CALCULUS

CALCULUS

Produced by the Consortium based at Harvard and funded by a National Science Foundation Grant. All proceeds from the sale of this work are used to support the work of the Consortium.

Deborah Hughes-Hallett
Harvard University

Andrew M. Gleason
Harvard University

Sheldon P. Gordon
Suffolk County Community College

David O. Lomen
University of Arizona

David Lovelock
University of Arizona

William G. McCallum
University of Arizona

Brad G. Osgood
Stanford University

Andrew Pasquale
Chelmsford High School

Jeff Tecosky-Feldman
Haverford College

Joe B. Thrash
University of Southern Mississippi

Karen R. Thrash
University of Southern Mississippi

Thomas W. Tucker
Colgate University

John Wiley & Sons, Inc.
New York Chichester Brisbane Toronto Singapore

Dedicated to: Duff Campbell, Kenny Ching, and Alex Kasman

Recognizing the importance of preserving what has been written, it is a policy of John Wiley & Sons, Inc. to have books of enduring value published in the United States printed on acid-free paper, and we exert our best efforts to that end.

All royalties from the sale of this book are assigned to the Core Calculus Consortium based at Harvard University, a project conducted pursuant to the National Science Foundation Prime Grant USE-8953923, to be used to further the mandates of the grant.

Photo Credit: Michael Halminski/SUPERSTOCK

Problems from *Calculus: The Analysis of Functions*, by Peter D. Taylor (Toronto: Wall & Emerson, Inc. 1992). Reprinted with permission of the publisher.

ISBN: 0-471-57723-5

Printed in the United States of America

10 9 8 7 6 5 4

Printed and bound by Courier Companies, Inc.

Preface

Calculus is one of the greatest achievements of the human intellect. Inspired by problems in astronomy, Newton and Leibniz developed the ideas of calculus 300 years ago. Since then, each century has demonstrated the power of calculus to illuminate questions in mathematics, the physical sciences, engineering, and the social and biological sciences.

Calculus has been so successful because of its extraordinary power to reduce complicated problems to simple rules and procedures. Therein lies the danger in teaching calculus: it is possible to teach the subject as nothing but the rules and procedures – thereby losing sight of both the mathematics and of its practical value. With the generous support of the National Science Foundation, our group set out to create a new calculus curriculum that would restore that insight. This book is part of that endeavor.

Basic Principles

Two principles guided our efforts. The first is our prescription for restoring the mathematical content to calculus:

> **The Rule of Three:** *Every topic should be presented geometrically, numerically and algebraically.*

We continually encourage students to think about the geometrical and numerical meaning of what they are doing. It is not our intention to undermine the purely algebraic aspect of calculus, but rather to reinforce it by giving meaning to the symbols. In the homework problems dealing with applications, we continually ask students what their answers mean in practical terms.

The second principle, inspired by Archimedes, is our prescription for restoring practical understanding:

> **The Way of Archimedes:** *Formal definitions and procedures evolve from the investigation of practical problems.*

Archimedes believed that insight into mathematical problems is gained by investigating mechanical or physical problems first.[1] For the same reason, our text is problem driven. Whenever possible, we start with a practical problem and derive the general results from it. By practical problems we usually, but not always, mean real world applications. These two principles have led

---

[1] ...I thought fit to write out for you and explain in detail ... the peculiarity of a certain method, by which it will be possible for you to get a start to enable you to investigate some of the problems in mathematics by means of mechanics. This procedure is, I am persuaded, no less useful even for the proof of the theorems themselves; for certain things first became clear to me by a mechanical method, although they had to be demonstrated by geometry afterwards because their investigation by the said method did not furnish an actual demonstration. But it is of course easier, when we have previously acquired, by the method, some knowledge of the questions, to supply the proof than it is to find it without any previous knowledge. From *The Method*, in *The Works of Archimedes* edited and translated by Sir Thomas L. Heath (Dover, NY)

to a dramatically new curriculum – more so than a cursory glance at the table of contents might indicate.

Technology

We have taken advantage of computers and graphing calculators to help students learn to think mathematically. Local linearity, for example, is much more easily seen by 'zooming in' on functions using a graphing calculator than by any other method. Furthermore, the ability to use technology effectively is itself of the greatest importance.

However, the book does not require any specific software or technology. Test sites have used the materials with graphing calculators, graphing software, and computer algebra systems. Any technology with the ability to graph functions and perform numerical integration will suffice.

What Student Background is Expected?

We have found this curriculum to be thought-provoking for well-prepared students while still accessible to students with weak algebra backgrounds. Providing numerical and graphical approaches as well as the algebraic gives students another way of mastering the material. This approach seems to enable at-risk students to persist, thereby lowering failure rates.

What's New?

When we designed this curriculum we started with a clean slate. We included some new topics, such as differential equations, and omitted some traditional topics whose inclusion we could not justify. In the process, we also changed the focus of certain topics. For example:

- **Chapter 1** introduces key properties of all the functions used in the text. Although the functions may be familiar, the emphasis on their graphical and numerical properties and on their practical uses is not. We introduce exponential functions at the earliest possible stage, since they are fundamental to the understanding of real-world processes.

- **Chapter 2** presents the key concept of the derivative according to the Rule of Three. We deliberately avoid differentiation formulas here in order to better develop the mathematical ideas.

- **Chapter 3** presents the key concept of the definite integral, along the same lines as Chapter 2. Some instructors using preliminary versions of the book have delayed covering Chapter 3 until after Chapter 5 without any difficulty.

- **Chapter 4** presents the symbolic approach to differentiation.

- **Chapter 5** presents applications of the derivative. It includes an investigation of parametrized families of functions according to the Way of Archimedes, using the graphing technology to observe basic properties and calculus to confirm them.

- **Chapter 6** treats the definite integral and gives an introduction to the antiderivative. We do not restrict our attention to functions that have closed-form antiderivatives. Instead, we emphasize the role of numerical integration as a basic tool. This chapter also includes a variety of techniques of integration. However, we have eliminated many of these methods and replaced them with the use of tables of integrals. While we do not specifically make use of computer algebra software, we certainly acknowledge that its existence changes the skills that students need to master.

- **Chapter 7** addresses applications of the definite integral. We emphasize the idea of subdividing the quantity at hand to produce Riemann sums which, in the limit, yield definite integrals. We avoid template problems and methods (the disk method, the washer method, etc.) and broaden the scope by including applications to economics and probability.

- **Chapter 8** presents an introduction to differential equations. We use the slope field to visualize the behavior of solutions of first-order differential equations. The emphasis is on qualitative solutions, modelling and interpretation. We include applications to population models (exponential and logistic) and predator-prey equations. There is also some material on second order differential equations with constant coefficients: the spring equation, both damped and undamped, and solutions using complex numbers.

- **Chapter 9** involves the approximation of functions. Our primary focus is on Taylor polynomials and series. This chapter also includes a brief introduction to Fourier analysis as a means of approximating a function globally, as opposed to locally with a Taylor series.

Supplementary Materials

- **Instructor's Manual** with teaching tips, calculator programs, some overhead transparency masters and sample exams.

- **Instructor's Solution Manual** with complete solutions to all problems.

- **Student's Solution Manual** with complete solutions to all odd-numbered problems.

- **Video for Instructors**, an orientation to teaching with the materials.

Our Experiences

In the process of developing the ideas incorporated in this book, we have been conscious of the need to test the ideas and the materials based on them thoroughly, in a wide variety of institutions serving many different types of students. Consortium members have used previous versions of the book for several years at large and small liberal arts colleges, at large and small public universities, at a two-year institution, and at a high school. During the 1991-92 academic year, we were assisted by colleagues at other schools around the country who class-tested the book and reported their experiences. We appreciate the valuable suggestions they made, which we have tried to incorporate into the present version of the text.

The institutions involved in this widespread testing include:

Arizona State University	Northern Arizona University
Brigham Young University	Pima County Community College
Cayuga Community College	San Francisco State University
DePaul University	Seattle Central Community College
The Evergreen State University	Shoreline Community College
Gettysburg College	St. Lawerence University
Middlesex County College	SUNY – Stony Brook
Millsaps College	University of Hartford
Monroe Community College	University of South Alabama
Nassau Community College	U.S. Merchant Marine Academy
New York Institute of Technology	West Valley College
Northeast Louisiana University	Western Washington University

Acknowledgements

First and foremost, we want to express our appreciation to the National Science Foundation for their faith in our ability to produce a revitalized calculus curriculum and, in particular, to Louise Raphael, John Kenelly, John Bradley, and James Lightbourne. We also want to thank the members of our Advisory Board, Lida Barrett, Bob Davis, John Dossey, Ron Douglas, Seymour Parter and Steve Rodi for their ongoing guidance and advice.

In addition, a host of other people around the country and abroad deserve our thanks for all that they did to help our project suceed. They include: Alice Adamson, Wayne Anderson, Ruth Baruth, Eric Belsley, Jeffrey Bergen, Maria Betkowski, Marty Betz, Peter and Helen Bing, Ross Biro, Sam Borah, Bill Bossert, Steve Boyd, Jackie Boyd-DeMarzio, Otto Bretscher, Carl Brettschneider, John Brillhart, Greg Brumfiel, Joan Carrafiello, Grace Cascio, Phil Cheifetz, Patrick Chiu, Jim Clay, David Cohen, Ralph Cohen, Bob Condon, Sterling G. Crossley, Caspar Curjel, Carolyn DeSilva, Carl DeVito, Bob Decker, Tom Dick, Kathleen Ann Drude, Neil Dummigan, Bill Dunn, Eugene Durenard, Ken Dutch, Wade Ellis, Boas Erez, Alice Essary, Sol Feferman, Hermann Flaschka, Dan Flath, Patti Frazier Lock, Lynn Garner, David Gay, Mike Goho, Randi Goldsmith, Craig Gordon, Florence Gordon, Ken Gordon, Danny Goroff, Robin Gottlieb, Milja Hakosalo, Joe Harris, Steve Harrison, Cymra Haskell, John Havlicek, Dudley Hershbach, JoEllen Hillyer, Andrew Ho, Luke Hunsberger, John Hutchinson, David-Olivier Jacquet-Chiffelle, Qin Jing, Mille Johnson, Yael Karshon, Matthias Kawski, Aaron King, Ed Kingham, Neal Koblitz, Bryna Kra, Donna Krawczyk, Robert Kuhn, Sylvain Laroche, Carl Leinbach, John Leonard, David Levermore, Harry Lewis, Ya-Yan Lu, Reginald Luke, Tom MacMahon, Dan Madden, Dean Madden, Andreu Mas-Colell, David Maslen, Steve Massaquoi, Warren May, Barry Mazur, Rafe Mazzeo, Dave Meredith, Bob Mizner, Jean Morris, David Mumford, Charlie Naffzinger, Lynn Narasimhan, Bridget Neale, Sam Nelson, Alan Newell, Fielding Norton, Huriye Önder, Cassie Osgood, Herbert and Ruth Osgood, Miles Osgood, Rebecca Osgood, Arnie Ostebee, Mike Pavloff, Barbara Peskin, Yionnis Petridis, Tony Phillips, Dan Pollack, Horacio Porta, John Prados, Amy Radunskaya, Wayne Raskind, Gabriella Ratay, Janet Ray, Tom Read, Marc Sanders, Wilfried Schmid, Marilyn Semrau, Dave Shao, Henry Shahrouz, Taka Shiota, Esther Silberstein, David Smith, Gary Snavely, Don Snow, Bob Speiser, Hal Stern, Fred Stevenson, Edith Stokey, Steve Strogatz, "Suds" Sudholz, Mary Sunseri, Peter Taylor, Ewart Thomas, Richard Thompson, Tom Timchek, Sue Tolman, Elias Toubassi, Eric Towne, Praja

Triverdi, Emily Tucker, Tom Tucker, Alan Tucker, Judy Turner, Jerry Uhl, Doug Ulmer, Bill Vélez, Faye Villalobos, Gary Walls, Charles Walter, Don Weitzman, Steve Wheaton, Jeff Wolcowitz, Andy Wolf, Janet Woodland, Debbie Yoklic, Lee Zia, Paul Zorn

Most of all, to the remarkable team that worked day and night to get the text into the computer (and out again), to get the solutions written and the pictures labeled: we greatly appreciate your ingenuity, energy and dedication. Thanks to: Reza Akhtar, Stefan Bilbao, Lori Bosch, Joshua Brandon, Barry Brent, Ruvim Breydo, Will Brockman, Duff Campbell, Kenny Ching, Eric Connally, Srdjan Divac, Stacy Evans, Jim Georges, Noah Graham, Christian Gromoll, Jeff Hass, Patricia Hersh, Adrian Iovita, Rajesh James, Sam Kaplan, Alex Kasman, Emmanuel Katz, Gabriel Katz, Mike Klucznik, Dimitri Kountourogiannis, Alex Mallozzi, Teresa Marrin, David Miller, Mike Mitzenmacher, Jean Morris, John Neale, Bill Niquette, Marc Nirenberg, Ed Park, Shankar Ramaswami, Maddy Rothberg, Jordan Samuels, Ayşegul Sener, Eugene Stern, Jeff Suzuki, Andras Szenes, Diane Tang, Sulian Tay, Alice Wang, Eric Wepsic, Gang Zhang, Ying Zhu.

Deborah Hughes-Hallett   David Lovelock   Jeff Tecosky-Feldman
Andrew M. Gleason   William G. McCallum   Joe B. Thrash
Sheldon P. Gordon   Brad G. Osgood   Karen R. Thrash
David O. Lomen   Andrew Pasquale   Thomas W. Tucker

To Students: How to Learn from this Book

- This book is probably very different from any other mathematics textbook that you have used, so it may be helpful to know about some of the differences in advance. This book emphasizes at every stage the *meaning* (in practical, graphical or numerical terms) of the symbols you are using. There is much less emphasis on "plug-and-chug" and using formulas, and much more emphasis on the interpretation of these formulas than you may expect. You will often be asked to explain your ideas in words or to explain an answer using graphs.

  Why does the book have this emphasis? Because *understanding* is the key to being able to remember and use your knowledge in other course and other fields. Much of the book is designed to help you gain such an understanding.

- The book contains the main ideas of calculus in plain English. Your success in using this book will depend on your reading, questioning, and thinking hard about the ideas presented. Although you may not have done this with other books, you should plan on reading the text in detail, not just the worked examples.

- There are very few examples in the text that are exactly like the homework problems. This means that you can't just look at a homework problem and search for a similar–looking "worked out" example. Success with the homework will come by grappling with the ideas of calculus.

- Many of the problems that we have included in the book are open-ended. This means that there may be more than one approach and more than one solution, depending on your analysis. Many times, solving a problem relies on common sense ideas that are not stated in the problem but which you will know from everyday life.

- This book assumes that you have access to a calculator or computer that can graph functions, find (approximate) roots of equations, and compute integrals numerically. There are many situations where you may not be able to find an exact solution to a problem, but you can use a calculator or computer to get a reasonable approximation. An answer obtained this way is usually just as useful as an exact one. However, the problem does not always state that a calculator is required, so use your judgement.

  If you mistrust technology, listen to this student, who started out the same way:

  Using computers is strange, but surprisingly beneficial, and in my opinion is what leads to success in this class. I have difficulty visualizing graphs in my head, and this has always led to my downfall in calculus. With the assistance of the computers, that stress was no longer a factor, and I was able to concentrate on the concepts behind the shapes of the graphs, and since these became gradually more clear, I got increasingly better at picturing what the graphs should look like. It's the old story of not being able to get a job without previous experience, but not being able to get experience without a job. Relying on the computer to help me avoid graphing, I was tricked into focusing on what the graphs meant instead of how to make them look right, and what graphs symbolize is the fundamental basis of this class. By being able to see what I was trying to describe and learn from, I could understand a lot more about the concepts, because I could change the conditions and see the results. For the first time, I was able to see how everything works together . . . .

That was a student at the University of Arizona who took calculus in Fall 1990, the first time we used the text. She was terrified of calculus, got a C on her first test, but finished with an A for the course.

- This book attempts to give equal weight to three methods for describing functions: graphical (a picture), numerical (a table of values) and algebraic (a formula). Sometimes you may find it easier to translate a problem given in one form into another. For example, you might replace the graph of a parabola with its equation, or plot a table of values to see its behavior. The best idea is to be flexible about your approach: if one way of looking at a problem doesn't work, try another.

- Students using this book have found discussing these problems in small groups very helpful. There are a great many problems which are not cut-and-dried; it can help to attack them with the other perspectives your colleagues can provide. If group work is not feasible, see if your instructor can organize a discussion session in which additional problems can be worked on.

- You are probably wondering what you'll get from the book. The answer is, if you put in a solid effort, you will get a real understanding of one of the most important accomplishments of the millennium – calculus – as well as a real sense of how mathematics is used in the age of technology.

Deborah Hughes-Hallett   David Lovelock   Jeff Tecosky-Feldman
Andrew M. Gleason   William G. McCallum   Joe B. Thrash
Sheldon P. Gordon   Brad G. Osgood   Karen R. Thrash
David O. Lomen   Andrew Pasquale   Thomas W. Tucker

Table of Contents:

1 A LIBRARY OF FUNCTIONS

1

1.1 What's a Function? 2

1.2 Linear Functions 9

1.3 Exponential Functions 19

1.4 The Number e 31

1.5 Power Functions 40

1.6 Inverse Functions 50

1.7 Logarithms 55

1.8 Natural Logs: Logs to Base e 63

1.9 New Functions from Old 68

1.10 The Trigonometric Functions 76

1.11 Polynomials and Rational Functions 88

1.12 Roots, Continuity, and Accuracy 96

1.13 Miscellaneous Exercises for Chapter 1 109

2 KEY CONCEPT: THE DERIVATIVE

119

2.1 How do we Measure Speed? 120

2.2 The Derivative 127

2.3 The Derivative Function 139

2.4 Interpretations of the Derivative 148

2.5 The Second Derivative 155

2.6 Approximations and Local Linearity 162

2.7 Notes on the Limit 168

2.8 Notes on Differentiability 171

2.9 Miscellaneous Exercises for Chapter 2 175

3 KEY CONCEPT: THE DEFINITE INTEGRAL   181

3.1 How Do We Measure Distance Traveled?   182

3.2 The Definite Integral   190

3.3 Interpretations of the Definite Integral   196

3.4 The Fundamental Theorem of Calculus   203

3.5 Further Notes on the Limit   211

3.6 Miscellaneous Exercises for Chapter 3   213

4 SHORT-CUTS TO DIFFERENTIATION   219

4.1 Formulas for Derivative Functions   220

4.2 Powers and Polynomials   223

4.3 The Exponential Function   233

4.4 The Product and Quotient Rules   239

4.5 The Chain Rule   245

4.6 The Trigonometric Functions   250

4.7 Applications of the Chain Rule   258

4.8 Implicit Functions   263

4.9 Notes on the Tangent Line Approximation   266

4.10 Miscellaneous Exercises for Chapter 4   271

5 USING THE DERIVATIVE   277

5.1 Maxima and Minima   278

5.2 Concavity and Inflection Points   289

5.3 Families of Curves: A Qualitative Study   300

5.4 Economic Applications: Marginality   311

5.5 Optimization   319

5.6 More Optimization: Introduction to Modeling   328

5.7 Newton's Method   337

5.8   Working Backwards: Antiderivatives   342

5.9   Notes on Equations of Motion: Why acceleration?   352

5.10 Miscellaneous Exercises for Chapter 5   355

6 THE INTEGRAL   361

6.1   The Definite Integral Revisited   362

6.2   Properties of the Definite Integral   366

6.3   Antiderivatives and the Fundamental Theorem   374

6.4   Integration by Substitution—Part I   383

6.5   Integration by Substitution—Part II   390

6.6   Integration by Parts   395

6.7   Tables of Integrals   403

6.8   Approximating Definite Integrals   414

6.9   Approximation Errors and Simpson's Rule   424

6.10 Improper Integrals   431

6.11 More on Improper Integrals   441

6.12 Notes on Constructing Antiderivatives   448

6.13 Miscellaneous Exercises for Chapter 6   456

7 USING THE DEFINITE INTEGRAL   463

7.1   The Definite Integral as an Area   464

7.2   Setting up Riemann Sums   469

7.3   Applications to Geometry   478

7.4   Applications to Physics   489

7.5   Applications to Economics   498

7.6   Probability and Distributions   507

7.7   Miscellaneous Exercises for Chapter 7   524

8 DIFFERENTIAL EQUATIONS 533

8.1 What is a Differential Equation? 534

8.2 Slope Fields 537

8.3 Euler's Method 545

8.4 Separation of Variables 549

8.5 Growth and Decay 555

8.6 Applications and Modeling 562

8.7 Models of Population Growth 573

8.8 Two Interacting Populations 579

8.9 Second Order Differential Equations: Oscillations 587

8.10 Damped Oscillations and Numerical Methods 599

8.11 Solving Second Order Differential Equations 605

8.12 Appendix: Complex Numbers 613

8.13 Miscellaneous Exercises for Chapter 8 619

9 APPROXIMATIONS 625

9.1 Taylor Polynomials 626

9.2 Taylor Series 635

9.3 Finding and Using Taylor Series 643

9.4 The Error in a Taylor Polynomial Approximation 651

9.5 Fourier Series 657

9.6 Miscellaneous Exercises for Chapter 9 666

Chapter 1

A LIBRARY OF FUNCTIONS

Functions are truly fundamental in mathematics. For example, in everyday language we say, "The price of a ticket is a function of where you sit," or "The speed of a rocket is a function of its payload." In each case, the word *function* expresses the idea that knowledge of one fact tells us another. In mathematics, the most important functions are those in which knowledge of one number tells us another number. If we know the length of the side of a square, its area is determined. If the circumference of a circle is known, its radius is determined.

Calculus starts with the study of functions. This chapter will lay the foundation for calculus by surveying the behavior of the most common functions, including powers, exponentials, logarithms, and the trigonometric functions. Besides the behavior of these functions, we will see ways of handling the graphs, tables and formulas that represent them.

1.1 What's a Function?

Let's look at an example. In the summer of 1990, the temperatures in Arizona reached an all-time high (so high, in fact, that some airlines decided it might be unsafe to land their planes there). The daily high temperatures in Phoenix for June 19–29 are given in Table 1.1.

Date: June	19	20	21	22	23	24	25	26	27	28	29
Temp(°F)	109	113	114	113	113	113	120	122	118	118	108

Table 1.1: Temperature in Phoenix, Arizona, June 1990

Although you may not have thought of something so unpredictable as temperature as being a function, the temperature *is* a function of date, because each day gives rise to one and only one high temperature. There is no formula for temperature (otherwise we would not need the weather bureau) but nevertheless the temperature does satisfy the definition of a function: Each date, t, has a unique high temperature, H, associated with it.

Definition of a Function

One quantity, H, is a **function** of another, t, if each value of t has a unique value of H associated with it. We say H is the *value* of the function or the *dependent variable* and t is the *argument* or *independent variable*. Alternatively, think of t as the *input* while H is the *output*. We write $H = f(t)$, where f is the name of the function.

The *domain* of a function is a set of possible values of the independent variable, while the *range* is the set of corresponding values of the dependent variable.

Functions play an important role in science. Frequently, one observes that one quantity is a function of another and then tries to find a reasonable formula to express this function. For example, before about 1590 there was no quantitative idea of temperature. Of course, people understood relative notions like warmer and cooler, and some absolute notions like boiling hot, freezing cold, or body temperature, but there was no numerical measure of temperature. It took the genius of Galileo to realize that the expansion of fluids as they warmed was the key to the measurement of temperature. He was the first to think of temperature as a function of fluid volume.

Finding a function which represents a given situation is called making a *mathematical model*. Such a model can throw light on the relationship between the variables and can thereby help us make predictions.

Representation of Functions: Tables, Graphs, and Formulas

Functions can be represented in at least three different ways: by tables, by graphs, and by formulas. For example, the function giving the temperatures in Phoenix, Arizona, as a function of time can be represented by the graphs in Figure 1.1 as well as by a table.

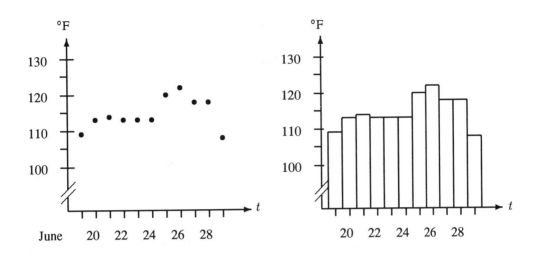

Figure 1.1: Phoenix Temperatures, June 1990

Other functions arise naturally as graphs. Figure 1.2 contains electrocardiogram (EKG) pictures showing the heartbeat patterns of two patients, one normal and one not. While it is possible to construct a formula to approximate an EKG function, this is seldom done. The pattern of repetitions is what a doctor needs to know, and these are much more easily seen from a graph than from a formula. However, each EKG represents a function showing electrical activity as a function of time.

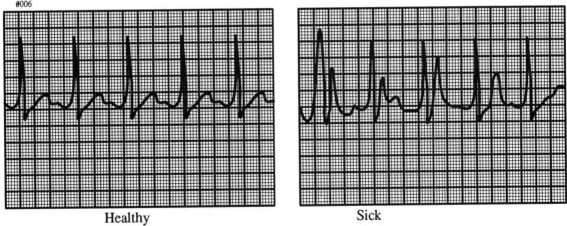

Healthy   Sick

Figure 1.2: EKG Readings on Two Patients

As another example of a function, consider the Snow Tree Cricket. Surprisingly enough, all such crickets chirp at essentially the same rate if they are at the same temperature. That means that the chirp-rate is a function of the temperature. In other words, if we know the temperature, we can determine the chirp-rate. Even more surprisingly, the chirp-rate, C, increases steadily with the temperature, T, and to a high degree of accuracy can be computed by the formula

$$C = 4T - 160.$$

The formula for C is written $C = f(T)$ to express the fact that we are thinking of C as a function of T. The graph of this function is in Figure 1.3.

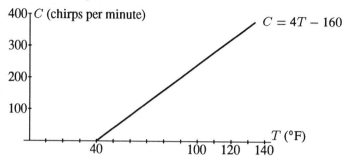

Figure 1.3: Temperature versus Cricket Chirp-rate

Examples of Domain and Range

Many functions have a natural domain. For example, we usually think of the domain of the function $f(x) = x^2$ as all real numbers, whereas the domain of the function $g(x) = \frac{1}{x}$ is all real numbers except zero. Sometimes however we may specify, or restrict, the domain. For example, if the function $f(x) = x^2$ is used to represent the area of a square of side x, we consider only non-negative values of x, and restrict the domain to non-negative numbers.

☐ **Example 1** Consider the function $C = f(T)$ giving chirp-rate as a function of temperature. We assume that this equation holds for all temperatures above that at which the predicted chirp-rate becomes negative, and up to the highest temperature ever recorded at a weather station, namely, 136°F. What is the domain of this function f?

Solution. If we consider

$$C = f(T) = 4T - 160$$

simply as a mathematical relationship between C and T, any T value is possible and the domain is all real T values. However, if we're thinking of it as a relationship between cricket chirps and temperature, then T cannot be less than 40°F, where C becomes negative. (See Figure 1.3.) In addition, we are told that the formula doesn't hold at temperatures above 136°. Thus we have

$$
\begin{aligned}
\text{Domain} &= \text{ all } T \text{ values between } 40°\text{F and } 136°\text{F} \\
&= \text{ all } T \text{ values with } 40 \le T \le 136.
\end{aligned}
$$

☐

☐ **Example 2** Find the range of the function f, given the domain from Example 1. In other words, find all possible values of the chirp rate, C, in the equation $C = f(T)$.

Solution. Again, if we consider $C = f(T)$ simply as a mathematical relationship, its range is all real C values. However, thinking of its meaning for crickets, the function will predict cricket chirps per minute between 0 (when $T = 40°F$) and 384 (when $T = 136°F$). Hence

$$
\begin{aligned}
\text{Range} \quad &= \quad \text{all } C \text{ values from 0 to 384} \\
&= \quad \text{all } C \text{ values with } 0 \leq C \leq 384.
\end{aligned}
$$

$\square$

So far we have used the temperature to predict the chirp-rate, and thought of the temperature as the *independent variable* and the chirp-rate as the *dependent variable*. However, we could do this backwards, and calculate the temperature from the chirp-rate. From this point of view, the temperature is dependent on the chirp-rate. Thus, which variable is dependent and which is independent may depend on your viewpoint.

Thinking of temperature as a function of chirp-rate would enable us (in theory, at least!) to use the chirp-rate instead of a thermometer to measure temperature. The way we actually do measure temperature is based on another such function: the relation between the height of the liquid in a thermometer and temperature. The height of the mercury is certainly a function of temperature; however, we always use this the other way round and determine the temperature from the height of the mercury, as suggested by Galileo.

Proportionality

A common functional relationship occurs when one quantity is *proportional* to another. For example, if apples are 60¢ a pound, we say the price you pay, p¢, is proportional to the weight you buy, w pounds, because

$$
p = f(w) = 60w.
$$

The area, A, of a circle is proportional to the square of the radius, r:

$$
A = f(r) = \pi r^2.
$$

In general, we say that

---

y **is proportional to** x if there is a constant k such that
$$
y = kx.
$$

We also say that one quantity is *inversely proportional* to another if one is proportional to the reciprocal of the other. For example, the speed, v, at which you make a 50-mile trip is inversely proportional to the time, t, taken, because v is proportional to $\frac{1}{t}$:

$$
v = \frac{50}{t} = 50\left(\frac{1}{t}\right).
$$

❖ Exercises for Section 1.1

1. Match the stories with three of the graphs in Figure 1.4 and write a story for the remaining graph.

    (a) I had just left home when I realized I had forgotten my books and so I went back to pick them up.

    (b) Things went fine until I had a flat tire.

    (c) I started out calmly, but sped up when I realized I was going to be late.

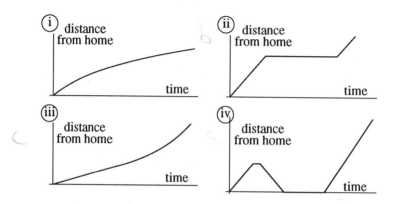

Figure 1.4: Which story goes with which graph?
(for Problem 1)

2. It warmed up throughout the morning, then suddenly got much cooler around noon when a storm came through. After the storm, it warmed up before cooling off at sunset. Sketch a possible graph of this day's temperature as a function of time.

3. Right after a certain drug is administered to a patient with a rapid heart rate, the heart rate plunges dramatically and then slowly rises again as the drug wears off. Sketch a possible graph of the heart rate against time from the moment the drug is administered.

4. When an impulse is sent down the spinal cord, a flood of sodium ions rush into a nerve cell. The sodium is then slowly pumped out again until the level is back to normal. Sketch a graph showing the number of sodium ions in the nerve cell as a function of time before and after an impulse.

5. To the right is a graph of electrical potential against time in a nerve cell. Describe in words how the electrical potential changes with time.

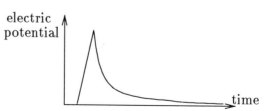

6. Generally, the more fertilizer that is used, the better the yield of the crop. However, if too much fertilizer is applied, the crops become poisoned and the yield goes down rapidly. Sketch a possible graph showing the yield of the crop as a function of the amount of fertilizer applied.

7. Describe what Figure 1.5 tells you about the assembly line whose productivity is represented as a function of the number of workers on that assembly line.

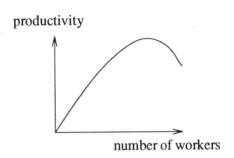

Figure 1.5: Productivity versus Number of Workers

8. A flight from Dulles Airport in Washington D.C., to LaGuardia Airport in New York City has to circle LaGuardia several times before being allowed to land. Plot a graph of distance from Washington against time from the moment of take-off until landing.

9. In her *Guide to Excruciatingly Correct Behavior*, Miss Manners states:

> There are three possible parts to a date of which at least two must be offered: entertainment, food and affection. It is customary to begin a series of dates with a great deal of entertainment, a moderate amount of food and the merest suggestion of affection. As the amount of affection increases, the entertainment can be reduced proportionately. When the affection has replaced the entertainment, we no longer call it dating. Under no circumstances can the food be omitted.

Based on this statement, sketch a graph showing entertainment as a function of affection, assuming the amount of food to be constant. Mark the point on the graph at which the relationship starts, as well as the point at which the relationship ceases to be called dating.

Problems 10 and 11 are about supply and demand curves. Economists are interested in how the quantity of a good which is manufactured and sold, q, depends on its price, p. They think of quantity as a function of price. However, for historical reasons, [1] the economists put price (the independent variable) on the vertical axis and quantity (the dependent variable) on the horizontal axis. There are two graphs in Figure 1.6 relating p and q: the *supply curve* which represents manufacturers' behavior and the *demand curve* which represents consumer behavior.

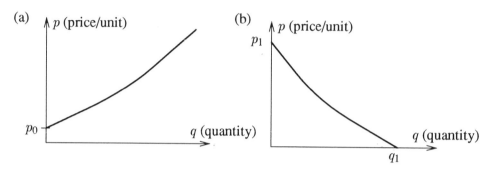

Figure 1.6: Supply and Demand: Which is Which?

10. One of the graphs in Figure 1.6 is a supply curve and the other is a demand curve. Which is which? Why?

11. The price p_0 in Figure 1.6(a) represents the price below which the manufacturers are unwilling to produce any of the good. What do the price p_1 and the quantity q_1 in Figure 1.6(b) represent in practical economic terms?

12. Specify the domain and range of the function $y = f(x)$ whose graph is shown to the right.

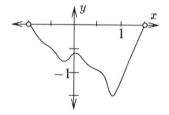

13. Use a calculator or computer graph of $f(x) = 13 - 20x - x^2 - 3x^4$ to determine:

    (a) the range of this function.

    (b) The number of zeros of this function.

---

[1]Originally, the economists thought of price as the dependent variable and put it on the vertical axis. Unfortunately, when the point of view changed, the axes did not.

14. The graph of $r = f(p)$ is given to the right.

    (a) What is the domain of f?

    (b) What is the range of f?

    (c) What values of r correspond to exactly one value of p?

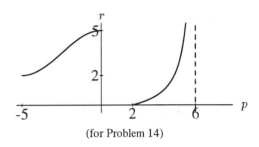

(for Problem 14)

15. If $y = f(x) = \dfrac{1}{\sqrt{4 - x^2}}$, what values of x do *not* determine a real value for y? For the same function f, solve $f(t) = 5$ for t.

16. When Galileo was formulating the laws of motion, he considered the motion of a body starting from rest and falling under gravity. He originally thought that the velocity of such a falling body was proportional to the distance it had fallen. What light does the first set of data below throw on this hypothesis of Galileo's? What alternative hypotheses is suggested by the two sets of data?

distance(feet)	0	1	2	3	4
velocity(ft/sec)	0	8	11.3	13.9	16

time(sec)	0	1	2	3	4
velocity(ft/sec)	0	32	64	96	128

1.2   Linear Functions

Probably the most commonly used functions are the *linear functions*. These are functions that represent a steady increase or a steady decrease. A function is linear if any change, or increment, in the independent variable causes a proportional change, or increment, in the dependent variable.

The Olympic Pole Vault

During the early years of the Olympics, the height of the winning pole vault increased approximately as shown in Table 1.2. Since the winning height increased regularly by 8 inches every four years, the height is a linear function of time over the period from 1900 to 1912. The height starts at 130 inches and increases at the equivalent of 2 inches every year, so if y is the height in inches, and t is the number of years since 1900, we can write

$$y = f(t) = 130 + 2t.$$

The 2 tells us the rate at which the height increases and is the *slope* of the line $f(t) = 130 + 2t$.

Year	Height (in inches)
1900	130
1904	138
1908	146
1912	154

Table 1.2: Olympic Pole Vault Records (Approximate)

You can visualize the slope in Figure 1.7 as the ratio:

$$\text{slope} = \frac{\text{rise}}{\text{run}} = \frac{8}{4} = 2.$$

Calculating the slope (rise/run) using any other two points on the line gives the same value. It is this fact—that the slope, or rate of change, is the same everywhere—that makes a line straight. For a function that is not linear, the rate of change will vary from point to point. Since $y = f(t)$ increases with t, we say that f is *an increasing function*.

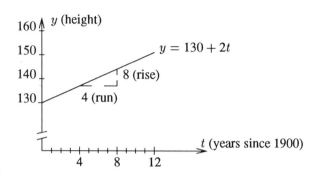

Figure 1.7: Olympic Pole Vault Records

What about the 130? This represents the initial height in 1900, when $t = 0$. Geometrically, the 130 is the *intercept* on the vertical axis.

You may wonder whether the linear trend continues beyond 1912. Not surprisingly, it doesn't exactly. The formula $y = 130 + 2t$ predicts the height in the 1988 Olympics would be 306″ or 25′6″, which is considerably higher than the actual value of 19′9″. In fact the height does increase at almost every Games, but not at a constant rate. Thus there is clearly a danger in *extrapolating* too far from the given data. You should also observe that the data in Table 1.2 is *discrete*, because it is given only at specific points (every four years). However, we have treated the variable t as though it were *continuous*, because the function $y = 130 + 2t$ makes sense for all values of t. The graph in Figure 1.7 is of the continuous function because it is a solid line, rather than four separate points representing the years in which the Olympics were held.

Linear Functions in General

A **linear function** has the form

$$y = f(x) = b + mx$$

and a graph where

m is the slope, or rate of change of y with respect to x

b is the vertical intercept, or value of y when x is zero.

Notice that if the slope is 0, we have $y = b$, a horizontal line.

To recognize that a function $y = f(x)$ given by a table of data is linear: look for differences in y values that are constant for equally spaced x values.

The slope of a linear function can be calculated from values of the function at two points, a and c, using the formula

$$m = \frac{\text{rise}}{\text{run}} = \frac{f(c) - f(a)}{c - a}.$$

The quantity $\frac{f(c)-f(a)}{c-a}$ is called a *difference quotient* because it is the quotient of two differences (see Figure 1.8). In Chapter 2, you will see that difference quotients play an important role in calculus.

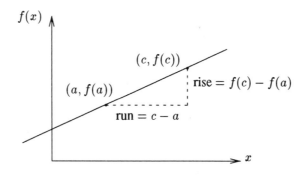

Figure 1.8: Difference Quotient $= \frac{f(c)-f(a)}{c-a}$

The Success of Search and Rescue Teams

Consider the problem of the "search and rescue" teams working to find lost hikers in remote areas in the West. To search for an individual, members of the search team separate and walk parallel to one another through the area to be searched. Experience has shown that the team's chance of finding a lost individual is related to the distance, d, by which team members are separated. The percentage found[1] for various separations are recorded in Table 1.3.

Separation distance: d feet	Percent found: $P\%$
20	90
40	80
60	70
80	60
100	50

Table 1.3: Separation of Searchers versus Success Rate

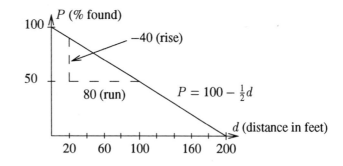

Figure 1.9: Separation of Searchers versus Success Rate

From the data in the table, you can see that as the separation distance decreases, a larger percentage is found. (This makes sense!) Since $P = f(d)$ decreases as d increases, we say P is a *decreasing function of d.* You can also see that for the data given, each 20 foot increase in distance

---

[1]Wartes, 1971.

causes the percentage found to drop by 10. This constant decrease in P as d increases by a fixed amount is a clear indication that the graph of P against d is a line. See Figure 1.9. Notice that the slope is $-\frac{40}{80} = -\frac{1}{2}$. The negative sign shows that P decreases as d increases. The slope is the rate at which P is increasing or decreasing, as d increases.

What about the vertical intercept? If $d = 0$, so the searchers are walking shoulder to shoulder, you'd expect everyone to be found, so $P = 100$. This is exactly what you get if the line is continued to the vertical axis (a decrease of 20 in d causes an increase of 10 in P). Therefore the equation of the line is

$$P = f(d) = 100 - \frac{1}{2}d.$$

What about the horizontal intercept? When $P = 0$ or $0 = 100 - \frac{1}{2}d$, then $d = 200$. The value $d = 200$ represents the separation distance at which, according to the model, no one is found. This is unreasonable, because even when the searchers are far apart, the search will sometimes be successful. What this suggests is that somewhere outside the data given, the linear relationship ceases to hold. As in the pole vault example, extrapolating too far beyond the given data may not give accurate answers.

Increasing versus Decreasing

A function f is **increasing** if the values of $y = f(x)$ increase as x increases; the function is **decreasing** if the values of $y = f(x)$ decrease as x increases.

The graph of an *increasing* function *climbs* as you move from left to right; the graph of a *decreasing* function *descends* as you move from left to right.

A Budget Constraint

An ongoing debate in the federal government concerns the allocation of money between defense and social programs. In general, the more that is spent on defense, the less that is available for social programs, and vice versa. Let's simplify the example to guns and butter. Assuming a constant budget, we will show that the relationship between the number of guns and the quantity of butter is linear. Suppose there is $12,000 to be spent and that it is to be divided between guns, costing $400 each, and butter, costing $2000 a ton. Suppose the number of guns bought is g, and the number of tons of butter is b. Then the amount of money spent on guns is $400g$ (because each one is $400), and the amount spent on butter is $2000b$. Assuming all the money is spent,

$$\left(\begin{array}{c}\text{amount spent} \\ \text{on guns}\end{array}\right) + \left(\begin{array}{c}\text{amount spent} \\ \text{on butter}\end{array}\right) = \$12{,}000$$

or

$$400g + 2000b = 12{,}000$$

or

$$g + 5b = 30.$$

This equation is the budget constraint. Its graph is the line shown in Figure 1.10, which can be found by plotting points. We will calculate the points at which the graph crosses the axes. If $b = 0$, then

$$g + 5(0) = 30 \quad \text{so} \quad g = 30.$$

If $g = 0$, then

$$0 + 5b = 30 \quad \text{so} \quad b = 6.$$

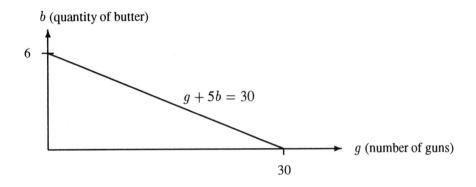

Figure 1.10: Budget Constraint

Since the number of guns bought determines the amount of butter bought (because all the money which doesn't go to guns goes to butter), b is a function of g. Similarly the amount of butter bought determines the number of guns, so g is a function of b. The budget constraint represents an *implicitly defined function*, because neither quantity is given explicitly in terms of the other. If we solve for g, giving

$$g = 30 - 5b$$

we have an *explicit* formula for g in terms of b. Similarly

$$b = \frac{30 - g}{5} \quad \text{or} \quad b = 6 - 0.2g$$

gives b as an explicit function of g. Since the explicit functions

$$g = 30 - 5b \quad \text{and} \quad b = 6 - 0.2g$$

are linear, the graph of the budget constraint must be a line.

Families of Linear Functions

Formulas like $f(x) = mx$, and $f(x) = b + mx$, containing constants such as m and b which can take on various values are said to define a *family of functions*. The constants m and b are called *parameters*. Each of the functions in this section belong to the family $f(x) = b + mx$.

Grouping functions into families which share important features is particularly useful for mathematical modeling. We often choose a family to represent a given situation on theoretical grounds, and then use data to determine the particular values of the parameters, the meaning of the parameters m and b are shown in Figures 1.11 and 1.12.

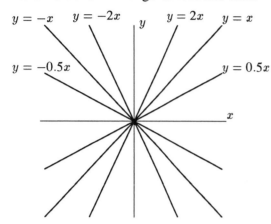

Figure 1.11: The family $y = mx$. $(b = 0)$

Figure 1.12: The family $y = b + x$. $(m = 1)$

❖ Exercises for Section 1.2

1. Find the slope and vertical intercept of the line whose equation is $2y + 5x - 8 = 0$.

2. Find the equation of the line through the point $(-1, 0)$ and $(2, 6)$.

3. Find the equation of the line with slope m through the point (a, c).

   point-slope formula $\quad y - y_1 = m(x - x_1)$

4. Recall that two lines are perpendicular if their slopes are negative reciprocals. Use this fact to find the equation of the line through the point $(2, 1)$ which is perpendicular to $f(x) = 5x - 3$.

5. Find the equations of the lines parallel to and perpendicular to the line $y + 4x = 7$ and through the point $(1, 5)$.

6. Estimate the slope of the line shown in Figure 1.13 and use the slope to find an equation for the line.

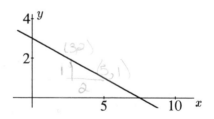

Figure 1.13: Note that the x and y scales are unequal. (For Problem 6)

7. Match the graphs in Fig. 1.14 with the equations below. (Note that the axes are not drawn to scale. The x-axes are all horizontal and the y-axes are all vertical.)

   (a) $y = x - 4$

   (b) $-2x + 3 = y$

   (c) $4 = y$

   (d) $y = -3x - 4$

   (e) $y = x + 5$

   (f) $y = \frac{1}{2}x$

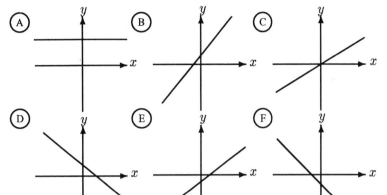

Figure 1.14: Match the graphs to the equations in Problem 7.

8. Match the graphs with the equations below. (Note that the axes are not drawn to scale. The x-axes are all horizontal and the y-axes are all vertical.)

   (a) $y = -2.72x$

   (b) $y = 0.01 + 0.001x$

   (c) $y = 27.9 - 0.1x$

   (d) $y = 0.1x - 27.9$

   (e) $y = -5.7 - 200x$

   (f) $y = \dfrac{x}{3.14}$

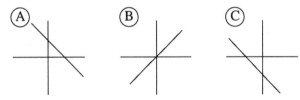

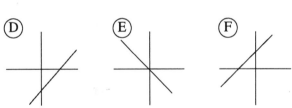

Figure 1.15: Match graphs to equations in Problem 8

9. Corresponding values of p and q are given in the table below.

   (a) Find q as a linear function of p.

   (b) Find p as a linear function of q.

p	1	2	3	4
q	950	900	850	800

10. A linear equation generated the values in the table to the right. Find the equation.

x	5.2	5.3	5.4	5.5	5.6
y	27.8	29.2	30.6	32.0	33.4

11. An equation of a line is $3x + 4y = -12$. Find the length of the portion of the line that lies between its x and y intercepts.

$$d = \sqrt{(x-x_1)^2 + (y-y_1)^2}$$

12. A car rental company offers cars at $40 a day and 15 cents a mile. Its competitor's cars are $50 a day and 10 cents a mile.

    (a) For each company, write a formula giving the cost of renting a car for a day as a function of the distance traveled.

    (b) On the same axes, sketch graphs of both functions.

    (c) How should you decide which company is cheapest?

13. Consider a graph of Fahrenheit temperature, F, against Celsius temperature, C, and assume that the graph is a line. You know that 212° Fahrenheit and 100° Celsius both represent the temperature at which water boils. Similarly, 32° Fahrenheit and 0° Centigrade both represent water's freezing point.

    (a) What is the slope of the graph?

    (b) What is the equation of the line?

    (c) Use the equation to find what temperature in Fahrenheit corresponds to 20° Centigrade.

    (d) What temperature is the same number of degrees in both Centigrade and Fahrenheit?

14. Suppose you are analyzing the cost of planting seed as a function of the number of acres sown. The cost of the equipment you need is called the *fixed cost* because you pay it regardless of the number of acres planted. The cost of supplies and labor varies with the number of acres planted and are called *variable costs*. Suppose the fixed costs are $10,000 and the variable costs are $200 per acre. Let C be the total cost, measured in thousands of dollars, and let x be the number of acres planted.

    (a) Find a formula for C as a function of x.

    (b) Sketch a graph of C against x.

    (c) Explain how you can visualize the fixed and variable costs on the graph.

15. Suppose you are driving at a constant speed from Chicago to Detroit, about 275 miles away. About 120 miles from Chicago you pass through Kalamazoo, Michigan. Sketch a graph of your distance from Kalamazoo as a function of time.

16. Hot peppers have been rated according to 'Scoville' units with a maximum human tolerance level of 14,000 Scovilles per dish. The West Coast Restaurant, known for spicy dishes, promises a daily special to satisfy the most avid spicy-dish fans. The restaurant imports Indian peppers rated at 1,200 Scovilles each and Mexican peppers with a Scoville rating of 900 each.

    (a) Determine the Scoville constraint equation relating the maximum number of Indian and Mexican peppers the restaurant should use for their specialty dish.

    (b) Solve the equation from (a) to show explicitly the number of Indian peppers needed in the hottest dishes as a function of the number of Mexican peppers.

17. You have a fixed budget of $\$k$ to spend on soda and suntan oil, which cost $\$p_1$ per liter and $\$p_2$ per liter respectively.

    (a) Write an equation expressing the relationship between the number of liters of soda and the number of liters of suntan oil that you can buy if you exhaust your budget. This is your *budget constraint*.

    (b) Graph the budget constraint, assuming that you can buy fractions of a liter. Label the intercepts.

    (c) Suppose your budget is suddenly doubled. Graph the new budget constraint on the same axes.

    (d) With a budget of $\$k$, the price of suntan oil suddenly doubles. Sketch the new budget constraint on the same axes.

18. Since the opening up of the West, the U.S. population has moved westward. To observe this, we look at the 'population center' of the U.S., which is the point at which the country would balance if it were a flat plate with no weight, and every person had equal weight. In 1790 the population center was east of Baltimore, Maryland. It has been moving westward ever since, and in 1990 it crossed the Mississippi river to Steelville, Missouri (southwest of St. Louis). During the second half of this century, the center of gravity has moved about 50 miles west every 10 years.

    (a) Express the approximate position of the center of gravity as a function of time, measured in years from 1990. Measure position westward from Steelville, along the line running through Baltimore.

    (b) If the distance from Baltimore to St. Louis is a bit over 700 miles, could the center of gravity have been moving at roughly the same rate for the last two centuries?

    (c) Could the function in (a) continue to apply for the next three centuries? Why or why not? [Hint: You may want to look at a map. Note that distances are in air miles and are not driving distances.]

19. For small changes in temperature, the formula for the expansion of a metal rod under a change in temperature is:

$$l - l_0 = al_0(t - t_0),$$

where l is the length of the object at temperature t, and l_0 is the initial length at temperature t_0, and a is a constant which depends on the type of metal.

(a) Express l as a linear function of t. Find the slope and y-intercept.
    [Hint: Treat the other quantities as constants.]

(b) Suppose you had a rod which was initially 100 cm long at 60° F and made of a metal with a equal to 10^{-5}. Write an equation giving the length of this rod at temperature t.

(c) What does the sign of the slope of the graph tell you about the expansion of a metal under a change in temperature?

20. When a cold yam is put into a hot oven to bake, the temperature of the yam rises. The rate, R (in degrees per minute), at which the temperature of the yam rises is governed by Newton's Law of Heating, which says that the rate is proportional to the temperature difference between the yam and the oven. If the oven is at 350°F, and the temperature of the yam is H°F,

(a) write a formula giving R as a function of H.

(b) Sketch the graph of R against H.

21. When a cup of hot coffee sits on the kitchen table, its temperature falls. The rate, R, at which its temperature changes is governed by Newton's Law of Cooling, which says that the rate is proportional to the temperature difference between the coffee and the surrounding air. Let's think of the rate, R, as a negative quantity because the temperature of the coffee is falling. If the temperature of the coffee is H°C, and the temperature of the room is 20°C,

(a) write a formula giving R as a function of H.

(b) Sketch a graph of R against H.

22. A body of mass m is falling downwards with velocity v. Newton's Second Law of Motion, $F = ma$, says that the net downward force, F, on the body is proportional to its downward acceleration, a. The net force, F, consists of the force due to gravity, F_g, which acts downwards, minus the air resistance, F_r, which acts upwards. The force due to gravity is mg, where g is a constant. Assume the air resistance is proportional to the velocity of the body.

(a) Write an expression for the net force, F, as a function of the velocity, v.

(b) Write a formula giving a as a function of v.

(c) Sketch a against v.

1.3   Exponential Functions

Population Growth

We will consider the data for the population of Mexico in the early 1980's in Table 1.4. To see how the population is growing, you might look at the increase in population from one year to the next as shown in the third column. If the population had been growing linearly, all the numbers in the third column would have been the same. But populations usually grow faster as they get bigger, because there are more people to have babies. So you shouldn't be surprised to see the numbers in the third column increasing.

Year	Population (in millions)	Change in Population (in millions)
1980	67.38	
		1.75
1981	69.13	
		1.80
1982	70.93	
		1.84
1983	72.77	
		1.89
1984	74.66	
		1.94
1985	76.60	
		1.99
1986	78.59	

Table 1.4: Population of Mexico (estimated), 1980–1986

Suppose we divide each year's population by the previous year's population. We get, approximately,

$$\frac{\text{Population in 1981}}{\text{Population in 1980}} = \frac{69.13 \text{ million}}{67.38 \text{ million}} = 1.026$$

$$\frac{\text{Population in 1982}}{\text{Population in 1981}} = \frac{70.93 \text{ million}}{69.13 \text{ million}} = 1.026$$

The fact that both calculations give 1.026 shows the population grew by about 2.6% between 1980 and 1981 *and* between 1981 and 1982. If you do similar calculations for other years, you will find the population grew by a factor of about 1.026, or 2.6%, every year. Whenever you have a constant growth factor (here 1.026), you have *exponential growth*. If t is the number of years since 1980,

when $t = 0$, population $= 67.38 = 67.38(1.026)^0$
when $t = 1$, population $= 69.13 = 67.38(1.026)^1$
when $t = 2$, population $= 70.93 = 69.13(1.026) = 67.38(1.026)^2$
when $t = 3$, population $= 72.77 = 70.93(1.026) = 67.38(1.026)^3$

and so t years after 1980, the population is given by

$$P = 67.38(1.026)^t.$$

This is an *exponential function* with base 1.026. It is called exponential because the variable, t, is in the exponent. The base represents the factor by which the population grows each year.

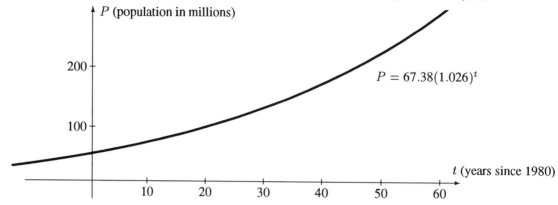

Figure 1.16: Population of Mexico (estimated): Exponential Growth

If we assume that the same formula will hold for the next 50 years or so, the population will have the shape shown in Figure 1.16. Clearly the function is increasing. Notice also that the graph grows faster and faster as time goes on. This behavior is typical of an exponential function. You should compare this with the behavior of a linear function which climbs at the same rate everywhere and so has a straight-line graph. Because this graph is bending upwards, we say it is *concave up*. Even exponential functions which climb slowly at first, such as this one, climb extremely quickly eventually. That is why exponential population growth is such a threat to the world.

Even if it represents reliable data, the smooth graph in Figure 1.16 is actually only an approximation to the true graph of the population of Mexico. Since we can't have fractions of people, the graph should really be jagged, jumping up or down by one each time someone is born or dies. However, with a population in the millions, the jumps are so small as to be invisible with the scale we are using. Therefore, the smooth graph is an extremely good approximation.

☐ **Example 1** Predict the population of Mexico in the year

    (a) 2007, (when $t = 27$)   (b) 2034, (when $t = 54$)   (c) 2061, (when $t = 81$).

Solution. Extrapolating so far into the future can be risky because it assumes that the population continues to grow exponentially at the same constant rate. (There could, for example, be a medical breakthrough that would increase the rate, or an epidemic that would decrease it.) Writing "$\approx$" to represent approximately equal, the model we are using predicts

    (a) $P = 67.38(1.026)^{27} = 134.74 \approx 67.38(2)$ million.

    (b) $P = 67.38(1.026)^{54} = 269.46 \approx 67.38(4)$ million.

    (c) $P = 67.38(1.026)^{81} = 538.85 \approx 67.38(8)$ million.

☐

If you look at the answers to Example 1, you will see something that may surprise you. After 27 years the population has doubled; after another 27 years (at $t = 54$), it has doubled again. Another 27 years later (when $t = 81$), the population has doubled yet again. As a result, we say that the *doubling time* of the population of Mexico is 27 years.

Every exponentially growing population has a fixed doubling time. The world's population currently has a doubling time of about 38 years. Notice what this means: if you live to be 76, the world's population is expected to quadruple in your lifetime.

Concavity

We have used the term concave up to describe the graph in Figure 1.16. In general:

> The graph of any function is **concave up** if it bends upwards, and **concave down** if it bends downwards.

A line is neither concave up nor concave down.

Musical Pitch

The pitch of a musical note is determined by the frequency of the vibration which causes it. Middle C on the piano, for example, corresponds to a vibration of 263 hertz (cycles per second). A note one octave above Middle C vibrates at 526 hertz, and a note two octaves above Middle C vibrates at 1052 hertz. See Table 1.5.

Number, n, of Octaves above Middle C	Number of hertz $(V = f(n))$
0	263
1	526
2	1052
3	2104
4	4208

Table 1.5: Pitch of Notes above Middle C

n	$V = 263 \cdot 2^n$
-3	$263 \cdot 2^{-3} = 263 \left(\frac{1}{2^3}\right) = 32.875$
-2	$263 \cdot 2^{-2} = 263 \left(\frac{1}{2^2}\right) = 67.75$
-1	$263 \cdot 2^{-1} = 263 \left(\frac{1}{2}\right) = 131.5$
0	$263 \cdot 2^0 = 263$

Table 1.6: Pitch of Notes below Middle C

Notice that

$$\frac{526}{263} = 2 \quad \text{and} \quad \frac{1052}{526} = 2 \quad \text{and} \quad \frac{2104}{1052} = 2$$

and so on. In other words, each value of V is twice the value before, so

$$f(1) = 526 = 263 \cdot 2 = 263 \cdot 2^1$$
$$f(2) = 1052 = 526 \cdot 2 = 263 \cdot 2^2$$
$$f(3) = 2104 = 1052 \cdot 2 = 263 \cdot 2^3.$$

In general

$$V = f(n) = 263 \cdot 2^n.$$

The base of 2 represents the fact that as we go up an octave, frequency of vibrations doubles. Indeed, our ears hear a note as one octave higher than another precisely because it vibrates twice as fast. For the negative values of n in Table 1.6, this function represents the octaves below Middle C. The notes on a piano are represented by values of n between -3 and 4, and the human ear finds values of n between -4 and 7 audible.

Although $V = f(n) = 263 \cdot 2^n$ only makes sense in musical terms for certain values of n, values of the function $f(x) = 263 \cdot 2^x$ can be calculated for all real x, and its graph has the typical exponential shape, as can be seen in Figure 1.17. It is concave up, climbing faster and faster as x increases.

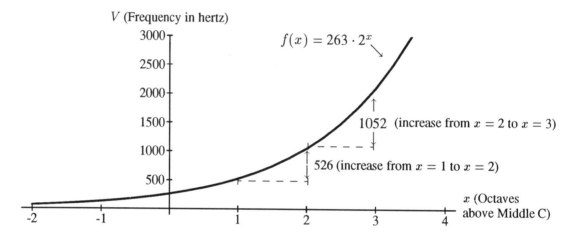

Figure 1.17: Pitch as a function of Number of Octaves above Middle C

Removal of Pollutants from Jet Fuel

Now we will look at an example in which a quantity is decreasing instead of increasing. Before kerosene can be used as jet fuel, federal regulations require that the pollutants in it be removed by passing the kerosene through clay. We will suppose the clay is in a pipe and that each foot of the pipe removes 20% of the pollutants that enter it. Therefore each foot leaves 80% of the pollution. If P_0 is the initial quantity of pollutant and $P = f(n)$ is the quantity left after n feet of pipe:

$$\begin{aligned} f(0) &= P_0 \\ f(1) &= (0.8)P_0 \\ f(2) &= (0.8)(0.8)P_0 = (0.8)^2 P_0 \\ f(3) &= (0.8)(0.8)^2 P_0 = (0.8)^3 P_0 \end{aligned}$$

and so, after n feet,

$$P = f(n) = P_0 (0.8)^n.$$

In this example, n must be nonnegative. However, the *exponential decay function*

$$P = f(x) = P_0(0.8)^x$$

makes sense for any real x. We'll plot it with $P_0 = 1$ in Figure 1.18; some values of the function are in Table 1.7.

x	$P = (0.8)^x$
-2	1.56
-1	1.25
0	1
1	0.8
2	0.64
3	0.51
4	0.41

Table 1.7: Values of Decay Function

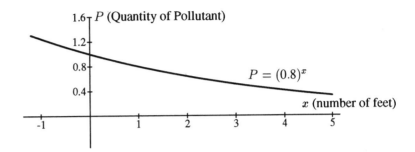

Figure 1.18: Pollutant Removal: Exponential Decay

Notice the way the function in Figure 1.18 is decreasing: each downward step is smaller than the last. This is because as the kerosene gets cleaner, there's less dirt to remove, and so each foot of clay takes out less pollutant than the previous one. Compare this to the exponential growth in Figures 1.16 and 1.17 on pages 20 and 22, where each step upward is larger than the last. Notice, however, that both graphs are concave up.

The General Exponential Function

> P is an **exponential function** of t with base a if
>
> $$P = P_0 a^t$$
>
> where P_0 is the initial quantity (when $t = 0$) and a is the factor by which P changes when t increases by 1.
> If $a > 1$, we have exponential growth; if $0 < a < 1$, we have exponential decay.

The largest possible domain for the exponential function is all real numbers, provided $a > 0$, $a \neq 1$. (Why do we not want $a < 0$, $a = 0$, or $a = 1$?)

> To recognize that a function $P = f(t)$ given by a table of data is exponential: look for ratios of P values that are constant for equally spaced t values.

In addition,

> the **doubling time** of an exponentially increasing population is the time for it to double; the **half-life** of an exponentially decaying quantity is the time for it to be reduced to half.

Drug Build-Up

Suppose that we want to model the amount of a certain drug in the body. Imagine that initially there is none, but that the quantity slowly starts to increase, via a continuous intravenous injection. As the quantity in the body increases, so does the rate at which the body excretes the drug, so that eventually the quantity levels off at a saturation value, S. The graph of quantity against time will look something like that in Figure 1.19.

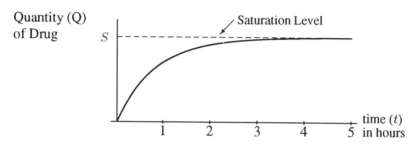

Figure 1.19: Build up of Drug in a Body

Notice that the quantity, Q, starts at zero, and increases towards S. We say that the line representing the saturation level is a *horizontal asymptote*, because the graph gets closer and closer to it as time increases. Since the rate at which the quantity of the drug increases slows as it approaches S, this graph is bending downward, i.e. it is *concave down*.

Suppose we want to make a mathematical model of this situation, that is, suppose we want to find a formula giving the quantity, Q, in terms of time, t. Making a mathematical model often involves looking at a graph and deciding what kind of function has that shape. The graph in Figure 1.19 looks like an exponential decay function, upside down. What actually decays is the difference between the saturation level S and the quantity Q in the blood. If we believe that the difference between the saturation level and the quantity in the body decreases by a factor of 30% every hour, then

$$\text{difference} = (\text{initial difference}) \cdot (0.3)^t$$

where t is in hours. Since the difference is $S - Q$, and the initial value of this difference is $S - 0 = S$:

$$S - Q = S \cdot (0.3)^t.$$

Solving for Q as a function of t gives

$$
\begin{aligned}
Q &= S - S \cdot (0.3)^t \\
Q &= f(t) = S \cdot \left(1 - (0.3)^t\right).
\end{aligned}
$$

This function is an upside down exponential because of the minus sign. As t gets larger, $(0.3)^t$ gets smaller and smaller. Using '→' to mean 'tends to', we can say $(0.3)^t \to 0$ as $t \to \infty$. This shows that

$$Q = S(1 - (0.3)^t) \to S(1 - 0) = S \quad \text{as} \quad t \to \infty$$

confirming that the graph of $Q = S(1 - (0.3)^t)$ has a horizontal asymptote at $Q = S$.

The Family of Exponential Functions

The formula $P = P_0 a^t$ gives a family of exponential functions with parameters P_0 (the initial quantity) and a (the base, or growth factor). The base is as important for an exponential function as the slope is for a linear function. We assume $a > 0$ and $a \neq 1$. Then the base tells you whether the function is increasing ($a > 1$) or decreasing ($a < 1$). Since a is the factor by which P changes when t is increased by 1, large values of a mean fast growth; values of a near 0 mean fast decay. See Figures 1.20 and 1.21.

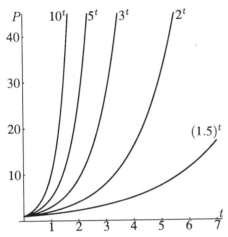

Figure 1.20: Exponential Growth: $P = a^t$, $a > 1$

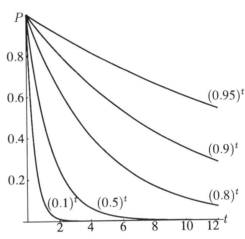

Figure 1.21: Exponential Decay: $P = a^t$, $0 < a < 1$

Alternate Formula for the Exponential Function

Exponential growth is often described in terms of percentages. For example, the population of Mexico is growing at 2.6% per year; in other words, the growth factor is $a = 1.026$. Similarly, each foot of clay removes 20% of the pollution from jet fuel, so the decay factor is $a = 1 - 0.20 = 0.8$. In general,

---

If r is the *growth* rate, then $a = 1 + r$, and

$$P = P_0 a^t = P_0 (1 + r)^t.$$

If r is the *decay* rate, then $a = 1 - r$, and

$$P = P_0 a^t = P_0 (1 - r)^t.$$

Note that $r = 0.05$ when the percentage growth rate is 5%.

☐ **Example 2** Suppose that $Q = f(t)$ is an exponential function of t. If $f(20.1) = 88.2$ and $f(20.3) = 91.4$:
(a) find the base, (b) find the percentage growth rate, (c) evaluate $f(21.4)$.

Solution.

(a) Let
$$Q = Q_0 a^t.$$

Then
$$88.2 = Q_0 a^{20.1} \quad \text{and} \quad 91.4 = Q_0 a^{20.3}.$$

Dividing gives
$$\frac{91.4}{88.2} = \frac{Q_0 a^{20.3}}{Q_0 a^{20.1}} = a^{0.2}.$$

Solve for the base, a:
$$a = \left(\frac{91.4}{88.2}\right)^{\frac{1}{0.2}} = 1.195.$$

(b) Since $a = 1.195$, the growth rate is $r = 0.195 = 19.5\%$.

(c) We want to find $f(21.4) = Q_0 a^{21.4} = Q_0 (1.195)^{21.4}$. First let's find Q_0 from the equation $91.4 = Q_0(1.195)^{20.3}$. Solving gives $Q_0 = 2.457$. Thus,
$$f(21.4) = 2.457(1.195)^{21.4} = 111.19.$$

❑

Reminder: Definition of Exponentials

$$a^0 = 1, \quad a^{-1} = \frac{1}{a}, \quad \text{and, in general, } a^{-x} = \frac{1}{a^x}$$

$$a^{\frac{1}{2}} = \sqrt{a}, \quad a^{\frac{1}{3}} = \sqrt[3]{a}, \quad \text{and, in general, } a^{\frac{1}{n}} = \sqrt[n]{a}.$$

Reminder: Rules for Manipulating Exponentials

1. $a^x \cdot a^t = a^{x+t}$ because, for example, $2^4 \cdot 2^3 = (2 \cdot 2 \cdot 2 \cdot 2) \cdot (2 \cdot 2 \cdot 2) = 2^7$

2. $\dfrac{a^x}{a^t} = a^{x-t}$ because, for example, $\dfrac{2^4}{2^3} = \dfrac{2 \cdot 2 \cdot 2 \cdot 2}{2 \cdot 2 \cdot 2} = 2^1$

3. $(a^x)^t = a^{xt}$ because, for example, $(2^3)^2 = 2^3 \cdot 2^3 = 2^6$

❖ Exercises for Section 1.3

1. The number of cancer cells grows slowly at first and then with increasing rapidity. Draw a possible graph of the number of cancer cells against time.

2. Each year, the world's annual consumption of electricity rises. In addition, each year the annual increase also rises. Sketch a possible graph of the annual world consumption of electricity as a function of time.

3. A drug is injected into a patient's bloodstream over a five minute interval. During this time, the quantity in the blood increases linearly. At five minutes, the injection is discontinued and after that the quantity decays exponentially. Sketch a graph of the quantity versus time.

4. When there are no other steroid hormones (for example, estrogen) in a cell, the rate at which steroid hormones diffuse into a cell is fast. The rate slows down as the amount in the cell builds up. Sketch a possible graph of the quantity of steroid hormone in the cell against time, assuming that initially there are no steroid hormones in the cell.

5. Each of the functions in the table to the right is increasing, but each increases in a different way. Which of the graphs below best fits each function?

t	$g(t)$	t	$h(t)$	t	$k(t)$
1	23	1	10	1	2.2
2	24	2	20	2	2.5
3	26	3	29	3	2.8
4	29	4	37	4	3.1
5	33	5	44	5	3.4
6	38	6	50	6	3.7

(a)   (b)   (c)

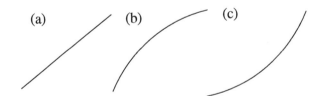

6. Each of the functions in the table to the right decreases, but each decreases in a different way. Which of the graphs below best fits each function?

x	$f(x)$	x	$g(x)$	x	$h(x)$
1	100	1	22.0	1	9.3
2	90	2	21.4	2	9.1
3	81	3	20.8	3	8.8
4	73	4	20.2	4	8.4
5	66	5	19.6	5	7.9
6	60	6	19.0	6	7.3

(a)   (b)   (c)

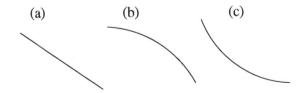

7. Match up the function values in the table to the right with the formulas

$$y = a(1.1)^s, \quad y = b(1.05)^s, \quad y = c(1.03)^s,$$

assuming a, b, and c are constants.

s	$h(s)$	s	$f(s)$	s	$g(s)$
2	1.06	1	2.20	3	3.47
3	1.09	2	2.42	4	3.65
4	1.13	3	2.66	5	3.83
5	1.16	4	2.93	6	4.02
6	1.19	5	3.22	7	4.22

For Problems 8–9, find a possible formula for the functions represented by the data.

8.

x	0	1	2	3
$f(x)$	4.30	6.02	8.43	11.80

9.

t	0	1	2	3
$g(t)$	5.50	4.40	3.52	2.82

Find possible equations for the graphs in Problems 10–13.

10.
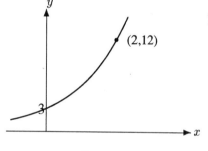

11.
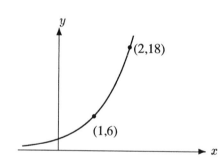

12.
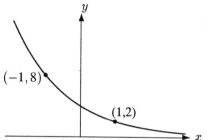

13.
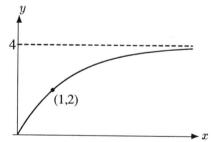

14. When the Olympic Games were held outside Mexico City in 1968, there was much discussion about the effect the high altitude (7340 feet) would have on the athletes. Assuming air pressure decays exponentially at 0.4% every 100 feet, by what percentage is air pressure reduced by moving from sea level to Mexico City?

15. A population is known to be growing exponentially. Estimate the doubling time of the population shown in Figure 1.22 and verify graphically that the doubling time is independent of where you start on the graph.

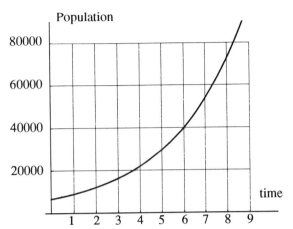

Figure 1.22: Is the doubling time the same no matter where you start? (For Problem 15)

16. A certain region has a population of 10,000,000 and an annual growth rate of 2%. Estimate the doubling time by trial and error.

17. A certain radioactive substance decays exponentially in such a way that after 10 years, 70% of the initial amount remains. Find an expression for the quantity remaining after any number of years t. How much will be present after 50 years? What is the half life? How long will it be before only 20% of the original amount is left? before only 10% is left? (Use trial and error where necessary.)

18. Assume that the median price, P, of a home rose from $50,000 in 1970 to $100,000 in 1990. Let t be the number of years since 1970.

    (a) Assume the increase in housing prices has been linear. Find an equation for the line representing price P in terms of t. Use this equation to complete column (a) of Table 1.8. Work with the price in $1000 units.

    (b) If instead the housing prices have been rising exponentially, determine an equation of the form $P = P_0 a^t$ which would represent the change in housing prices from 1970-1990 and complete column (b) of Table 1.8.

    (c) On the same set of axes, sketch the functions represented in column (a) and column (b) of Table 1.8.

t	a) Linear Growth Price in 1000's of $	b) Exponential Growth Price in 1000's of $
0	50	50
10	75	70.5
20	100	100
30	125	140
40	150	200

Table 1.8: The cost of a home
(for Problem 18)

Countries with very high inflation rates often publish monthly rather than yearly inflation figures, because monthly figures are less alarming. Problems 19– 21 involve such high rates, which are called *hyperinflation*.

19. In 1989, U.S. inflation was 4.6% a year. In 1989, Argentina had an inflation rate of about 33% a month.

   (a) What is the yearly equivalent of Argentina's 33% monthly rate?

   (b) What is the monthly equivalent of the U.S. 4.6% yearly rate?

20. Between December 1988 and December 1989, Brazil's inflation rate was 1290% a year. (This means that between 1988 and 1989, prices increased by a factor of 13.90.)

   (a) What would an article which cost 1000 cruzados (the Brazilian currency unit) in 1988 cost in 1989?

   (b) What was Brazil's monthly inflation rate during this period?

21. During 1988, Nicaragua's inflation rate averaged 1.3% a day. This means that, on average, prices went up by 1.3% from one day to the next.

   (a) By what factor did Nicaraguan prices increase each month during 1988?

   (b) What was Nicaragua's annual inflation rate during 1988?

1.4   The Number e

Many applications of exponential functions are expressed in terms of the base e, where $e \approx 2.718\ldots$, rather than the seemingly more natural bases 2 or 10. The fact that most calculators have a key for e^x is good evidence that people find the exponential function with this base useful. At the moment we will look at the behavior of exponential functions involving e, and in Chapter 4 we will see that they are used because some formulas are simpler when written with base e than with any other base.

The Family of Exponential Functions with Base e

By sketching the graph of $y = e^x$ we see that for each positive value of y there is a unique value of x such that $y = e^x$. Thus, for any positive number a, we can write

$$a = e^k$$

for some number k. (In fact, k is called the *natural logarithm* of a.) If $a > 1$, k is positive, and if $0 < a < 1$, k is negative. Thus the function representing an exponentially growing population can be rewritten as

$$P = P_0 a^t = P_0 (e^k)^t = P_0 e^{kt}$$

with k positive. When $0 < a < 1$,

$$a = e^{-k}$$

with k another positive constant. Thus if Q is a quantity that is decaying exponentially and Q_0 is the initial quantity, at time t we will have

$$Q = Q_0 a^t = Q_0 (e^{-k})^t = Q_0 e^{-kt} = \frac{Q_0}{e^{kt}}.$$

Since e^{kt} is now in the denominator, Q will decrease as time goes on—as you'd expect if Q is decaying.

Any **exponential growth** function can be written in the form

$$P = P_0 e^{kt}$$

and any **exponential decay** function can be written as

$$Q = Q_0 e^{-kt}$$

where P_0 and Q_0 are the initial quantities and k is positive.

We say that P and Q are growing or decaying at a *continuous rate* of k.
(Note that $k = 0.02$ if the continuous growth rate is 2%.)

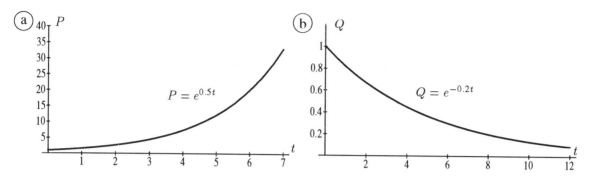

Figure 1.23: (a) An Exponential Growth Function (b) An Exponential Decay Function

☐ **Example 1**   Sketch the graphs of $P = e^{0.5t}$ and $Q = e^{-0.2t}$.

Solution. The graph of $P = e^{0.5t}$ is in Figure 1.23(a). Notice that the graph is the same shape as the previous exponential growth curves: increasing and concave up. Indeed, the families a^t, $a > 1$, and e^{kt}, $k > 0$, are one and the same. The graph of $Q = e^{-0.2t}$ is in Figure 1.23(b); it too has the same shape as other exponential decay functions. The families a^t, $0 < a < 1$, and e^{kt}, $k < 0$, are one and the same.   ☐

You can see that the use of e does not change the properties of the exponential function. Any exponential growth or decay equation that can be written involving e can also be written using any other base, and vice versa. However, there are a great many calculations which are much easier with e than with any other base.

Radioactive Decay

Radioactive decay is a process which is usually modeled by an exponential function with base e. You have seen that a population which grows exponentially has a doubling time, and that a radioactive material has a half-life, which is the time it takes for half the material to decay.

☐ **Example 2**   Carbon-14, used by archaeologists for carbon dating, is radioactive and decays according to the equation

$$Q = Q_0 e^{-0.000121t} \quad (t \text{ in years}).$$

Confirm that the half-life of Carbon-14 is about 5730 years.

Solution. Using a calculator we'll check that the quantity Q remaining is reduced by half when $t = 5730$:

$$Q = Q_0 e^{-0.000121(5730)} \approx (0.50)Q_0 \leftarrow \begin{cases} \text{half the} \\ \text{original} \\ \text{quantity} \end{cases}$$

After another 5730 years (i.e., after a total of 11460 years), the quantity remaining should be reduced by half again, and we see that it is:

$$Q = Q_0 e^{-0.000121(11460)} \approx (0.25)Q_0 \leftarrow \begin{cases} \text{one-quarter} \\ \text{the original} \\ \text{quantity} \end{cases}$$

☐

Compound Interest

If you have some money, you may decide to invest it to earn interest. The interest can be paid in many different ways - for example, once a year or many times a year. If the interest is paid more frequently than once per year there is a benefit to the investor since the interest earns interest. This effect is called *compounding*. You may have noticed banks offering accounts that differ both in interest rates and in compounding methods. Some offer interest compounded annually, some offer quarterly and others daily compounding. Some even offer continuous compounding.

What is the difference between a bank account advertising 8% compounded annually (once per year) and one offering 8% compounded quarterly (four times per year)? In both cases 8% is an annual rate of interest. The 8% *compounded annually* means that at the end of each year, 8% of the current balance is added. This is equivalent to multiplying the current balance by 1.08. Thus, if $100 is deposited, the balance, B, in dollars will be

$$B = 100(1.08) \qquad \text{after one year,}$$
$$B = 100(1.08)^2 \qquad \text{after two years,}$$
$$B = 100(1.08)^t \qquad \text{after } t \text{ years.}$$

The 8% *compounded quarterly* means that interest is added four times per year (every three months) and that $\frac{8}{4} = 2\%$ of the current balance is added each time. Thus, if $100 is deposited, at the end of one year, four compoundings will have taken place and the account will contain $100(1.02)^4$. Thus the balance, B, in dollars will be

$$B = 100(1.02)^4 \qquad \text{after one year,}$$
$$B = 100(1.02)^8 \qquad \text{after two years,}$$
$$B = 100(1.02)^{4t} \qquad \text{after } t \text{ years.}$$

Note that 8% is *not* the rate used for each three month period; the annual rate is divided into four 2% payments. Calculating the total balance after one year under each method shows that

$$\text{annual compounding:} \qquad B = 100(1.08) = 108.00$$
$$\text{quarterly compounding:} \qquad B = 100(1.02)^4 = 108.24$$

Thus, more money is earned from quarterly compounding, because the interest earns interest as the year goes by. In general, the more often interest is compounded, the more money will be earned (although the increase may not be very large).

We can measure the effect of compounding by introducing the notion of *effective annual yield*. Since $100 invested at 8% compounded quarterly grows to $108.24 by the end of one year, we say that the *effective annual yield* in this case is 8.24%. We now have two interest rates which describe the same investment: the 8% compounded quarterly, and the 8.24% effective annual yield. Banks call the 8% the *annual percentage rate*, or *APR*. We may also call the 8% the *nominal rate* (nominal means 'in name only'). However, it is the effective yield which tells you exactly how much interest the investment really pays: to compare two bank accounts, simply compare the effective annual yields. The next time that you walk by a bank, look at the advertisements, which should (by law) include both the APR, or nominal rate, and the effective annual yield. We will often abbreviate *annual percentage rate* by *annual rate*.

Using the Effective Annual Yield

☐ **Example 3** Which is better: Bank X paying a 7% annual rate compounded monthly or Bank Y offering a 6.9% annual rate compounded daily?

Solution. We will find the effective annual yield for each bank.

Bank X: There are 12 interest payments in a year, each payment being $\frac{0.07}{12} = 0.005833$ times the current balance. If the initial deposit were \$100 then the balance B will be

$$
\begin{aligned}
B &= 100(1.005833) && \text{after one month,} \\
B &= 100(1.005833)^2 && \text{after two months,} \\
B &= 100(1.005833)^t && \text{after } t \text{ months.}
\end{aligned}
$$

To find the effective annual yield we need to look at one year, or 12 months, which gives $B = 100(1.005833)^{12} = 100(1.072286)$, so the effective annual yield $\approx 7.23\%$.

Bank Y: There are 365 interest payments in a year (assuming it is not a leap-year), each being $\frac{0.069}{365} = 0.000189$ times the current balance. Then the balance B in the account is

$$
\begin{aligned}
B &= 100(1.000189) && \text{after one day,} \\
B &= 100(1.000189)^2 && \text{after two days,} \\
B &= 100(1.000189)^t && \text{after } t \text{ days.}
\end{aligned}
$$

so at the end of one year we have multiplied the initial deposit by

$$(1.000189)^{365} = 1.071413$$

so the effective annual yield for Bank Y $\approx 7.14\%$.

Comparing effective annual yields for the banks, we see that Bank X is offering a better investment, by a small margin. ☐

☐ **Example 4** If \$1000 is invested in each bank in Example 3, write an expression for the balance in each bank after t years.

Solution. For Bank X, the effective annual yield $\approx 7.23\%$, so after t years the balance in dollars will be

$$B = 1000(1.0723)^t.$$

For Bank Y, the effective annual yield $\approx 7.14\%$, so after t years the balance in dollars will be

$$B = 1000(1.0714)^t.$$

(Again, we are ignoring leap years.) ☐

In general:

If interest at an annual rate of r is compounded n times a year, then $\frac{r}{n}$ times the current balance is added n times a year. Thus the effective annual yield is $(1 + \frac{r}{n})^n - 1$.

With an initial deposit of $\$P$, the balance t years later is

$$B = P\left(1 + \frac{r}{n}\right)^{nt}$$

Note that r is the nominal rate, and that $r = 0.05$ when the annual rate is 5%.

Increasing the Frequency of Compounding: Continuous Compounding

☐ **Example 5** Find the effective annual yield for a 7% annual rate compounded
    a) 1000 times/year and   b) 10,000 times/year.

Solution.
a)
$$\left(1 + \frac{0.07}{1000}\right)^{1000} \approx 1.0725056$$

giving an effective annual yield of about 7.25056%.

b)
$$\left(1 + \frac{0.07}{10000}\right)^{10000} \approx 1.0725079$$

giving an effective annual yield of about 7.25079%.   ❏

You can see that there's not a great deal of difference between compounding 1000 times each year (about three times per day) and 10,000 times each year (about 30 times per day). What happens if we compound more often still? Every minute? Every second? You may be surprised that the effective annual yield does not increase indefinitely, but tends to a finite value. The benefit of increasing the frequency of compounding becomes negligible beyond a certain point.

For example, if you were to compute the effective annual yield on a 7% investment compounded n times per year for values of n larger than 100,000, you would find that

$$\left(1 + \frac{0.07}{n}\right)^n \approx 1.0725082.$$

So the effective annual yield is about 7.25082%. Even if you take $n = 1,000,000$ or $n = 10^{10}$, the effective annual yield will not change appreciably. The 7.25082% is an upper bound which is approached as the frequency of compounding increases.

When the effective annual yield is at this upper bound, we say that the interest is being *compounded continuously*. (The word 'continuously' is used because the upper bound is approached by compounding more and more frequently.) Thus, when a 7% nominal annual rate is compounded so frequently that the effective annual yield is 7.25082%, we say that the 7% is compounded *continuously*. This represents the most one can get from a 7% nominal rate.

Where does the number e fit in?

It turns out that e is intimately connected to continuous compounding. To see this, use your calculator to check that $e^{0.07} \approx 1.0725082$, which is the same number we obtained when we compounded 7% a large number of times. So you have discovered that for very large n:

$$\left(1 + \frac{0.07}{n}\right)^n \approx e^{0.07}.$$

As n gets larger, the approximation gets better and better, which we write as

$$\left(1 + \frac{0.07}{n}\right)^n \longrightarrow e^{0.07}$$

meaning that as n increases, the value of $\left(1 + \frac{0.07}{n}\right)^n$ gets closer and closer to $e^{0.07}$.

Thus, if $\$P$ is deposited at an annual rate of 7% compounded continuously, the balance, $\$B$, will be

$$
\begin{aligned}
B &= P(1.0725082) = Pe^{0.07} && \text{after one year,} \\
B &= P(1.0725082)^2 = P(e^{0.07})^2 = Pe^{(0.07)2} && \text{after two years,} \\
B &= P(1.0725082)^t = P(e^{0.07})^t = Pe^{0.07t} && \text{after } t \text{ years.}
\end{aligned}
$$

Generalizing, we conclude that

If interest on an initial deposit of $\$P$ is *compounded continuously* at an annual rate of r, the balance t years later can be calculated using the formula

$$B = Pe^{rt}.$$

(Again, r is the nominal rate, and $r = 0.05$ when the annual rate is 5%.)

When solving a problem involving compound interest, it is important to be clear whether interest rates are nominal rates or effective yields, as well as whether compounding is continuous or not.

☐ **Example 6** Find the effective annual yield of a 6% annual rate, compounded continuously.

Solution. In one year, an investment of P becomes $Pe^{0.06}$. Using a calculator, we see that

$$Pe^{0.06} = P(1.0618365)$$

So the affective annual yield is about 6.18%. ☐

☐ **Example 7** Suppose you want to invest money in a certificate of deposit (CD) for your child's education. You need it to be worth $12,000 in 10 years. How much should you invest if the CD pays interest at a 9% annual rate compounded quarterly? Continuously?

Solution. Suppose you invest $\$P$ initially. A 9% annual rate compounded quarterly has an effective annual yield given by $(1 + \frac{0.09}{4})^4 = 1.0930833$ or 9.30833%. So after 10 years you will have

$$P(1.0930833)^{10} = 12000.$$

Thus you should invest

$$P = \frac{12000}{(1.0930833)^{10}} = \frac{12000}{2.4351885} = 4927.75$$

On the other hand, if the CD pays 9% compounded continuously, after 10 years you will have

$$Pe^{(0.09)10} = 12000.$$

So you would need to invest

$$P = \frac{12000}{e^{(0.09)10}} = \frac{12000}{2.4596031} = 4878.84$$

Notice that to achieve the same result, continuous compounding requires a smaller initial investment than quarterly compounding. This is to be expected since the effective annual yield is higher for continuous than for quarterly compounding. ❑

❖ Exercises for Section 1.4

1. (a) A population grows according to the equation $P = P_0 e^{kt}$ (with P_0, k constants). Find the population as a function of time if it grows at a continuous rate of 2% a year and starts at 1 million.

    (b) Plot a graph of the population you found in (a) against time.

2. Air pressure, P, decreases exponentially with the height above the surface of the earth, h:

$$P = P_0 e^{-1.2 \times 10^{-4} h}$$

where P_0 is the air pressure at sea level and h is in meters.

    (a) If you go to the top of Mount McKinley (height 6,198 meters or 20,320 feet), what is the air pressure, as a fraction of the pressure at sea level?

    (b) The maximum cruising altitude of an ordinary commercial jet is around 12,000 meters (about 40,000 feet). At that height, what is the air pressure, as a fraction of the sea level value?

3. Under certain circumstances, the velocity, V, of a falling raindrop is given by

$$V = V_0(1 - e^{-t}),$$

where t is time and V_0 is a positive constant.

    (a) Sketch a rough graph of V against t, for $t \geq 0$.

    (b) What does V_0 represent?

4. One of the main contaminants of a nuclear accident, such as that at Chernobyl, is Strontium-90, which decays exponentially at a continuous rate of approximately 2.47% per year. Preliminary estimates after the Chernobyl disaster suggested that it would be about 100 years before the region would again be safe for human habitation. What percent of the original Strontium-90 would still remain by this time?

5. Use a graph of $y = \left(1 + \dfrac{0.07}{x}\right)^x$ to find the value of $\left(1 + \dfrac{0.07}{x}\right)^x$ as $x \to \infty$. Confirm that the value you get is $e^{0.07}$.

6. If you deposit $10,000 in an account earning interest at an 8% annual rate compounded continuously, how much money is in the account after five years?

7. (a) Find the effective annual yield for a 5% annual interest rate compounded
   i) 1000 times/year,   ii) 10,000 times/year,   iii) 100,000 times/year.

   (b) Look at the sequence of answers in a), and deduce the effective annual yield for a 5% annual rate compounded continuously.

   (c) Compute $e^{0.05}$. How does this confirm your answer to b)?

8. (a) Find $\left(1 + \dfrac{0.04}{n}\right)^n$ for $n = 10000$, and $100,000$, and $1,000,000$. Use the results to deduce the effective annual yield of a 4% annual rate compounded continuously.

   (b) Confirm your answer by computing $e^{0.04}$.

9. Use the number e to find the effective annual yield of a 6% annual rate, compounded continuously.

10. What is the effective annual yield of an investment paying at a 12% annual rate, compounded continuously?

11. Explain how you can match the interest rates a)-e) with the effective annual yields A)-E) without doing any calculations.

   (a) 5.5% annual rate, compounded continuously.

   (b) 5.5% annual rate, compounded quarterly.

   (c) 5.5% annual rate, compounded weekly.

   (d) 5% annual rate, compounded yearly.

   (e) 5% annual rate, compounded twice a year.

$$A)\,5\% \quad B)\,5.06\% \quad C)\,5.61\% \quad D)\,5.651\% \quad E)\,5.654\%.$$

12. When you rent an apartment, you are often required to give the landlord a security deposit which is returned if you leave the apartment undamaged. In Massachusetts, the landlord is required to pay the tenant interest on the deposit once a year, at a 5% annual rate, compounded

annually. The landlord, however, may invest the money at a higher (or lower) interest rate. Suppose the landlord invests a $1000 deposit at a

(a) 6% annual rate, compounded continuously

(b) 4% annual rate, compounded continuously.

In each case, how much net interest does the landlord make or lose each year? (Give your answer to the nearest cent.)

13. Refer to the newspaper article in Figure 1.24 and fill in the three blanks. (For the last blank, assume the interest has been compounded yearly, and give your answer in dollars.)

213 Years After Loan, Uncle Sam Is Dunned
By LISA BELKIN

Special to The New York Times

SAN ANTONIO, May 26 — More than 200 years ago, a wealthy Pennsylvania merchant named Jacob DeHaven lent $450,000 to the Continental Congress to rescue the troops at Valley Forge. That loan was apparently never repaid.

So Mr. DeHaven's descendants are taking the United States Government to court to collect what they believe they are owed. The total: ____ in today's dollars if the interest is compounded daily at 6 percent, the going rate at the time. If compounded yearly, the bill is only ____.

Family Is Flexible

The descendants say that they are willing to be flexible about the amount of a settlement and that they might even accept a heartfelt thank you or perhaps a DeHaven statue. But they also note that interest is accumulating at ____ a second.

Figure 1.24: From the New York Times, May 27, 1990.
(for Problem 13)

14. (a) The Banque Nationale du Zaïre pays 100% nominal interest on deposits, compounded monthly. You invest 1 million zaïre. (The "zaïre" is the unit of currency of the Republic of Zaïre.) How much money do you have after one year?

(b) How much money do you have after one year if you invest 1 million zaïre with interest compounded

   i. daily,   ii. hourly,   iii. each minute.

(c) Does this amount increase without bound as interest is compounded more and more often, or does it level off? If it levels off, provide a close "upper" estimate for the total after one year.

1.5 Power Functions

The family of power functions is an important family of functions. A power function is one in which the dependent variable is proportional to a power of the independent variable. For example, the area, A, of a square of side s is given by

$$A = f(s) = s^2.$$

The volume, V, of a sphere of radius r is

$$V = f(r) = \frac{4}{3}\pi r^3.$$

Both of these are *power functions*. So is the function which describes how the gravitational attraction of the Earth varies with distance. If g is the force of gravitational attraction on a unit mass at a distance r from the Earth, Newton's Inverse Square Law of Gravitation says that

$$g = \frac{k}{r^2} \qquad \text{or} \qquad g = kr^{-2}$$

where k is a positive constant.

In general, a **power function** has the form

$$y = f(x) = kx^p$$

where k and p are any constants.

In this section we will compare various power functions with one another and with the exponential functions. Since $y = f(x) = mx$ (with m a constant) is also a power function (because $x = x^1$, the first power), linear functions are included in the comparison too. We will look at where such functions are increasing, where they are decreasing, their concavity, what kind of symmetry they have, and their behavior near the origin. In addition, we'll often be interested in the comparative growth rates of these functions, meaning which grows most rapidly for large values of x.

Positive Integral Powers: $y = x$, $y = x^2$, $y = x^3$,...

First, we'll look at functions of the form $f(x) = x^n$, with n a positive integer. In Figure 1.25 you will observe that the graphs of these functions fall into two groups: the odd powers and the even powers. All the odd powers, x, x^3, x^5, are increasing everywhere and their graphs are symmetric about the origin. All odd powers above $n = 1$ have a 'bend' or 'seat' at the origin. The even powers, on the other hand, are first decreasing and then increasing, making them '$\bigcup$'-shaped with symmetry about the y-axis. The even powers are concave up everywhere, while the odd ones (greater than 1) are concave down for negative x and concave up for positive x. All odd and even powers, however, go through the points $(0, 0)$ and $(1, 1)$.

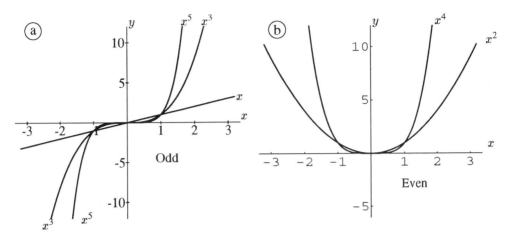

Figure 1.25: Comparison of Odd and Even Powers of x

Looking at Figure 1.26 shows that the higher the power of x, the faster the function climbs. (This is true for non-integral powers of x too, which we usually consider only for $x \geq 0$.) For large values of x (in fact, for all $x > 1$), $y = x^5$ is above $y = x^4$, which is above $y = x^3$, and so on. This is because if $x = 10$, for example, 10^5 is bigger than 10^4, which is bigger than 10^3. Not only are the higher powers larger, but they are *much* larger. As x gets larger (written as $x \to \infty$), any positive power of x completely swamps all lower powers of x. We say that, as $x \to \infty$, higher powers of x *dominate* lower powers.

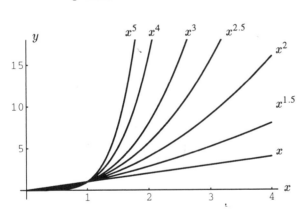

Figure 1.26: Powers of x: Which is Largest for Large Values of x?

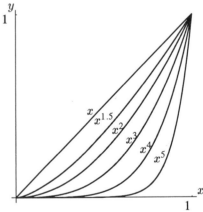

Figure 1.27: Between 0 and 1: Small Powers of x Dominate

The close-up view near the origin in Figure 1.27 shows an entirely different story. For x between 0 and 1, the order is reversed: x^3 is bigger than x^4, which is bigger than x^5. (Try $x = 0.1$ to confirm this.) The fact that higher powers of x climb faster is true for large values of x but not for small. For big values of x, the highest powers are largest; for values of x near zero, smaller powers dominate.

☐ **Example 1** On a graphing calculator or computer, plot

1. $y = x$, $y = x^3$, $y = x^6$, $y = x^9$

2. $y = x$, $y = x^4$, $y = x^7$, $y = x^{10}$

first for $-5 \leq x \leq 5$, $-100 \leq y \leq 100$, and then for $-1.2 \leq x \leq 1.2$, $-2 \leq y \leq 2$. Observe the general shape of these functions: (a) Do the odd powers still have the same general shape? What about the even powers? (b) Which function is largest for big x? For x near 0? Is this what you expected?

Zero and Negative Integral Powers: $y = x$, $y = x^{-1}$, $y = x^{-2}, \ldots$

The function $y = x^0 = 1$ has a graph that is a horizontal line. Rewriting

$$y = x^{-1} = \frac{1}{x} \quad \text{and} \quad y = x^{-2} = \frac{1}{x^2}$$

makes it easier to see that as x increases, the denominators increase and the functions decrease. The graphs of $y = x^{-1}$ and $y = x^{-2}$ have both x and y axes as asymptotes. See Figure 1.28. For $x > 1$, the graph of $y = x^{-2}$ is below that of $y = x^{-1}$ and both must stay below $y = x^0 = 1$.

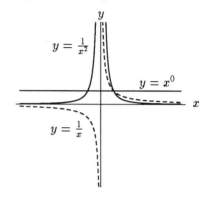

Figure 1.28: Comparison of Zero and Negative Powers of x

❏ **Example 2** Plot a graph of pressure against volume for a fixed quantity of gas at a constant temperature. Use the fact that pressure is inversely proportional to volume.

Solution. Think of a fixed quantity of air, for example inside a cylinder in a car engine. If the volume of the air is decreased (by moving the pistons), the pressure of the air will be increased. Conversely, if the volume is increased, the pressure will decrease. For an ideal gas Boyle's Law gives the exact relationship between pressure and volume, provided the temperature is constant. It says pressure times volume is constant, or

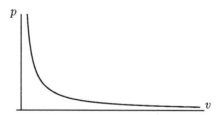

Figure 1.29: Graph of pressure against volume: $p = \frac{k}{v}$

$$pv = k$$

with k positive. So

$$p = \frac{k}{v}.$$

The relationship $p = k/v$ is equivalent to saying p is inversely proportional to v. When v is large, p is small, and when v is small, p is large. For any (positive) value of k, the graph has the shape shown in Figure 1.29. Both axes are asymptotes, showing that as the volume tends to infinity, the pressure tends to zero, and vice versa. The shape is known as a rectangular *hyperbola*. The power function $p = \frac{k}{v} = kv^{-1}$ differs from exponential decay in being undefined for $v = 0$, so this graph does not cross the vertical axis. In addition, it approaches the horizontal axis more slowly than the exponential. ☐

☐ **Example 3** Quantum mechanics predicts that the force between two gas molecules has two components: an attractive force which is proportional to r^{-7} (where r is the distance between the molecules) and a repulsive force proportional to r^{-13}. How does the net force vary with r?

Solution. A repulsive force is usually considered to be positive, while an attractive force is negative. Thus we write $F = -ar^{-7} + br^{-13}$ where a, b are positive constants. If r is very small, r^{-13} is larger than r^{-7} so $F \approx br^{-13}$ and the net force is repulsive. (This occurs when the molecules are so close together that the electrons surrounding the nuclei repel one another.) For larger r, r^{-7} is larger than r^{-13} and $F \approx -ar^{-7}$, making the net force attractive. As $r \to \infty$, all the forces die away to zero. ☐

Positive Fractional Powers: $y = x^{1/2}$, $y = x^{1/3}$, $y = x^{3/2}$, $\ldots$

The function giving the side of a square, s, in terms of its area, A, involves a root, or fractional power:

$$s = \sqrt{A} = A^{1/2}.$$

Similarly, the average number of species found on an island varies with the size of the island. If N is the number of species and A is the area of the island, observations have shown[2] that approximately

$$N = k\sqrt[3]{A} = kA^{1/3}$$

where k is a constant depending on the region of the world in which the island is found.

We will now look at functions of the form $y = x^{m/n} = \sqrt[n]{x^m}$. Since some fractional powers such as $x^{1/2}$ involve roots and are defined only for positive x and 0, we frequently restrict the domain of positive fractional powers of x to $x \geq 0$.

Figure 1.30 shows that for large x (in fact, all $x > 1$), the graph of $y = x^{1/2}$ is below the graph of $y = x$ and $y = x^{1/3}$ is below $y = x^{1/2}$. This is reasonable since, for example, $10^{1/2} = \sqrt{10} \approx 3.16$ and $10^{1/3} = \sqrt[3]{10} \approx 2.15$, so $10^{1/3} < 10^{1/2} < 10$. Between $x = 0$ and $x = 1$, the situation is reversed, and $y = x^{1/3}$ is on top. (Why?) Not surprisingly, $y = x^{3/2}$ is between $y = x$ and $y = x^2$ for all x.

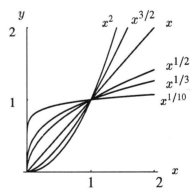

Figure 1.30: Comparison of Some Fractional Powers of x

---

[2]*Scientific American*, September 1989, p. 112.

☐ **Example 4** Do some calculations using specific values of x to verify that $y = x^{1/3}$ is above $y = x^{1/2}$ and that $y = x^{1/2}$ is above $y = x$ for $0 < x < 1$.

The other important feature to notice about the graphs of $y = x^{1/2}$ and $y = x^{1/3}$ is that they are bending in the opposite direction to the graphs of $y = x^2$ and x^3. For example, the graph of $y = x^2$ is climbing faster and faster as x increases; it is concave up. On the other hand, the graphs of $y = x^{1/2}$ and $y = x^{1/3}$ are climbing slower and slower. They are concave down. Despite this, all these functions do become infinitely large as x increases.

What Effect do Coefficients Have?

We know that $x^2 < x^3$ for all $x > 1$. But which is larger: $50x^2$ or x^3? Eventually $50x^2 < x^3$ too. In fact $50x^2 < x^3$ for all $x > 50$. See Figure 1.31. The effect of the factor of 50 is to change the point at which the graphs cross. However it does not change which power of x ends up on top: provided the coefficients are positive, as $x \to \infty$, the higher power is always larger eventually.

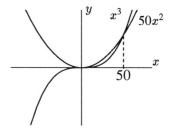

Figure 1.31: Comparison of $y = 50x^2$ and $y = x^3$

☐ **Example 5** Which of $f(x) = 100x^2$ and $g(x) = 0.1x^3$ is larger as $x \to \infty$?

Solution. Since $x \to \infty$, we are looking at large positive values of x, where the highest power will eventually dominate. Thus $g(x) = 0.1x^3$ will be larger. See Figure 1.32. ☐

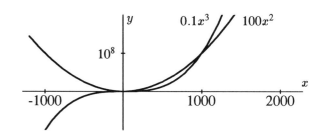

Figure 1.32: Solution to Example 5.

☐ **Example 6** Sketch a global view of $f(x) = -x^3$, $g(x) = 40x^4$, $h(x) = -0.1x^5$. Which function has the largest positive values as $x \to \infty$? As $x \to -\infty$?

Solution. As $x \to \infty$, $g(x) = 40x^4$ is the only one which is positive. As $x \to -\infty$, $h(x) = -0.1x^5$ is eventually (for $x < -400$) above the other functions. See Figure 1.33.

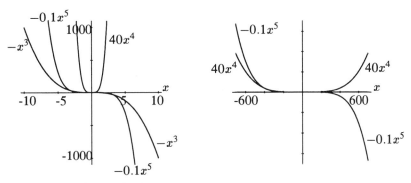

Figure 1.33: Solution Graphs for Example 6.

Exponentials and Power Functions: Which Dominate?

In everyday language, 'exponential' is often used to imply very fast growth. But do exponential functions always grow faster than power functions?

Let's consider $y = 2^x$ and $y = x^3$. The close-up, or local view, in Figure 1.34(a) shows that between $x = 2$ and $x = 5$, the graph of $y = 2^x$ lies below the graph of $y = x^3$. But the more global, or far-away, view in Figure 1.34(b) shows that the exponential function $y = 2^x$ eventually overtakes $y = x^3$. And Figure 1.34(c), which gives a truly global view, shows that, for large x, x^3 is insignificant compared to 2^x. Indeed, 2^x is growing so much faster than x^3 that it appears almost vertical—in comparison to the more leisurely climb of x^3.

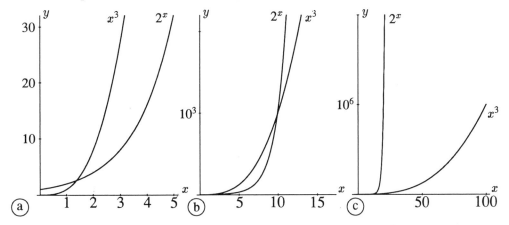

Figure 1.34: Comparison of $y = 2^x$ and $y = x^3$

Example 7 By trial and error, use a calculator to find to two decimal places the point near $x = 10$ at which $y = 2^x$ and $y = x^3$ cross.

Solution. $9.935 < x < 9.94$, so $x \approx 9.94$.

In fact, *every* exponential growth function eventually dominates *every* power function. Although an exponential function may be below a power function for some values of x, if you look at large enough x values, a^x (with $a > 1$) will eventually dominate x^n, no matter what n is. Two more examples are in Figure 1.35 and Table 1.9.

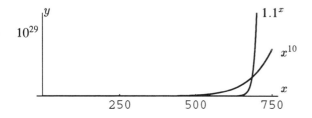

Figure 1.35: Comparison of $y = x^{10}$ and $y = 1.1^x$

x	x^{100}	1.01^x
10^4	10^{400}	1.6×10^{43}
10^5	10^{500}	1.4×10^{432}
10^6	10^{600}	2.4×10^{4321}

Table 1.9: Comparison of x^{100} and 1.01^x

Can you guess what happens in the case of negative powers and negative exponents? For example, consider $y = 2^{-x}$ and $y = x^{-2}$. Since $y = 2^{-x} = \frac{1}{2^x}$ and $y = x^{-2} = \frac{1}{x^2}$, knowing that 2^x is eventually larger than x^2 tells you that 2^{-x} is eventually smaller than x^{-2}. Hence $y = 2^{-x}$ is eventually below $y = x^{-2}$. See Figure 1.36. This behavior is also typical: every exponential decay function will eventually approach 0 faster than every power function with a negative exponent.

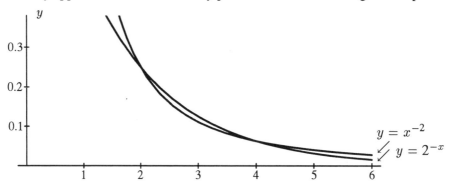

Figure 1.36: Comparison of $y = 2^{-x}$ and $y = x^{-2}$

❖ Exercises for Section 1.5

1. Simplify each of the following:

   (a) $8^{\frac{2}{3}}$

   (b) $9^{(-\frac{3}{2})}$

2. Sketch graphs of $y = x^{\frac{1}{2}}$ and $y = x^{\frac{2}{3}}$ on the same axes. Which is larger as $x \to \infty$?

3. Sketch a graph of $y = x^{-4}$.

4. Consider the graph in Figure 1.37

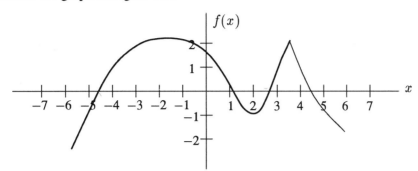

Figure 1.37: Graph for Problem 4

(a) How many zeros does this function have? Approximately where are they?

(b) Give approximate values for $f(2)$ and $f(4)$.

(c) Is the function increasing or decreasing at $x = -1$? How about at $x = 3$?

(d) Is the graph concave up or concave down at $x = 2$? How about at $x = -4$?

(e) List all intervals (approximately) on which the function is increasing.

5. (a) What happens to $y = x^4$ as $x \to \infty$? As $x \to -\infty$?

(b) What happens to $y = -x^7$ as $x \to \infty$? As $x \to -\infty$?

6. By hand, sketch global pictures of $f(x) = x^3$ and $g(x) = 20x^2$ on the same axes. Which is larger as $x \to \infty$?

7. By hand, sketch pictures of $f(x) = x^5$, $g(x) = -x^3$, and $h(x) = 5x^2$ on the same axes. Which has larger positive values as $x \to \infty$? As $x \to -\infty$?

8. (a) Use a graphing calculator (or a computer) to plot the graphs of x^3, x^4 and x^5 on the interval $-0.1 \le x \le 0.1$. Determine an appropriate range for y so that all powers will be distinguishable in the viewing rectangle.

(b) Plot the same graphs for $-100 \le x \le 100$ and determine an appropriate range for y.

9. Use a graphing calculator to find the point(s) of intersection of the functions $y = (1.06)^x$ and $y = 1 + x$.

10. For what values of x is $4^x > x^4$?

11. For what values of x is $3^x > x^3$? (How can you deal with the fact that the graphs of 3^x and x^3 are very close together?)

12. According to the April 1991 issue of *Car and Driver*, an Alfa Romeo going at 70 mph requires 177 feet to stop. Assuming that the stopping distance is proportional to the square of velocity, find the stopping distance required by an Alfa Romeo going at 35 mph and at 140 mph (its top speed).

13. Poiseuille's Law gives the rate of flow, R, of a gas through a cylindrical pipe in terms of the radius of the pipe, r, for a fixed drop in pressure. Assume a constant drop in pressure throughout the remainder of this problem.

    (a) Determine a formula for Poiseuille's Law, given that the rate of flow is proportional to the fourth power of the radius.

    (b) If R is 400 cm^3/sec. in a pipe of radius 3 cm. for a certain gas, determine an explicit formula for the rate of flow of that gas through a pipe of radius r cm.

    (c) What is the rate of flow of the same gas through a pipe with a 5 cm. radius?

14. The values of three functions are given in Table 1.10. One function is exponential, one is of the form $y = at^2$ and one is of the form $y = bt^3$. Which is which?

t	$f(t)$	t	$g(t)$	t	$h(t)$
2.0	4.40	1.0	3.00	0.0	2.04
2.2	5.32	1.2	5.18	1.0	3.06
2.4	6.34	1.4	8.23	2.0	4.59
2.6	7.44	1.6	12.29	3.0	6.89
2.8	8.62	1.8	17.50	4.0	10.33
3.0	9.90	2.0	24.00	5.0	15.49

Table 1.10: Which is exponential, which quadratic, and which cubic?

15. Data from three functions are contained in Table 1.11. (The numbers have been rounded to two decimal places.) Two are power functions and one is an exponential. One of the power functions is a quadratic and one a cubic. Which is which?

x	$f(x)$	x	$g(x)$	x	$k(x)$
8.4	5.93	5.0	3.12	0.6	3.24
9.0	7.29	5.5	3.74	1.0	9.01
9.6	8.85	6.0	4.49	1.4	17.66
10.2	10.61	6.5	5.39	1.8	29.19
10.8	12.60	7.0	6.47	2.2	43.61
11.4	14.82	7.5	7.76	2.6	60.91

Table 1.11: Which is exponential and which are power functions?

16. Owing to improved seed types and new agricultural techniques, the grain production of a region has been increasing. Over a 20-year period, production (in millions of tons) was as follows:

'70	'75	'80	'85	'90
5.35	5.90	6.49	7.05	7.64

At the same time the population (in millions) was:

'70	'75	'80	'85	'90
53.2	56.9	60.9	65.2	69.7

(a) Fit a linear or exponential function to each set of data. (Pick whichever type of function fits best.)

(b) If this region was more or less exactly self-supporting in this grain in 1970, was it self-supporting between 1970 and 1990? What are your predictions for the future if the trends continue?

17. Use a graphing calculator (or a computer) to graph $y = x^4$ and $y = 3^x$. Determine the appropriate domains and ranges to give the pictures in Figure 1.38.

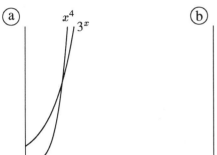 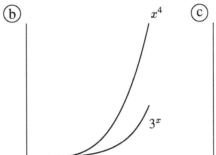 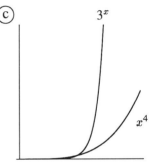

Figure 1.38: The effect of scale on a graph.

(for Problem 17)

18. Using a calculator, plot $y = x^2$ and $y = 0.01e^{0.01x}$ for

(a) $0 \le x \le 1500$, $0 \le y \le 225$. What do you notice? Why?

(b) $0 \le x \le 1500$, $0 \le y \le 2,250,000$. What do you notice? Why?

19. What domains and ranges would make the graphs of $y = x^2$ and $y = 0.01e^{0.01x}$ look like each of the graphs in Figure 1.39?

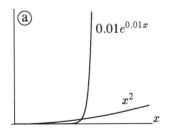 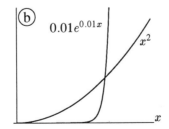 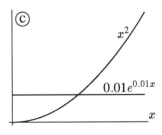

Figure 1.39: What scales are these?

(for Problem 19)

1.6 Inverse Functions

From Distance to Time and Back

On August 18, 1989, Arturo Barrios of Mexico set a world record in the 10,000 meter run with a time of 27 minutes and 8.23 seconds. His times in seconds at 2000 meter intervals are recorded in Table 1.12. This is an abbreviated table of values for the function f, where $f(d)$ is the number of seconds Barrios took to complete the first d meters of the race. For example, Barrios ran the first 4000 meters in 650.1 seconds, so $f(4000) = 650.1$. The function f is useful to athletes planning to compete with Barrios.

d (meters)	$f(d)$ (seconds)
0	0.0
2000	325.9
4000	650.1
6000	975.5
8000	1307.0
10000	1628.23

Table 1.12: Barrios' running time

t (seconds)	$f^{-1}(t)$ (meters)
0.0	0
325.9	2000
650.1	4000
975.5	6000
1307.0	8000
1628.23	10000

Table 1.13: Distance run by Barrios

Let us now change our point of view, and ask for distances rather than for times. Table 1.12 is still helpful, although we must read it backwards. If we ask how far Barrios ran during the first 650.1 seconds of his race, the answer is clearly 4000 meters. Going backwards like this from numbers of seconds to numbers of meters sounds like a function, and it is. However, it is not f. The new function is called the inverse function of f, and is denoted by f^{-1}. Thus $f^{-1}(t)$ is the number of meters that Barrios ran during the first t seconds of his race. Table 1.13 is a brief table of values for f^{-1}.

The two functions f and f^{-1} contain the same information, but they package it differently. For example, the fact that Barrios ran the first 6000 meters in 975.5 seconds can be expressed with either f or f^{-1}:

$$f(6000) = 975.5 \quad \text{or} \quad f^{-1}(975.5) = 6000.$$

The difference is that the independent variable for f is the dependent variable for f^{-1}, and vice versa. The domains and ranges of f and f^{-1} are also interchanged. The domain of f is all distances d such that $0 \le d \le 10000$, which is the same as the range of f^{-1}. The range of f is all times t, such that $0 \le t \le 1628.23$, which is the same as the domain of f^{-1}.

Definition of an Inverse Function

Not every function f has an inverse, a problem we will discuss later in this section. But when an inverse exists,

$$\boxed{f^{-1}(t) = d \quad \text{means} \quad f(d) = t}$$

The notation f^{-1} to denote an inverse function is perhaps unfortunate, as it is easy to confuse with a reciprocal, which the inverse function is not. However, there's no changing such well-established notation!

☐ **Example 1** The temperature at which water boils decreases as altitude increases, a fact important to cooks. Let $f(h)$ be the boiling point ($°C$) of water at an altitude of h meters above sea level during standard atmospheric conditions. What is the meaning in practical terms of $f^{-1}(90)$ and of $f^{-1}(90) = 3000$? Evaluate $f^{-1}(100)$.

Solution. The function f goes from altitude to temperature, so f^{-1} goes back from temperature to altitude. Thus $f^{-1}(90)$ is the altitude in meters at which the boiling point of water is $90°C$. The equation $f^{-1}(90) = 3000$ means that the boiling point of water is $90°C$ at an altitude of 3000 meters. The equation $f(3000) = 90$ has the same meaning. Since the boiling point of water is $100°C$ at sea level (where the altitude is 0 meters), we must have $f^{-1}(100) = 0$. ☐

Formulas for Inverse Functions

If a function is defined by a formula, it is sometimes possible to find a formula for the inverse function as well. In Section 1.1, we looked at the Snow Tree Cricket, whose chirp-rate, C, in chirps per minute is approximated by the formula

$$C = f(T) = 4T - 160$$

where T is the temperature in $°F$. So far we have used this formula to predict the chirp rate from the temperature. But it is perfectly possible to use this formula backwards to calculate the temperature from the chirp-rate.

☐ **Example 2** Find the formula for the function giving temperature in terms of the number of cricket chirps per minute; i.e., find the inverse function f^{-1} such that

$$T = f^{-1}(C).$$

Solution. We know $C = 4T - 160$. We solve for T giving

$$T = \frac{C}{4} + 40$$

so

$$f^{-1}(C) = \frac{C}{4} + 40.$$

  ☐

When Does a Function have an Inverse?

Some functions do not have an inverse. However, if a function has an inverse, we say it is *invertible*. Let us look at some examples.

Consider the flight of the Mercury spacecraft "Freedom 7" that carried the first American, Alan Shepard, Jr., into space in May 1961. The spacecraft rose to an altitude of 116 miles, then came down into the sea for a total of 15 minutes flight time. The function $f(t)$ giving the altitude in miles at t minutes after liftoff does not have an inverse. To see why not, try to decide on a value for $f^{-1}(100)$. Clearly, $f^{-1}(100)$ should tell us the time when the altitude of the spacecraft was 100 miles. However, there are two such times, one when the spacecraft was ascending and one when it was descending. Since functions must take single definite values, there is no inverse function f^{-1}. (Actually, there are two functions g and h that are partial substitutes for the nonexistent inverse function of f. Namely, $g(x)$ equals the number of minutes after liftoff when the spacecraft was at an altitude of x miles and ascending, and $h(x)$ equals the number of minutes after liftoff when the spacecraft was at an altitude of x miles and descending.)

For an abstract example, let $f(x) = x^2$ be the squaring function. Since $f(3) = 9$ and $f(-3) = 9$, it is not possible to decide whether $f^{-1}(9)$ should equal 3 or -3. There is no inverse function for this f. Partial substitutes for the nonexistent inverse are the two functions $g(x) = \sqrt{x}$ and $h(x) = -\sqrt{x}$. The reason the Barrios-time and chirp-rate functions do have inverses is that they both increase everywhere. Thus, each distance d corresponds to some unique time t, and any biologically reasonable number of chirps, C, corresponds to some particular temperature, T.

A function does not have to be increasing everywhere to have an inverse, but the graph in Figure 1.40 does suggest when an inverse will exist. The original function, f, takes you from x to y, as shown in Figure 1.40. Since having an inverse means there is a function going from y to x, the crucial question is whether you can get back. In other words, does each y value corresponds to a unique x value? If so, there's an inverse; if not, there isn't. Put geometrically,

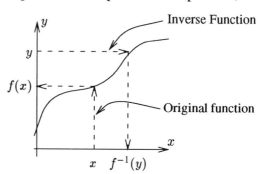

Figure 1.40: Why an Increasing Function has an Inverse

> Any function whose graph is cut by any horizontal line at most once has an inverse.

Thus, for example, the function $f(x) = x^2$ does not have an inverse because many horizontal lines cut the parabola twice.

Graphs of Inverse Functions

The graph of $f(x) = x^3$ shows that this function is increasing everywhere and so has an inverse. To find the inverse, we solve

$$y = x^3$$

for x, giving

$$x = \sqrt[3]{y}.$$

Thus the inverse function is

$$f^{-1}(y) = \sqrt[3]{y}$$

or, if we want to call the variable x:

$$f^{-1}(x) = \sqrt[3]{x}.$$

Remember the graphs of $y = x^3$ and $y = x^{1/3}$ (shown again in Figure 1.41)? Notice that these graphs are the reflections of one another in the line $y = x$. For example, $(8,2)$ is on the graph of $y = x^{\frac{1}{3}}$ because $2 = 8^{1/3}$ and $(2,8)$ is on the graph of $y = x^3$ because $8 = 2^3$. The points $(8,2)$ and $(2,8)$ are reflections of one another in the line $y = x$.

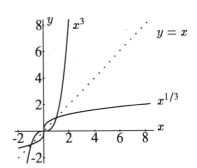

Figure 1.41: Comparison of $y = x^3$ and $y = x^{1/3}$

In general: The graph of f^{-1} is the reflection of the graph of f in the line $y = x$.

❖ Exercises for Section 1.6

1. Let $f(x)$ equal the temperature (°F) when the column of mercury in a particular thermometer is x inches long. What is the meaning of $f^{-1}(75)$ in practical terms?

For Problems 2–6, decide whether the function f is invertible.

2. $f(d)$ is the number of gallons of gasoline in your automobile when the odometer reads d miles.

3. $f(t)$ equals the number of customers in Macy's department store at t minutes past noon on December 18, 1991.

4. $f(x)$ is the volume in liters of x kilograms of water at 4°C.

5. $f(w)$ is the cost in cents of mailing a letter that weighs w grams.

6. $f(n)$ is the number of students in your calculus class whose birthday is on the n-th day of the year.

7. Write a table of values for f^{-1}, where f is as given below. The domain of f is the integers from 1 to 7. State the domain of f^{-1}.

x	1	2	3	4	5	6	7
$f(x)$	3	−7	19	4	178	2	1

8. The function $f(x) = x^3 + x + 1$ is invertible. Use a graphing calculator to give an approximate value for $f^{-1}(20)$.

For Problems 9–11, use a graphing calculator or computer to sketch the graphs of the following functions and decide whether or not they are invertible.

9. $f(x) = x^2 + 3x + 2$

10. $f(x) = x^3 - 5x + 10$

11. $f(x) = x^3 + 5x + 10$

12. The function $f(x) = x^2 + 5$ is not invertible. Suggest two functions that could act as partial substitutes for the nonexistent inverse function of f.

13. The cost of producing q articles is given by the function

$$C = f(q) = 100 + 2q.$$

(a) Find a formula for the inverse function.

(b) Explain in practical terms what the inverse function tells you.

14. A kilogram is about 2.2 pounds.

(a) Write a formula for the function, f, which gives an object's mass in kilograms, k, as a function of its weight in pounds, p.

(b) Find a formula for the inverse function of f. What does this inverse function tell you, in practical terms?

15. Suppose f is invertible and increasing, what can you say about whether its inverse is increasing or decreasing?

16. If a function f is invertible and concave up, what can you say about the concavity of its inverse?

17. Figure 1.42 is the graph of the function f, where $f(t)$ is the number (in millions) of motor vehicles registered in the world in the year t. (In 1988, one third of the registered vehicles were in the United States.)

    (a) Is f invertible? Explain.

    (b) What is the meaning of $f^{-1}(400)$ in practical terms? Evaluate $f^{-1}(400)$.

    (c) Sketch the graph of f^{-1}.

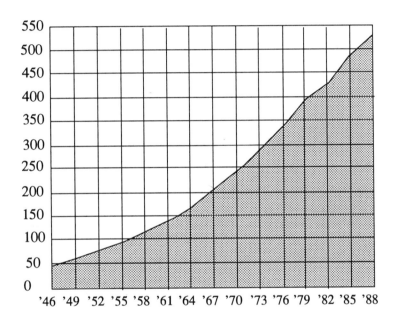

Figure 1.42: World population of motor vehicles[3]

1.7   Logarithms

In Section 1.3, we set up a function approximating the population of Mexico (in millions) as:

$$P = f(t) = 67.38(1.026)^t$$

where t is the number of years since 1980. Writing the function this way shows that we are thinking of the population as a function of time, and that we believe the population to be 67.38 million in 1980 and to grow by 2.6% every year.

    Now suppose that instead of calculating the population, we want to find when the population is expected to reach 100 million. This means we want to find the value of t for which

$$100 = f(t) = 67.38(1.026)^t.$$

---

[3]From "Energy for Motor Vehicles" by D. Blerics and P. Walzer, *Scientific American*, Sept. 1990.

Since the exponential function is always increasing and is eventually more than 100, there's exactly one value of t making $P = 100$. How should we find it? A reasonable way to start is trial-and-error: taking $t = 10$ and $t = 20$ we get

$$P = f(10) = 67.38(1.026)^{10} \approx 87.1\ldots \quad \text{(so } t = 10 \text{ is too small)}$$
$$P = f(20) = 67.38(1.026)^{20} \approx 112.58\ldots \quad \text{(so } t = 20 \text{ is too large)}$$

Some more experimenting leads to

$$P = f(15) = 67.38(1.026)^{15} \approx 99.0$$
$$P = f(16) = 67.38(1.026)^{16} \approx 101.6$$

so t is between 15 and 16. In other words, the population is projected to reach 100 million sometime during 1995.

Although it is always possible to approximate t by trial-and-error like this, it would clearly be better to have a formula that gives t in terms of P. We need the *logarithm*.

Definition and Properties of Logs to Base 10:

Special notations have been invented for the inverses of the most basic functions. The inverse function of 10^x is written $\log_{10} x$

$$\boxed{\log_{10} x = c \quad \text{means} \quad 10^c = x.}$$

We call 10 the *base*. In other words,

$$\boxed{\text{the } logarithm \text{ to base 10 of } x \text{ is the power of 10 you need to get } x.}$$

Thus, for example, $\log_{10} 1000 = \log_{10} 10^3 = 3$ since 3 is the power of 10 needed to get 1000. Similarly, $\log_{10}(0.1) = -1$ because $0.1 = \frac{1}{10} = 10^{-1}$. No power of 10 is negative or 0, and therefore

$$\boxed{\log_{10} x \text{ is not defined if } x \text{ is negative or 0.}}$$

(Try to find $\log_{10}(-3)$ on your calculator. What happens?)

We will use $\log x$ to mean $\log_{10} x$, which is the notation your calculator uses. In the next section we will use logarithms to base e, which are written in their own notation, 'ln'. In working with logarithms, you will need to recall:

Rules for Manipulating Logarithms

1. $\log(AB) = \log A + \log B$

2. $\log\left(\frac{A}{B}\right) = \log A - \log B$

3. $\log(A^p) = p\log A$

4. $\log(10^x) = x$

5. $10^{\log x} = x$

In addition, remember that $\log 1 = 0$ because $10^0 = 1$.

The fact that $\log x$ is the power of 10 giving x justifies the identities 4 and 5.

Solving Equations using Logarithms :

Logs are frequently useful when we have to solve for unknown exponents, as in the next examples.

☐ **Example 1** Find t such that $2^t = 7$.

Solution. First, notice that we expect t to be between 2 and 3 (because $2^2 = 4$ and $2^3 = 8$). To find t exactly, take logs to base 10:

$$\log(2^t) = \log 7.$$

Then use the third log rule which says $\log(2^t) = t\log 2$:

$$t\log 2 = \log 7.$$

Using a calculator to find the logs gives

$$t(0.301) = 0.845$$

so

$$t = \frac{0.845}{0.301} \approx 2.81.$$

☐

☐ **Example 2** Solve $100 = 67.38(1.026)^t$ for t, using logs.

Solution. Dividing both sides of the equation by 67.38, we get

$$1.484 = \frac{100}{67.38} = (1.026)^t.$$

Now take logs and use the fact that $\log(A^t) = t\log A$

$$\log(1.484) = \log(1.026^t),$$
$$\log(1.484) = t\log(1.026).$$

Then, using a calculator to find the logs, we get

$$0.1714 = 0.0111t$$

Solving this equation gives

$$t \approx 15.4$$

which is between $t = 15$ and $t = 16$, as we had found at the beginning of this section. ❑

❑ **Example 3** Find a formula for the inverse of the function

$$P = f(t) = 67.38(1.026)^t.$$

Solution. We must find a formula expressing t as a function of P. Take logs:

$$\log P = \log(67.38(1.026)^t).$$

Since $\log(AB) = \log A + \log B$,

$$\log P = \log 67.38 + \log((1.026)^t).$$

Now use $\log(A^t) = t \log A$:

$$\log P = \log 67.38 + t \log 1.026.$$

Now solve for t in two steps, using a calculator at the final stage:

$$t \log 1.026 = \log P - \log 67.38$$

$$t = \frac{\log P}{\log 1.026} - \frac{\log 67.38}{\log 1.026} = 89.7 \log P - 164.0$$

Thus,

$$f^{-1}(P) = 89.7 \log P - 164.0.$$

Note that

$$f^{-1}(100) = 89.7(\log 100) - 164.0 = (89.7)(2) - 164.0 = 15.4,$$

which agrees with the result of Example 2. ❑

❑ **Example 4** Find the half-life of the decaying exponential $P = P_0(0.8)^x$ that modeled the removal of pollutants in jet fuel on page 22. What does your answer mean in practical terms?

Solution. We seek the x such that

$$P = \frac{1}{2}P_0.$$

Thus we must solve the equation

$$\frac{1}{2}P_0 = P_0(0.8)^x$$

Dividing both sides by P_0 leaves

$$(0.8)^x = \frac{1}{2}.$$

Takings logs gives

$$x(\log 0.8) = \log(\frac{1}{2})$$

$$-0.097x = -0.30$$

$$x = 3.1$$

The half-life is 3.1. In practical terms, this tells us that forcing kerosene through 3.1 feet of clay pipe removes half of its impurities. □

Logs can't always be used to find exponents. The next example involves *transcendental* equations which contain both polynomials and exponentials and which can only be solved numerically or by using a graph.

□ **Example 5** Solve (a) $1 + x = 2^x$, (b) $1 + x = 5^x$.

Solution.

(a) Taking logs is of no help, because it gives an equation that is no easier to solve than the original:

$$\log(1 + x) = \log(2^x) = x \log 2 = (0.301)x$$

The next possibility is to guess, and you can check that $x = 0$ and $x = 1$ both satisfy the equation. Looking at the graphs of $y = 1 + x$ and $y = 2^x$ in Figure 1.43(a) shows that there are no other solutions.

(b) Logs are no help here either, and guessing isn't quite as easy. Again, $x = 0$ is a solution, and Figure 1.43(b) shows that there is another solution to the left of $x = 0$. By zooming in on the graph, or by trial and error, we find $x \approx -0.65$ is also a solution.

□

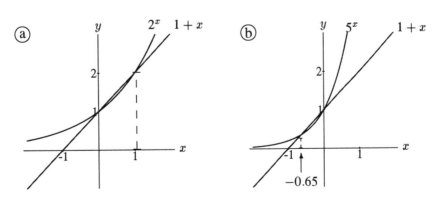

Figure 1.43: Graphs showing Solutions of (a) $1 + x = 2^x$ and (b) $1 + x = 5^x$
(for Example 5)

The Graph of log x

x	$\log x$
0	undefined
1·	0
2	0.3
3	0.5
4	0.6
⋮	⋮
10	1·

x	10^x
0	1
1	10
2	100
3	10^3
4	10^4
⋮	⋮
10	10^{10}

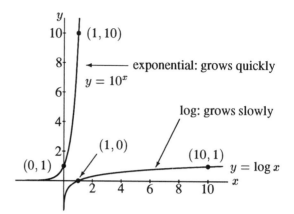

Table 1.14: $\log x$ and 10^x

Figure 1.44: Graphs of $\log x$ and 10^x

☐ **Example 6**   Graph $f(x) = \log x$ and compare it to the graph of $g(x) = 10^x$.

Solution. The values of $\log x$ in Table 1.14 were found using a calculator. Because no power of 10 gives 0, $\log 0$ is undefined. See Figure 1.44 for the graphs.   ☐

You will notice that one of the big differences between $g(x) = 10^x$ and $f(x) = \log x$ is that the exponential function grows extremely quickly while the log function grows extremely slowly. Of course these two facts are directly related. Since $\log x$ is the power of 10 you need to get x, and you only need a relatively small power of 10 to get a pretty large x, the log doesn't grow very fast. Indeed to make $\log x$ large, you will need a gigantic value of x. However, albeit slowly, the log function does go to infinity as x increases.

The log function is not defined for $x \leq 0$, and has a vertical asymptote at $x = 0$. As x gets closer to 0 from the right (written $x \to 0^+$), the log graph drops towards $-\infty$. Check this on a calculator by finding $\log(0.1), \log(0.01), \log(0.001)$ and so on. In addition, you should notice that the log function crosses the x-axis at $x = 1$, because $10^0 = 1$ so $\log(1) = 0$.

Since $f(x) = \log x$ and $g(x) = 10^x$ are inverse functions, the graphs of the two functions are reflections of one another in the line $y = x$. See Figure 1.44. For example,

$$\log(10) = 1 \quad \text{so } (10, 1) \text{ is on the log graph}$$

and this means that

$$10^1 = 10 \quad \text{so } (1, 10) \text{ is on the exponential graph.}$$

The points $(10, 1)$ and $(1, 10)$ are reflections of one another in the line $y = x$. A similar argument shows that if (a, b) is on the exponential graph, (b, a) is on the log graph.

Comparison between Logarithms and Power Functions

There is a superficial resemblance between the graphs of $y = x^{\frac{1}{3}}$ and $y = \log x$, shown in Figure 1.45. The resemblance is that both $x^{\frac{1}{3}}$ and $\log x$ are increasing and concave down for positive x. But note carefully the differences. The graph of $x^{\frac{1}{3}}$ includes the origin $(0,0)$, whereas the graph of $\log x$ never reaches the vertical axis $x = 0$, which is an asymptote. Another difference is the rate of growth of the two functions for large x. The calculations $\log 1,000,000 = 6$ and $\log 10,000,000 = 7$ make it pretty clear that the logarithm climbs slowly as x increases beyond 1. You won't be surprised to discover, therefore, that the logarithm climbs slower than any positive power of x. In fact, x^p will dominate $A \log x$ for large x no matter what the values of $p > 0$ and $A > 0$. Figure 1.46 compares $x^{\frac{1}{3}}$ and $100 \log x$; Table 1.15 compares $x^{0.001}$ and $1000 \log x$.

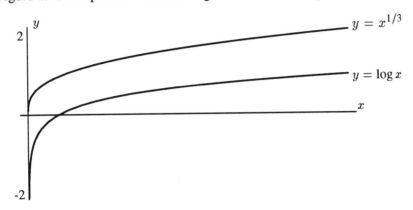

Figure 1.45: Comparison of $y = \log x$ and $y = x^{1/3}$

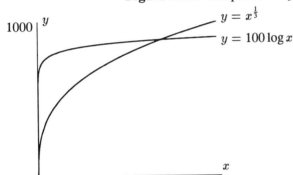

Figure 1.46: Comparison of $y = x^{\frac{1}{3}}$ and $y = 100 \log x$

x	$x^{0.001}$	$1000 \log x$
10^{5000}	10^5	5×10^6
10^{6000}	10^6	6×10^6
10^{7000}	10^7	7×10^6

Table 1.15: Comparison of $x^{0.001}$ and $1000 \log x$

❖ Exercises for Section 1.7

1. Construct a table of values for $x = 1, 2, \ldots, 10$ to compare the values of $f(x) = \log x$ and $g(x) = \sqrt{x}$. Round to two decimal places. Use these values to graph both functions.

For Problems 2 – 3, plot a graph of the given function on a calculator or computer. Describe and explain what you see.

2. $y = \log 10^x$

3. $y = 10^{\log x}$

For Problems 4–11, solve for t using logs. (In Problems 4–7, you can check your answer using a graphing calculator or computer if you like.)

4. $5^t = 7$

5. $2 = (1.02)^t$

6. $7 \cdot 3^t = 5 \cdot 2^t$

7. $5.02\,(1.04)^t = 12.01\,(1.03)^t$

8. $a = b^t$

9. $P = P_0\, a^t$

10. $Q = Q_0\, a^{nt}$

11. $P_0\, a^t = Q_0\, b^t$

For Problems 12–19, simplify as much as possible.

12. $\log A^2 + \log B - \log A - \log B^2$

13. $2 \log \alpha - 3 \log B - \dfrac{\log \alpha}{2}$

14. $\dfrac{\log A^2 - \log A}{\log B - \frac{1}{2} \log B}$

15. $\log(10^{x+7})$

16. $10^{\log A^2}$

17. $10^{2 \log Q}$

18. $10^{-\log P}$

19. $10^{-(\log B)/2}$

20. Find the equation of the line l in Figure 1.47.

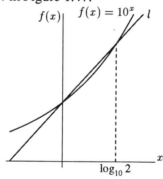

Figure 1.47: (For Problem 20)

21. What is the doubling time of prices which are increasing by 5% a year?

22. In 1980, there were about 170 million vehicles (cars and trucks) and about 227 million people in the United States. If the number of vehicles was growing at 4% a year, while the population was growing at 1% a year, in what year was there, on average, one vehicle per person?

23. The population of a region is growing exponentially. If there were 40,000,000 people in 1980 ($t = 0$) and 56,000,000 in 1990, find an expression for the population at any time t. What would you predict for the year 2000? What is the doubling time?

24. (a) Find the doubling time for annual growth rates of 3%, 4%, and 5%.

    (b) Does the change in doubling time suggest that the doubling time D is a linear function of the growth rate k? If so, estimate the slope and find the equation of the line. If not, what type of function might it be?

25. The half-life of a certain radioactive substance is 12 days. If there are 10.32 grams initially:

    (a) Write an equation to determine the amount, A, of the substance as a function of time.

    (b) When will the substance be reduced to one gram?

26. Owing to an innovative rural public health program, infant mortality in Senegal, West Africa, is being reduced at a rate of 10% per year. How long will it take for infant mortality to be reduced by 50%?

27. Assume that the rate of inflation continues at the current (1991) rate of 4.6% a year. If postage stamps go up, on average, at the rate of inflation, when will it cost $1 to mail a letter?   [Note: The cost of mailing a letter became 29¢ in 1990.]

28. Find the inverse function to $p(t) = (1.04)^t$.

For Problems 29-34, imagine the graph of $y = 2^x$ on a scale of one x unit per inch and one y unit per inch, where the origin of the coordinate system is on the Earth. For approximately what x value is the point $(x, 2^x)$

29. in the Moon? (239,000 miles from Earth)

30. In the Sun? (93,000,000 miles from Earth)

31. In Proxima Centauri, the nearest star except for the Sun? (1.3 parsecs from Earth, where 1 parsec = 1.917×10^{13} miles)

32. At the center of the Milky Way? (10^4 parsecs from Earth)

33. In Andromeda Galaxy? (6.7×10^5 parsecs from Earth)

34. At the most distant object known in the Universe, a Quasar? (5×10^9 parsecs from Earth)

1.8   Natural Logs: Logs to Base e

Many naturally occurring phenomena are modeled using an exponential function with base e. For example, we saw in Section 1.4 that population growth may be written in the form

$$P = P_0 e^{kt}, \quad k > 0$$

while radioactive decay is described by

$$Q = Q_0 e^{-kt}, \quad k > 0$$

Logarithms to base e are called *natural logarithms* and have their own notation. The natural logarithm of x, written $\ln x$, is defined to be the inverse function of e^x:

$$\ln x = \log_e x = c \quad \text{means} \quad e^c = x$$

so

$$\ln x \quad \text{is the power of} \quad e \quad \text{needed to get} \quad x.$$

The rules for manipulating natural logs are similar to those for logs to base 10, and $\ln x$ is not defined when x is negative or 0.

Rules For Manipulating Natural Logarithms

1. $\ln(AB) = \ln A + \ln B$

2. $\ln\left(\dfrac{A}{B}\right) = \ln A - \ln B$

3. $\ln(A^p) = p \ln A$

4. $\ln e^x = x$

5. $e^{\ln x} = x$

In addition, $\ln 1 = 0$ because $e^0 = 1$.

☐ **Example 1**

Use the $\boxed{\text{LN}}$ button on a calculator to plot a graph of $f(x) = \ln x$ for $0 < x \leq 10$. Observe that the graph of $y = \ln x$, like the graph of $y = \log x$, climbs very slowly as x increases.

☐ **Example 2** The release of chlorofluorocarbons used in air conditioners and, to a lesser extent, in household sprays (hair spray, shaving cream, etc.) destroys the ozone in the upper atmosphere. At the present time, the amount of ozone, Q, is decaying exponentially at a continuous yearly rate of 0.25%. At this rate, how long will it take for half the ozone to disappear?

Solution. If Q_0 is the initial quantity of ozone, then

$$Q = Q_0 e^{-0.0025t}.$$

We want to find T, the value of t making $Q = \frac{Q_0}{2}$, so

$$\frac{Q_0}{2} = Q_0 e^{-0.0025T}.$$

Taking natural logs yields

$$-0.0025T = \ln\left(\frac{1}{2}\right) = -0.6931,$$

so

$$T \approx 277 \text{ years.}$$

☐

In Example 2 the growth rate was given. However, in many situations where we expect to find exponential growth the rate is not given. To find it, we must know the quantity at two different times, and then solve for the growth rate, as in the next example.

☐ **Example 3** The population of Kenya was 19.5 million in 1984 and 21.2 million in 1986. Assuming it increases exponentially, find a formula for the population of Kenya as a function of time.

Solution. If we measure the population, P, in millions and time, t, in years since 1984, we can say

$$P = P_0 e^{kt} = 19.5 e^{kt}$$

where $P_0 = 19.5$ is the initial value of P. We find k by using the fact that $P = 21.2$ when $t = 2$, so

$$21.2 = 19.5 e^{k \cdot 2}.$$

To find k, we divide both sides by 19.5, giving

$$1.087 = \frac{21.2}{19.5} = e^{2k}.$$

Now take natural logs:

$$\ln(1.087) = \ln(e^{2k}).$$

Using a calculator, and the fact that $\ln(e^{2k}) = 2k$, this becomes

$$0.0834 = 2k.$$

So

$$k \approx 0.042 \quad \text{or about} \quad 4.2\%.$$

Therefore

$$P = 19.5 e^{0.042t}.$$

☐

This calculation shows that Kenya's population is growing continuously at 4.2% a year, currently the fastest in the world. We chose to express Kenya's population in the form

$$P = P_0 e^{kt}$$

and find k. However, we could equally well have written

$$P = P_0 a^t$$

and solved for a. An exponential problem of this kind can be done with or without using e.

❖ Exercises for Section 1.8

1. Construct a table of values for $x = 1, 2, \ldots, 10$ to compare the values of $f(x) = \log x$ and $g(x) = \ln x$. Round to two decimal places. Use these values to graph both functions.

For Problems 2–3, plot a graph of the given function on a calculator or computer. Describe and explain what you see.

2. $y = \ln e^x$

3. $y = e^{\ln x}$

4. (a) Evaluate the quantity $\dfrac{\ln x}{\log x}$ for several values of x (say $x = 0.5, 5, 10, 100$). What do you notice?

    (b) Plot a graph of $y = \dfrac{\ln x}{\log x}$ on a computer or calculator. What do you see? How does it relate to your results in (a) ?

For Problems 5–8, solve for t using natural logs.

5. $a = be^t$

6. $P = P_0 e^{kt}$

7. $ae^{kt} = e^{bt}$,   where $k \neq b$

8. $ce^{-\alpha t} = be^{-\gamma t/n}$, where $\alpha n \neq \gamma$

9. In 1986 the population of Costa Rica was 2.5 million. Assuming the population is growing at a continuous rate of 2.8% a year,

    (a) Express P as a function of t

    (b) In what year will the population reach 5 million? 10 million? 20 million?

10. The air in a factory is being filtered so that the quantity of a pollutant, P (measured in mg/liter), is decreasing according to the equation $P = P_0 e^{-kt}$ where t represents time in hours . If 10% of the pollution is removed in the first five hours,

    (a) What percentage of the pollution is left after 10 hours?

    (b) How long will it take before the pollution is reduced by 50%?

    (c) Plot a graph of pollution against time. Show the results of your calculations on the graph.

    (d) Give a qualitative argument to explain why the quantity of pollutant might decrease in this way.

11. If an exponentially growing bacteria colony doubles every five hours, how long will it take for the number of bacteria to triple?

12. If $1000 is invested at 7% interest compounded continuously, when will the investment be worth $2000? What rate of interest would be required for the investment to double in value in five years?

13. Radioactive Carbon-14 is present in all living tissues and is maintained at a constant level as long as a plant or animal is alive. Once an organism is dead, however, the radioactive Carbon-14 is no longer replenished and the quantity remaining after t years is given by the equation

$$Q = Q_0 e^{-0.000121t}$$

   (a) A skull uncovered at an archeological dig has 15% of the original amount of Carbon-14 present. Estimate its age.

   (b) Show how you can calculate the half-life of Carbon-14 from this equation.

14. Geological dating of rocks is done using Potassium-40 rather than Carbon-14 because potassium has a longer half-life. The potassium decays to argon, which remains trapped in the rocks and can be measured; thus the original quantity of potassium can be calculated. If the half-life of Potassium-40 is $1.28 \cdot 10^9$ years, find a formula giving the quantity, P, of potassium remaining as a function of time in years, assuming the initial quantity is P_0.

15. Find the inverse function to $f(t) = 50e^{0.1t}$.

16. Define $f(x) = \dfrac{1}{1 + e^{-x}}$.

   (a) Is f increasing or decreasing?

   (b) Explain why f is invertible and find a formula for $f^{-1}(x)$.

   (c) What is the domain of f^{-1}?

   (d) Sketch the graphs of f and f^{-1}, on the same axes, and explain their relationship.

1.9 New Functions from Old

Shifts and Stretches

The graph of a constant multiple of a given function is easy to visualize: each y value is 'stretched' or 'shrunk' by that multiple. For example, consider the function $f(x)$ and its multiples $g(x) = 3f(x)$ and $h(x) = -2f(x)$, whose graphs are in Figure 1.48. The factor 3 in the function $g(x)$ stretches each $f(x)$ value by multiplying it 3; the factor -2 in $h(x)$ stretches $f(x)$ by multiplying by 2 and reflects it in the x-axis. You can think of the multiples of a given function as a family of functions.

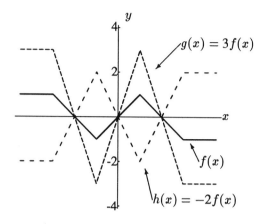

Figure 1.48: The family of multiples of the function $f(x)$

It is also easy to create families of functions by shifting graphs. For example, $y - 4 = x^2$ is the same as $y = x^2 + 4$ which is the graph of $y = x^2$ shifted up by 4, and $y = (x - 2)^2$ is the graph of $y = x^2$ shifted right by 2. See Figure 1.49.

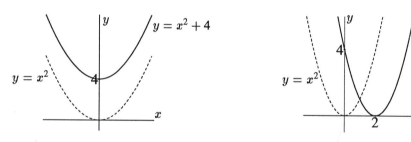

Figure 1.49: Graphs of $y = x^2$ with $y = x^2 + 4$ and $y = (x - 2)^2$

In general:

- Multiplying a function by a constant 'stretches' or 'shrinks' its graph vertically; a negative sign 'flips' it.

- Replacing y by $(y - k)$ moves a graph up by k (down if k is negative).

- Replacing x by $(x - h)$ moves a graph to the right by h (to the left if h is negative).

Sums of Functions

The right-hand graph in Figure 1.50[4] shows the number of students majoring in science and engineering, broken down into men and women. The top line represents the total; the upper shaded area represents the women and the bottom represents the men. It is really a picture of three functions: the number of men majoring in science and engineering, $m(t)$, the number of women majoring in science and engineering, $w(t)$, and the total number of students, $n(t)$, where t is the year. The men's and women's graphs are shown separately to the left.

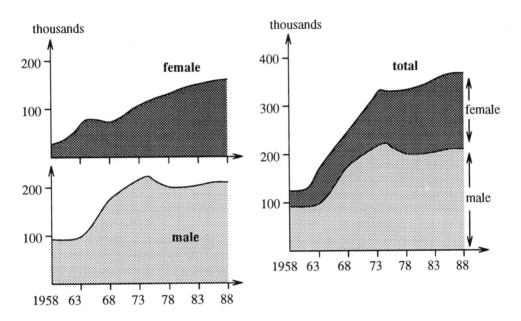

Figure 1.50: Numbers of Students Majoring in Science and Engineering (1958-1988)

Now

$$\text{total number} = (\#\text{ men}) + (\#\text{ women})$$

so, in functional notation

$$n(t) = m(t) + w(t).$$

Figure 1.50 makes it clear that the graph of the sum function is obtained by stacking the other graphs one on top of the other. Let's look at the same idea in a more mathematical setting.

❏ **Example 1**   Graph the function $y = 2x^2 + \dfrac{1}{x}$ for $x > 0$.

Solution. We graph both of the functions $y = 2x^2$ and $y = \dfrac{1}{x}$ separately in Figure 1.51. Now imagine the corresponding y values stacked on the top of one another, and you will get the graph of $y = 2x^2 + \dfrac{1}{x}$. See Figure 1.52.

---

[4]National Science Foundation

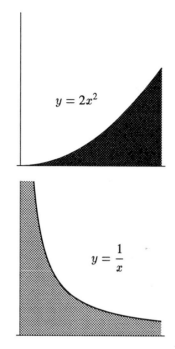

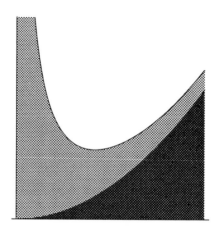

Figure 1.51: Graphs of $y = 2x^2$ and $y = \dfrac{1}{x}$

Figure 1.52: Graph of $2x^2 + \dfrac{1}{x}$

Notice that, for x near 0, the graph of $y = 2x^2 + \dfrac{1}{x}$ looks much like the graph of $y = \dfrac{1}{x}$ (because $\dfrac{1}{x}$ is much larger than $2x^2$). For large x, the graph of $y = 2x^2 + \dfrac{1}{x}$ looks like the graph of $y = 2x^2$.

Composite Functions

If oil is spilled from a tanker, the area of the oil slick will grow with time. Suppose that the oil slick is always a perfect circle. (In practice, this does not happen because of winds and tides and the location of the coastline.) The area of the slick is a function of its radius

$$A = f(r) = \pi r^2.$$

Now the radius is also a function of time, because the radius increases as more oil spills. Thus the area, being a function of the radius, is also a function of time. If, for example, the radius is given by

$$r = g(t) = 1 + t$$

then area is given as a function of time by substitution

$$A = \pi r^2 = \pi(1 + t)^2.$$

We say that A is a *composite function* or 'function of a function', which is written

$$A = \underbrace{f(g(t))}_{\substack{\text{composite function;} \\ f \text{ is outside function} \\ g \text{ is inside function}}} = \pi(g(t))^2 = \pi(1+t)^2.$$

Here the *inside function*, g, represents the calculation which is done first, and the *outside function*, f, represents the calculation done second. Look at the example and think about how you would calculate A using the formula $\pi(1+t)^2$. For any given t, the first step is to find $1+t$, and the second step is to square and multiply by π. So the first step corresponds to the inside function $g(t) = 1+t$, and the second step corresponds to the outside function $f(r) = \pi r^2$.

☐ **Example 2** Express each of the following functions as a composition:

    (a) $h(t) = (1+t^3)^{27}$ (b) $k(x) = \log x^2$ (c) $l(x) = \log^2 x$ (d) $n(y) = e^{-y^2}$

Solution. In each case think about how you would calculate a value of the function. The first stage of the calculation will give you the inside function and the second stage will give you the outside one.

    (a) The first stage is cubing and adding one, so the inside function is $g(t) = 1 + t^3$. The second stage is taking the 27th power, so the outside function is $f(y) = y^{27}$. Then

$$f(g(t)) = f(1+t^3) = (1+t^3)^{27}.$$

In fact there are lots of different answers, with $g(t) = t^3$ and $f(y) = (1+y)^{27}$ being another possibility.

    (b) By convention, $\log x^2$ means $\log(x^2)$, so evaluating this function involves squaring x first and then taking the log. So if $g(x) = x^2$ is the inside and $f(y) = \log y$ is the outside, then $f(g(x)) = \log x^2$.

    (c) By convention, $\log^2 x$ means $(\log x)^2$, so this function involves taking the log first and then squaring. Using the same definitions of f and g as in (b), namely $f(x) = \log x$ and $g(t) = t^2$, the composition is $g(f(x)) = (\log x)^2$. By evaluating the functions in (b) and (c) for $x = 2$, say, giving $\log(2^2) = 0.602$, and $\log^2(2) = 0.091$, you can see that the order in which you compose two functions does make a difference.

    (d) To calculate e^{-y^2} you will square y, take the negative, and then take e to that power. So $g(y) = -y^2$ and $f(z) = e^z$. Then $f(g(y)) = e^{-y^2}$. Alternatively, you could take $g(y) = y^2$ and $f(z) = e^{-z}$.

☐

Odd and Even Functions

There is a certain symmetry apparent in the graphs of $f(x) = x^2$ and $g(x) = x^3$ in Figure 1.53. Namely, for each point (x, x^2) on the graph of f, the point $(-x, x^2)$ is also on the graph; while for each point (x, x^3) on the graph of g, the point $(-x, -x^3)$ is also on the graph (see Figure 1.53). The graph of $f(x) = x^2$ is therefore symmetric about the y-axis, while the graph of $g(x) = x^3$ is symmetric about the origin. We say that the squaring function is *even* and the cubing function is *odd*. The names come from the fact that the even powers are even functions and odd powers are odd functions.

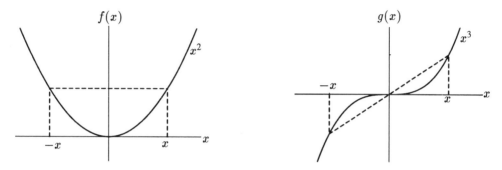

Figure 1.53: $f(x) = x^2$ is even and $g(x) = x^3$ is odd

In general, for any function f,

$$
\begin{array}{llll}
f \text{ is an } \textbf{even} \text{ function} & \text{if} & f(-x) = f(x) & \text{for all } x \\
f \text{ is an } \textbf{odd} \text{ function} & \text{if} & f(-x) = -f(x) & \text{for all } x
\end{array}
$$

You should be aware that many functions do not have any symmetry, and so are neither even nor odd.

❖ Exercises for Section 1.9

1. (a) Write an equation for a graph obtained by vertically stretching the graph of $y = x^2$ by a factor of 2 followed by a vertical upward shift of one unit. Sketch the graph.

    (b) What is the equation if the order of the transformations in part (a) is interchanged?

    (c) Are the two graphs the same? Explain the effect of reversing the order of transformations.

2. If $f(x) = 2x + 3$ and $g(x) = \log x$, find

    (a) $g(f(x))$ (b) $f(g(x))$ (c) $f(f(x))$

3. What is the difference (if any) between $\ln(\ln(x))$ and $\ln^2(x)$ $(= (\ln x)^2)$?

4. If $h(x) = x^3 + 1$, and $g(x) = \sqrt{x}$, find

    (a) $g(h(x))$ (b) $h(g(x))$ (c) $h(h(x))$ (d) $g(x) + 1$ (e) $g(x + 1)$

5. If $f(t) = (t+7)^2$ and $g(t) = \dfrac{1}{t+1}$, find

    (a) $f(g(t))$ (b) $g(f(t))$ (c) $f(t^2)$ (d) $g(t-1)$

6. If $f(t) = 2^t$ and $g(t) = t^2$, find

    (a) $f(g(t))$ (b) $g(f(t))$ (c) $g(g(t))$ (d) $f(f(t))$

7. If $f(z) = 10^z$ and $g(z) = \log(z)$, find

    (a) $f(g(100))$ (b) $g(f(3))$ (c) $f(g(x))$ (d) $g(f(x))$

For Problems 8–11, let $m(z) = z^2$. Find and simplify the following quantities:

8. $m(z+1) - m(z)$ 10. $m(z) - m(z-h)$

9. $m(z+h) - m(z)$ 11. $m(z+h) - m(z-h)$

For Problems 12–15 determine functions f and g such that $h(x) = f(g(x))$. [There is more than one correct answer. Do not choose $f(x) = x$ or $g(x) = x$.]

12. $h(x) = x^3 + 1$ 14. $h(x) = \ln^3 x$

13. $h(x) = (x+1)^3$ 15. $h(x) = \ln(x^3)$

16. By translating the graph of $y = x^3$, find a cubic polynomial with a graph similar to the graph to the right.

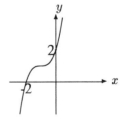

For the functions whose graphs are in Problems 17 and 18 sketch graphs of

    (a) $y = 2f(x)$ (b) $y = f(x+1)$ (c) $y = f(x) + 1$

Where possible, identify x and y intercepts and horizontal and vertical asymptotes.

17.

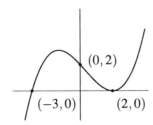

18.

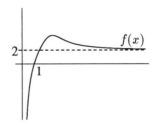

For the functions whose graphs are in Problems 19 and 20, sketch graphs of

(a) $y = -f(x)$ (b) $y = f(x) + 2$ (c) $y = f(x + 2)$

Identify the horizontal and vertical asymptotes.

19.

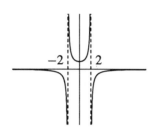

20.

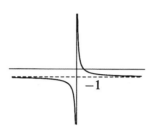

21. If $f(x)$ and $g(x)$ are odd, what about $h(x) = f(x) + g(x)$? What about $k(x) = f(x)g(x)$?

22. What symmetries do the graphs of even and odd functions have?

23. For which positive integers n is $f(x) = x^n$ even, odd?

24. Which polynomials are even, odd? Are there polynomials which are neither?

25. Are $\dfrac{1}{x}$, $\ln|x|$, e^x even, odd, or neither?

For Problems 26–28, use the graph of $y = f(x)$ in Figure 1.54, to sketch the graph of:

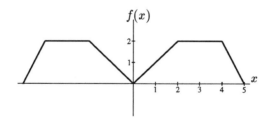

Figure 1.54: This function is even.

26. $y = 2f(x)$

27. $y = 2 - f(x)$

28. $y = \dfrac{1}{f(x)}$

For Problems 29–34, suppose that f and g are given by the graphs below:

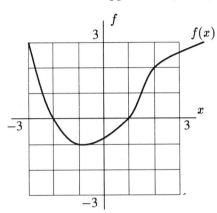

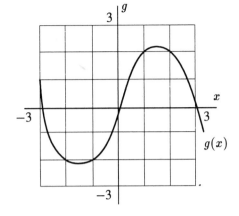

29. Find $f(g(1))$.

30. Find $g(f(2))$.

31. Find $f(f(1))$.

32. Sketch a graph of $f(g(x))$.

33. Sketch a graph of $g(f(x))$.

34. Sketch a graph of $f(f(x))$.

35. Complete Table 1.16 to show values for functions f, g and h, given the following conditions:

   (a) f is symmetric about the y-axis.

   (b) g is symmetric about the origin.

   (c) h is the composition of f with g [i.e. $h(x) = g(f(x))$].

x	$f(x)$	$g(x)$	$h(x)$
-3	0	0	
-2	2	2	
-1	2	2	
0	0	0	
1			
2			
3			

Table 1.16: For Problem 35

36. Complete the graphs of $f(x)$ and $g(x)$ for $-10 \le x \le 10$ given that $f(x)$ is even and $g(x)$ is odd.

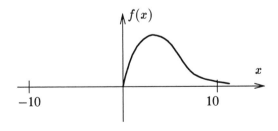

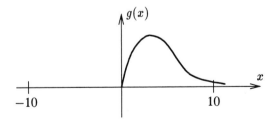

37. Suppose the graph of $f(x)$ is given to the right. Sketch the graphs of $g(x) = f(2x)$ and $h(x) = f(x/2)$. If a is constant, how is the graph of $g(x) = f(ax)$ related to the graph of $f(x)$?

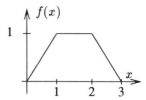

1.10 The Trigonometric Functions

Trigonometry originated as part of the study of triangles. Indeed, the name 'tri-gon-o-metry' means the measurement of three-cornered figures. Thus the first definitions of the trigonometric functions were in terms of triangles. The trigonometric functions can also be defined using the unit circle, a definition that makes them periodic, or repeating. Many naturally occurring processes are also periodic. The water level in a tidal basin, the blood pressure in a heart, an alternating current and the position of the air molecules transmitting a pure musical note all fluctuate regularly. Such phenomena can be represented by trigonometric functions.

We will use only the three trigonometric functions found on a calculator: the sine, the cosine, and the tangent.

Radians

When trigonometry is used to represent oscillations, the angles are usually measured in radians. The formulas of calculus, as you will see, come out much more neatly in radians than in degrees.

> A *radian* is defined to be the angle at the center of a unit circle which cuts off an arc of length 1, measured counter-clockwise. See Figure 1.58(a).

An angle of two radians cuts off an arc of length 2. A negative angle, such as $-1/2$ radians, cuts off an arc of length $1/2$, but measured clockwise. See Figure 1.58(b). It is useful to think of

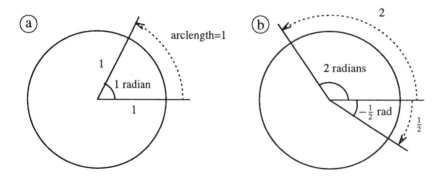

Figure 1.58: Radians Defined Using Unit Circle

angles as rotations, since then we can make sense of angles of over $360°$. Since one full rotation of

$360°$ cuts off an arc of length 2π, the perimeter of the circle, it follows that:

$$360° = 2\pi \text{ radians} \quad \text{so} \quad 180° = \pi \text{ radians.}$$

Since 1 radian $= \frac{180°}{\pi}$, one radian is about $60°$. The word radians is often dropped, so if an angle or rotation is referred to without units, it is understood to be in radians.

Arclength

Radians are also useful to compute the arclength of a sector of a circle. If the circle
has radius r, and the sector has angle θ,
as in Figure 1.59, then we have

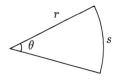

$$\boxed{\text{Arclength} = s = r\theta.}$$

Figure 1.59: Arclength of a Sector of a Circle

The Sine and Cosine

The two basic trigonometric functions—the sine and cosine—are defined using a unit circle. In Figure 1.60, a length t is measured counter-clockwise around the circle from the point $(1,0)$ to the point P, giving an angle, or rotation, of t radians at the origin. Then if P has coordinates (x, y):

$$\cos t = x \quad \text{and} \quad \sin t = y.$$

Since the equation of the unit circle is $x^2 + y^2 = 1$, we have the fundamental identity

$$\boxed{\cos^2 t + \sin^2 t = 1.}$$

As t increases and P moves around the circle, the values of $\sin t$ and $\cos t$ oscillate up and down, and eventually repeat as P moves through points where it has been before. If t is negative, the length is measured clockwise around the circle.

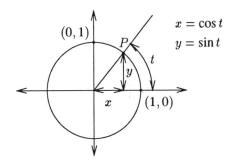

Figure 1.60: The Unit Circle

The graphs of the sine and cosine are in Figure 1.61. Notice that the sine is an odd function while the cosine is even. The maximum and minimum values of the sine and cosine are $+1$ and -1, because those are the maximum and minimum values of y and x on the unit circle. After the point P has moved around the complete circle once, the values of $\cos t$ and $\sin t$ start to repeat; we say the functions are *periodic*. We define:

The *amplitude* of an oscillation is half the distance between maximum and minimum values.

The *period* of an oscillation is the time period needed for the oscillation to execute one complete cycle.

The amplitude of $\cos t$ and $\sin t$ is 1, and the period is 2π. Why 2π? Because that's the value of t when the point P has gone exactly once around the circle. (The circle has radius 1 and perimeter $2\pi \cdot 1 = 2\pi$.)

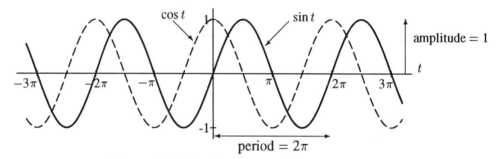

Figure 1.61: Graphs of $\cos t$ and $\sin t$

In looking at Figure 1.61 you undoubtedly noticed that the sine and cosine graphs are exactly the same shape, only shifted horizontally. Since the cosine is the sine shifted $\pi/2$ to the left,

$$\cos t = \sin(t + \pi/2).$$

This says that the cosine of any number is the same as the sine of the number $\pi/2$ further to the right—which is exactly what the graphs say. Since the sine is the cosine shifted $\pi/2$ to the right,

$$\sin t = \cos(t - \pi/2).$$

We say that the *phase difference* between $\sin t$ and $\cos t$ is $\pi/2$. (Why does this make sense from the definitions of the sine and cosine on the unit circle?)

To describe arbitrary amplitudes and periods, we use functions of the form

$$f(t) = A \sin Bt \qquad \text{and} \qquad g(t) = A \cos Bt,$$

where A is the amplitude and $2\pi/B$ is the period.

☐ **Example 1** Find and show on a graph the amplitude and period of

(a) $y = 5 \sin 2t$ (b) $y = -5 \sin\left(\dfrac{t}{2}\right)$ (c) $y = 1 + 2 \sin t$

Solution.

(a) From Figure 1.62, you can see that the amplitude of $y = 5 \sin 2t$ is 5 because the factor of 5 stretches the oscillations up to 5 and down to -5. The period of $y = \sin 2t$ is π, because when t changes from 0 to π, $2t$ changes from 0 to 2π, so the sine function will have gone through one complete oscillation.

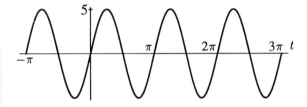

Figure 1.62: Graph of $y = 5 \sin 2t$

(b) Figure 1.63 shows that the amplitude of $y = -5 \sin\left(\frac{t}{2}\right)$ is again 5, because the negative sign reflects the oscillations in the t-axis but does not change how far up or down they go. The period of $y = -5 \sin\left(\frac{t}{2}\right)$ is 4π because when t changes from 0 to 4π, $\frac{t}{2}$ changes from 0 to 2π, so the sine function goes through one complete oscillation.

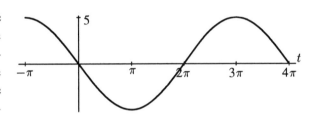

Figure 1.63: Graph of $y = -5 \sin\left(\frac{t}{2}\right)$

(c) The 1 shifts the graph $y = 2 \sin t$ up by 1. Since $y = 2 \sin t$ has an amplitude of 2 and a period of 2π, the graph of $y = 1 + 2 \sin t$ goes up to 3 and down to -1, and has a period of 2π. See Figure 1.64. Thus $y = 1 + 2 \sin t$ also has amplitude 2 and period 2π.

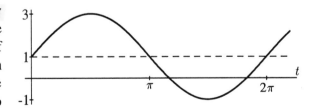

Figure 1.64: Graph of $y = 1 + 2 \sin t$

☐

☐ **Example 2** Find functions describing the oscillations in Figure 1.65.

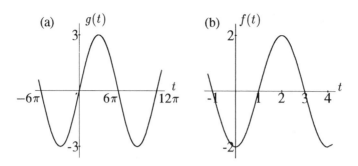

Figure 1.65: What are these functions?

Solution. The function in graph (a) looks like a sine function of amplitude 3, so $A = 3$. Since the function has executed one full period between $t = 0$ and $t = 12\pi$, a change of 12π in t goes to the sine function as 2π. This means $B \cdot 12\pi = 2\pi$ so $B = 1/6$. Therefore $g(t) = 3\sin(t/6)$ has the graph shown.

The function in graph (b) looks like a cosine function with amplitude 2 which completes one period between $t = 0$ and $t = 4$. Therefore a change of 4 in t goes to the cosine function as 2π, so $B \cdot 4 = 2\pi$ or $B = \pi/2$. The function $2\cos(\pi t/2)$ has the right amplitude and period but has a vertical intercept of 2, instead of -2. Therefore $f(t) = -2\cos(\pi t/2)$ has the graph shown. ❏

❑ **Example 3**   On February 10, 1990, high tide in Boston was at midnight. The water level at high tide was 9.9 feet; later, at low tide, it was 0.1 feet. Assuming the next high tide is exactly 12 hours later and that the height of the water is given by a sine or cosine curve, find a formula for water level as a function of time.

Solution.   Suppose the height of the water level is y feet, and let t be the time measured in hours from midnight. Then the oscillations are to have amplitude 4.9 feet ($= \frac{9.9 - 0.1}{2}$) and period 12, so $12B = 2\pi$ and $B = \frac{\pi}{6}$. Since the water is highest at midnight, when $t = 0$, the oscillations are best represented by a cosine, because the cosine is at its maximum at the beginning of the cycle. See Figure 1.66. The oscillations about the average are

$$\text{oscillation} = 4.9\cos\left(\frac{\pi}{6}t\right).$$

Since the average depth of the water was 5 feet, we want a cosine curve shifted up by 5, which we get by adding 5:

$$y = 5 + 4.9\cos\left(\frac{\pi}{6}t\right).$$

❑

❑ **Example 4**   Of course, there's something wrong with the assumption that the next high tide will be at noon. If so, the high tide would always be at noon or midnight, instead of progressing slowly through the day, as in fact it does. The interval between successive high tides actually averages

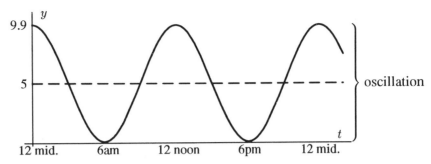

Figure 1.66: The Tide in Boston on February 10, 1990

about 12 hours 24 minutes. Using this, give a more accurate formula for the height of the water as a function of time.

Solution. The period is 12 hours 24 minutes $= 12.4$ hours, so $B = 2\pi/12.4$ giving

$$y = 5 + 4.9 \cos\left(\frac{2\pi}{12.4}t\right) = 5 + 4.9\cos(0.507t).$$

$\square$

The Tangent Function

If t is any number with $\cos t \neq 0$, we define the tangent function by

$$\tan t = \frac{\sin t}{\cos t}.$$

Figure 1.60 on page 77 shows the geometrical meaning of the tangent function: $\tan t$ is the slope of the line through the origin $(0,0)$ and the point $P = (\cos t, \sin t)$ on the unit circle.

The tangent function is undefined wherever $\cos t = 0$, namely at $t = \pm\frac{\pi}{2}, \pm\frac{3\pi}{2}, \ldots$, and has a vertical asymptote at each of these points. Now think about where $\tan t$ will be positive and where it will be negative: it will be positive where $\sin t$ and $\cos t$ have the same sign. The graph of the tangent is in Figure 1.67.

Notice that the tangent function has period π, because it repeats every π units. Does it make sense to talk about the amplitude of the tangent function? Not if we're thinking of the amplitude as a measure of the size of the oscillation, because the tangent becomes infinitely large near each vertical asymptote. We can still multiply the tangent by a constant, but that constant no longer represents an amplitude.

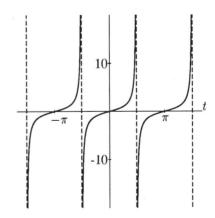

Figure 1.67: The Tangent Function: $f(t) = \tan t$

The Inverse Trigonometric Functions

On occasion, you may need to find a number with a given sine. For example, you might want to find x such that

$$\sin x = 0$$

or

$$\sin x = 0.3.$$

The first of these equations can be solved by inspection; the solution is $x = 0, \pm\pi, \pm 2\pi, \ldots$. Solving the second equation requires a calculator but it also has infinitely many solutions. For each equation, we pick out the only solution between $-\frac{\pi}{2}$ and $\frac{\pi}{2}$ as the preferred solution. For example, the preferred solution to $\sin x = 0$ is $x = 0$. We define the inverse sine, written arcsin or $\sin^{-1}$, as the function which gives the preferred solution.

> For $-1 \le y \le 1$
> $$\arcsin y = x$$
> means
> $$\sin x = y \quad \text{and} \quad -\frac{\pi}{2} \le x \le \frac{\pi}{2}.$$

Note that calculators have the inverse sine written as $\boxed{\sin^{-1}}$ and this function usually appears as the second function on the sine key.

Thus the arcsine is the inverse function to the piece of the sine function having domain $\left[-\frac{\pi}{2}, \frac{\pi}{2}\right]$. See Table 1.17 and Figure 1.68.

x	$\sin x$	x	$\sin^{-1} x$
$-\dfrac{\pi}{2}$	-1.000	-1.000	$-\dfrac{\pi}{2}$
-1.0	-0.841	-0.841	-1.0
-0.5	-0.479	-0.479	-0.5
0.0	0.000	0.000	0.0
0.5	0.479	0.479	0.5
1.0	0.841	0.841	1.0
$\dfrac{\pi}{2}$	1.000	1.000	$\dfrac{\pi}{2}$

Table 1.17: Values of $\sin x$ and $\sin^{-1} x$

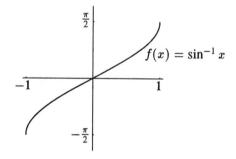

Figure 1.68: The Arcsine Function

The inverse tangent, written arctan or $\tan^{-1}$, is the inverse function for the piece of the tangent having domain $\left[-\frac{\pi}{2}, \frac{\pi}{2}\right]$. By definition

> $$\arctan y = x$$
> means
> $$\tan x = y \quad \text{and} \quad -\frac{\pi}{2} < x < \frac{\pi}{2}.$$

On a calculator, the inverse tangent is represented by $\boxed{\tan^{-1}}$. The graph of the arctangent is in Figure 1.69.

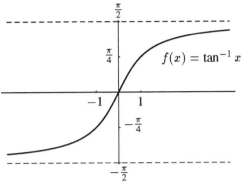

Figure 1.69: The Arctangent Function

Further Examples

☐ **Example 5**   Explain why the function

$$f(x) = (0.9)^x \sin x \quad \text{for } 0 \le x \le 20$$

represents an oscillation which is dying out, called a *damped oscillation*.

Solution. You can think of $f(x)$ as a sine function with an amplitude of $(0.9)^x$, which decays as x increases. Figure 1.70 shows that $f(x)$ is an oscillation which dies away with time. This is the kind of motion you'd expect from a weight on a spring bobbing up and down, or any other kind of oscillation dying out because of friction.   ☐

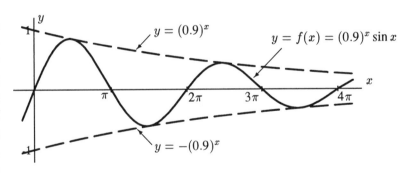

Figure 1.70: $f(x) = (0.9)^x \sin x$

☐ **Example 6**   Graph $f(x) = \sin x \sin(20x)$ on a calculator or computer. Describe what you see.

☐ **Example 7**   Plot and explain the graph of the function $f(x) = x + \sin x$.

Solution. The graphs of $y = x$ and $y = \sin x$ are shown in Figure 1.71(a). Imagine the corresponding y values stacked vertically and you will get the graph of $f(x) = x + \sin x$ in Figure 1.71(b).   ☐

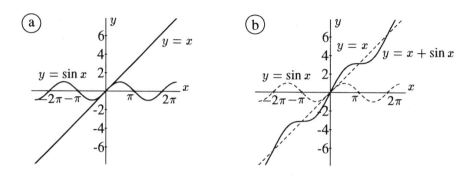

Figure 1.71: Graphs of $y = x$ and $y = \sin x$ and their sum, $y = x + \sin x$

❖ Exercises for Section 1.10

For Problems 1–8, draw the angle using a ray through the origin, and determine whether the sine, cosine, and tangent of that angle are positive, negative, zero, or undefined.

1. $\frac{\pi}{4}$
2. $\frac{\pi}{2}$
3. 3π
4. $\frac{4\pi}{3}$
5. $-\frac{\pi}{12}$
6. $-\frac{7\pi}{6}$
7. 4
8. -1

Given that $\sin \frac{\pi}{12} = 0.258$ and $\cos \frac{\pi}{5} = 0.809$, compute (without using a calculator) the quantities in Problems 9–18. You may want to draw a picture showing the angles involved, and then check your answers on a calculator.

9. $\sin \frac{11\pi}{12}$
10. $\cos\left(-\frac{\pi}{5}\right)$
11. $\sin \frac{13\pi}{12}$
12. $\cos \frac{4\pi}{5}$
13. $\sin \frac{23\pi}{12}$
14. $\cos\left(-\frac{4\pi}{5}\right)$
15. $\sin \frac{37\pi}{12}$
16. $\cos\left(-\frac{11\pi}{5}\right)$
17. $\cos \frac{\pi}{12}$
18. $\sin \frac{\pi}{5}$

19. Use the solution to Example 3 in Section 1.10 to estimate the water level in Boston Harbor at 3 am, 4 am, and 5 pm.

20. What is the period of the Earth's revolution around the sun?

21. What is the approximate period of the moon's revolution around the Earth?

22. What is the period of the Earth's rotation about its axis joining the North and South Poles?

23. What is the period of the motion of the minute hand of a clock?

24. An LP record rotates $33\frac{1}{3}$ times in a minute. What is the period of its motion?

25. A record turns at a rate of $33\frac{1}{3}$ revolutions per minute. What is the equivalent rate measured in radians per second?

26. When a car's engine makes less than about 200 revolutions per minute, it stalls. What is the period of the rotation of the engine when it is about to stall?

27. What is the difference between $\sin x^2$ and $\sin^2 x = (\sin x)^2$ and $\sin(\sin x)$? Express each of the three as a composition.

28. (a) Match the functions f, g, h, k, whose values are given in the table below, with the functions whose formulas are:

   (i) $\omega = 1.5 + \sin t$, (iii) $\omega = -0.5 + \sin t$,

   (ii) $\omega = 0.5 + \sin t$, (iv) $\omega = -1.5 + \sin t$.

t	$\omega = f(t)$	t	$\omega = g(t)$	t	$\omega = h(t)$	t	$\omega = k(t)$
6.0	−0.78	3.0	1.64	5.0	−2.46	3.0	0.64
6.5	−0.28	3.5	1.15	5.1	−2.43	3.5	0.15
7.0	0.16	4.0	0.74	5.2	−2.38	4.0	−0.26
7.5	0.44	4.5	0.52	5.3	−2.33	4.5	−0.48
8.0	0.49	5.0	0.54	5.4	−2.27	5.0	−0.46

   (b) Looking at the table, what is the relationship between the values of $g(t)$ and $k(t)$? Explain this relationship using the formulas you chose for g and k.

   (c) Using the formulas you chose for g and h, explain why all the values of g are positive, while all the values of h are negative.

For Problems 29–34, sketch graphs of the functions. What are their amplitudes and periods?

29. $y = 3 \sin x$

30. $y = 3 \sin 2x$

31. $y = -3 \sin 2\theta$

32. $y = 4 \cos 2x$

33. $y = 4 \cos(\tfrac{1}{2}t)$

34. $y = 5 - \sin 2t$

35. Identify the trigonometric functions whose graphs are in Figure 1.72.

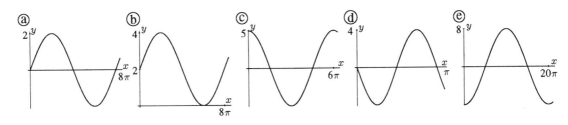

Figure 1.72: Identify these functions.

(for Problem 35)

36. Determine possible formulas for the trigonometric functions graphed in Figure 1.73.

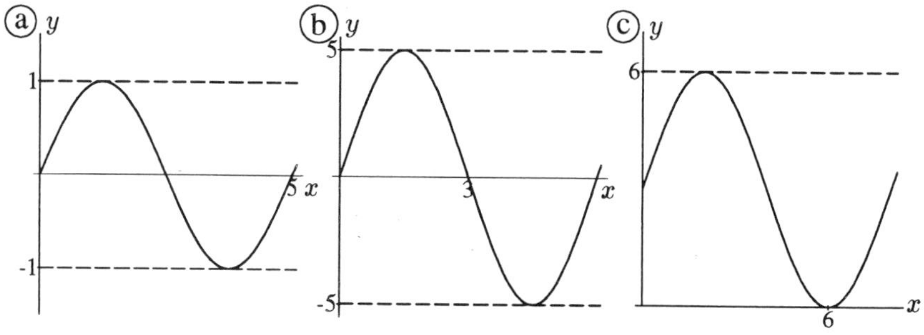

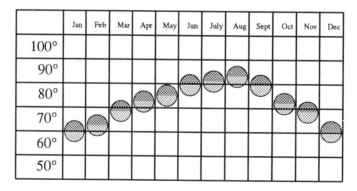

Figure 1.73: What are these functions?

(for Problem 36)

37. Using a calculator or computer, estimate all points of intersection of the graphs of $f(x) = x + \sin x$ and $g(x) = x^3$. How do you know you have found all of them?

38. The Visitors Guide to St. Petersburg, Florida, contains the chart shown in Figure 1.74 to advertise their good weather. Fit a trigonometric function approximately to the data. The independent variable should be time in months. In order to do this, you will need to find the amplitude and period of the data, and when the maximum occurs. [There are many possible answers to this problem, depending on how you read the graph.]

	Jan	Feb	Mar	Apr	May	Jun	July	Aug	Sept	Oct	Nov	Dec
100°												
90°					●	●	●	●				
80°			●	●	●				●	●		
70°	●	●	●							●	●	
60°	●											●
50°												

Figure 1.74: "St. Petersburg...where we're famous for our wonderful weather and year-round sunshine."

(Reprinted with permission)

39. The voltage, V of an electrical outlet in your home is given as a function of time, t (in seconds), by $V = V_0 \cos(120\pi t)$.

    (a) What is the period of the oscillation?

    (b) What does V_0 represent?

    (c) Sketch the graph of V against t. Label the axes.

40. (a) Using a calculator set in radians, make a table of values, to two decimal places, of $f(x) = \arcsin x$, for $x = -1, -0.8, -0.6, \ldots, 0, \ldots, 0.8, 1$ (arcsin = $\boxed{\sin^{-1}}$ on most calculators).

   (b) Sketch $f(x) = \arcsin x$. Mark on the graph the domain and range of f.

41. (a) Using a calculator set in radians make a table of values, to two decimal places, of $g(x) = \arccos x$, for $x = -1, -0.8, -0.6, \ldots, 0, \ldots, 0.8, 1$ (arccos = $\boxed{\cos^{-1}}$ on most calculators.)

   (b) Sketch the graph of $g(x) = \arccos x$.

   (c) From your graph, what are the domain and range of the arccosine?

   (d) Why is the domain of the arccosine the same as the domain of the arcsine?

   (e) Why is the range of the arccosine *not* the same as the range of the arcsine? To answer this, look at how the domain of the original sine function was restricted to construct the arcsine. Why can't the domain of the cosine be restricted in exactly the same way to construct the arccosine? How should the domain of the cosine be restricted?

42. From the data in Table 1.18, determine a possible formula for each function. Explain your reasoning. [Note: Before checking the values of a trigonometric function, be certain your calculator is in radian mode.]

x	$f(x)$	$g(x)$	$h(x)$	$F(x)$	$G(x)$	$H(x)$
−5	−10	20	25	0.958924	0.544021	2.958924
−4.5	−9	19	20.25	0.97753	−0.412118	2.97753
−4	−8	18	16	0.756802	−0.989358	2.756802
−3.5	−7	17	12.25	0.350783	−0.656987	2.350783
−3	−6	16	9	−0.14112	0.279415	1.85888
−2.5	−5	15	6.25	−0.598472	0.958924	1.401528
−2	−4	14	4	−0.909297	0.756802	1.090703
−1.5	−3	13	2.25	−0.997495	−0.14112	1.002505
−1	−2	12	1	−0.841471	−0.909297	1.158529
−0.5	−1	11	0.25	−0.479426	−0.841471	1.520574
0	0	10	0	0	0	2
0.5	1	9	0.25	0.479426	0.841471	2.479426
1	2	8	1	0.841471	0.909297	2.841471
1.5	3	7	2.25	0.997495	0.14112	2.997495
2	4	6	4	0.909297	−0.756802	2.909297
2.5	5	5	6.25	0.598472	−0.958924	2.598472
3	6	4	9	0.14112	−0.279415	2.14112
3.5	7	3	12.25	−0.350783	0.656987	1.649217
4	8	2	16	−0.756802	0.989358	1.243198
4.5	9	1	20.25	−0.97753	0.412118	1.02247
5	10	0	25	−0.958924	−0.544021	1.041076

Table 1.18: Mystery Functions
(for Problem 42)

43. Two trigonometric functions each have period π and their graphs intersect at $x = 3.64$. Without assuming anything else about these two functions,

    (a) can you say whether these two graphs intersect at any smaller positive x value? If so, what is it?

    (b) Find an x value greater than 3.64 at which these graphs intersect.

    (c) Find a negative x value at which these graphs intersect.

44. (a) Use a graphing calculator or computer to find the period of $2 \sin \theta + 3 \cos 2\theta$.

    (b) Explain your answer, using the fact that the period of $\sin \theta$ is 2π and the period of $\cos 2\theta$ is π.

45. (a) Use a graphing calculator or computer to find the period of $2 \sin 3\alpha + 3 \cos \alpha$.

    (b) What is the period of $\sin 3\alpha$? of $\cos \alpha$?

    (c) Use your answers to (b) to explain your answer to (a).

46. (a) Use a graphing calculator or computer to find the period of $2 \sin 3t + 3 \cos 2t$.

    (b) Give your answer as a multiple of π.

    (c) Find the periods of $\sin 3t$ and $\cos 2t$ and use them to explain your answer to (a).

1.11 Polynomials and Rational Functions

Polynomials

Some of the best-known functions for which there are formulas are the polynomials:

$$y = p(x) = a_n x^n + a_{n-1} x^{n-1} \cdots + a_0.$$

The shape of the graph of a polynomial depends on its degree, n, as shown in Figure 1.75. These graphs correspond to a positive coefficient for x^n; a negative coefficient flips the graph over. Notice that the graph of the quadratic "turns around" once, the cubic "turns around" twice, and the quartic "turns around" three times. In general, an nth degree polynomial "turns around" at most $n - 1$ times (where n is a positive integer), but there may be fewer turns.

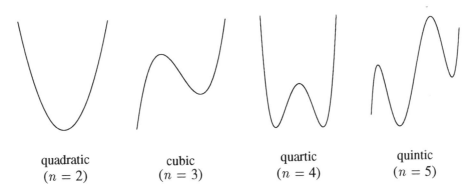

quadratic
($n = 2$)

cubic
($n = 3$)

quartic
($n = 4$)

quintic
($n = 5$)

Figure 1.75: Graphs of Typical Polynomials of Degree n

☐ **Example 1** Find possible formulas for the polynomials whose graphs are in Figure 1.76.

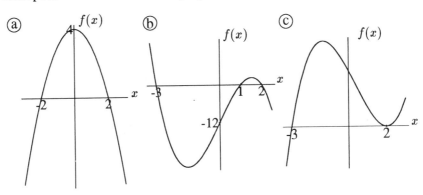

Figure 1.76: Mystery Polynomials

Solution.

(a) This graph appears to be a parabola, turned upside down and moved up by 4, so

$$f(x) = -x^2 + 4.$$

The minus sign turns the parabola upside down and the $+4$ moves it up by 4. You should notice that this formula does give the correct x-intercepts since $0 = -x^2 + 4$ has solutions $x = \pm 2$.

You can also solve this problem by looking at the x-intercepts first, which tell you $f(x)$ must have factors of $(x + 2)$ and $(x - 2)$, so

$$f(x) = k(x + 2)(x - 2).$$

To find k, use the fact that the graph has a y-intercept of 4, so $f(0) = 4$, giving:

$$4 = k(0 + 2)(0 - 2)$$

so $k = -1$. Therefore $f(x) = -(x + 2)(x - 2)$, which multiplies out to $-x^2 + 4$. Note that

$$f(x) = 4 - \frac{x^4}{4}$$

also fits the requirements, but its 'shoulders' are sharper. There are many possible answers to these questions.

(b) This looks like a cubic with factors $(x + 3)$, $(x - 1)$ and $(x - 2)$, one for each intercept:

$$f(x) = k(x + 3)(x - 1)(x - 2).$$

Since the y-intercept is -12:

$$-12 = k(0 + 3)(0 - 1)(0 - 2)$$

so $k = -2$ and

$$f(x) = -2(x + 3)(x - 1)(x - 2).$$

(c) This also looks like a cubic, but with a *double zero* at $x = 2$ and a single zero at $x = -3$, giving

$$f(x) = k(x + 3)(x - 2)^2.$$

The repeated factor, $(x - 2)^2$, comes from the double zero. Here's why: Imagine two zeros close together, say at $x = 2$ and $x = 2.1$, giving

$$f(x) = k(x + 3)(x - 2)(x - 2.1).$$

Now let the zero at $x = 2.1$ move towards the zero at $x = 2$, and you will have a factor of $(x - 2)^2$.

Another thing to notice here is that when $x > 2$, the factor $(x - 2)^2$ is positive, and when $x < 2$, $(x - 2)^2$ is still positive. This means that $f(x)$ doesn't change sign near $x = 2$. Does this agree with the given graph? Compare with the behavior near the single zero, where f does change sign.

You cannot find k, as no coordinates are given for points off of the x-axis. Inserting any positive value of k will stretch the graph but not change the zeros and therefore will still work.

◻

☐ **Example 2** Using a calculator or computer, sketch graphs of $y = x^4$ and $y = x^4 - 15x^2 - 15x$ for $-4 \le x \le 4$ and for $-20 \le x \le 20$. Set the y range to $-100 \le y \le 100$ for the first domain, and to $-100 \le y \le 200,000$ for the second. What do you observe?

Solution.

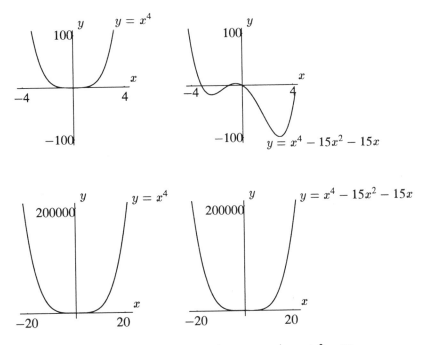

Figure 1.77: Local and Global Views of $y = x^4$ and $y = x^4 - 15x^2 - 15x$

From the graphs in Figure 1.77 you can see that close up (for $-4 \le x \le 4$) the graphs look different; from far away, however, they are almost indistinguishable. The reason is that the leading terms (those with the highest power of x) are the same, namely x^4, and for large values of x, the leading term dominates the other terms.

Looked at numerically in Table 1.19, the differences in the values of the two functions, although large, are tiny compared with the vertical scale (0 to 200,000) and so can't be seen.

x	$y = x^4$	$y = x^4 - 15x^2 - 15x$	difference
-20	160,000	154,300	5700
-15	50,625	47,475	3150
15	50,625	47,025	3600
20	160,000	153,700	6300

Table 1.19: Numerical Values of $y = x^4$ and $y = x^4 - 15x^2 - 15x$

▢

□ **Example 3**   If a tomato is thrown vertically up into the air at time $t = 0$ with velocity 48 feet per second, its distance, y, above the surface of the earth at time t (in seconds) is given by the equation

$$y = -16t^2 + 48t.$$

Sketch a graph of position against time, and mark on the graph the coordinates of the points corresponding to instants when the tomato (a) hits the ground, and (b) reaches its highest point.

Solution. This graph is a parabola opening downwards. Its intercepts are given by

$$0 = -16t^2 + 48t$$

or

$$0 = 16t(t - 3) \quad \text{so } t = 0, 3 \text{ seconds.}$$

The tomato's journey occurs between $t = 0$ and $t = 3$. The point representing the instant when the tomato hits the ground is where $y = 0$ and $t = 3$. Its highest point occurs halfway through the journey at $t = 1.5$, when $y = -16(1.5)^2 + 48(1.5) = 36$ feet. See Figure 1.78. You should notice that although the graph is arch-shaped, the tomato in fact moves vertically up and down, and will land at exactly the spot from which you threw it.

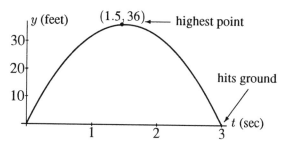

Figure 1.78: The Height of the Tomato

Rational Functions

Rational functions are those of the form

$$f(x) = \frac{p(x)}{q(x)}$$

where p and q are polynomials. Their graphs often have vertical asymptotes where the denominator is zero. If the denominator is nowhere zero there are no vertical asymptotes.

☐ **Example 4** Plot and explain the graph of $y = \dfrac{x^2 - 4}{x^2 - 1}$.

Solution. Factoring gives

$$y = \frac{x^2 - 4}{x^2 - 1} = \frac{(x + 2)(x - 2)}{(x + 1)(x - 1)}$$

so $x = \pm 1$ are vertical asymptotes. If $y = 0$ then $(x + 2)(x - 2) = 0$ or $x = \pm 2$; these are the x-intercepts. Note that the vertical asymptotes arise from zeros of the denominator, while zeros of the numerator give rise to x-intercepts. Substituting $x = 0$ gives $y = 4$; this is the y-intercept. To see what happens as $x \to \pm\infty$, look at the y values in Table 1.20. Notice that, in this example, positive and negative x's give the same y value.

x	$y = \dfrac{x^2-4}{x^2-1}$
±10	0.969697
±100	0.999700
±1000	0.999997

Table 1.20: Values of $y = \frac{x^2-4}{x^2-1}$

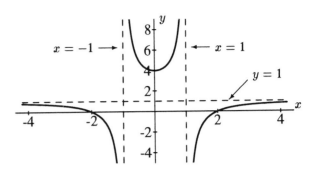

Figure 1.79: Graph of the Function $y = \frac{x^2-4}{x^2-1}$

Clearly y is getting closer to 1 as x gets large positively or negatively. This can also be seen by realizing that as $x \to \pm\infty$, only the highest powers of x really matter. For large x, the 4 and the 1 are insignificant compared to x^2, so

$$y = \frac{x^2 - 4}{x^2 - 1} \approx \frac{x^2}{x^2} = 1.$$

Therefore the horizontal asymptote is $y = 1$. Since, for $x > 1$, the denominator is positive and the numerator is less than the denominator, the graph lies *below* its asymptote. (Why doesn't the graph lie below $y = 1$ when $-1 < x < 1$?) See Figure 1.79. ☐

❖ Exercises for Section 1.11

1. Assume that each of the graphs in Figure 1.80 is of a polynomial. For each graph:

   (i) What is the minimum possible degree of the polynomial?

   (ii) Is the leading coefficient of the polynomial positive or negative? (The leading coefficient is the coefficient of the highest power of x.)

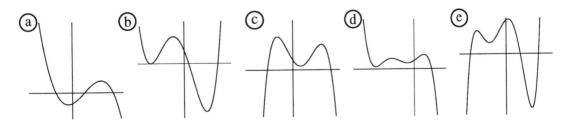

Figure 1.80: What degree polynomials have graphs like these?

(for Problem 1)

2. Describe the *end behavior* of each of the functions below. Specifically:

   • As $x \to +\infty$, does $f(x) \to +\infty$ or $-\infty$? Why?
   • As $x \to -\infty$, does $f(x) \to +\infty$ or $-\infty$? Why?

   (a) $f(x) = -3x^3 + 70x^2 - 20$

   (b) $f(x) = 20x^4 + 3x^3 + x^2 + 1000$

   (c) $f(x) = -3x^4 + 20x^3 - 5x^2 + x - 20$

   (d) $f(x) = x^4 + x^5$

   (e) $f(x) = 4x^4 + 5x^5 - 6x^6$.

For Problems 3–6, sketch graphs of the following polynomials:

3. $f(x) = (x+2)(x-1)(x-3)$

4. $f(x) = 5(x^2 - 4)(x^2 - 25)$

5. $f(x) = -5(x^2 - 4)(25 - x^2)$

6. $f(x) = 5(x-4)^2(x^2 - 25)$

For each of the rational functions, in Problems 7–10, find asymptotes, end behavior of the function, and the behavior of the function near any vertical asymptotes. Then, use this information to sketch a graph. Check your work by looking at the actual graph of each on a computer or graphing calculator.

7. $y = \dfrac{1 - 4x}{2x + 2}$

8. $y = \dfrac{3x^3 + 2x^2 - 11}{x^3 + 8}$

9. $y = \dfrac{x^2 + 2x + 1}{x^2 - 4}$

10. $y = \dfrac{1 - x^2}{x - 2}$

11. If $f(x) = ax^2 + bx + c$, what do you know about the values of a, b, and c if:

    (a) $(1, 1)$ is on the graph of $f(x)$?

    (b) $(1, 1)$ is the vertex of the graph of $f(x)$? (You may want to use the fact that the equation for the axis of symmetry of the parabola of $y = ax^2 + bx + c$ is $x = -\frac{b}{2a}$.)

    (c) The y-intercept of the graph is $(0, 6)$?

    (d) Find a quadratic function that satisfies all three conditions.

12. A pomegranate is thrown from ground level straight up into the air at time $t = 0$ with velocity 64 feet per second. Its height at time t will be $f(t) = -16t^2 + 64t$. Find the time it hits the ground and the instant that it reaches its highest point. What is the maximum height?

13. The height of an object above the ground at time t is given by

$$s = v_0 t - \frac{1}{2}gt^2$$

where v_0 represents the initial velocity and g is a constant called the acceleration due to gravity.

    (a) At what height is the object initially?

    (b) How long is the object in the air before it hits the ground?

    (c) When will the object reach its maximum height?

    (d) What is that maximum height?

14. Determine cubic polynomials to represent the graphs in Figure 1.81.

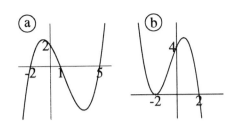

Figure 1.81: Can you build these cubic polynomials?

(for Problem 14)

15. (a) Find a possible formula for each of the graphs shown in Figure 1.82.

   (b) Using the graphs, read off approximate intervals on which the function is increasing and on which it is decreasing.

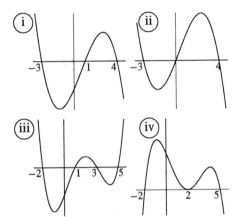

Figure 1.82: What polynomials could these graphs represent?

(for Problem 15)

16. The graph of a rational function $y = f(x)$ is given in Fig. 1.83.

   If $f(x) = \frac{g(x)}{h(x)}$ with $g(x)$ and $h(x)$ both quadratic functions, give possible formulas for $g(x)$ and $h(x)$. [There are many possible answers.]

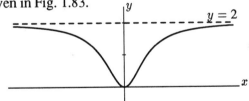

Figure 1.83: The graph of a rational function.

(for Problem 16)

For Problems 17–20, find all polynomials, p, of degree ≤ 2 which satisfy the given conditions.

17. $p(0) = p(1) = p(2) = 1$

18. $p(0) = p(1) = 1$ and $p(2) = 2$

19. $p(0) = p(1) = 1$

20. $p(0) = p(1)$

21. Newton's Second Law of Motion, $F = ma$, tells us that the net force, F, on a train of mass m is proportional to its acceleration, a. Suppose that the only forces are those of the engine, which exerts a constant force, F_E, in the direction of motion, and the wind resistance, which exerts a force proportional to the square of the train's velocity, v, but in the opposite direction.

    (a) Write a formula giving a as a function of v.

    (b) Sketch a graph of a against v.

22. The rate, R, at which a population in a confined space increases is proportional to the product of the current population, P, and the difference between the *carrying capacity*, L, and the current population. (The carrying capacity is the maximum population the environment can sustain.)

    (a) Write R as a function of P.

    (b) Sketch R as a function of P.

1.12   Roots, Continuity, and Accuracy

It is often necessary to find the zeros of a polynomial or the point of intersection of two curves. So far, you have probably used algebraic methods, such as the quadratic formula, or trial and error to solve such problems. Unfortunately, however, mathematicians' search for "closed form" solutions to equations, such as the quadratic formula, has not been all that successful. The formulas for the solutions to third and fourth degree equations are so complicated that you'd never want to use them. Early in the 19th century, it was proved that there is no algebraic formula for the solutions to fifth degree equations and higher. For non-polynomial equations, the situation is no better: most equations cannot be solved by using a formula.

Thus, we can seldom find closed form solutions. However, we can still find roots of equations, provided we use approximate methods, not formulas. In this section we will discuss three ways to find roots: algebraic, graphical and numerical. Of these, only the algebraic gives closed form, or exact, solutions.

First, let's get some terminology straight. Given the equation $x^2 = 4$, we call $x = -2$ and $x = 2$ the *roots, or solutions, of the equation*. If we are given the function $f(x) = x^2 - 4$, then -2 and 2 are called the *zeros of the function*; that is, the zeros of the function f are the roots of the equation $f(x) = 0$.

The Algebraic Viewpoint: Roots by Factoring

If the product of two numbers is zero, then one or the other must be zero, i.e. if $AB = 0$ then $A = 0$ or $B = 0$. This simple observation lies behind finding roots by factoring. You have probably spent a lot of time factoring polynomials. Here you will also have to factor expressions involving trigonometric and exponential functions.

❏ **Example 1** Find the roots of $x^2 - 7x = 8$.

Solution. Rewrite the equation as $x^2 - 7x - 8 = 0$. Then factor the left side: $(x + 1)(x - 8) = 0$. By our observation about products, either $x + 1 = 0$ or $x - 8 = 0$ so the roots are $x = -1$ and $x = 8$. ❏

❏ **Example 2** Find the roots of $\dfrac{1}{x} - \dfrac{x}{(x + 2)} = 0$.

Solution. Rewrite the left side with a common denominator: $\dfrac{x + 2 - x^2}{x(x + 2)} = 0$. Whenever a quotient is zero, the numerator must be zero, i.e., if $\dfrac{A}{B} = 0$ then $A = 0$. Thus $x + 2 - x^2 = 0$. Now factor:

$$x + 2 - x^2 = (-1)(x^2 - x - 2) = (-1)(x - 2)(x + 1) = 0.$$

We conclude that $x - 2 = 0$ or $x + 1 = 0$, so 2 and -1 are the roots. They can be checked by substitution. ❏

❏ **Example 3** Find the roots of $e^{-x} \sin x - e^{-x} \cos x = 0$.

Solution. Factor the left side: $e^{-x}(\sin x - \cos x) = 0$. The factor e^{-x} is never zero; it is impossible to raise e to a power and get zero. Therefore, the only possibility is that $\sin x - \cos x = 0$. This equation is equivalent to $\sin x = \cos x$. If we divide both sides by $\cos x$, we get

$$\frac{\sin x}{\cos x} = \frac{\cos x}{\cos x} \quad \text{so} \quad \tan x = 1.$$

The roots of this equation are just

$$\cdots \frac{-7\pi}{4}, \frac{-3\pi}{4}, \frac{\pi}{4}, \frac{5\pi}{4}, \frac{9\pi}{4}, \frac{13\pi}{4} \cdots .$$

❏

Warning: Factoring only works when one side of the equation is 0. It is not true that if, say, $AB = 7$ then $A = 7$ or $B = 7$. For example, you *cannot* solve $x^2 - 4x = 2$ by factoring $x(x - 4) = 2$ and then assuming that either x or $x - 4$ equals 2.

The problem with factoring is that factors are not easy to find. For example, the quadratic equation $x^2 - 4x - 2 = 0$ does not factor, at least not into "nice" factors with integer coefficients. For the general quadratic equation:

$$ax^2 + bx + c = 0$$

there is the quadratic formula for the roots:

$$x = \frac{-b \pm \sqrt{b^2 - 4ac}}{2a}.$$

Thus the roots of $x^2 - 4x - 2 = 0$ are $\frac{4 \pm \sqrt{24}}{2}$, or $2 + \sqrt{6}$ and $2 - \sqrt{6}$.

Notice that in each of these examples, we have found the roots exactly.

The Graphical Viewpoint: Roots by 'Zooming'

To find the roots of an equation $f(x) = 0$, it helps to draw the graph of f. The roots of the equation, that is the zeros of f, are *the values of x where the graph of f crosses the x-axis*. Even a very rough sketch of the graph can be useful in determining how many zeros there are and their approximate values. If you have a computer or graphing calculator, then finding solutions by graphing is the easiest method, especially if you use the "zoom" feature. However, a graph can never tell you the exact value of a root, but only an approximate one.

☐ **Example 4**   Find the roots of $x^3 - 4x - 2 = 0$.

Solution.

The left side does not factor with integer coefficients, so we cannot easily find the roots by algebra. We know the graph of $f(x) = x^3 - 4x - 2$ will have the usual cubic shape; see Figure 1.84. There are clearly three roots: one between $x = -2$ and $x = -1$, another between $x = -1$ and $x = 0$, and a third between $x = 2$ and $x = 3$. Zooming in on the largest root with a graphing calculator or computer shows that it lies in the following interval:

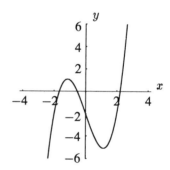

Figure 1.84: The cubic $f(x) = x^3 - 4x - 2$

$$2.213 < x < 2.215.$$

Thus the root is $x = 2.21$, accurate to two decimal places. Zooming in on the other two roots shows them to be $x = -1.68$ and $x = -0.54$, accurate to two decimal places.   ☐

Useful trick: Suppose you want to solve the equation $\sin x - \cos x = 0$ graphically. Instead of graphing $f(x) = \sin x - \cos x$, you may find it easier to rewrite the equation as $\sin x = \cos x$, and graph $g(x) = \sin x$ and $h(x) = \cos x$. (After all, you already know what these two graphs look like. See Figure 1.85.) The roots of the original equation are then precisely the x-coordinates of the points of intersection of the graphs of the two functions $g(x)$ and $h(x)$.

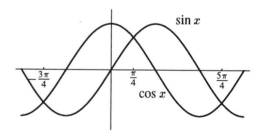

Figure 1.85: Finding Roots of $\sin x - \cos x = 0$.

☐ **Example 5**   Find the roots of $2 \sin x - x = 0$.

Solution. Rewrite the equation as $2 \sin x = x$ and graph both sides. Since $g(x) = 2 \sin x$ is always between -2 and 2, there are no roots of $2 \sin x = x$ for $x > 2$ or for $x < -2$. Thus we need only consider the graphs between -2 and 2 (or between $-\pi$ and π, which makes graphing the sine function easier). Figure 1.86 shows the graphs. There are three points of intersection: one appears to be at $x = 0$, one between $x = 1.5$ and $x = 3$, and one between $x = -1.5$ and $x = -3$. You can tell that $x = 0$ is the exact value of one root because it satisfies the original equation exactly. Zooming in shows that there is a second root $x \approx 1.9$ and the third root is $x \approx -1.9$ by symmetry. $\square$

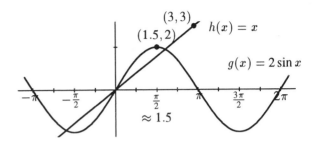

Figure 1.86: The graphs of $g(x) = 2 \sin x$ and $h(x) = x$.

Continuity of a Function

How much do we know about a function if we only know its values at a few points? Can we find its zeros, at least approximately? The answer must be yes at least for some functions, since we have been finding zeros with a calculator, which graphs by plotting just a finite number of points.

Let's return to our example of finding the roots of the cubic equation $f(x) = x^3 - 4x - 2 = 0$. From the graph in Figure 1.84 we can see there is a root somewhere between $x = 2$ and $x = 3$ because the function changes sign between these points: $f(2) = -2$ and $f(3) = 13$.

In general, if we find an interval on which $f(x)$ changes sign, can we be sure that $f(x)$ has a zero there? To be certain, we need to know that the graph of the function has no breaks or jumps in it. Otherwise the graph could "jump" across the x-axis, changing sign, but not creating a zero. For example, $f(x) = \frac{1}{x}$ has opposite signs at $x = -1$ and $x = 1$ but no zeros at all because of the break at $x = 0$. See Figure 1.87. To avoid this problem, we consider only intervals where our functions are *continuous*.

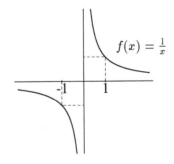

Figure 1.87: No zero although $f(-1)$ and $f(1)$ have opposite signs

What Does it Mean for a Function to be Continuous?

Graphically

A function is said to be continuous on an interval if its graph has no breaks, jumps, or holes in that interval. Alternatively, a continuous function has a graph which can be drawn without lifting the pencil from the paper.

Example: Figure 1.87 shows that $f(x) = \frac{1}{x}$ is not continuous on any interval containing the origin because of the break at $x = 0$, where f is not defined.

Example: The function $g(\theta) = \frac{\sin\theta}{\theta}$ is graphed to the right. Since g is not defined at $\theta = 0$, the graph has a hole there. This function is not continuous on any interval containing the origin.

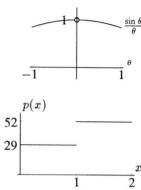

Example: Suppose $p(x)$ is the price of the mailing a first class letter weighing x ounces. Since it costs 29¢ up to the first ounce, and 52¢ between the first and second, the graph (shown to the right) is a series of steps. This function is not continuous on any interval containing a positive integer.

Numerically

A function is continuous if nearby values of the independent variable give nearby values of the function. In practical work, continuity is important because it means that we don't have to know the independent variable exactly in order to have a pretty good idea of the value of a function.

Example: If $f(x) = x^2$, then $f(2) = 4$, $f(2.01) = 4.0401$, and $f(1.99) = 3.9601$. Thus, values of x near 2 give values of $f(x)$ near $f(2)$.

Example: If $g(\theta) = \sin\theta$, then $\sin 1 = 0.841$, $\sin 1.01 = 0.847$, and $\sin 0.99 = 0.836$, so values of θ near 1 give values of the sine near $\sin 1$.

Example: If $p(x)$ is the cost of mailing a first class letter weighing x ounces, then $p(0.99) = p(1) = 29$¢, while $p(1.01) = 52$¢, because as soon as you get over 1 ounce, the price goes up to 52¢. So, values of x near 1 do not necessarily give values of $p(x)$ near $p(1)$.

Which Functions are Continuous?

Requiring functions to be continuous is not very restrictive, as any function whose graph is a single curve over any interval on which it is defined is continuous. For example, polynomials are continuous everywhere (meaning for $-\infty < x < \infty$), and rational functions are continuous on any interval in which their denominators are not zero. Exponential functions are continuous everywhere, as are the sine and cosine. All the new functions that we created by composing continuous functions are continuous too.

▢ **Example 6**   What does the following table of values tell you about the zeros of $f(x) = \cos x - 2x^2$?

x	0	0.2	0.4	0.6	0.8	1.0
$f(x)$	1.00	0.90	0.60	0.11	−0.58	−1.46

Solution. You can conclude that $f(x)$ has at least one zero in the interval $0.6 < x < 0.8$ since $f(x)$ is continuous and changes from positive to negative on that interval. From the graph of $f(x)$ in Figure 1.88, it seems that there is in fact only one zero in the entire interval $0 \le x \le 1$, but one cannot be sure of that just from the given table of values. ☐

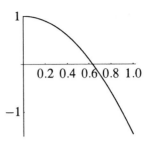

Figure 1.88: Graph of $f(x) = \cos x - 2x^2$.

Boundedness of a Function

Knowing how big or how small a function gets can sometimes be useful, especially when you can't easily find exact values of the function. You can say, for example, that $\sin x$ stays between -1 and 1 for all x, and that $2\sin x + 10$ stays between 8 and 12 for all x. But 2^x is not confined between any two numbers, because 2^x will exceed any number you can name if x is large enough. We say that $\sin x$ and $2\sin x + 10$ are *bounded* functions, and that 2^x is an *unbounded* function.

> A function f is *bounded* on an interval if there are numbers L and U such that
>
> $$L \le f(x) \le U$$
>
> for all x in the interval. Otherwise, f is *unbounded* on the interval.
>
> If $L \le f(x)$ for all x in the interval, then L is a *lower bound for* f on the interval.
> If $f(x) \le U$ for all x in the interval, then U is an *upper bound for* f on the interval.

☐ **Example 7** Which of the following are bounded?

(a) x^3 on $-\infty < x < \infty$; on $0 \le x \le 100$.

(b) $\dfrac{2}{x}$ on $0 < x < \infty$; on $1 \le x < \infty$.

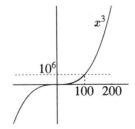

Figure 1.89: Is x^3 bounded?

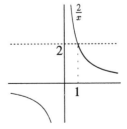

Figure 1.90: Is $\dfrac{2}{x}$ bounded?

Solution.

(a) The graph of x^3 in Figure 1.89 shows that x^3 will exceed any number no matter how large if x is big enough, so x^3 does not have an upper bound on $-\infty < x < \infty$. Therefore x^3 is unbounded on $-\infty < x < \infty$. But on the interval $0 \leq x \leq 100$, x^3 stays between 0 (a lower bound) and $100^3 = 1{,}000{,}000$ (an upper bound). Therefore x^3 is bounded on the interval $0 \leq x \leq 100$. Observe that upper and lower bounds, when they exist, are not unique. For example, -100 is another lower bound and $2{,}000{,}000$ another upper bound for x^3 on $0 \leq x \leq 100$.

(b) $2/x$ is unbounded on $0 < x < \infty$ since it has no upper bound there. But $0 \leq 2/x \leq 2$ for $1 \leq x < \infty$, so $2/x$ is bounded on $1 \leq x < \infty$. See Figure 1.90.

❑

The Numerical Viewpoint: Roots by Bisection

Let's go back to the problem of finding the root of $f(x) = x^3 - 4x - 2 = 0$ between 2 and 3. To locate the root, we "zero in" on it by evaluating the function at the midpoint of the interval, $x = 2.5$. Since $f(2) = -2$, $f(2.5) = 3.625$ and $f(3) = 13$, the function changes sign between $x = 2$ and $x = 2.5$, so the root is between these points. Now we look at $x = 2.25$.

$f(2.25) = 0.39$	There is a root between 2 and 2.25, so we look at 2.125
$f(2.125) = -0.90$	There is a root between 2.125 and 2.25, so we look at 2.1875
$f(2.1875) = -0.28$	There is a root between 2.1875 and 2.25, ...

and so on. See Figure 1.91. If we want to locate the root more precisely, we continue the process. (You may want to round the decimals as you work.) The intervals containing the root are in Table 1.21 and show that the root is $x = 2.21$ to two decimal places.

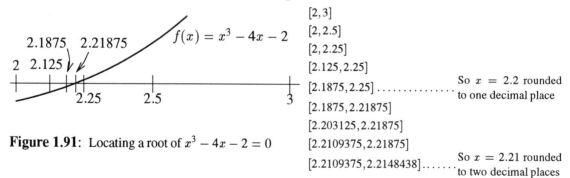

Figure 1.91: Locating a root of $x^3 - 4x - 2 = 0$

$[2,3]$
$[2,2.5]$
$[2,2.25]$
$[2.125,2.25]$
$[2.1875,2.25]$ .... So $x = 2.2$ rounded to one decimal place
$[2.1875,2.21875]$
$[2.203125,2.21875]$
$[2.2109375,2.21875]$
$[2.2109375,2.2148438]$ .... So $x = 2.21$ rounded to two decimal places

Table 1.21: Intervals containing root to $x^3 - 4x - 2 = 0$ (Note that $[2,3]$ means $2 \leq x \leq 3$)

This method of finding roots is called the **Bisection Method**.

- To solve an equation $f(x) = 0$ using the bisection method, we need two starting values for x, say $x = a$ and $x = b$, such that $f(a)$ and $f(b)$ have opposite signs and f is continuous on $[a, b]$.

- Evaluate f at the midpoint of the interval $[a, b]$, and decide in which half-interval the root lies.

- Repeat, using the new half-interval instead of $[a, b]$.

There are some problems with the bisection method:

1. The function may not change signs near the root. For example, $f(x) = x^2 - 2x + 1 = 0$ has a root at $x = 1$, but $f(x)$ is never negative because $f(x) = (x - 1)^2$ and a square cannot be negative. See Figure 1.92.

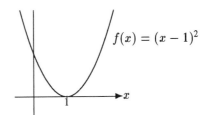

$$f(x) = (x - 1)^2$$

Figure 1.92: f doesn't change signs

2. The function f must be continuous between the starting values $x = a$ and $x = b$.

3. If there is more than one root between the starting values $x = a$ and $x = b$, the method will find only one of the roots. For example, if we had tried to solve $x^3 - 4x - 2 = 0$ starting at $x = -12$ and $x = 10$, the bisection method would zero in on the root between $x = -2$ and $x = -1$, not the root between $x = 2$ and $x = 3$ that we found earlier. (Try it! Then see what happens if you use $x = -10$ instead of $x = -12$.)

4. The bisection method is slow and not very efficient. Applying bisection three times in a row only traps the root in an interval $(\frac{1}{2})^3 = \frac{1}{8}$ as large as the starting interval. Thus, if we initially know that a root is between, say, 2 and 3, then we would need to apply the bisection method at least four times to know the first digit after the decimal point. Also, the midpoint between a and b is often not the best value to try next. For example, with $f(x) = x^3 - 4x - 2 = 0$, since $f(2) = -2$ and $f(3) = 13$, you may suspect that the root is nearer to 2 than 3. In fact, $2\frac{2}{15}$ would be a good next try. Can you see why?

There are much more powerful methods of finding roots available, such as Newton's Method and the secant method. They are also more complicated, but they avoid some of these difficulties.

☐ **Example 8** Find all the roots of $xe^x = 5$ to at least one decimal place.

Solution. If we rewrite the equation as $e^x = 5/x$ and graph both sides as in Figure 1.93, it is clear that there is exactly one root and it is somewhere between 1 and 2. Table 1.22 shows the intervals obtained by the bisection method. After five iterations we have the root trapped between 1.3125 and 1.34375, so we can say $x = 1.3$ is our root to one decimal place. ☐

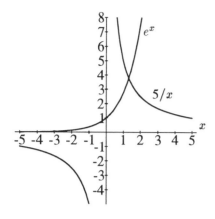

Figure 1.93: Intersection of $y = e^x$
and $y = 5/x$

Interval Containing Root
$[1, 2]$
$[1, 1.5]$
$[1.25, 1.5]$
$[1.25, 1.375]$
$[1.3125, 1.375]$
$[1.3125, 1.34375]$

Table 1.22: Bisection Method for
$f(x) = xe^x - 5 = 0$ (Note that $[1, 2]$
means the interval $1 \leq x \leq 2$)

Iteration

Both zooming in and bisection as discussed above are examples of *iterative* methods, in which a sequence of steps is repeated over and over again, using the results of one step as the input for the next. We can use the method to locate a root to any degree of accuracy. In bisection, each iteration traps the root in an interval that is half the length of the previous one. Each time you zoom in on a calculator, you bracket the root in a smaller interval – how much smaller depends on the settings on your calculator.

Accuracy and Error

In the previous discussion, we used the phrase "accurate to 2 decimal places". For an iterative process where we get closer and closer estimates for some quantity, we take a common sense approach to accuracy: we watch the numbers carefully, and when a digit stays the same for a few iterations, we assume it has stabilized and is correct, especially if the digits to the right of that digit also stay the same. For example, suppose 2.21429 and 2.21431 are two successive estimates for a zero of $f(x) = x^3 - 4x - 2$. Since these two estimates agree to the third digit after the decimal point, we probably have at least 3 decimal places correct.

There is a problem with this, however. Suppose we are finding a root whose true value is 1, and the estimates are converging to the value from below, say 0.985, 0.991, 0.997 and so on. In this case, not even the first decimal place is "correct" even though the difference between the estimates and the actual answer is very small – much less than 0.1. To avoid this difficulty, we say that an estimate a for some quantity r is *accurate to p decimal places* if the error, which is the absolute value of the difference between a and r, or $|r - a|$, is as follows:

Accuracy to p decimal places	means	Error less than
$p = 1$		0.05
2		0.005
3		0.0005
$\vdots$		$\vdots$
n		$0.\underbrace{000\ldots0}_{n}5$

This is the same as saying that r must lie in an interval of length twice the maximum error, centered on a. For example, if a is accurate to 1 decimal place, r must lie in the following interval:

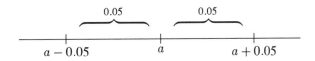

Since both the graphing calculator and the bisection method give us an interval in which the root is trapped, this definition of decimal accuracy is a natural one for these processes.

☐ **Example 9** Suppose the numbers $\sqrt{10}$, $\frac{22}{7}$, and 3.14 are given as approximations to $\pi = 3.1415\ldots$. To how many decimal places is each approximation accurate?

Solution. Using $\sqrt{10} = 3.1622\ldots$,

$$|\sqrt{10} - \pi| = |3.1622\ldots - 3.1415\ldots| = 0.0206\ldots < 0.05,$$

so $\sqrt{10}$ is accurate to one decimal place. Similarly, using $\frac{22}{7} = 3.1428\ldots$,

$$\left|\frac{22}{7} - \pi\right| = |3.1428\ldots - 3.1415\ldots| = 0.0013\ldots < 0.005,$$

so $\frac{22}{7}$ is accurate to two decimal places. Lastly,

$$|3.14 - 3.1415\ldots| = 0.0015\ldots < 0.005,$$

so 3.14 is accurate to two decimal places. ☐

Warning:

- Saying an approximation is accurate to, say, 2 decimal places does *not* guarantee that its first two decimal places are "correct", i.e., that the two digits of the approximation are the same as the corresponding two digits in the true value. For example, an approximate value of 5.997 is accurate to 2 decimal places if the true value is 6.001, but neither of the digit 9's in the approximation agree with the digit 0's in the true value (nor does the digit 5 agree with the digit 6).

- You should notice that the 3 decimal places of accuracy refers to the number of digits that have stabilized in the root, r. It does *not* refer to the number of digits of $f(r)$ that are zero. For example, Table 1.21 on page 102 shows that $x = 2.2$ is a root of $f(x) = x^3 - 4x - 2 = 0$, accurate to one decimal place. Yet, $f(2.2) = -0.152$, so $f(2.2)$ does not have one zero after the decimal point. Similarly, $x = 2.21$ is the root accurate to two decimal places, but $f(2.21) = -0.046$ does not have two zeros after the decimal point.

☐ **Example 10** Is $x = 2.2143$ a zero of $f(x) = x^3 - 4x - 2$ accurate to four decimal places?

Solution. We want to know whether r, the exact value of the zero, lies in the interval

$$2.2143 - 0.00005 < r < 2.2143 + 0.00005$$

or

$$2.21425 < r < 2.21435.$$

Since $f(2.21425) < 0$ and $f(2.21435) > 0$ the zero does lie in this interval, and so $r = 2.2143$ is accurate to four decimal places. ☐

How to Write a Decimal Answer

The graphing calculator and bisection method naturally give an interval for a root or zero. Other numerical techniques, however, do not give a pair of numbers bounding the true value, but rather a single number near the true value. What should you do if you want to give a single number, rather than an interval, for an answer?

When giving that single number as an answer and interpreting it, be careful about giving rounded answers. For example, suppose you have computed a value to be 0.84 and you know the true value is between 0.81 and 0.87. It would be wrong to round 0.84 to 0.8 and say that the answer is 0.8 accurate to one decimal place; the true value could be 0.86, which is not within 0.05 of 0.8. The right thing to say is that the answer is 0.84 accurate to one decimal place. Similarly, to give an answer accurate to, say, 2 decimal places, you may have to show 3 or more decimal places in your answer.

❖ Exercises for Section 1.12

For Problems 1–11, determine the roots or points of intersection to one decimal place accuracy.

1. (a) The root of $x^3 - 3x + 1 = 0$ between 0 and 1

    (b) The root of $x^3 - 3x + 1 = 0$ between 1 and 2

    (c) The third root of $x^3 - 3x + 1 = 0$

2. The root of $x^4 - 5x^3 + 2x - 5 = 0$ between -2 and -1

3. The root of $x^5 + x^2 - 9x - 3 = 0$ between -2 and -1

4. The largest real root of $2x^3 - 4x^2 - 3x + 1 = 0$

5. All real roots of $x^4 - x - 2 = 0$

6. All real roots of $x^5 - 2x^2 + 4 = 0$

7. The first positive root of $x \sin x - \cos x = 0$

8. The first positive point of intersection between $y = 2x$ and $y = \tan x$

9. The first positive point of intersection between $y = \frac{1}{2^x}$ and $y = \sin x$

10. The point of intersection between $y = e^{-x}$ and $y = \ln x$

11. All roots of $\cos t = t^2$

12. Find the largest zero of
$$f(x) = 10xe^{-x} - 1$$
to two decimal places, using the bisection method. Make sure that you demonstrate that your approximation is as good as you claim.

13. (a) Find the first positive value of x where the graphs of $f(x) = \sin x$ and $g(x) = 2^{-x}$ intersect.

    (b) Repeat with $f(x) = \sin 2x$ and $g(x) = 2^{-x}$.

14. Use a graphing calculator to sketch $y = 2 \cos x$ and $y = x^3 + x^2 + 1$ on the same set of axes. Find the positive zero of $f(x) = 2 \cos x - x^3 - x^2 - 1$. Your best friend claims there is one more real zero. Is your friend correct? Explain.

15. Use the table below to investigate the zeros of the function
$$f(\theta) = (\sin 3\,\theta)(\cos 4\,\theta) + 0.8$$
in the interval $0 \le \theta \le 1.8$.

θ	0	0.2	0.4	0.6	0.8	1.0	1.2	1.4	1.6	1.8
$f(\theta)$	0.80	1.19	0.77	0.08	0.13	0.71	0.76	0.12	-0.19	0.33

(a) Decide how many zeros the function has in the interval $0 \le \theta \le 1.8$.

(b) Locate each zero, or a small interval containing each zero.

(c) Are you sure you have found all the zeros in the interval $0 \le \theta \le 1.8$? Graph the function on a calculator or computer to decide.

16. (a) Use Table 1.23 to locate approximately the solution(s) to

$$(\sin 3x)(\cos 4x) = \frac{x^3}{\pi^3}$$

in the interval $1.07 \le x \le 1.15$. Give an interval of length 0.01 in which each solution lies.

(b) Make an estimate for each solution accurate to two decimal places.

x	x^3/π^3	$(\sin 3x)(\cos 4x)$
1.07	0.0395	0.0286
1.08	0.0406	0.0376
1.09	0.0418	0.0442
1.10	0.0429	0.0485
1.11	0.0441	0.0504
1.12	0.0453	0.0499
1.13	0.0465	0.0470
1.14	0.0478	0.0417
1.15	0.0491	0.0340

Table 1.23: Find the intersection of these two curves

17. Sketch the graphs of three different functions $f(x)$ that are continuous on $0 \le x \le 1$ and have the table of values given. The first function is to have exactly one zero in $[0, 1]$, the second is to have at least two zeros in the interval $[0.6, 0.8]$, and the third is to have at least two zeros in the interval $[0, 0.6]$.

x	0	0.2	0.4	0.6	0.8	1.0
$f(x)$	1.00	0.90	0.60	0.11	-0.58	-1.46

18. (a) With your calculator in radian mode, take the arctangent of 1 and multiply that number by 4. Now, take the arctangent of the result and multiply *it* by four. Continue this process ten times or so and record each result as in the table to the right. At each step, you get 4 arctan of the result of the previous step.

1
3.14159...
5.05050...
5.50129...
$\vdots$

(b) Your table allows you to find a solution of the equation

$$4 \arctan x = x.$$

Why, and what is that solution?

(c) What does your table in (a) have to do with Figure 1.94 (which is not to scale)? [Hint: The coordinates of P_0 are (1,1). Find the coordinates of P_1, P_2, P_3....]

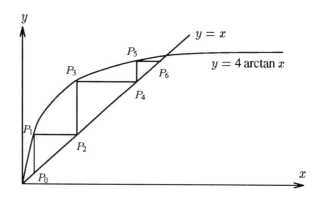

Figure 1.94: $y = x$ and $y = 4 \arctan x$

(for Problem 18)

(d) In (a), what happens if you start with an "initial guess" of 10? of -10? What types of behavior do you observe (i.e., for which initial guesses is the sequence increasing, decreasing; does the sequence approach a limit)? Explain your answers graphically (as in (c)).

Note: The method used in this problem is another example of *iteration* [i.e., repetition of the same process with the output of one step used as input to the next].

19. Using radians, apply the iteration method of Problem 18 to the equation

$$\cos x = x.$$

Represent your results graphically, as in Figure 1.94.

1.13   Miscellaneous Exercises for Chapter 1

1. A car starts out slowly and then goes faster and faster until a tire blows out. Sketch a possible graph of the distance the car has traveled as a function of time.

2. Having left home in a hurry, I'd only gone a short distance when I realized I hadn't turned off the washing machine and so I went back to do so. I then set out again immediately. Sketch my distance from home as a function of time.

3. The graph in Figure 1.95 shows how the usage of household gas (for cooking, etc) varies with the time of day in Ankara, the capital city of Turkey. Explain why the graph has the shape it does.

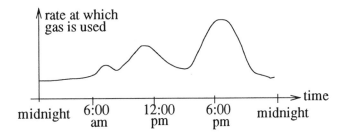

Figure 1.95: Gas usage in Ankara, Turkey

4. Sketch a possible graph for a function that is decreasing everywhere, concave up for negative x and concave down for positive x.

5. Table 1.24 gives the average temperature in Wallingford, Connecticut, for the first ten days in March 1990.

   (a) Over which intervals was the average temperature increasing? Decreasing?

   (b) Find a pair of consecutive intervals over which the average temperature was increasing at a decreasing rate. Find another pair of consecutive intervals over which the average temperature was increasing at an increasing rate.

Date	Average Temp. (°F)
March 1	42°
March 2	42°
March 3	34°
March 4	25°
March 5	22°
March 6	34°
March 7	38°
March 8	40°
March 9	49°
March 10	49°

Table 1.24: Temperature in Wallingford, March 1–10, 1990

(for Problem 5)

6. For tax purposes, you may have to report the value of your assets, such as cars or refrigerators. The value you report depreciates, or drops, with time. The idea is that a car you originally paid $10,000 for may be worth only $5000 a few years later. The simplest way to calculate the value of your asset is using straight line depreciation, which assumes that the value is a linear function of time. If a $950 refrigerator depreciates completely in seven years, find a formula for its value as a function of time.

7. An airplane uses a fixed amount of fuel for take-off, a (different) fixed amount for landing, and a fixed amount per mile when it is in the air. How does the total quantity of fuel required depend on the length of the trip? Write a formula for the function involved. Explain the meaning of the constants in your formula.

8. Sketch reasonable graphs for the following. Pay particular attention to the concavity of the graphs and explain your reasoning.

   (a) The total revenue generated by a car rental business, plotted against the amount spent on advertising.

   (b) The temperature of a cup of hot coffee standing in a room, plotted as a function of time.

9. If $f(x) = x^2 + 1$, find:
   (a) $f(t+1)$ (b) $f(t^2+1)$ (c) $f(2)$ (d) $2f(t)$ (e) $(f(t))^2 + 1$

10. For $g(x) = x^2 + 2x + 3$ determine:
    (a) $g(2+h)$ (b) $g(2)$ (c) $g(2+h) - g(2)$

11. For $f(n) = 3n^2 - 2$ and $g(n) = n + 1$, determine:

   (a) $f(n) + g(n)$

   (b) $f(n)g(n)$

   (c) the domain of $\dfrac{f(n)}{g(n)}$.

   (d) $f(g(n))$

   (e) $g(f(n))$

12. Estimate all real zeros of the following polynomials, accurate to 2 decimal places:

   (a) $f(x) = x^3 - 2x^2 - x + 3$

   (b) $f(x) = x^3 - x^2 - 2x + 2$

13. Sketch the graph of a function defined for $x \geq 0$ with all of the following properties. [There are lots of possible answers.]

   (a) $f(0) = 2$;

   (b) $f(x)$ is increasing for $0 \leq x < 1$;

   (c) $f(x)$ is decreasing for $1 < x < 3$;

   (d) $f(x)$ is increasing for $x > 3$;

   (e) $f(x) \to 5$ as $x \to \infty$.

14. When a new product is advertised, more and more people try it. However, the rate at which new people try it slows as time goes on.

   (a) Sketch a graph of the total number of people who have tried such a product against time.

   (b) What do you know about the concavity of the graph?

15. Consider the data in Table 1.25.

   (a) For what x-values given in the table is $f(x)$ at its maximum?

   (b) Over which subinterval (i.e. between which consecutive x-values) is the value of f increasing fastest? Decreasing fastest?

   (c) Determine two different subintervals where it appears likely that there would be a value of x such that $f(x) = 0$.

   (d) Determine consecutive subintervals over which f is increasing at an increasing rate.

x	$f(x)$
-2.5	76.96
-2	15.04
-1.5	5.31
-1	8.14
-0.5	87.04
0	29.16
0.5	24.64
1	8.5
1.5	-12.96
2	-27.44
2.5	-16.64
3	43.74

Table 1.25: Where is this function increasing at an increasing rate?

16. (a) Use the data from Table 1.26 to determine a formula of the form

$$Q = Q_0 e^{rt}$$

which would give the number of rabbits, Q, at time t (in months).

(b) What is the approximate doubling time for this population of rabbits?

(c) Use your equation to predict when the rabbit population will reach 1000 rabbits.

t	# of rabbits
0	25
1	43
2	75
3	130
4	226
5	391

Table 1.26: Rapid rabbit reproduction
(for Problem 16)

17. If you need $20,000 in your bank account in 6 years, how much must be deposited now? (Assume an annual interest rate of 10%, compounded continuously.)

18. What nominal annual interest rate has an effective annual yield of 5%?

19. What is the effective annual yield for a nominal annual interest rate of 8%?

20. Different kinds of the same element (called different *isotopes*) can have very different half-lives. The decay of Plutonium-240 is described by the formula

$$Q = Q_0 e^{-0.00011t}$$

while the decay of Plutonium-242 is described by

$$Q = Q_0 e^{-0.0000018t}.$$

Find the half-lives of Plutonium-240 and Plutonium-242.

21. An animal skull still has 20% of the Carbon-14 present when the animal died. If the half-life of Carbon-14 is 5730 years, find the approximate age of the skull.

22. Suppose prices are increasing by 0.1% a day.

(a) By what percent do prices increase a year?

(b) Looking at your answer to (a), guess the approximate doubling time of prices increasing at this rate. Check your guess.

23. In the period from 1984 to 1989, Argentina's inflation rate averaged about 450% a year.

(a) What was the average daily inflation rate in Argentina during this period?

(b) Approximately how long did it take for prices to double in Argentina during this period?

24. Each planet moves around the sun in an elliptical orbit. The orbital period, T, of a planet is the time it takes the planet to go once around the sun. The semi-major axis of each planet's orbit is the average of the largest and the smallest distances between the planet and the sun. Kepler discovered that the period of a planet is proportional to the $\frac{3}{2}$ power of its semi-major axis. What is the orbiting period (in days) of Mercury, the closest planet to the sun, with a semi-major axis of 58 million km? What is the period (in years) of Pluto, the farthest planet, with semi-major axis of 6000 million km? The semi-major axis of the Earth is 150 million km. (What is its period?)

25. Under certain circumstances, the velocity, V, of a falling raindrop is given by

$$V = V_0 \left(1 - e^{-\frac{t}{t_0}} \right)$$

where t is time, V_0 and t_0 are positive constants.

(a) Sketch rough graphs of V as a function of t, for $t \geq 0$, and $t_0 = 1, 2, 3$.

(b) How does increasing t_0 change the shape of the graph?

26. Find possible formulas for the functions in Figure 1.96 using exponentials. [There may be more than one answer.]

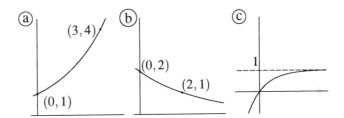

Figure 1.96: What are these functions?

(for Problem 26)

27. Find equations whose graphs would resemble those in Figure 1.97.

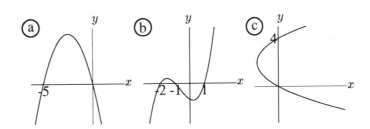

Figure 1.97: Find polynomials whose graphs would look like these. (For Problem 27)

28. (a) Consider the functions shown in Figure 1.98(a). Find the coordinates of C.

    (b) Consider the functions shown in Figure 1.98(b). Find the coordinates of C in terms of b.

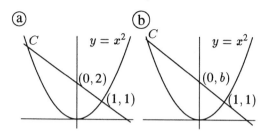

Figure 1.98: Find the coordinates of C as a function of b. (For Problem 28)

29. Consider the table:

t	0	1	2	3	4
$\sin\left(t^2\right)$	0	0.84	-0.76	0.41	-0.29

    (a) What does it tell you about the number of roots of
$$\sin\left(t^2\right) = 0$$
    in the interval $0 \le t \le 4$? Approximately where do they lie?

    (b) What does a graphing calculator or computer tell you about the number of roots the equation has in the interval $0 \le t \le 4$?

    (c) Use the calculator or computer to estimate each of these roots to one decimal place.

    (d) Explain why the smallest positive root is $\sqrt{\pi}$.

    (e) Find an exact, symbolic expression (like $\sqrt{\pi}$) for each of the other positive roots you found.

30. Using a calculator or computer, sketch a graph of $y = \sin x$, $y = 0.4$ and $y = -0.4$.

    (a) From the graph estimate, to one decimal place, all the solutions of $\sin x = 0.4$ with $-\pi \le x \le \pi$.

    (b) Use a calculator to find $\arcsin(0.4)$. What is the relation between $\arcsin(0.4)$ and each of the solutions you found in (a)?

    (c) Estimate all the solutions to $\sin x = -0.4$ with $-\pi \le x \le \pi$. (Again, to one decimal place.)

    (d) What is the relation between $\arcsin(0.4)$ and *each* of the solutions you found in (c)?

31. (a) Use a graphing calculator or computer to find the period of $3 \sin \alpha + 4 \cos \frac{1}{3}\alpha$.

    (b) What is the period of $\sin \alpha$? of $\cos \frac{1}{3}\alpha$?

    (c) Use your answers to (b) to explain your answer to (a).

32. (a) Use a graphing calculator or computer to find the period of $2 \sin 4x + 3 \cos 2x$.

   (b) Give your answer in an exact form (as a multiple of π).

   (c) Find the periods of $\sin 4x$ and $\cos 2x$ and use them to explain your answer to (a).

33. Match the formulas with the graphs in Figure 1.99:

   (a) $y = 1 - 2^{-x}$ (f) $y = x^3 - x^2 - x + 1$
   (b) $y = \log(x + 1)$ (g) $y = 2^{-x} \sin x$
   (c) $y = 2 \cos x$ (h) $y = 1 + \cos x$
   (d) $y = 1 - x^2$ (i) $y = \arctan x$
   (e) $y = \tan x$

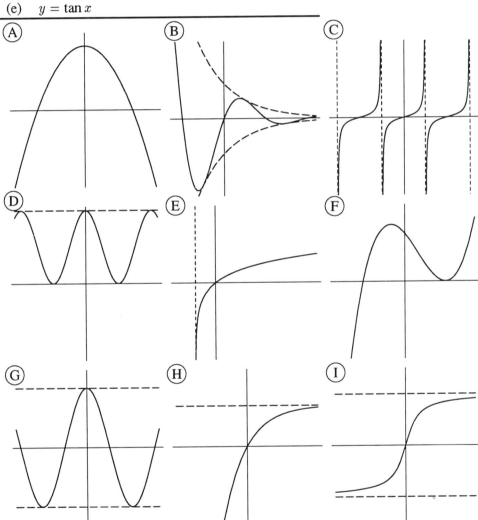

Figure 1.99: Graphs of Functions

(for Problem 33)

34. Each of the functions described by the data in Table 1.27 is increasing over its domain, but each increases in a different way. Which of the graphs in Fig. 1.100 best fits each function?

x	$f(x)$	x	$g(x)$	x	$h(x)$
1	1	3.0	1	10	1
2	2	3.2	2	20	2
4	3	3.4	3	28	3
7	4	3.6	4	34	4
11	5	3.8	5	39	5
16	6	4.0	6	43	6
22	7	4.2	7	46.5	7
29	8	4.4	8	49	8
37	9	4.6	9	51	9
47	10	4.8	10	52	10

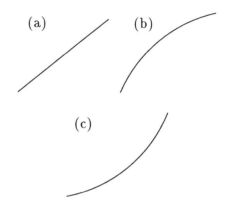

Table 1.27: Lists of data from increasing functions.

(for Problem 34)

Figure 1.100: Plots of three increasing functions

(for Problem 34)

35. Given the graph of $y = h(x)$ to the right,

(a) sketch a graph of:

i. $y = h^{-1}(x)$

ii. $y = \dfrac{1}{h(x)}$

(b) What becomes of the asymptote when you sketch the inverse function?

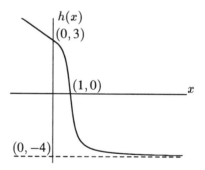

36. Find reasonable formulas for the graphs in Figure 1.101.

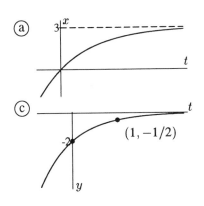
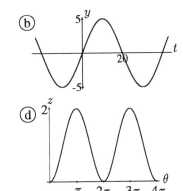

Figure 1.101: What formulas define these graphs?

(for Problem 36)

37. (a) What effect does the transformation

$$y = p(x) \qquad \text{to} \qquad y = p(1 + x)$$

have on the graph of $p(x)$?

(b) If p is a polynomial of degree ≤ 2 such that

$$p(x) = p(1 + x),$$

what can you say about p?

38. For each of the two conditions below, find all polynomials, p, of degree ≤ 2 which satisfy the condition for all x.

(a) $p(x) = p(-x)$
(b) $p(2x) = 2p(x)$

39. A *catalyst* in a chemical reaction is a substance which speeds up the reaction but which does not itself change. If the product of a reaction is itself a catalyst, the reaction is said to be *autocatalytic*. Suppose the rate, r, of a particular autocatalytic reaction is proportional to the quantity of the original material remaining times the quantity of product, p, produced. If the initial quantity of the original material is A, and the amount remaining is $A - p$,

(a) express r as a function of p.
(b) What is the value of p when the reaction is proceeding fastest?

40. Glucose is fed by intravenous injection at a constant rate, k, into a patient's bloodstream. Once there, the glucose is broken down chemically at a rate proportional to the amount of glucose present. If R is the net rate at which the quantity, G, of glucose in the blood is increasing,

(a) Write a formula giving R as a function of G.
(b) Sketch a graph of R against G.

41. A fish population is reproducing at an annual rate equal to 5% of the current population, P. Meanwhile, fish are being caught by fishermen at a constant rate, Y (measured in fish per year).

    (a) Write a formula for the rate, R, at which the fish population is increasing as a function of P.

    (b) Sketch a graph of R against P.

42. A tomato is dropped off of a 55 foot tall building. Sketch a graph of its *velocity* as a function of its height above the ground. [Hint: The domain of the function is 0 feet through 55 feet. Don't worry about concavity.]

43. Let $S(x)$ be the number of sunlight hours on a cloudless June 21, as a function of the latitude x (measured in degrees).

    (a) What is $S(0)$? [Hint: Latitude 0° is the equator.]

    (b) Let x_0 be the latitude of the Arctic Circle ($x_0 \approx 66\frac{1}{2}°$). In the Northern Hemisphere, $S(x)$ is given by:

    $$S(x) = \begin{cases} a + b \arcsin\left(\frac{\tan x}{\tan x_0}\right), & \text{for } 0 \le x \le x_0 \\ 24, & \text{for } x_0 \le x \le 90 \end{cases}$$

    for some constants a and b. Find a and b.

    (c) Compute $S(x)$ for Tucson, Arizona ($x = 32°13'$) and Walla Walla, Washington ($46°4'$).

    (d) Graph $S(x)$, for $-90° \le x \le 90°$, using a graphing calculator or computer. Is the graph smooth?

44. Figure 1.102, is the graph of the function f, where $f(t)$ is the depth in meters below the Atlantic Ocean floor where t million-year-old rock can be found.

    (a) Evaluate $f(15)$ and say what it means in practical terms.

    (b) Is f invertible? Explain.

    (c) Evaluate $f^{-1}(120)$ and say what it means in practical terms.

    (d) Sketch a graph of f^{-1}.

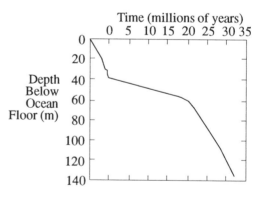

Figure 1.102: Sedimentation Rates[5], Hole 522

---

[5]Data of Dr. Murlene Clark based on core samples drilled by the research ship *Glomar Challenger*, taken from Initial Reports of the Deep Sea Drilling Project

Chapter 2

KEY CONCEPT: THE DERIVATIVE

We begin this chapter by investigating the problem of speed: how can we measure the speed of a moving object at a given instant in time? Or, more fundamentally, what do we mean by the term speed? By carefully examining some examples, we'll come up with a definition of speed that will have wide-ranging implications, not just for the speed problem, but also for measuring any rate of change. Our journey will lead us to the key concept of *derivative*, which forms the basis for our study of calculus.

The derivative can be interpreted geometrically, as the slope of a curve, and physically, as a rate of change. Because derivatives can be used to represent everything from fluctuations in interest rates to the rate at which fish are dying and gas molecules are moving, they have applications throughout the sciences.

2.1   How do we Measure Speed?

Let's take a few minutes to set the stage for our discussion of speed. The notion of speed and, in particular, the speed of an object at an instant in time, is subtle and a bit tricky to define precisely. Consider the statement: "At the instant it crossed the finish line, the horse was traveling at 42 mph." How can such a claim be substantiated? A photograph taken at that instant will show the horse motionless—it is no help at all. There is some paradox in trying to quantify the property of motion at a particular instant in time, since by focusing on a single instant you stop the motion!

A similar difficulty arises whenever we attempt to measure the rate of change of anything—for example, oil leaking out of a damaged tanker. The statement "One hour after the ship's hull ruptured, oil was leaking at a rate of 200 barrels per second" seems not to make sense. You could argue that at any given instant *no* oil is leaking.

Problems of motion were of central concern to Zeno and other philosophers as early as the 5th century BC. The modern approach, made famous by Newton's calculus, is to stop looking for a simple notion of speed at an instant, and instead to look at speed over small intervals containing the instant. This method sidesteps the philosophical problems mentioned earlier, but brings along new ones of its own.

The time is ripe for an example! We shall illustrate the ideas discussed above by an idealized experiment, called a thought experiment. It is idealized in the sense that we assume that we can make measurements of distance and time as accurately as we wish. In fact, the numbers we shall use come from a mathematical formula, not from real measurements, but that doesn't matter for our purposes.

A Thought Experiment: Average and Instantaneous Velocity

We shall look at the speed or velocity of a small object (say a grapefruit) that is thrown high in the air at $t = 0$ seconds. The behavior of the grapefruit is not surprising: it goes up, slows down, reverses direction, falls down and finally "*Splat!*" Informally, we can say that the grapefruit leaves the thrower's hand at high speed, slows down until it reaches its peak and then gradually speeds up in the downward direction. But suppose that we would like to be more precise in our determination of the speed, say at $t = 1$ second. We'll assume that we can measure the height of the grapefruit above the ground at any time t; we'll think of the height, y, as a function of time. See Table 2.1.

t(secs)	0	1	2	3	4	5	6
y(feet)	6	90	142	162	150	106	30

Table 2.1: Height of the Grapefruit

"*Splat!*" comes sometime between 6 and 7 seconds. The numbers show the behavior noted above: in the first second, the grapefruit travels $90 - 6 = 84$ feet, while in the second second it travels only $142 - 90 = 52$ feet. Hence the grapefruit traveled faster over the first interval, $0 < t < 1$, than the second interval, $1 < t < 2$. Since speed = distance/time, we say the average speed over the interval $0 < t < 1$ is 84 ft/sec and the average speed over the next interval is 52 ft/sec.

Velocity versus Speed: From now on, we will make a distinction between velocity and speed. Suppose an object moves along a line. If we pick one direction to be positive, the *velocity* is positive if it is in the same direction, and negative if it is in the opposite direction. *Speed* is the magnitude of the velocity and so is always positive or zero. In general, we define

> the *average velocity* of an object over the interval $a < t < b$ is the net change in position during the interval divided by the change in time, $b - a$.

☐ **Example 1** Compute the average velocity of the grapefruit over the interval $4 < t < 5$. What is the significance of the sign of your answer?

Solution. During this interval, the grapefruit moves $(106 - 150) = -44$ feet. Therefore the average velocity is -44 ft/sec. The negative sign means the height is decreasing and the grapefruit is moving downward. ☐

☐ **Example 2** Compute the average velocity of the grapefruit over the interval $1 < t < 3$.

Solution. Average velocity $= (162 - 90)/(3 - 1) = \frac{72}{2} = 36$ ft/sec. ☐

The average velocity is a useful concept since it gives a rough idea of the behavior of the grapefruit: if two identical grapefruits are hurled into the air, and one has an average velocity of 10 ft/sec over the interval $0 < t < 1$ while the second has an average velocity of 100 ft/sec over the same interval, clearly the second one was thrown harder.

But average velocity over an interval doesn't solve the problem of measuring the velocity of the grapefruit at *exactly* $t = 1$ second. To get closer to an answer to that question, we'll have to look at what happens near $t = 1$ in more detail. Suppose we are given the data[1] in Figure 2.1, showing the average velocity over small intervals on either side of $t = 1$.

Notice that the average velocity before $t = 1$ is slightly more than the average velocity after $t = 1$. We would expect to define the velocity *at* $t = 1$ to be between these two average velocities. As the size of the interval shrinks, the values of the velocity before $t = 1$ and the velocity after $t = 1$ get closer together. By the third example, both velocities are 68.0 ft/sec, (to one decimal place), so we will define the velocity at $t = 1$ to be 68.0 ft/sec (to one decimal place).

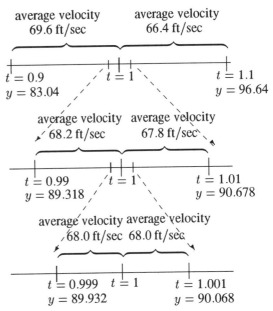

Figure 2.1: Average Velocity of Grapefruit

[1]The data is in fact calculated from the formula $y = 6 + 100t - 16t^2$.

Of course, if we showed more decimal places, average velocities before and after $t = 1$ would no longer agree in the third example. To calculate the velocity at $t = 1$ to more decimal places, you should take smaller and smaller intervals on either side of $t = 1$, until the average velocities agree to as many decimal places as you want. The velocity at $t = 1$ is then defined to be this common average velocity.

Defining Instantaneous Velocity using the Idea of a Limit

When you take smaller and smaller intervals, it turns out that the average velocities are always just above or just below 68 ft/sec. It seems natural, then, to define velocity at the instant $t = 1$ to be 68 feet/sec. This is called the *instantaneous velocity* at this point, and its definition depends on our being adequately convinced that smaller and smaller intervals will provide average speeds that come arbitrarily close to 68. Modern mathematics has a name for this process: it is called *taking the limit*. Thus we can state

> the *instantaneous velocity* of an object at time t is given by the limit of the average velocity over the interval as the interval shrinks around t.

You should be sure that you see how we have replaced the original difficulty of computing velocity at a point by a search for an argument to convince ourselves that the average velocities do approach a number as the time intervals shrink in size. In a sense we have traded one hard question for another, since we don't yet have any idea how to be certain what number the average velocities are approaching. In our thought experiment, the number seemed to be exactly 68, but what if it were 68.000001? How can we be sure that we have taken small enough intervals?

For most practical purposes, it is not likely to be important whether the velocity is exactly 68 or 68.000001. Showing that the limit is exactly 68 requires more precise knowledge of how the velocities were calculated and of the limiting process; we will see this in Chapter 4.

Visualizing Velocity: Slope of Curve

We have just seen how to estimate velocity numerically. Now we will see how to visualize velocity using a graph of height. Let's go back to the grapefruit. Suppose that Figure 2.2 shows the height of the grapefruit plotted against time. (Note that this is not a picture of the grapefruit's path, which is straight up-and-down.)

How can we visualize the average velocity on this graph? Suppose $y = s(t)$. Let's consider the interval $1 < t < 2$ and the expression

$$\text{average velocity} = \frac{\text{distance moved, } 1 < t < 2}{\text{time elapsed}} = \frac{s(2) - s(1)}{2 - 1} = \frac{142 - 90}{1} = 52 \text{ ft/sec.}$$

Now $s(2) - s(1)$ is the change in height over the interval, or the distance moved, and it is marked vertically in Figure 2.2. The 1 in the denominator is the time elapsed and is marked horizontally in Figure 2.2. Therefore,

$$\text{average velocity} = \frac{\text{distance moved}}{\text{time elapsed}} = \text{slope of line joining } BC.$$

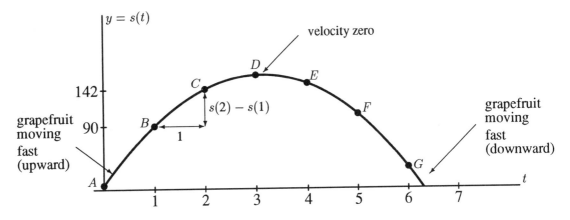

Figure **2.2**: The Height, y, of the Grapefruit at Time t

A similar argument shows that

> the *average velocity* over any interval is the slope of the line joining the points on the graph of $s(t)$ corresponding to the endpoints of the interval.

The next question is how to visualize the velocity at an instant. Let's think about how we found the instantaneous velocity. We took average velocities across smaller and smaller intervals ending at the point $t = 1$. Two such velocities are represented by the slopes of the lines in Figure 2.3. As the length of the interval shrinks, the slope of the line gets closer to the slope of the curve at $t = 1$.

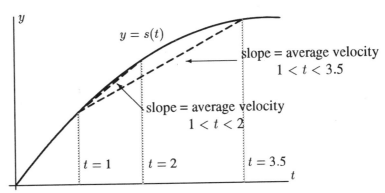

Figure **2.3**: Average Velocities over Small Intervals

The cornerstone of the idea is the fact that, on a very small scale, most functions look almost linear. Think of taking the graph of a function near a point and enlarging the region with a microscope. See Figure 2.4. The higher the magnification, the more the curve will appear to be a straight line. In other words, if you repeatedly magnify a section of the curve centered on a point of interest, the section of curve will eventually look like a straight line. We call the slope of this magnified line *the slope of the curve* at the point (see Figure 2.4).

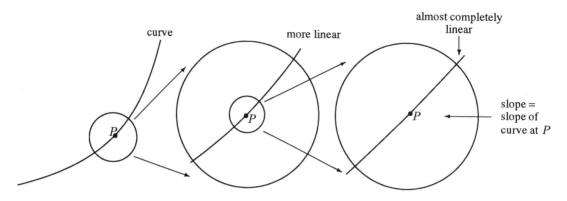

Figure 2.4: Finding the Slope of the Curve at a Point by Successive Magnifications

Therefore the slope of the magnified line is the instantaneous velocity. Thus,

the *instantaneous velocity* is *the slope of the curve* at the point.

Look back at the graph of the grapefruit's height as a function of time in Figure 2.2. If you think of the velocity at any point as the slope of the curve there, you can see how the grapefruit's velocity varies during its journey. At points A and B the curve has a large positive slope, indicating that the grapefruit is traveling up rapidly. Point D is almost at the top: the grapefruit is slowing as it reaches the peak. At the peak, the slope of the curve is zero: the fruit has slowed to a stop in preparation for its return to Earth. At point E, the curve has a small negative slope, indicating a slow velocity of descent. Finally, the slope of the curve at point G is large and negative, indicating a large downward velocity that is responsible for the *"Splat"*.

❐ **Example 3**   At the top of the grapefruit's path, its velocity is zero. Does this mean the grapefruit has stopped?

Solution. At the top of its path the grapefruit changes direction—from moving upward to moving downward—and at the moment it makes this change, the grapefruit's instantaneous velocity is zero. Deciding whether the grapefruit has stopped depends on what you think 'stopped' means. Most people think an object has stopped only if it has zero velocity for some interval of time; with this definition, our grapefruit has not stopped, but only had zero instantaneous velocity for one moment. ❑

The Precise Definition of Velocity and the Notation for Limit

To recap the results so far, we found that a natural way to define the velocity of an object at an instant t was to look at the average velocity over smaller and smaller intervals around t, and see to what number those average velocities converge. Let us introduce a formal expression for this idea. As before, let s be the name for the function of time that gives the position (or height) y of the grapefruit at time t, so $y = s(t)$. Then we need to write down an expression that gives the average velocity over an interval. If the interval is $a < t < b$, then

$$\text{Average Velocity} = \frac{\text{change in position}}{\text{change in time}} = \frac{s(b) - s(a)}{b - a}.$$

If we're interested in looking for instantaneous velocity at $t = a$, then we'd like to look at smaller and smaller intervals near $t = a$. We will consider intervals of the form $a < t < a + h$, where h is the size of the interval. Then, over the interval $a < t < a + h$,

$$\text{average velocity} = \frac{s(a + h) - s(a)}{h}.$$

We get a similar formula when $h < 0$. Now remember in the previous section we agreed that the instantaneous velocity is the number that the average velocities approach as the intervals decrease in size, that is, as h becomes smaller. Thus the instantaneous velocity at $t = a$ is

$$\text{the limit as } h \text{ approaches } 0 \text{ of } \quad \frac{s(a + h) - s(a)}{h}.$$

One final small change is for the sake of economy: instead of writing the phrase "limit as h approaches 0 of", we'll just use

$$\lim_{h \to 0}$$

Now we can exhibit our definition in all of its finery. We write

$$\left(\begin{array}{c} \text{Instantaneous Velocity} \\ \text{at } t = a \end{array} \right) = \lim_{h \to 0} \frac{s(a + h) - s(a)}{h}.$$

This expression forms the foundation of the rest of calculus. Be sure that you are not confused by the notation and recognize it for what it is: the number that the average velocities approach as the intervals shrink. To find that number, the limit, we look at intervals of smaller and smaller, but never zero, length. You should realize that we haven't introduced any new ideas in this definition; we have simply found a compact way to write the ideas developed previously.

❖ Exercises for Section 2.1

1. A car drives at a constant speed. Sketch a graph of the distance the car has traveled as a function of time.

2. A car drives at an increasing speed. Sketch a possible graph of the distance the car has traveled as a function of time.

3. A car starts at a high speed and its speed then decreases slowly. Sketch a graph of the distance the car has traveled as a function of time.

4. A bicyclist pedals at a fairly constant rate, with evenly spaced intervals of coasting. Sketch a graph of the distance she has traveled as a function of time.

5. Find the average velocity over the interval $0 \leq t \leq 0.2$ and estimate the velocity at $t = 0.2$ of a car whose position, s, is given by

t(secs)	0	0.2	0.4	0.6	0.8	1.0
s(feet)	0	0.5	1.8	3.8	6.5	9.6

6. A ball is tossed into the air from a bridge and its height y (in feet) above the ground, t seconds after it is thrown, is given by

$$y = f(t) = -16t^2 + 50t + 36.$$

   (a) How high above the ground is the bridge?
   (b) What is the average velocity of the ball for the first second?
   (c) Approximate the velocity of the ball at $t = 1$ second.
   (d) Graph the function f and determine the maximum height the ball will reach. What should the velocity be at the time the ball is at the peak?
   (e) Use the graph to decide at what time, t, the ball reaches its maximum height.

7. Match the points labeled on the curve with the slopes given below.

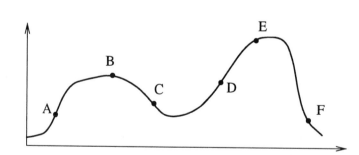

slope	point
-2	
-1	
0	
$\frac{1}{2}$	
1	
3	

8. For the function shown in Figure 2.5, at what labeled points is the slope of the curve positive? negative? Which labeled point has the most positive slope? The most negative (i.e. negative and with the largest magnitude)?

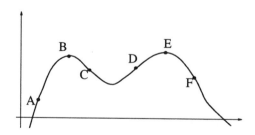

Figure 2.5: Function for Problem 8

9. For the graph $y = f(x)$ shown in Figure 2.6, arrange the following numbers in ascending (i.e. smallest to largest) order:

- the slope of the curve at A
- the slope of the curve at B
- the slope of the curve at C
- the slope of the line AB
- the number 0
- the number 1

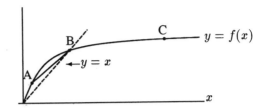

Figure 2.6: Put the numbers in ascending order.
(for Problem 9)

10. Suppose a particle is moving along a straight line and that $s = f(t)$ represents the distance of the particle from a point as a function of time, t. Sketch a possible graph for f if the average velocity of the particle between $t = 2$ and $t = 6$ is the same as the instantaneous velocity at $t = 5$.

2.2   The Derivative

Average Rate of Change

Now we will apply the analysis of Section 2.1 to any function $y = f(x)$ — not necessarily height as a function of time. In the case of height, we looked at

$$\frac{s(a + h) - s(a)}{h},$$

which is the change in height divided by the length of a a small time interval. Now we'll consider

$$\frac{f(a + h) - f(a)}{h}$$

for some function f. This ratio is called the *difference quotient*. What does it represent? The numerator, $f(a + h) - f(a)$, measures the change in f during the interval from a to $a + h$. Thus

$$\left(\begin{array}{c} \text{Average Rate of Change} \\ \text{of } f \text{ over the interval} \\ \text{from } a \text{ to } a + h \end{array} \right) = \frac{f(a + h) - f(a)}{h}$$

This ratio compares the change in the value of the function, $f(x)$, with h, the change in x. If a small change in x produces a large change in $f(x)$, this ratio will be large; conversely if a small change in x produces an even smaller change in $f(x)$, the ratio will be small. Although the interval is not necessarily a time interval any more, we still talk about the *average rate of change* of f over the interval. If we want to emphasize the independent variable, we talk about the average rate of change of f *with respect to x*.

For example, consider the function which gives the radius of a sphere in terms of the volume of the sphere. Think of blowing air into a balloon: as you change the volume of air the radius changes. You've probably noticed that a balloon seems to blow up faster at the start and then slows down as you blow more air into it. What you're seeing is the rate of change of the radius with respect to change in volume. You can compute average rates of change of any function over any interval exactly as we did above. When the balloon is small, a small change in volume produces a relatively large change in its radius. As the balloon grows, the same small change in volume (a puff of air) produces a much smaller change in the radius.

☐ **Example 1** The function r for the radius of a sphere in terms of its volume is given by the formula

$$r(V) = \left(\frac{3V}{4\pi}\right)^{1/3}.$$

Calculate the average rate of change of r with respect to V over the intervals $0.5 < V < 1$ and $1 < V < 1.5$.

Solution.

$$\left(\begin{array}{c} \text{Average rate of change} \\ \text{of radius for } 0.5 < V < 1 \end{array}\right) = \frac{r(1) - r(0.5)}{0.5} = 2\left(\left(\frac{3}{4\pi}\right)^{1/3} - \left(\frac{1.5}{4\pi}\right)^{1/3}\right) \approx 0.26$$

$$\left(\begin{array}{c} \text{Average rate of change} \\ \text{of radius for } 1 < V < 1.5 \end{array}\right) = \frac{r(1.5) - r(1)}{0.5} = 2\left(\left(\frac{4.5}{4\pi}\right)^{1/3} - \left(\frac{3}{4\pi}\right)^{1/3}\right) \approx 0.18$$

As you can see, the rate does decrease as the volume increases. ☐

The average rate of change of a function during an interval is not at all the same as the absolute change: absolute change is just the difference in the values of f at the ends of the interval (i.e. $f(a+h) - f(a)$), while the rate of change is the absolute change divided by the size of the interval (i.e. $\frac{f(a+h)-f(a)}{h}$). The rate of change tells how quickly (or slowly) the function changes from one end of the interval to the other, relative to the size of the interval. It is often a more useful thing to know; for example, if someone offers you a $150 return on a $100 investment, you will want to know how long it is going to take to make that money. Just knowing the absolute change in your money $50, is not enough, but knowing the rate of change (i.e. $50, divided by the time it takes to make it), will help you decide whether or not to make the investment.

Instantaneous Rate of Change: the Derivative

We can also define the *instantaneous rate of change* of a function at a point. We simply mimic what we did for velocity, namely, look at the average rate of change over smaller and smaller intervals.

By taking the limit:

$$\begin{array}{c} \text{Rate of Change of } f \\ \text{at the point } a \end{array} = \lim_{h \to 0} \frac{f(a+h) - f(a)}{h}.$$

This number is so important that it is given its own name, *the derivative of f at a*, and is denoted by $f'(a)$. Thus we define the

Derivative of f at a :

$$f'(a) = \lim_{h \to 0} \frac{f(a+h) - f(a)}{h}.$$

❏ **Example 2**   By choosing small values for h, estimate the instantaneous rate of change of the radius of a sphere with respect to change in volume at $V = 1$.

Solution. With $h = 0.01$ and $h = -0.01$,

$$\frac{r(1.01) - r(1)}{0.01} \approx 0.2061 \quad \text{and} \quad \frac{r(0.99) - r(1)}{-0.01} \approx 0.2075.$$

With $h = 0.001$ and $h = -0.001$,

$$\frac{r(1.001) - r(1)}{0.001} \approx 0.2067 \quad \text{and} \quad \frac{r(0.999) - r(1)}{-0.001} \approx 0.2069.$$

The values of these difference quotients suggest that the limit is between 0.2067 and 0.2069. Rounding off suggests the value is about 0.207; taking other h values confirms this. So we say

$$\left(\begin{array}{c} \text{Instantaneous rate of change} \\ \text{of radius with respect to} \\ \text{change in volume at } V = 1 \end{array} \right) \approx 0.207.$$

❏

In the previous example we got an approximation to the instantaneous rate of change, or derivative, by substituting in small values of h. In the next example we will compute the exact value of the derivative.

◻ **Example 3** Find the derivative of the function $f(x) = x^2$ at the point $x = 1$.

Solution. We need to look at

$$f'(1) = \lim_{h \to 0} \frac{f(1 + h) - f(1)}{h}.$$

This is the same as

$$\lim_{h \to 0} \frac{(1 + h)^2 - 1^2}{h} = \lim_{h \to 0} \frac{(1 + 2h + h^2) - 1}{h} = \lim_{h \to 0} \frac{2h + h^2}{h}.$$

Now choose several small values for h (e.g. $0.1, 0.01, 0.001$), compute $\frac{2h+h^2}{h}$ for these and see if you can guess the limit.

An alternative approach is to divide by h in the expression $\frac{2h+h^2}{h}$, a valid operation since the limit only examines values of h close to but not equal to zero. We get

$$\lim_{h \to 0} \frac{h(2 + h)}{h} = \lim_{h \to 0} (2 + h).$$

This limit is pretty clearly equal to 2, so $f'(1) = 2$; i.e. at $x = 1$ the rate of change of x^2 is 2. ◻

Graphically, what does it mean that the derivative is 2? Since the derivative is the rate of change, it means that for small changes in x, near $x = 1$, the change in $f(x) = x^2$ is about twice as big. As an example, if x changes from 1 to 1.1, a net change of 0.1, $f(x)$ should change by about 0.2. Figure 2.7 shows this geometrically.

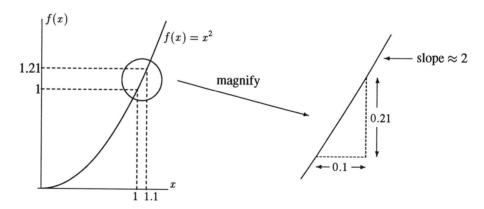

Figure 2.7: Graph of $f(x) = x^2$ near $x = 1$ has slope ≈ 2.

Table 2.2 shows the derivative of $f(x) = x^2$ numerically. Notice that near $x = 1$, every time the value of x increases by 0.001, the value of x^2 increases by approximately 0.002. Thus near $x = 1$ the graph is approximately linear with slope $\frac{0.002}{0.001} = 2$.

x	x^2	Difference in x^2 values
0.998	0.996004	
		0.001997
0.999	0.998001	
		0.001999
1.000	1.000000	
		0.002001
1.001	1.002001	
		0.002003
1.002	1.004004	
↑		↑
x increments 0.001		all approximately 0.002

Table 2.2: Values of $f(x) = x^2$ near $x = 1$

Visualizing the Derivative: Slope of Curve and Slope of Tangent

As with velocity, we can visualize the derivative $f'(a)$ as the slope of the graph of f at a. In addition, there is another way to think of $f'(a)$. Consider the difference quotient $\frac{f(a+h)-f(a)}{h}$. The numerator, $f(a+h) - f(a)$, is the vertical distance marked in Figure 2.8, and h is the horizontal distance, so the

$$\text{average rate of change of } f = \frac{f(a+h) - f(a)}{h} = \text{slope of line } AB.$$

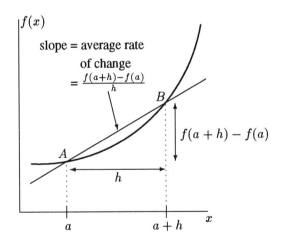

Figure 2.8: Visualizing the Average Rate of Change of f

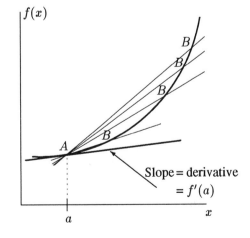

Figure 2.9: Visualizing the Instantaneous Rate of Change of f

As h becomes smaller, the line AB approaches the tangent line to the curve at A. See Figure 2.9.

Thus

$$\left(\begin{array}{c} \text{instantaneous} \\ \text{rate of change of } f \end{array} \right) = \lim_{h \to 0} \frac{f(a+h) - f(a)}{h} = \text{slope of tangent at } A.$$

Therefore,

---

the derivative can be thought of as

- the slope of the curve at A

- the slope of the tangent line at A.

The slope interpretation is often useful in gaining rough information about the derivative, as the following examples show.

☐ **Example 4** Is the derivative of $\sin x$ at $x = \pi$ positive or negative?

Solution. Looking at a graph of $\sin x$ in Figure 2.10, we see that a tangent line drawn at $x = \pi$ has negative slope, so the derivative is negative. (Remember, x is in radians.)

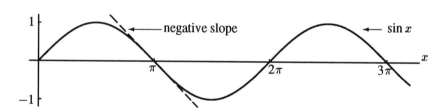

Figure 2.10: Tangent line to $\sin x$ at $x = \pi$

☐

Remember that if you zoom in on the graph of a function $y = f(x)$ at the point where $x = a$, you will usually find that the graph looks more and more like a straight line with slope $f'(a)$.

☐ **Example 5** By zooming in on the point $(0, 0)$ on the graph of the sine function, estimate the derivative of $\sin x$ at $x = 0$, with x in radians.

Solution. Figure 2.11 shows successive graphs of $\sin x$, with smaller and smaller scales. On the interval $-0.1 \le x \le 0.1$, the graph looks like a straight line of slope 1. Thus the derivative of $\sin x$ at $x = 0$ is about 1.

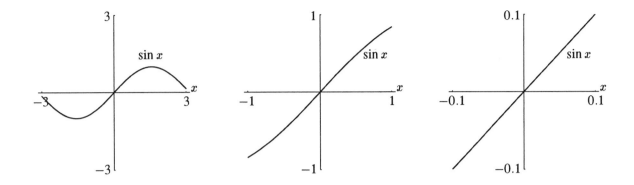

Figure 2.11: Zooming in on the graph of sin x near $x = 0$.

Later we will show that the derivative of sin x at 0 is 1. (See Problems 32 and 33 on page 257.) From now on we will use this fact.

❏ **Example 6**   Use the tangent line at $x = 0$ to estimate values of sin x near $x = 0$.

Solution. The previous example shows that in the neighborhood of $x = 0$, the graph of $y = \sin x$ looks like the graph of the straight line $y = x$; we can use this to estimate values of sin x when x is close to 0. For example, the point on the straight line $y = x$ with x-coordinate 0.32 is $(0.32, 0.32)$. Since the line is close to the graph of $y = \sin x$, we estimate that $\sin 0.32 \approx 0.32$. (See Figure 2.12.) Checking on the calculator, we find that $\sin 0.32 = 0.3146$, so our estimate is quite close. Notice that the graph leads you to expect that the real value of $\sin 0.32$ is slightly less than 0.32.   ❏

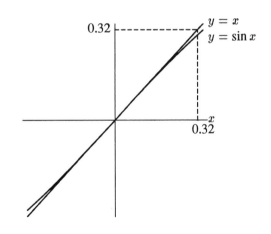

Figure 2.12: Approximating $y = \sin x$ by $y = x$.

Why Do We Use Radians And Not Degrees?

The fact that the derivative of sin x at $x = 0$ turns out to be 1, when x is in radians, is what makes us decide to use radians. If we had done the previous example in degrees, the derivative of sin x would have turned out to be a much messier number. (See Problem 10, page 138.)

☐ **Example 7** Estimate the derivative of $f(x) = 2^x$ at $x = 0$ graphically and numerically.

Solution. Graphically: If you try to draw a tangent line at $x = 0$ to the exponential curve in Figure 2.13, you'll see that it must have positive slope. Since the slope of the line BA is $\frac{2^0 - 2(-1)}{0 - (-1)} = \frac{1}{2}$ and the slope of the line AC is $\frac{2^1 - 2^0}{1 - 0} = 1$, we know the derivative is between $\frac{1}{2}$ and 1.

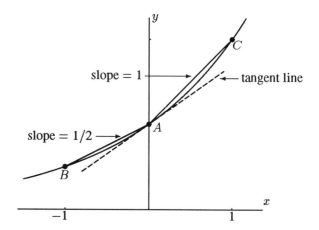

Numerically: To find the derivative at $x = 0$, we need to look at values of the difference quotient

$$\frac{f(0 + h) - f(0)}{h} = \frac{2^h - 2^0}{h} = \frac{2^h - 1}{h}$$

for small h. Table 2.3 shows some values of 2^h together with values of the difference quotients.

Figure 2.13: Graph of $y = 2^x$ showing the derivative at $x = 0$

h	2^h	difference quotient: $\frac{2^h - 1}{h}$
−0.0003	0.999792077	0.693075
−0.0002	0.999861380	0.693099
−0.0001	0.999930688	0.693123
0	1	
0.0001	1.00006932	0.693171
0.0002	1.00013864	0.693195
0.0003	1.00020797	0.693219

Table 2.3: Numerical Values for Difference Quotients of 2^x near $x = 0$

As h approaches 0 from the negatives, the difference quotient approaches a value that is slightly greater than 0.6931, and similarly as h approaches 0 from the positives. So the derivative, or slope of the tangent, is a bit above 0.6931 at $x = 0$.

The graph suggests that difference quotients calculated with negative h's are smaller than the derivative, while those calculated with positive h's should be larger. Thus the derivative is between 0.693123 and 0.693171. ☐

☐ **Example 8** Find an approximate equation for the tangent line to $f(x) = 2^x$ at $x = 0$.

Solution. From the previous example, we know the slope of the tangent line is about 0.693. Since we also know the line has y-intercept 1, its equation is

$$y = 0.693x + 1.$$

☐

☐ **Example 9** Find the derivative of $f(x) = x^2 + 1$ at $x = 3$ algebraically. Find the equation of the tangent line to f at $x = 3$.

Solution. We need to look at the difference quotient and take the limit as h approaches zero. The difference quotient is

$$\frac{f(3 + h) - f(3)}{h} = \frac{(3 + h)^2 + 1 - 10}{h} = \frac{9 + 6h + h^2 - 9}{h} = \frac{6h + h^2}{h}.$$

Since we are not going to let $h = 0$, we can divide by h in the last expression to get $6 + h$. Now the limit as h goes to 0 of $6 + h$ is clearly 6. So,

$$f'(3) = \lim_{h \to 0} \frac{6h + h^2}{h} = \lim_{h \to 0} (6 + h) = 6.$$

Thus we know that the slope of the tangent line at $x = 3$ is 6. Since $f(3) = 10$, the tangent passes through $(3, 10)$ so the equation of the tangent line is

$$y - 10 = 6(x - 3) \quad \text{or} \quad y = 6x - 8.$$

☐

☐ **Example 10** The table below shows values of $f(x) = x^3$ near $x = 2$, rounded to three decimal places. Use it to estimate $f'(2)$.

x	1.998	1.999	2.000	2.001	2.002
x^3	7.976	7.988	8.000	8.012	8.024

Solution. The derivative, $f'(2)$, is the rate of change of x^3 at $x = 2$. Notice that each time x changes by 0.001 in the table, the value of x^3 changes by 0.012. Therefore, we estimate

$$f'(2) = \left(\begin{array}{c} \text{Rate of Change} \\ \text{of } f \text{ at } x = 2 \end{array} \right) \approx \frac{0.012}{0.001} = 12.$$

The function values in the table look exactly linear because they have been rounded. For example, the exact value of x^3 when $x = 2.001$ is 8.012006001, not 8.012. Thus the table can tell us only that the derivative is approximately 12. Showing that the derivative is exactly 12 (which it is) requires algebra. ☐

❖ Exercises for Section 2.2

1. (a) Make a table of values, rounded to two decimal places, for $f(x) = \log x$ (that is, log base 10) where $x = 1, 1.5, 2, 2.5, 3$. Use this information to answer the following questions.

    (b) Find the average rate of change of $f(x)$ between $x = 1$ and $x = 3$.

    (c) Use average rates of change to approximate the instantaneous rate of change of $f(x)$ at $x = 2$.

2. Sketch a rough graph of $f(x) = \sin x$ and use the graph to decide whether the derivative of $f(x)$ at $x = 3\pi$ is positive or negative. Explain your reasons briefly.

3. On a sketch of $y = f(x)$ similar to that in Fig. 2.14, mark lengths that represent the following quantities. (Pick any convenient x and assume $h > 0$.)
    (a) $f(x)$
    (b) $f(x + h)$
    (c) $f(x + h) - f(x)$
    (d) h
    (e) Using your answers to (a)–(d), show how you can represent the quantity $\frac{f(x+h)-f(x)}{h}$ as the slope of a line on the graph.

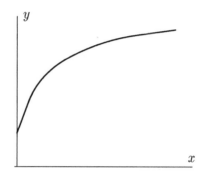

Figure 2.14: Represent $\frac{f(x+h)-f(x)}{h}$ on this graph

(for Problem 3)

4. On a sketch of $y = f(x)$ similar to that in Fig. 2.15, mark lengths that represent the following quantities. (Pick any convenient x and assume $h > 0$.)
    (a) $f(x)$
    (b) $f(x + h)$
    (c) $f(x + h) - f(x)$
    (d) h
    (e) Using your answers to (a)–(d), show how you can represent the quantity $\frac{f(x+h)-f(x)}{h}$ as the slope of a line on the graph.

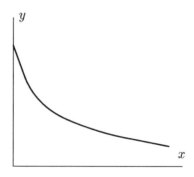

Figure 2.15: Represent $\frac{f(x+h)-f(x)}{h}$ as a slope of a line on this graph

(for Problem 4)

5. Show how you can represent the following
   on a sketch in Figure 2.16:

   (a) $f(4)$

   (b) $f(4) - f(2)$

   (c) $\dfrac{f(5) - f(2)}{5 - 2}$

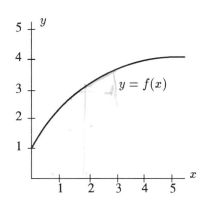

Figure 2.16: Function for Problems 5–8

6. Consider the function $y = f(x)$ shown in Figure 2.16. Decide which of each of the following pairs of numbers is larger. Explain your answer.

   (a) $f(3)$ or $f(4)$?

   (b) $f(3) - f(2)$ or $f(2) - f(1)$?

   (c) $\dfrac{f(2) - f(1)}{2 - 1}$ or $\dfrac{f(3) - f(1)}{3 - 1}$?

7. Suppose $y = f(x)$ graphed in Figure 2.16 represents the cost of manufacturing x kilograms of a chemical. Then $\dfrac{f(x)}{x}$ represents the average cost of producing 1 kilogram when x kilograms are made. This problem asks you to visualize these averages graphically.

   (a) Show how to represent $\dfrac{f(4)}{4}$ as the slope of a line.

   (b) Which is larger, $\dfrac{f(3)}{3}$ or $\dfrac{f(4)}{4}$?

8. With the function f given by Figure 2.16, arrange the following quantities in ascending order:

$$0, \quad 1, \quad f'(2), \quad f'(3), \quad f(3) - f(2)$$

9. Refer to the graph of the function k to the right:

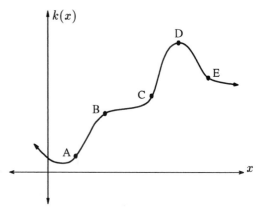

   (a) Between which pair of consecutive points is the average rate of change of k greatest?

   (b) Between which pair of consecutive points is the average rate of change of k closest to zero?

   (c) Between which two pairs of consecutive points are the average rates of change of k about equal?

10. (a) Estimate $f'(0)$ if $f(x) = \sin x$ with x in degrees.

    (b) In Example 5, page 132, you found that the derivative of $\sin x$ at $x = 0$ was 1. Why do you get a different result here?

11. If $f(x) = x^3 + 4x$, estimate $f'(3)$ using a table similar to that in Example 10 on page 135.

12. For $g(x) = x^5$, use tables similar to that in Example 10 on page 135 to estimate $g'(2)$ and $g'(-2)$. What relationship do you notice between $g'(2)$ and $g'(-2)$? Explain geometrically why this must occur.

13. For the function $f(x) = \log x$, estimate $f'(1)$. From the graph of $f(x)$, would you expect your estimate to be greater than or less than $f'(1)$?

14. Estimate the derivative of $f(x) = x^x$ at $x = 2$.

15. For $y = f(x) = 3x^{3/2} - x$, use your calculator to construct a graph of $y = f(x)$, for $0 \le x \le 2$. From your graph, estimate $f'(0)$ and $f'(1)$.

16. (a) Use your calculator to approximate the derivative of the hyperbolic sine function (written sinh) at the points 0, 0.3, 0.7, and 1.

    (b) Can you find a relation between the values of this derivative and the values of the hyperbolic cosine?

17. Let $f(x) = \ln(\cos x)$. Use your calculator to approximate the instantaneous rate of change of f at the point $x = 1$. Do the same thing for $x = \frac{\pi}{4}$. (Note: Be sure that your calculator is set in *radians*.)

18. (a) If f is even and $f'(10) = 6$, what must $f'(-10)$ equal?

    (b) If f is any even function and $f'(0)$ exists, what must $f'(0)$ equal?

19. If g is an odd function and $g'(4) = 5$, what must $g'(-4)$ equal?

20. (a) Sketch graphs of the functions $f(x) = \frac{1}{2}x^2$ and $g(x) = f(x) + 3$ on the same set of axes. What can you say about the slopes of the tangent lines to the two graphs at the point $x = 0$? $x = 2$? $x = x_0$?

    (b) Show that adding a constant value, C, to any function does not change the value of the slope of its graph at any point. (Hint: Let $g(x) = f(x) + C$ and calculate the difference quotients for f and g.)

2.3 The Derivative Function

In the last section we looked at the derivative of a function at a fixed point. Now we'll consider what happens at a variety of points and see that, in general, the derivative takes on different values at different points.

First, remember that the derivative of a function at a point tells you the rate at which the value of the function is changing at that point. Geometrically, if you "zoom in" on a point in the graph until it looks like a straight line, the slope of that line is the derivative at that point. Equivalently, you can think of the derivative as the slope of the tangent line at the point, because as you "zoom in" the curve and the tangent line become indistinguishable.

☐ **Example 1** Estimate the derivative of the function given by the graph in Figure 2.17 at $x = -2, -1, 0, 1, 2, 3, 4, 5$.

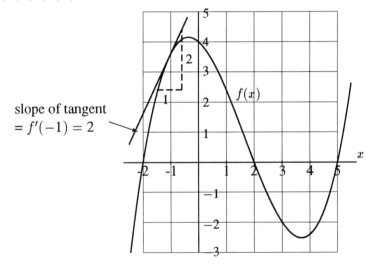

Figure 2.17: Estimating the Derivative Graphically as the Slope of the Tangent Line

Solution. From the graph you can estimate the derivative at any point by placing a straight edge so that it forms the tangent at that point, and then using the grid squares to estimate the slope of the straight edge. (For example, the tangent at $x = -1$ is drawn in Figure 2.17, and has a slope of about 2, so $f'(-1) \approx 2$.) Notice that the slope at $x = -2$ is positive and fairly large; the slope at $x = -1$ is positive but smaller. At $x = 0$, the slope is negative; by $x = 1$, it has become more negative, and so on. Some estimates of the derivative are listed in Table 2.4. You should check these values yourself. Are they reasonable? Is the derivative positive where you expect? Negative?

x	-2	-1	0	1	2	3	4	5
Derivative at x	6	2	-1	-2	-2	-1	1	4

Table 2.4: Estimated Values of Derivative of Function in Figure 2.17

The important point to notice is that for every x value, there's a corresponding value of the derivative. The derivative, therefore, is itself a function of x. For any function f, we define

the **Derivative Function, f',** by

$$f'(x) = \left(\begin{array}{c} \text{Rate of Change of } f \\ \text{at } x \end{array} \right) = \lim_{h \to 0} \frac{f(x+h) - f(x)}{h}.$$

For every x value for which this limit exists, we say f is *differentiable* at that x value. If the limit exists for all x in the domain of f, we say f is *differentiable everywhere*. The functions we shall deal with will be differentiable at every point in their domain, except perhaps for a few isolated points.

Alternative notation: If $y = f(x)$, we can write the derivative as $\dfrac{dy}{dx}$.

The reason for the $\dfrac{dy}{dx}$ notation will be explained more fully on page 150.

Finding the Derivative of a Function Given Graphically

☐ **Example 2** Sketch the graph of the derivative of the function shown in Figure 2.17 of Example 1.

Solution. Table 2.4 gives some values of this derivative which we can plot. However, it is a good idea first to identify some of the key features of the derivative graph from the graph of the original function. For example, we can see from Figure 2.17 that the function is increasing from -2 to a little after -0.5. Thus the derivative is positive in this interval, and so we must draw the graph of f' above the x-axis from -2 to about -0.5. Between -0.5 and about 3.7, the function is decreasing, so the derivative is negative and its graph must be below the x-axis. Beyond 3.7 the function is increasing, so the derivative is positive again and the graph of f' should be above the axis. Somewhere in the region where the derivative is negative, it is going to reach its lowest point; this will be at the point where the graph of the original function is decreasing most steeply. From Figure 2.17 we see that this occurs a little before $x = 2$, where the slope is slightly steeper than -2. Thus we should make our derivative graph have a minimum of slightly below -2 a little before $x = 2$. With this in mind, and using the data in Table 2.4, we obtain Figure 2.18, which shows a graph of the derivative (the gray line), along with the original function (black).

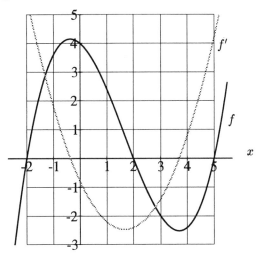

Figure 2.18: Function (black) and Derivative (gray) from Example 2

You should check for yourself that this graph of f' makes sense. Notice that at the points where f has large upward slope, such as $x = -2$, the graph of the derivative is way above the x-axis, which is as it should be, since the value of the derivative should be large there. On the other hand, at points where the slope is gentle, the graph of f' is close to the x-axis, since the derivative is small. ◻

Finding the Derivative of a Function Given Numerically

If we are given a table of function values instead of a graph, we can estimate values of its derivative.

◻ **Example 3** Suppose Table 2.5 gives values of $c(t)$, the concentration (in mg/cc) of a drug in the bloodstream at time t (in mins). Construct a table of estimated values for $c'(t)$, the rate of change of $c(t)$.

t (min)	0	0.1	0.2	0.3	0.4	0.5	0.6	0.7	0.8	0.9	1.0
$c(t)$ (mg/cc)	0.84	0.89	0.94	0.98	1.00	1.00	0.97	0.90	0.79	0.63	0.41

Table 2.5: Concentration as a function of time

Solution. We want to estimate the derivative of c using the values in the table. To do this, we have to assume that the data points are close enough together that there are no big changes in concentration between them. From the table, we can see that the concentration is increasing between $t = 0$ and $t = 0.4$, so we'd expect a positive derivative there. However, the increase is quite slow, so we would expect the derivative to be small. The concentration doesn't change between 0.4 and 0.5, so we expect the derivative to be 0 there. From $t = 0.5$ to $t = 1$, the concentration starts to decrease, and the rate of decrease gets larger and larger, so we would expect the derivative to be negative and of greater and greater magnitude.

Using the data in the table, we can estimate the derivative using the difference quotient

$$\frac{c(t+h) - c(t)}{h}$$

with $h = 0.1$. For example,

$$c'(0) \approx \frac{c(0.1) - c(0)}{0.1} = \frac{0.89 - 0.84}{0.1} = 0.5 \text{ mg/cc/min}.$$

Thus we estimate that

$$c'(0) \approx 0.5.$$

Similarly, we get the estimates

$$c'(0.1) \approx \frac{c(0.2) - c(0.1)}{0.1} = \frac{0.94 - 0.89}{0.1} = 0.5$$

$$c'(0.2) \approx \frac{c(0.3) - c(0.2)}{0.1} = \frac{0.98 - 0.94}{0.1} = 0.4$$

$$c'(0.3) \approx \frac{c(0.4) - c(0.3)}{0.1} = \frac{1.00 - 0.98}{0.1} = 0.2$$

$$c'(0.4) \approx \frac{c(0.5) - c(0.4)}{0.1} = \frac{1.00 - 1.00}{0.1} = 0.0$$

and so on. These values are tabulated in Table 2.6. Notice that the derivative has small positive values up until $t = 0.4$, and then gets more and more negative, as we expected. The slopes are shown on the graph of $c(t)$ in Figure 2.19.

t	$c'(t)$
0	0.5
0.1	0.5
0.2	0.4
0.3	0.2
0.4	0.0
0.5	−0.3
0.6	−0.7
0.7	−1.1
0.8	−1.6
0.9	−2.2

Table 2.6: Values of Derivative

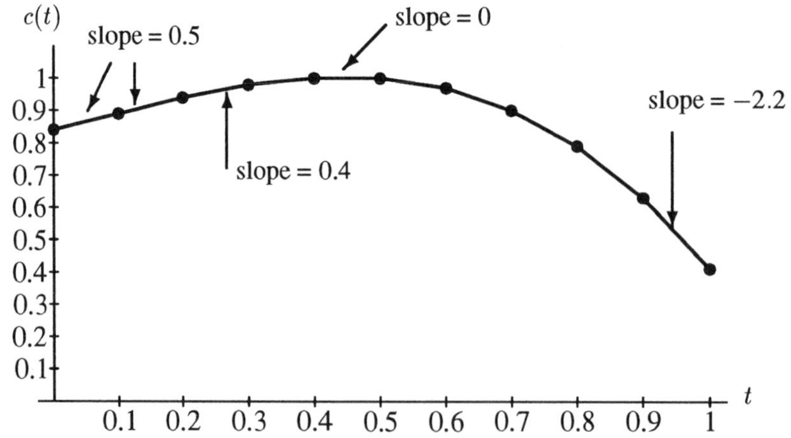

Figure 2.19: Graph of Function in Example 3

Other Ways We Could Calculate the Derivative Numerically

In the previous example, our estimate for the derivative at 0.2 used the point to the right: we found the average rate of change between $t = 0.2$ and $t = 0.3$. However, we could equally well have gone to the left and used the rate of change between $t = 0.1$ and $t = 0.2$ to approximate the derivative at 0.2. For a more accurate result, we could average these slopes and say

$$c'(0.2) \approx \frac{1}{2}\left(\begin{array}{c}\text{slope to left}\\\text{of 0.2}\end{array} + \begin{array}{c}\text{slope to right}\\\text{of 0.2}\end{array}\right) = \frac{0.5 + 0.4}{2} = 0.45.$$

Each of these methods of approximating the derivative gives a reasonable answer. For convenience, unless there is a reason to do otherwise, we will estimate the derivative by going to the right.

Finding the Derivative of a Function Given by a Formula

What if we are given a formula for f: can we come up with a formula for f'? Using the definition of the derivative, we often can, as shown in the next example. Indeed, much of the power of calculus depends on our ability to find formulas for the derivatives of all the functions in our library. This is done in detail in Chapter 4.

☐ **Example 4** Find a formula for the derivative of $f(x) = x^2$.

Solution. Before computing the formula for $f'(x)$ algebraically, let's try to guess the formula by looking for a pattern in the values of $f'(x)$. Table 2.7 contains values of $f(x) = x^2$ (rounded to three decimals) which we can use to estimate the values of $f'(1)$, $f'(2)$ and $f'(3)$. Near $x = 1$, x^2 increases by about 0.002 each time x increases by 0.001, so

x	x^2 (approx)	x	x^2 (approx)	x	x^2 (approx)
0.999	0.998	1.999	3.996	2.999	8.994
1.000	1.000	2.000	4.000	3.000	9.000
1.001	1.002	2.001	4.004	3.001	9.006
1.002	1.004	2.002	4.008	3.002	9.012

Table 2.7: Values of $f(x) = x^2$ near $x = 1$, $x = 2$, $x = 3$ (rounded to three decimals)

$$f'(1) \approx \frac{0.002}{0.001} = 2.$$

Similarly

$$f'(2) \approx \frac{0.004}{0.001} = 4$$

$$f'(3) \approx \frac{0.006}{0.001} = 6.$$

Knowing the value of f' at specific points can never tell us the formula for f', but it certainly can be suggestive: knowing $f'(1) \approx 2$, $f'(2) \approx 4$, $f'(3) \approx 6$ certainly suggests that $f'(x) = 2x$. Now we'll show that $f'(x) = 2x$ is the correct formula algebraically. The derivative is calculated by forming the difference quotient and taking the limit as h goes to zero. The difference quotient is

$$\frac{f(x + h) - f(x)}{x} = \frac{(x + h)^2 - x^2}{h} = \frac{x^2 + 2xh + h^2 - x^2}{h} = \frac{2xh + h^2}{h}.$$

Since h never actually reaches zero, we can divide it out in the last expression to get $2x + h$. The limit of this as h goes to zero is $2x$, so

$$f'(x) = \lim_{h \to 0} (2x + h) = 2x.$$

☐

What Does the Derivative Tell Us?

When f' is positive, the tangent is sloping up; when f' is negative, the tangent is sloping down. If $f' = 0$, then the tangent is horizontal everywhere and so f is constant. Thus, the sign of f' tells us whether f is increasing or decreasing. Specifically

> if $f' > 0$ on an interval   then f is *increasing* over that interval
>
> and
>
> if $f' < 0$ on an interval   then f is *decreasing* over that interval
>
> and
>
> if $f' = 0$ on an interval   then f is *constant* over that interval.

Moreover, the magnitude of the derivative gives us the magnitude of the rate of change; so if f' is large (positive or negative), then the graph of f will be steep (up or down), whereas if f' is small the graph of f will slope gently.

With this in mind, you can deduce a lot about the behavior of a function from the behavior of its derivative.

☐ **Example 5** Suppose the derivative of f is the spike function illustrated in Figure 2.20. What can you say about the graph of f itself?

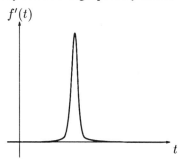

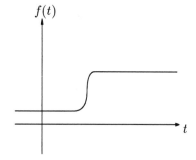

Figure 2.20: Spike Derivative Function
(for Example 5)

Figure 2.21: 'Step' function
(Possible Solution to Example 5)

Solution. On intervals where $f' = 0$, f is not changing at all, and is therefore constant. On the small interval where $f' > 0$, f is increasing; at the point where f' hits the top of its spike, f is increasing quite sharply. So f should be constant for a while, have a sudden increase, and then be constant again. A possible graph for f is shown in Figure 2.21. ☐

❖ Exercises for Section 2.3

For Problems 1–6, sketch a graph of the derivative function of each of the given functions.

1.

2.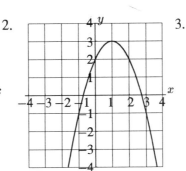

3.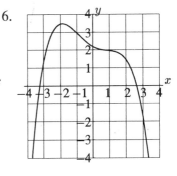

4.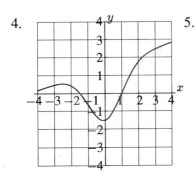

5.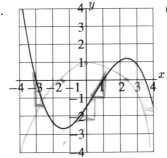

6.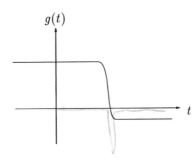

7. What would the derivative of the step function in Figure 2.22 look like?

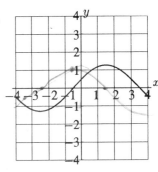

Figure 2.22: Downward step function (for Problem 7)

8. (a) Sketch a smooth curve whose slope is both increasing gradually and everywhere positive.

   (b) Sketch a smooth curve whose slope is both decreasing gradually and everywhere positive.

   (c) Sketch a smooth curve whose slope is both increasing (i.e., becoming less and less negative) gradually and everywhere negative

   (d) Sketch a smooth curve whose slope is both decreasing (i.e., becoming more and more negative) gradually and everywhere negative

9. Given the numerical table shown, find approximate values for the derivative of $f(x)$ at each of the x values given. Where is the rate of change of $f(x)$ positive? where negative? Where does the rate of change of $f(x)$ seem to be greatest?

x	0	1	2	3	4	5	6	7	8
$f(x)$	18	13	10	9	9	11	15	21	30

10. Suppose $f(x) = \frac{1}{3}x^3$. Make tables similar to those on page 143 for Example 4 to estimate $f'(2)$, $f'(3)$, and $f'(4)$. What do you notice? Can you guess a formula for $f'(x)$?

11. If $g(t) = t^2 + t$, use tables similar to those on page 143 for Example 4 to estimate $g'(1)$, $g'(2)$, and $g'(3)$. Use these to guess a formula for $g'(t)$.

12. Draw a possible graph of $y = f(x)$ given the following information about its derivative.

  (a) $f'(x) > 0$ on $1 < x < 3$

  (b) $f'(x) < 0$ for $x < 1$ and $x > 3$

  (c) $f'(x) = 0$ at $x = 1$ and $x = 3$

13. Draw a possible graph of $y = f(x)$ given the following information about its derivative:

  (a) $f'(x) > 0$ for $x < -1$

  (b) $f'(x) < 0$ for $x > -1$

  (c) $f'(x) = 0$ at $x = -1$

14. In the graph of f to the right, at which of the labeled x values is

  (a) $f(x)$ greatest?

  (b) $f(x)$ least?

  (c) $f'(x)$ greatest?

  (d) $f'(x)$ least?

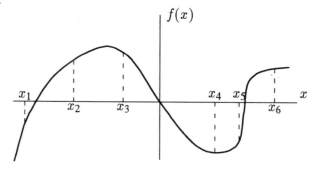

15. (a) On the same set of axes, graph $f(x) = \sin x$ and $g(x) = \sin 2x$ from $x = 0$ to $x = 2\pi$.

  (b) On a second set of axes, sketch the graphs of $f'(x)$ and $g'(x)$ and compare them. (In producing these graphs be very careful when comparing the slope of $f(x)$ and $g(x)$ at each point.)

For Problems 16–21, first sketch the graph of $f(x)$, and then use this graph to sketch the graph of $f'(x)$.

16. $f(x) = x(x - 1)$

17. $f(x) = 5x$

18. $f(x) = \cos x$

19. $f(x) = e^x$

20. $f(x) = \log x$

21. $f(x) = \dfrac{1}{x^3}$

For Problems 22–27, sketch the graph of $y = f'(x)$ for the function given.

22.

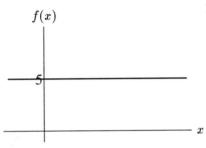

23.

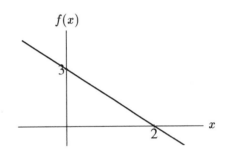

24.

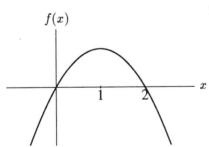

25.

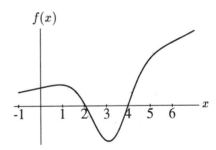

26.

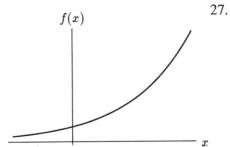

27.

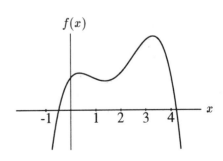

28. The population of a herd of deer is given by

$$P(t) = 4000 + 500\sin(2\pi t - \frac{\pi}{2})$$

where t is measured in years.

(a) How does this population vary with time? Sketch a graph of $P(t)$ for 1 year.

(b) Use the graph to decide when in the year the population is a maximum. What is that maximum? How about a minimum?

(c) Use the graph to decide when the population is growing fastest. When is it decreasing fastest?

(d) Estimate roughly how fast the population is changing on the first of July.

29. Show that if $f(x)$ is an even function then $f'(x)$ is odd.

30. Show that if $g(x)$ is an odd function then $g'(x)$ is even.

31. To give a patient an antibiotic slowly, the drug is injected into the muscle. (For example, penicillin for venereal disease is administered this way.) The quantity of the drug in the bloodstream starts out at zero, increases to a maximum and then decays to zero again.

(a) Sketch a possible graph of the quantity of the drug in the bloodstream as a function of time. Mark the time at which the drug is a maximum by t_0.

(b) Describe in words how the rate at which the drug is entering or leaving the blood changes over time. Sketch a graph of this rate, marking t_0 on the time axis.

2.4 Interpretations of the Derivative

We have already seen how the derivative can be interpreted as a slope and as a rate of change. A good test of whether you really understand the definition of the derivative is to see if you can use it to gain new interpretations. In what follows we will give a few examples of how to do this. The point of these examples is not to make a catalogue of interpretations, but to illustrate the process of obtaining them.

Acceleration

We started this chapter by showing how velocity could be calculated as a rate of change of position with respect to time. If $s(t)$ measures the distance an object has moved from a reference point along a straight line, then, at the instant when $t = a$:

$$\left(\begin{array}{c}\text{instantaneous velocity}\\ \text{at } t = a\end{array}\right) = s'(a) = \lim_{h\to 0}\frac{s(a+h) - s(a)}{h}.$$

Now, *acceleration* is the rate of change of velocity, $v(t)$, with respect to time, so we can define

$$\left(\begin{array}{c}\text{average acceleration}\\\text{from } c \text{ to } c + h\end{array}\right) = \frac{v(c + h) - v(c)}{h}$$

and

$$\left(\begin{array}{c}\text{instantaneous acceleration}\\\text{at } t = c\end{array}\right) = v'(c) = \lim_{h \to 0} \frac{v(c + h) - v(c)}{h}.$$

> If the terms 'velocity' or 'acceleration' are used alone, they are assumed to be instantaneous.

☐ **Example 1** Consider an accelerating sports car which goes from 0 mph to 60 mph in 5 seconds. Its velocity is given in Table 2.8, converted from miles per hour to feet per second so that all time measurements are in seconds. (Note: 1 mph = 22/15 ft/sec.) Find the average acceleration of the car over each of the first two seconds.

time, t, in sec	0	1	2	3	4	5
velocity, v, in ft/sec	0	30	52	68	80	88

Table 2.8: Velocity of Sports Car

Solution. To measure the average acceleration over an interval, we calculate the average rate of change of velocity over the interval. The units of acceleration are ft/sec per second, or $\frac{\text{ft/sec}}{\text{sec}}$, written ft/sec^2.

$$\left(\begin{array}{c}\text{average acceleration}\\\text{for } 0 < t < 1\end{array}\right) = \frac{\text{change in velocity}}{\text{time}} = \frac{30 - 0}{1} = 30 \text{ ft/sec}^2.$$

$$\left(\begin{array}{c}\text{average acceleration}\\\text{for } 1 < t < 2\end{array}\right) = \frac{52 - 30}{2 - 1} = 22 \text{ ft/sec}^2.$$

☐

☐ **Example 2** The graph in Figure 2.23 shows the velocity of a sports car as a function of time. Estimate its acceleration when $t = 1$.

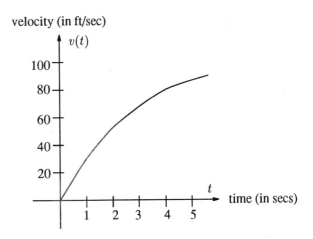

Figure 2.23: Velocity of Sports Car

Solution. The acceleration at $t = 1$ is the derivative $v'(1)$, the slope of the tangent to the velocity curve at $t = 1$. (See Figure 2.24.) Thus, an estimate for

$$\left(\begin{array}{c} \text{Acceleration} \\ \text{at } t = 1 \end{array} \right) = v'(1) = \frac{28}{1}$$

$$= 28 \text{ ft/sec}^2.$$

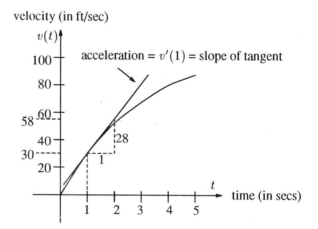

Figure 2.24: Estimating Acceleration from Velocity Graph

Why the Alternative Derivative Notation is Useful

So far we have used the notation f' for the derivative of a function f. The alternative notation for the derivative can be very helpful in interpreting the meaning of the derivative. If

$$y = f(x),$$

then the older notation is to write

$$\frac{dy}{dx} = f'(x)$$

for the derivative. The d in the old notation is not to be interpreted as multiplying the x or the y; rather, dx and dy are single entities. The dx represents what we called h before, namely a small change in x. Then dy represents the corresponding small change in y. In other words, if x is increased from x to $x + dx$, then y changes to $y + dy$, so that

$$y + dy = f(x + dx).$$

The ratio of dy to dx is then the ratio of the change in y to the change in x, which is nothing more than the rate of change or derivative; hence the notation. In the definition of the derivative, you don't get an exact value for the derivative unless you let h tend to zero. In this context, that would mean letting dx, and hence dy, tend to zero. Even if dx and dy individually are infinitesimally small, their ratio, $\frac{dy}{dx}$, is usually meaningful. It was the confusion arising from trying to think about infinitesimal quantities that eventually forced people to come up with the precise definition we gave earlier; but the old notation is still a useful way of thinking about the derivative when you are trying to understand what it means in a particular context.

For example, recall that if $s = f(t)$ is the position of a moving object at time t, then $v = f'(t)$ is the velocity of the object at time t. Writing

$$v = \frac{ds}{dt}$$

directly reminds you of this fact, since it shows a distance, ds, over a time, dt, and we know that distance over time is velocity. Similarly, we recognize

$$\frac{dy}{dx} = f'(x)$$

as the slope of the graph of $y = f(x)$ by remembering that slope is vertical rise, dy, over horizontal run, dx.

The disadvantage of this notation is that it is rather awkward if you want to specify the x value at which you are evaluating the derivative. To specify $f'(2)$, for example, we have to write

$$\left. \frac{dy}{dx} \right|_{x=2}$$

More Interpretations of the Derivative

☐ **Example 3** The cost C (in dollars) of building a house A square feet in area is given by the function $C = f(A)$. What is the practical interpretation of the function $f'(A)$?

Solution. In the alternative notation,

$$f'(A) = \frac{dC}{dA}.$$

This is a cost divided by an area, so it is measured in dollars per square foot. The dC is the extra cost of building an extra dA square feet of house. Thus $\frac{dC}{dA}$ is the additional cost per square foot. So if you are planning to build a house roughly A square feet in area, $f'(A)$ is the cost per square foot of the *extra* area involved in building a slightly larger house, and is called the *marginal cost*. The marginal cost is not necessarily the same thing as the average cost per square foot for the entire house, since once you are already set up to build a large house, the cost of adding a few square feet could be comparatively small. ☐

☐ **Example 4** The cost of extracting T tons of ore from a copper mine is $C = f(T)$ dollars. What does it mean to say that $f'(2000) = 100$?

Solution. In the alternative notation,

$$f'(2000) = \left. \frac{dC}{dT} \right|_{T=2000}.$$

Since C is measured in dollars and T is measured in tons, $\frac{dC}{dT}$ must be measured in dollars per ton. So the statement

$$\left. \frac{dC}{dT} \right|_{T=2000} = 100$$

says that when 2000 tons of ore have already been extracted from the mine, the cost of extracting the next one is approximately $100. In other words, after 2000 tons have been removed, extraction costs are $100 per ton. Another way of saying this is that it costs about $100 to extract the 2000th or 2001st ton of ore. Note that this may well be different from the cost of extracting the 10th ton, which is likely to be more accessible. ☐

Making Use of the Units of Measurement

The previous two examples illustrated another point: if you need to interpret in practical terms the meaning of a derivative, it helps to think about what the units of measurement are. This sort of reasoning can be turned around if you are given a quantity which seems as if it ought to be interpreted as the derivative of *something*, but you can't quite say what.

☐ **Example 5** You are told that water is flowing through a pipe at a rate of 10 cubic feet per second. Interpret this rate as the derivative of some function.

Solution. You might think at first that the statement has something to do with the velocity of the water, but in fact a flow rate of 10 cubic feet per second could be achieved both with very slowly moving water through a large pipe, or very rapidly moving water through a narrow pipe. If we look at the units — cubic feet per second — we realize that we are being given the rate of change of a quantity measured in cubic feet. But a cubic foot is a measure of volume, so we are being told the rate of change of a volume. If you imagine all the water that is flowing past ending up in a tank somewhere and let $V(t)$ be the volume of the tank at time t, then we are being told that the rate of change of $V(t)$ is 10, or:

$$V'(t) = \frac{dV}{dt} = 10$$

☐

❖ Exercises for Section 2.4

1. Consider the graph shown in Figure 2.25.

   (a) If $f(t)$ gives the position of a particle at time t, list the points at which the particle has zero velocity.

   (b) If we now suppose that $f(t)$ is the *velocity* of a particle at time t, what is the significance of the points listed in your answer to part (a)?

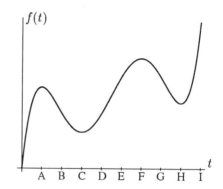

Figure 2.25: One graph to think of in two ways.

(for Problem 1)

2. If $\lim_{x \to \infty} f(x) = 50$ and $f'(x)$ is positive for all x, what is $\lim_{x \to \infty} f'(x)$? (Assume this limit exists.)

3. After investing $1000 at 7% for t years, your balance is $\$B$, where $B = f(t)$. What are the units of $\dfrac{dB}{dt}$? What is the financial interpretation of $\dfrac{dB}{dt}$?

4. An economist is interested in how the price of a certain commodity affects its sales. Suppose that at a price of $\$p$, a quantity, q, of the commodity is sold. If $q = f(p)$, explain in economic terms the meaning of the statements $f(10) = 240,000$ and $f'(10) = -29,000$.

5. The temperature, T, in degrees Fahrenheit of a cold yam placed in a hot oven is given by $T = f(t)$, where t is the time in minutes since the yam was put in the oven.

   (a) What is the sign of $f'(t)$? Why?

   (b) What is the practical interpretation of the statement $f'(20) = 2$?

6. Investing $1000 at an annual interest rate of $r\%$, compounded continuously, for 10 years gives you a balance of $\$B$, where $B = g(r)$. What is a financial interpretation of the statements

   (a) $g(5) \approx 1649$?

   (b) $g'(5) \approx 165$?

7. If $g(v)$ is the fuel efficiency of a car going at v miles per hour (i.e., $g(v) =$ the number of miles per gallon at v mph), what is the practical meaning of the statement

$$g'(55) = -0.54?$$

8. Let P be the total petroleum reservoir on Earth in the year t. (In other words, P represents the total quantity of petroleum, including what's not yet discovered, on Earth at time t.) Assume that no new petroleum is being made. What is the meaning of $\frac{dP}{dt}$? What is its sign? How would you set about estimating this derivative in practice? What would you need to know to make such an estimate?

9. (a) If you jump out of an airplane without a parachute, you will fall faster and faster until wind resistance causes you to approach a steady velocity, called a *terminal* velocity. Sketch a graph of your velocity against time.

   (b) Explain the concavity of your graph.

   (c) Assuming wind resistance to be negligible at $t = 0$, what natural phenomenon is represented by the slope of the graph at $t = 0$?

10. A company's revenue from car sales, C (measured in thousands of dollars), is a function of advertising expenditure, a, also measured in thousands of dollars. Suppose $C = f(a)$.

    (a) What does the company hope is true about the sign of f'?

    (b) What does the statement $f'(100) = 2$ mean in practical terms? How about $f'(100) = 0.5$?

    (c) Suppose the company plans to spend $100 thousand on advertising. What does the fact that $f'(100) = 2$ or $f'(100) = 0.5$ tell us about the plan? Should more or less than $100 thousand be spent on advertising?

11. Let $P(x)$ be the number of people in the U.S. of height $\leq x$ inches. What is the meaning of $P'(66)$? What are its units? Estimate $P'(66)$ (using common sense). Is $P'(x)$ ever negative? [Hint: You may want to approximate $P'(66)$ by a difference quotient, using $h = 1$. Also, you may use the fact that the U.S. population is about 250 million, and that $66'' = 5'6''$.]

12. Census figures for the U.S. population (in millions) are listed in Table 2.9. Since population varies with time, there is a function, f, such that $P = f(t)$. Assume that f is increasing (as the values in the table suggest). Then f is invertible.

    (a) What is the meaning of $f^{-1}(100)$?

    (b) What does the derivative of $f^{-1}(P)$ at $P = 100$ represent? What are its units?

    (c) Estimate $f^{-1}(100)$.

    (d) Estimate the derivative of $f^{-1}(P)$ at $P = 100$.

Year	Population	Year	Population
1790	3.9	1890	62.9
1800	5.3	1900	76.0
1810	7.2	1910	92.0
1820	9.6	1920	105.7
1830	12.9	1930	122.8
1840	17.1	1940	131.7
1850	23.1	1950	150.7
1860	31.4	1960	179.0
1870	38.6	1970	205.0
1880	50.2	1980	226.5

Table 2.9: U.S. Population (in millions) 1790–1980 (for Problems 12, 13 and 14)

13. Table 2.9 gives the U.S. population figures between 1790 and 1980.

    (a) Estimate the rate of change of the population for the years 1900, 1945 and 1980.

    (b) When, approximately, was the population growth maximal?

    (c) Estimate the U.S. population in 1956.

    (d) Based on the data from the table, what would you predict for the 1990 census?

14. (a) In Problem 13, we thought of the U.S. population as a smooth function of time. To what extent is this justified? What happens if we zoom in at a point of the graph? What about events like the Louisiana Purchase? Or the moment of your birth?

    (b) What do we in fact mean by the rate of change of the population for a particular time t?

    (c) Give another example of a real world function which is not smooth, but is usually treated as such.

2.5   The Second Derivative

Since the derivative is itself a function, we can calculate its derivative. For a function f, the derivative of its derivative is called the *second derivative*, and written f'' (read "f double-prime"). If $y = f(x)$, the second derivative can also be written as $\dfrac{d^2y}{dx^2}$, which means $\dfrac{d}{dx}\left(\dfrac{dy}{dx}\right)$, the derivative of $\dfrac{dy}{dx}$.

What Does the Second Derivative Tell Us?

Recall that the derivative of a function tells you whether a function is increasing or decreasing:

if $f' > 0$ on an interval, then f is increasing on that interval,

if $f' < 0$ on an interval, then f is decreasing on that interval.

Since f'' is the derivative of f',

if $f'' > 0$ on an interval, then f' is increasing on that interval,

if $f'' < 0$ on an interval, then f' is decreasing on that interval.

So the question becomes: What does it mean for f' to be increasing or decreasing? First let's look at the cases when f' is increasing. In both cases, the curve is bending upwards or is *concave up*.

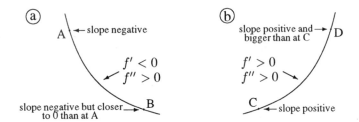

Figure 2.26: Meaning of f'': the slope increases from left to right so f'' is positive and f is concave up.

In the cases where f' is decreasing, shown in Figures 2.27(a) and 2.27(b), the graphs are bending downwards and are *concave down*.

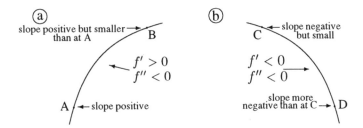

Figure 2.27: Meaning of f'': the slope decreases from left to right so f'' is negative and f is concave down

Thus

$f'' > 0$	means	f' is increasing	so the graph of f is concave up on that interval,
$f'' < 0$	means	f' is decreasing	so the graph of f is concave down on that interval.

☐ **Example 1**   For the functions whose graphs are given in Figure 2.28, decide where their second derivatives are positive and where they are negative.

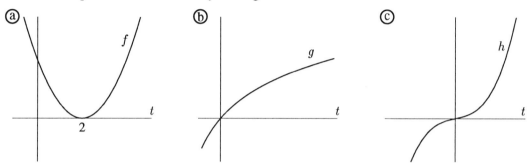

Figure 2.28: What Sign do the Second Derivatives Have?

Solution. From the graphs it appears that:

(a) $f'' > 0$ everywhere because the graph of f is concave up everywhere.

(b) $g'' < 0$ everywhere because the graph is concave down everywhere.

(c) $h'' > 0$ for $t > 0$ because the graph of h is concave up there; $h'' < 0$ for $t < 0$ because the graph of h is concave down there.

☐

Interpretation of the Second Derivative as a Rate of Change

If we think of the derivative as a rate of change, then the second derivative is a rate of change of a rate of change. If the second derivative is positive, the rate of change is increasing; if the second derivative is negative, the rate of change is decreasing.

The second derivative is often a matter of practical concern. In 1985, a newspaper headline reported the Secretary of Defense as saying that Congress and the Senate had cut the defense budget. As his opponents pointed out, however, Congress had merely cut the rate at which the defense budget was increasing[2]. In other words, the derivative of the defense budget was still positive (the budget was increasing), but the second derivative was negative (the budget's rate of increase had slowed).

---

[2]In the <u>Boston Globe</u>, March 13, 1985, Rep. William Gray (D-Pa) was reported as saying: "It's confusing to the American people to imply that Congress threatens national security with reductions when you're really talking about a reduction in the increase."

❏ **Example 2**   A population, P, growing in a confined environment often follows a logistic growth curve, like the graph shown in Figure 2.29. Describe how the rate at which the population is increasing changes over time. What is the practical interpretation of t_0 and L?

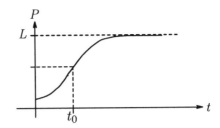

Figure 2.29: Logistic Growth Curve

Solution. Initially, the population is increasing, and at an increasing rate. Thus initially $\frac{dP}{dt}$ is increasing and so $\frac{d^2P}{dt^2} > 0$. At t_0, the rate at which the population is increasing is a maximum. Thus at time t_0 the population is growing fastest. Beyond t_0, the rate at which the population is growing is decreasing, and so $\frac{d^2P}{dt^2} < 0$. At t_0, the concavity changes from positive to negative, and $\frac{d^2P}{dt^2} = 0$. The quantity L represents the limiting value of the population which is approached as t tends to infinity. L is called *carrying capacity* of the environment and represents the maximum population that the environment can support. ❏

❏ **Example 3**   The table below shows the number of abortions per year, A, performed in the U.S. in year t (as reported to the Centers for Disease Control). Suppose these data points lie on a smooth curve $A = f(t)$.

Year (t)	1972	1976	1980	1985
Number of Abortions (A)	586,760	988,267	1,297,606	1,328,570

(a) Estimate $\dfrac{dA}{dt}$ for the time intervals shown between 1972 and 1985.

(b) What can you say about the sign of $\dfrac{d^2A}{dt^2}$ during the period 1972-85?

Solution.

(a) For each time interval, we can calculate the average rate of change of the number of abortions per year over this interval. For example, between 1972 and 1976

$$\frac{dA}{dt} \approx \left(\begin{array}{c} \text{Average Rate} \\ \text{of Change} \end{array} \right) = \frac{988267 - 586760}{1976 - 1972} \approx 100,377.$$

Values of $\frac{dA}{dt}$ are:

Time	'72–'76	'76–'80	'80–'85
Average Rate of Change, $\frac{dA}{dt}$	100,377	77,335	6,193

(b) Since the values of $\frac{dA}{dt}$ are decreasing, $\frac{d^2A}{dt^2}$ must be negative. This is confirmed by the graph in Figure 2.30, which is concave down.

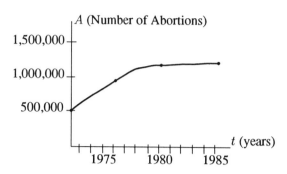

Figure 2.30: How the Rate of Reported Abortions in the U.S. is Changing with Time

Velocity and Acceleration

The velocity, v, of a moving body is the rate at which the position, s, of the body is changing with respect to time:

$$v = \frac{ds}{dt}.$$

The acceleration of a moving body describes how fast the velocity is changing with time, so

$$a = \frac{dv}{dt}$$

or, since the velocity is itself a derivative,

$$a = \frac{dv}{dt} = \frac{d^2s}{dt^2}.$$

Example 4 A particle is moving along a straight line. If its distance, s, to the right of a fixed point is given by Figure 2.31, estimate:

(a) When the particle is moving to the right and when it is moving to the left.

(b) When the particle has positive acceleration and when it has negative acceleration.

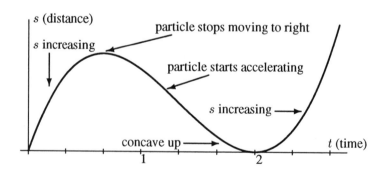

Figure 2.31: Distance of Particle to Right of Fixed Point

Solution.

(a) The particle is moving to the right whenever s is increasing. From the graph, this appears to be for $0 < t < 2/3$ and $t > 2$. For $2/3 < t < 2$, the value of s is decreasing so the particle is moving to the left.

(b) The particle has positive acceleration whenever the curve is concave up, which appears to be for $t > 4/3$. The particle has negative acceleration when the curve is concave down, for $t < 4/3$.

❑

❖ Exercises for Section 2.5

1. If f'' is positive on an interval, then f' is _____ on that interval, and f is _____ on that interval.
   If f'' is negative on an interval, then f' is _____ on that interval, and f is _____ on that interval.

2. (a) Sketch a curve whose first and second derivatives are everywhere positive.

   (b) Sketch a curve whose second derivative is everywhere negative, but whose first derivative is everywhere positive.

   (c) Sketch a curve whose second derivative is everywhere positive, but whose first derivative is everywhere negative.

   (d) Sketch a curve whose first and second derivatives are everywhere negative.

3. (a) Sketch a smooth curve whose slope is both positive and increasing at first, but later on positive and decreasing.

   (b) Sketch the graph of the first derivative of the curve in (a).

   (c) Sketch the graph of the second derivative of the curve in (a).

4. 'Winning the war on poverty' has been described cynically as slowing the rate at which people are slipping below the poverty line. Assuming that this is happening,

   (a) sketch a graph of the total number of people in poverty against time.

   (b) If N is the number of people below the poverty line at time t, what are the signs of $\dfrac{dN}{dt}$ and $\dfrac{d^2N}{dt^2}$?

5. In economics "total utility" refers to the total satisfaction from consuming some commodity. Here is a quote, from Samuelson's *Economics*:

> As you consume more of the same good, the total (psychological) utility increases. However,...,with successive new units of the good, your total utility will grow at a slower and slower rate because of a fundamental tendency for your psychological ability to appreciate more of the good to become less keen.

   (a) Sketch the total utility as a function of the number of units consumed.

   (b) In terms of derivatives, what is Samuelson telling us?

6. Let $P(t)$ represent the price of a share of stock of a corporation at time t. What does each of the following statements tell us about the signs of the first and second derivatives of $P(t)$?

   (a) "The price of the stock is rising faster and faster."

   (b) "The price of the stock is close to bottoming out."

7. An industry is being charged by the Environmental Protection Agency (EPA) with dumping unacceptable levels of toxic pollutants in a lake. Before going to court, an engineering firm has been asked to measure the rate at which pollutants are being discharged into the lake. Measurements are made daily over a several-month period.

   Suppose the engineers produce a graph similar to either Figure 2.32(a) or to Figure 2.32(b). For each case, give an idea of what argument the EPA might make in court and of what the industry's defense might be.

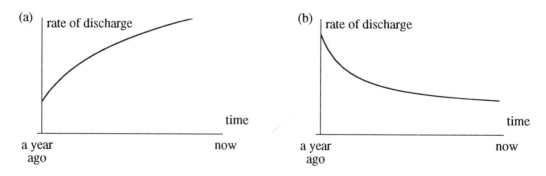

Figure 2.32: Discharge of pollutants

(for Problem 7)

8. IBM-Peru uses second derivatives to compare the relative successes of various advertising campaigns. They assume that all campaigns produce some increase in sales. If a graph of sales against time shows a positive second derivative during a new advertising campaign, what does that suggest to IBM management? Why? What does a negative second derivative during a campaign suggest?

9. Given the following data:

x	0	0.2	0.4	0.6	0.8	1.0
$f(x)$	3.7	3.5	3.5	3.9	4.0	3.9

   (a) Estimate $f'(0.6)$ and $f'(0.5)$.

   (b) Estimate $f''(0.6)$.

   (c) Where do you think the maximum and minimum values of f occur in the interval $0 \le x \le 1$?

10. The graph of f' (not f) is given in Figure 2.33. At which of the marked values of x is

   (a) $f(x)$ greatest?

   (b) $f(x)$ least?

   (c) $f'(x)$ greatest?

   (d) $f'(x)$ least?

   (e) $f''(x)$ greatest?

   (f) $f''(x)$ least?

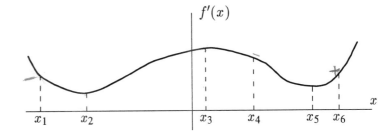

Figure 2.33: Note that this is a graph of f' against x

(for Problem 10)

11. Which of the points labeled by a letter in Figure 2.34 have

   (a) f' and f'' non-zero and of the same sign?

   (b) At least two of f, f', f'' equal to zero?

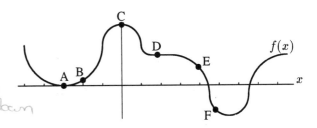

Figure 2.34: Graph for Problem 11

2.6 Approximations and Local Linearity

As you know, near any point, the graphs of most functions look almost like a line whose slope is the derivative of the function at the point. If we are studying a complicated function over a small interval, we can approximate the graph of the function by this line. This property of most functions is called *local linearity*—local because it holds only in a small region of the domain of the function.

If we consider the exact values of a function over a very small interval, you know that they look almost linear. The purpose of this section is to work backwards, and use local linearity to calculate approximate values for the function, using the derivative. First, as a reminder, look at the values of $f(x) = x^2$ in Table 2.10. Although $f(x) = x^2$ is definitely *not* a linear function, in the table of values it looks *almost* linear. For a constant change in x of 0.001, there is a nearly constant change in $f(x)$ of 0.002. Thus, near $x = 1$, the x^2 function appears nearly linear with a slope of 2. In the next example we use local linearity to estimate values of $f(x)$.

x	x^2
1	1
1.001	1.002001
1.002	1.004004
1.003	1.006009
1.004	1.008016
1.005	1.010025

Table 2.10: Table of Values for $f(x) = x^2$

❑ **Example 1** Use the fact that if $f(x) = x^2$, then $f'(1) = 2$ to estimate the value of $(1.0013)^2$.

Solution. Since we have observed that x^2 looks linear near $x = 1$, its graph looks like that in Figure 2.35. Near $x = 1$, the slope is 2, so

$$\frac{\text{change in } f(x)}{0.0013} \approx 2 \quad \text{or} \quad \text{change in } f(x) \approx 2(0.0013) = 0.0026.$$

This means that in moving from P to Q, x changes by 0.0013 and $f(x)$ changes by approximately 0.0026, from 1 to 1.0026. So

$$f(1.0013) = (1.0013)^2 \approx 1.0026.$$

❑

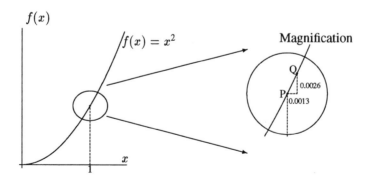

Figure 2.35: A portion of $f(x) = x^2$, magnified until almost straight

☐ **Example 2**   Check the approximation $(1.0013)^2 \approx 1.0026$ with a calculator. How large is the error? Is the approximation an over- or under- estimate?

Solution. $(1.0013)^2 = 1.00260169$, so the error is 0.00000169. The approximation is smaller than the true value.   ☐

As another example, look at the square root function. It too looks locally linear near any point. In Table 2.11 are values near $x = 4$. Here you can observe that every change in x of 0.001 gives rise to a change in $f(x)$ of about 0.00025 (Look at the table and check this!). Thus we have locally linear behavior with a slope of about $0.00025/0.001 = 1/4$.

x	$f(x) = \sqrt{x}$
4	2.0000000
4.001	2.0002500
4.002	2.0005000
4.003	2.0007500
4.004	2.0010000
4.005	2.0012496

Table 2.11: Table of Values for $f(x) = \sqrt{x}$

☐ **Example 3**   Use the fact that the derivative of $\sqrt{x}$ at $x = 4$ is $\frac{1}{4}$ to estimate $\sqrt{4.006}$.

Solution. Since 4 and 4.006 differ by 0.006, the values of $f(x)$ will differ by about $(0.006)(\frac{1}{4}) = 0.0015$. So $\sqrt{4.006} \approx 2 + 0.0015 = 2.0015$.   ☐

When we estimate the value of $\sqrt{4.006}$ from the value of $\sqrt{4}$, we say we have *extrapolated* from 4 to 4.006.

☐ **Example 4**   Use a calculator to see how good the estimate $\sqrt{4.006} \approx 2.0015$ is. How large is the error? Is the approximation too large or too small?

Solution. $\sqrt{4.006} \approx 2.0014994$, so the error is 0.0000006. The approximation is too large.   ☐

> This method is called *local linearization* or the *tangent line approximation* of a function. We are using the fact that near the point of contact, a curve and its tangent line are not very far apart. The values we have been calculating are coordinates of points on the tangent line.

Of course, there's no reason to expect that the curve will look like the tangent line if we go too far away and usually it doesn't. For example, if you use the same linearization we used above to find $\sqrt{16}$, we get $\sqrt{16} \approx 2 + 12(\frac{1}{4}) = 5$. As you can see, this is not so good! (See Figure 2.36.) The problem is that we have traveled too far from the place where the curve looks like a line with slope 1/4. In general, extrapolating too far from known data is dangerous.

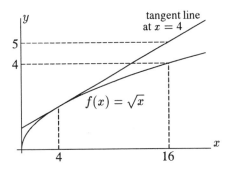

Figure 2.36: Local Linearization: Approximating $f(x) = \sqrt{x}$ by its Tangent Line at $x = 4$

☐ **Example 5** Local linearization will give values too small for the function x^2 and too large for the function $\sqrt{x}$. Draw pictures to explain why. Can you formulate a conjecture that tells when the tangent line approximation will be too large or too small?

Solution. The graph of x^2 is concave up and lies above its tangent line; therefore the linearization will always be too small. The graph of $\sqrt{x}$ is concave down and lies below its tangent line and therefore the linearization will be too large. ☐

Of course, the main use of this method of approximation is not for quantities like $(1.0013)^2$ and $\sqrt{4.006}$, which are much better found using a calculator. The real purpose of such approximations is to make estimates which cannot easily be made in other ways, such as in the following example.

☐ **Example 6** Climbing health care costs have been a source of concern for some time. The data[3] in Table 2.12 shows the average yearly per capita (i.e. per person) health care expenditures for various years since 1970. Use this data to estimate average per capita expenditures in 1988 and 1995.

Year	Per capita expenditure ($)
1970	349
1975	591
1980	1055
1985	1596
1987	1987[4]

Table 2.12: Health Care Costs

Solution. Health care costs were clearly increasing throughout the last twenty years. Between 1985 and 1987 they increased $(1987 - 1596)/2 = \$195.50$ per year. To make estimates beyond 1987 we will assume that costs continue to climb at the same rate. Therefore,

$$\text{Costs in 1988} = \text{Costs in 1987} + \text{Change in Costs}$$
$$\approx \$1987 + \$195.50 = \$2182.50.$$

Since 1995 is 8 years beyond 1987,

$$\text{Costs in 1995} \approx \$1987 + \$195.50(8) = \$3551.$$

☐

You should realize that the estimate for 1988 in the above example is much more likely to be close to the true value than the estimate for 1995. The further we extrapolate from the given data, the more errors we are likely to introduce. It is unlikely the rate of change of health care costs will stay at $195.50/year all the way until 1995.

Graphically what we have done is to extend the line joining the points for 1985 and 1987 to make projections for the future. (See Figure 2.37.) You might be concerned that we only used the last two pieces of data to make the estimates. Isn't there valuable information to be gained from the rest of the data? Yes, indeed—though there's no fixed way of taking this information into account. You might look at the rate of change for the years before 1985 and take an average. Alternatively,

---

[3]Dept. of Health and Human Services, Universal Almanac 1990, John Wright, ed. (Andrews & McNeal).

[4]Yes, this number is correct. The expenditures in 1987 were $1987 per person.

you might fit a line to all the data and use that. However, in each of these methods you are estimating a rate of change of health care costs with time. In other words, you are estimating the derivative of the cost function.

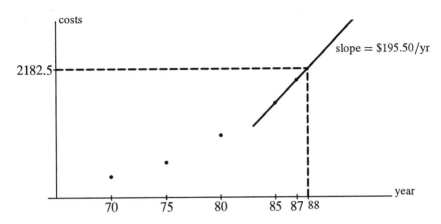

Figure 2.37: Graph of Health Care Costs

❖ Exercises for Section 2.6

1. Find the equation of the tangent line to $f(x) = \sqrt{x}$ at $x = 4$. Confirm that the point used in the approximation of $\sqrt{4.006}$ lies on this line.

2. Use the fact that the derivative of $f(x) = x^3$ is 12 at $x = 2$ to fill in approximate values of x^3 near 2 in the table below. Check your answers by finding x^3 exactly with a calculator.

x	2.000	2.001	2.002	2.003	2.004	2.005
x^3	8.000	8.012	8.024			

3. Use a calculator to construct a table of values for $\tan x$, rounded to three decimals, for $x = 0.80, 0.81, 0.82,..., 0.90$. Is the table approximately linear?

4. Given the following data:

x	0	0.2	0.4	0.6	0.8	1.0
$f(x)$	3.7	3.5	3.5	3.9	4.0	3.9

(a) Estimate an equation of the tangent line to $y = f(x)$ at $x = 0.6$.

(b) Using this equation, estimate $f(0.7)$, $f(1.2)$, $f(1.4)$. Which of these estimates do you feel most confident about? Why?

5. Given the following data about a function, f:

x	6.5	7.0	7.5	8.0	8.5	9.0
$f(x)$	10.3	8.2	6.5	5.2	4.1	3.2

   (a) Estimate $f'(7.0)$, $f'(8.5)$ and $f'(6.75)$.

   (b) Estimate the rate of change of f' at $x = 7$.

   (c) Find, approximately, an equation of the tangent line at $x = 7$.

   (d) Estimate $f(6.8)$.

6. If a child were to step off from the edge of a spinning merry-go-round, she would move in the direction of the line tangent to the path of the merry-go-round at the point where she steps off. (See Fig. 2.38). Suppose a merry-go-round of radius 10 is centered at the origin and the child steps off from the point $(6, 8)$. Considering the child's horizontal motion only, and not her height above the ground, find the equation of the line along which she will move. [Hint: We are considering the child's horizontal motion only (not vertical). Estimate the derivative of $y = \sqrt{100 - x^2}$ at $x = 6$ and use that value as the slope of your tangent line.]

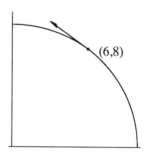

Figure 2.38: Stepping off a merry-go-round.

(for Problem 6)

7. The child in problem 6 moved along the path of the tangent line to the circle she had been following. This phenomenon, a result of *inertia*, would occur regardless of the original path of an object—i.e., the momentum of an object is always along the tangent line to the path it is following.

   Suppose a man was riding "The Quadratic Express", a train from $(-2, 4)$ to $(2, 4)$ along a track which followed the curve $y = x^2$. If he happened to be riding on top of the train and stepped off just as he was passing the origin, along what line would he move? What if he stepped off at $(1, 1)$? (Again, we are considering horizontal motion only).

8. In Example 6, page 164, we assumed that health care costs were increasing linearly. We will now assume that health care costs are increasing exponentially.

   (a) Fit an exponential function to the data for 1985 and 1987 in Table 2.12, and use this function to predict health care costs in 1988 and 1995.

   (b) How close are the linear and exponential predictions for 1988? For 1995? How will the two predictions compare further into the future?

9. The number of hours, H, of daylight in Madrid as a function of date is approximated by the formula

$$H = 12 + 2.4\sin(0.0172(t - 80))$$

where t is the number of days since the start of the year. Fig. 2.39 shows a one-month portion of the graph of H against t.

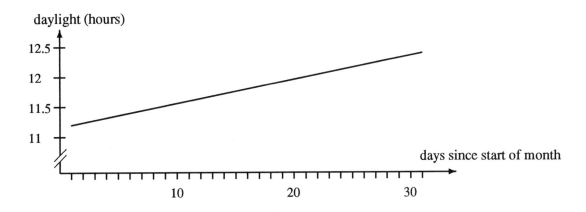

Figure 2.39: Amount of Daylight in Madrid
(for Problem 9)

(a) Comment on the shape of the graph. Why does it look like a straight line?

(b) What month does this graph show? How do you know?

(c) What is the approximate slope of this line? What does the slope represent in practical terms?

10. Suppose you put a yam in a hot oven, maintained at a constant temperature of 200° C. As the yam picks up heat from the oven, its temperature rises.[3]

(a) Draw a possible graph of the temperature T of the yam against time t (minutes) since it is put into the oven. Explain any interesting features of the graph, and in particular explain its concavity.

(b) Suppose that, at $t = 30$, the temperature T of the yam is 120° and increasing at the (instantaneous) rate of 2°/min. Using this information, plus what you know about the shape of the T graph, estimate the temperature at time $t = 40$.

(c) Suppose in addition I tell you that at $t = 60$, the temperature of the yam is 165°. Can you improve your estimate of the temperature at $t = 40$?

(d) Assuming all the data given so far, estimate the time at which the temperature of the yam is 150°.

---

[3]From *Calculus: The Analysis of Functions*, by Peter D. Taylor (Toronto: Wall & Emerson, Inc., 1992)

11. (a) Sketch a smooth curve $y = f(x)$ satisfying the following:
- $|f'(x)| \leq 0.5$ for all x.
- $f'(x) = 0.5$ for at least three different values of x.
- $f'(x) = -0.5$ for exactly one value of x.

(b) Pick an arbitrary point somewhere in the middle of the curve you drew above and label it "A". Sketch a line whose slope equals 0.5 and which passes through the point A. Sketch a line whose slope equals -0.5 and which passes through A.

(c) If f is any function satisfying the conditions above, and $f(a) = 1$ for some a, what is the *maximum* possible value of $f(a + 3)$? What is the *minimum* possible value of $f(a + 2)$?

2.7   Notes on the Limit

Our initial consideration of the speed problem has taken us a long way: we have ended up with a notion of instantaneous rate of change for any function. As mentioned previously, we have not really simplified the difficulties in measurement of such a quantity—we have just neatly tucked all of them into the notion of limit. Let's look a bit harder at the limit concept.

The expression

$$\frac{f(a + h) - f(a)}{h}$$

is called the difference quotient of f, and as we've seen, it measures the average rate of change of f over the interval $a < x < a + h$. As the interval shrinks, that is, when h gets smaller and smaller, these average rates may begin to approach a particular number. If there is a number L, say, such that we can make the average rate as close to L as we please simply by choosing a small enough interval, then L is said to be the *limit* of the difference quotients as h approaches 0. In symbols,

$$L = \lim_{h \to 0} \frac{f(a + h) - f(a)}{h}.$$

The thing to note here is that we are asserting that as h gets small, the difference quotient approaches L. We are not saying *anything* about what happens when h *equals* 0. Indeed when $h = 0$, the difference quotient becomes $0/0$, which is meaningless. The limit was introduced just to avoid such trouble. More generally, for any function g, we define

$$\lim_{h \to c} g(h) = L$$

to mean that $g(h)$ can be made as close to L as we like by choosing h close enough to c.

Limits have lots of uses in calculus—finding the instantaneous rate of change is just one of them. Actually, you can investigate the limit of any function at any point you choose. For example, the value of

$$\lim_{h \to 2} h^2$$

is 4. It's 4 because we can make h^2 as close to 4 as we like by taking h sufficiently close to 2. (Look at 1.9^2, 1.99^2, 1.999^2, and 2.1^2, 2.01^2, 2.001^2, to convince yourself.) Since the limit doesn't "ask" what happens *at* $h = 2$, it is not sufficient to simply put in 2 to find the answer. The limit measures behavior of a function *near* a point, not *at* the point.

Not every limit statement is as transparent as $\lim_{h \to 2} h^2 = 4$. For example, how would you find

$$\lim_{h \to 0} (1 + h)^{\frac{1}{h}}?$$

You might let $h = 0.01$, 0.001, 0.0001, and so on. Try it—the result is remarkable! Notice that substituting $h = 0$ makes the expression undefined, so the only way to investigate this is to use values near but not equal to zero.

☐ **Example 1** Use a graph to find $\lim_{\theta \to 0} \dfrac{\sin \theta}{\theta}$ for θ in radians.

Solution. Figure 2.40 shows that as θ gets closer to 0, $\dfrac{\sin \theta}{\theta}$ gets close to 1, so

$$\lim_{\theta \to 0} \frac{\sin \theta}{\theta} = 1.$$

☐

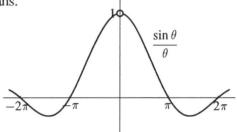

Figure 2.40: Graph of $\dfrac{\sin \theta}{\theta}$

One- and Two-Sided Limits

One other technical comment about limits is needed. When we write

$$\lim_{h \to 2} f(h),$$

we want this to represent the number that $f(h)$ approaches as h gets close to 2 *from both sides*. We examine values of $f(h)$ as h slides in to 2 from values greater than 2 (e.g. 2.1, 2.01, 2.001) and values less than 2 (e.g. 1.9, 1.99, 1.999). If $f(h)$ approaches the same number as we slide in from both sides, then that number is called the limit of f as h approaches 2. It is almost always the case that these "left" and "right" limits are the same. If they are not, we say the limit does not exist.

If, for some reason, we want h to approach 2 only through values greater than 2, we write

$$\lim_{h \to 2^+} f(h)$$

for the number that $f(h)$ approaches (assuming such a number exists). Similarly,

$$\lim_{h \to 2^-} f(h)$$

denotes the number (if it exists) obtained by letting $h \to 2$ from below. We call $\lim_{h \to 2^+}$ a *right-hand limit* and $\lim_{h \to 2^-}$ a *left-hand limit*.

Limits and Continuity

In Section 1.12, we saw how to recognize whether a function is continuous from its graph: the graph should have no jumps, breaks, or holes. We also saw that, numerically, a function is continuous if nearby values of the independent variable gave nearby values of the function. Suppose $f(x)$ is a continuous function of x on some interval, and $x = a$ is in that interval. Then if x is close to a, we know that $f(x)$ is close to $f(a)$. In fact, the closer x gets to a, the closer $f(x)$ gets to $f(a)$. Thus, as $x \to a$, the limit of $f(x)$ must be $f(a)$. Mathematicians use this idea and the limit notation to give a more formal definition of continuity than we have been using.

> A function f is **continuous** at $x = a$ if
> $$\lim_{x \to a} f(x) = f(a).$$

Note that to calculate $\lim_{h \to 2} h^2$ you were told not simply to substitute 2—although, as you have probably noticed, substituting 2 does give the right answer. The reason is that *if* you know the function $f(h) = h^2$ is continuous (which it is), then you *can* calculate the limit by substitution. In fact, the criterion for continuity is that the limit and the value of the function at the point are the same.

❖ Exercises for Section 2.7

1. How shall we define 0^0? Using the idea that $0^x = 0$ for all positive x, we might say that $0^0 = 0$. But using the idea that $x^0 = 1$ for all positive x, we might say that $0^0 = 1$. Let's decide to define 0^0 as
$$0^0 = \lim_{x \to 0} x^x.$$
Calculate this limit either by evaluating x^x for smaller and smaller positive values of x ($x = 0.1, 0.01, 0.001, \ldots$) or by zooming on the graph of $y = x^x$ near $x = 0$.

Use a graph to evaluate each of the limits in Problems 2–4.

2. $\lim_{\theta \to 0} \dfrac{\sin 2\theta}{\theta}$,   ($\theta$ in radians)

3. $\lim_{\theta \to 0} \dfrac{\cos \theta - 1}{\theta}$,   ($\theta$ in radians)

4. $\lim_{x \to 0} x \ln |x|$

5. Calculate
$$\lim_{x \to 0} (1 + x)^{\frac{1}{x}}$$
using the graph of $y = (1 + x)^{\frac{1}{x}}$. You should recognize the answer you get. What is it?

6. Using a graph, calculate
$$\lim_{x \to 0} (1 + 2x)^{\frac{1}{x}}.$$

2.8   Notes on Differentiability

A function is said to be **differentiable** at any point at which it has a derivative. Most of the functions we will be dealing with will have a derivative at every point; they are said to be differentiable everywhere.

How do you know whether a function is differentiable?

If a function has a derivative at a point, its graph must have a tangent line there. The slope of the tangent line is the derivative. Thus

- A function is differentiable at a point if when you "zoom in," or magnify, the graph at that point, you see a non-vertical straight line.

Occasionally—but not often—we will meet a function which fails to have a derivative at a few points; For example, a discontinuous function whose graph has a break at some point cannot have a derivative at that point. Most of the functions we will deal with are continuous and have graphs with no breaks, jumps, or holes.

Are there continuous functions which are not differentiable at some point?

The answer is yes; the points of non-differentiability are where the graph of the function does not look like a nonvertical straight line when you zoom in. Two of the ways in which a function can fail to be differentiable at a point are if it has a graph with:

- a sharp "corner" at that point, or
- a vertical tangent line.

Look at the hypothetical function in Figure 2.41. There is a problem at A because the graph has

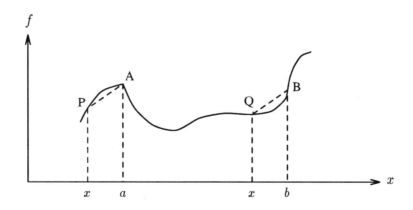

Figure 2.41: A function which is not differentiable at A and B

a "corner" there. As x approaches a from the left, the slope of the line joining P to A converges to some positive number. As x approaches a from the right, the slope of the line joining P to A converges to some negative number. Thus the slopes approach different numbers as you approach $x = a$ from different sides. Therefore the function is not differentiable at $x = a$.

At B, there is no sharp corner, but as x approaches b from the left or right, the slope of the lines joining B to Q do not converge—they just keep growing larger and larger. This reflects the fact that the graph has a vertical tangent at B. However, since the slope of a vertical line is not defined, the function is not differentiable at $x = b$ either. The function appears to be differentiable, however, at all points except $x = a$ and $x = b$.

If the graph of f looks more and more straight and nonvertical as the region near x is magnified, then f will possess a derivative there. On the other hand, if no amount of magnification will cause a "corner" to straighten out at some point, or if it has a vertical tangent line at that point, then f does not have a derivative there.

The Absolute Value Function

The best-known function with a 'corner' that can't be straightened out is the absolute value function

$$f(x) = |x| = \begin{cases} x & \text{if } x \geq 0, \\ -x & \text{if } x < 0. \end{cases}$$

The graph of this function is in Figure 2.42.

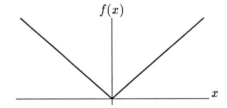

Figure 2.42: Absolute Value Function: $f(x) = |x|$

☐ **Example 1** For the function $f(x) = |x|$, try to compute the derivative at $x = 0$. Is f differentiable there?

Solution. To find the slope at $x = 0$, we want to look at

$$\lim_{h \to 0} \frac{|h| - 0}{h} = \lim_{h \to 0} \frac{|h|}{h}.$$

As h approaches 0 from the right, since h is always positive, $|h| = h$, and the ratio is always 1. On the other side, as h approaches 0 from the left, then $|h| = -h$, and the ratio is -1. These limits correspond to the fact that the slope of the right hand part of the graph is 1, and the slope of the left hand part is -1. Since the limits are different on each side, the limit of the difference quotient fails to exist. Thus the absolute value function is not differentiable at $x = 0$. If you imagine a high magnification of the curve near $x = 0$, you will always see the corner there. Try it on a graphing calculator or see Figure 2.42. By the way, this function has a derivative everywhere *except* at $x = 0$.
☐

❏ **Example 2**   Look at the graph of $f(x) = (x^2 + 0.0001)^{1/2}$ shown in Figure 2.43. The graph of f appears to have a sharp corner at $x = 0$. Does f have a derivative at $x = 0$?

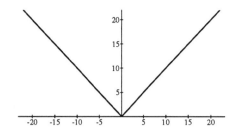

Figure 2.43: Graph of $f(x) = (x^2 + 0.0001)^{1/2}$

Solution. We want to look at
$$\lim_{h \to 0} \frac{(h^2 + 0.0001)^{1/2} - (0.0001)^{1/2}}{h}.$$

As $h \to 0$ from positive or negative numbers, the difference quotient approaches 0 (try evaluating it for $h = 0.001, 0.0001$, etc). So there is a derivative at $x = 0$ (namely 0). How can this be if f has a corner at $x = 0$?

The answer lies in the fact that what appears to be a corner is in fact smooth—under higher magnification the graph of f will look more and more like a straight line with slope 0! If you have a graphing calculator, use it to 'zoom in' on the graph at $x = 0$. If not, look at Figure 2.44.   ❏

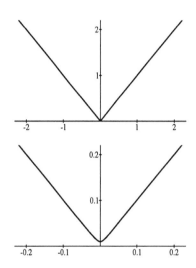

Figure 2.44: Magnification of $f(x) = (x^2 + 0.0001)^{1/2}$

❏ **Example 3**   Investigate the differentiability of $f(x) = x^{1/3}$ at $x = 0$.

Solution. This function is perfectly smooth at $x = 0$ (no lumps or bumps) but appears to have a vertical tangent there. Zoom in using a graphing calculator or see Figure 2.45. Looking at the difference quotient at $x = 0$ we see

$$\lim_{h \to 0} \frac{(0 + h)^{1/3} - 0^{1/3}}{h} = \lim_{h \to 0} \frac{h^{1/3}}{h} = \lim_{h \to 0} \frac{1}{h^{2/3}}.$$

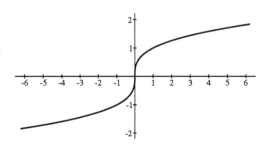

Figure 2.45: $f(x) = x^{1/3}$
(for Example 3)

As $h \to 0$ the denominator becomes small, so the fraction grows without bound. Hence the function fails to have a derivative at $x = 0$.   ❏

☐ **Example 4** Consider the function given by the formulas

$$f(x) = \begin{cases} x + 1 & \text{when} \quad x \leq 1 \\ 3x - 1 & \text{when} \quad x > 1 \end{cases}$$

Draw the graph of f. Is f continuous? Is f differentiable at $x = 1$? Explain.

Solution. The graph in Figure 2.46 has no breaks in it and therefore the function is continuous. However, the graph has a 'corner' at $x = 1$ which no amount of magnification will remove. To the left of $x = 1$, the slope is 1; to the right of $x = 1$, the slope is 3. Thus the function is not differentiable at $x = 1$. This function is called *piecewise linear* because each part of it is linear . Such functions are often used to make approximations, because although they are not smooth, they are easy to work with. ☐

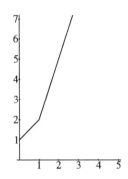

Figure 2.46: Function for Example 4

A great deal of interest has been sparked in the last few years in the study of curves which do not possess derivatives *anywhere*. These curves arise in the modeling of natural processes that are random and chaotic, such as the path of a water molecule in a glass of water. As the molecule bounces off of its neighbors haphazardly, it traces out a path with many jagged, nondifferentiable corners. Although the path may be smooth between collisions, it can be modeled effectively by a curve that is not smooth anywhere.

❖ Exercises for Section 2.8

Are the functions in Problems 1–3 differentiable at $x = 0$?

1. $f(x) = (x + |x|)^2 + 1$

2. $f(x) = \begin{cases} x \sin(\frac{1}{x}) + x & \text{if } x \neq 0 \\ 0 & \text{for } x = 0 \end{cases}$

3. $f(x) = \begin{cases} x^2 \sin(\frac{1}{x}) + 1 & \text{if } x \neq 0 \\ 1 & \text{for } x = 0 \end{cases}$

4. The acceleration due to gravity, g, varies with height above the surface of the Earth. If you go down below the surface of the Earth, g varies in a different way. It can be shown that, as a function of r, the distance from the center of the Earth, g is given by

$$g = \begin{cases} \dfrac{GMr}{R^3} & \text{for } r < R \\[2mm] \dfrac{GM}{r^2} & \text{for } r \geq R \end{cases}$$

where R is the radius of the Earth, M is the mass of the Earth and G is the gravitational constant.

(a) Is g a continuous function of r? Explain your answer.

(b) Sketch a graph of g against r.

(c) Is g a differentiable function of r? Explain your answer.

5. Sometimes, odd behavior can be hidden beneath the surface of a rather normal-looking function. Consider the following function:

$$f(x) = \begin{cases} 0 & \text{if } x < 0. \\ x^2 & \text{if } x \geq 0. \end{cases}$$

(a) Sketch a graph of this function. Does it have any vertical segments or corners? Is it differentiable everywhere? If so, sketch the derivative f' of this function. (Hint: You may want to use the result of Example 4 on page 143.)

(b) Is the derivative differentiable everywhere? If not, at what point(s) is it not differentiable? Draw the second derivative wherever it exists. Is the second derivative differentiable? Continuous?

2.9 Miscellaneous Exercises for Chapter 2

1. Using a calculator or computer, sketch a graph of $f(x) = x \sin x$ on the interval $-10 \leq x \leq 10$.

(a) How many zeros does $f(x)$ have on this interval?

(b) Is f increasing or decreasing at $x = 1$? at $x = 4$?

(c) On which interval is the average rate of change greater: $0 \leq x \leq 2$ or $6 \leq x \leq 8$?

(d) Is the instantaneous rate of change greater at $x = -9$ or $x = 1$?

2. Let $f(x) = x^2$. In Section 2.2 we calculated an estimate of $f'(2)$ by tabulating values of x^2 near $x = 2$ (see Table 2.2). Do the same for $x = -1$ and for $x = 5$. In each case estimate the derivative at that point. Recalling from Example 4 of Section 2.3 that $f'(2) \approx 4$, and $f'(3) \approx 6$, guess a general formula for the derivative of $f(x) = x^2$.

3. Construct tables of values, rounded to three decimals, for $f(x) = x^3$ near $x = 1$, near $x = 3$ and near $x = 5$, as in Problem 2. In each case estimate the derivatives $f'(1)$, $f'(3)$, and $f'(5)$. Then guess a general formula for $f'(x)$.

4. For $f(x) = \ln x$, construct tables, rounded to four decimals, near $x = 1$, $x = 2$, $x = 5$ and $x = 10$ as in Problem 2. Use the tables to estimate $f'(1)$, $f'(2)$, $f'(5)$ and $f'(10)$. Then guess a general formula for $f'(x)$.

For Problems 5–6, sketch a graph of the derivative function of each of the given functions.

5.

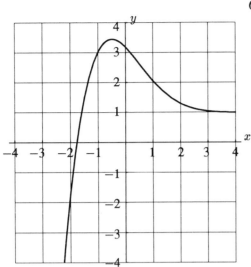

6.

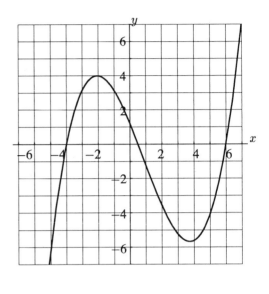

Sketch the graphs of the derivatives of the functions shown in Problems 7–11. Be sure your sketches are consistent with the important features of the original functions.

7.

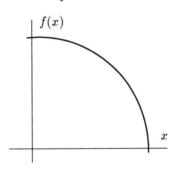

8.

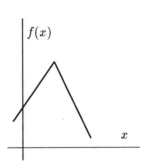

9.

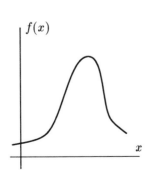

10.

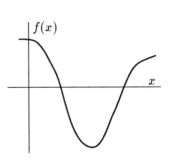

11.

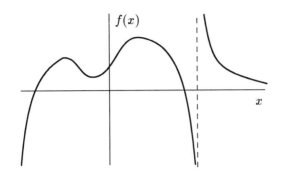

12. Draw a possible graph of $y = f(x)$ given the following information about its derivative:

   (a) For $x < -2$, $f'(x) > 0$ and the derivative is increasing.

   (b) For $-2 < x < 1$, $f'(x) > 0$ and the derivative is decreasing.

   (c) At $x = 1$, $f'(x) = 0$.

   (d) For $x > 1$, $f'(x) < 0$ and the derivative is decreasing (getting more and more negative).

13. Students were asked to evaluate $f'(4)$ from the following table which shows the values of the function f:

x	1	2	3	4	5	6
$f(x)$	4.2	4.1	4.2	4.5	5.0	5.7

   • Student A estimated the derivative as $f'(4) \approx \frac{f(5)-f(4)}{5-4} = 0.5$.

   • Student B estimated the derivative as $f'(4) \approx \frac{f(4)-f(3)}{4-3} = 0.3$.

   • Student C suggested that they should split the difference and estimate the average of these two results, i.e, $f'(4) \approx \frac{1}{2}(0.5 + 0.3) = 0.4$.

   (a) Sketch the graph of f and indicate how the three estimates are represented on the graph.

   (b) Explain which answer is likely to be best.

   (c) Use Student C's method to find an algebraic formula which approximates $f'(x)$ using increments of size h.

14. Each of the graphs in Figure 2.47 shows the position of a particle moving along the x-axis as a function of time, $0 \le t \le 5$. The vertical scales of each of the graphs is the same. During this time interval, which particle has

   (a) constant velocity
   (b) the greatest initial velocity
   (c) the greatest average velocity
   (d) zero average velocity
   (e) zero acceleration
   (f) positive acceleration throughout.

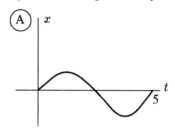

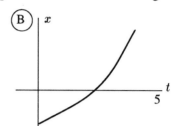

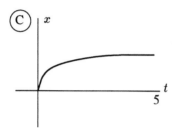

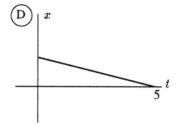

Figure 2.47: Position Graphs

15. Let $g(x) = \sqrt{x}$ and $f(x) = kx^2$, where k is a constant.

   (a) Find the slope of the tangent line to the graph of g at the point $(4, 2)$.
   (b) Find the equation of this tangent line.
   (c) If the graph of f contains the point $(4, 2)$, find k.
   (d) Where does the graph of f intersect the tangent line found above?

16. A circle with center at the origin and radius of length $\sqrt{19}$ has equation $x^2 + y^2 = 19$. Graph the circle.

   (a) Just from looking at the graph, what can you say about the slope of the line tangent to the circle at the point $(0, \sqrt{19})$? What about the slope of the tangent at $(\sqrt{19}, 0)$?
   (b) Estimate the slope of the tangent to the circle at the point $(2, -\sqrt{15})$ by graphing the tangent carefully at that point.
   (c) Use the result of part (b) and the symmetry of the circle to find slopes of the tangents drawn to the circle at $(-2, \sqrt{15})$, $(-2, -\sqrt{15})$ and $(2, \sqrt{15})$.

17. A person with a certain liver disease first exhibits larger and larger concentrations of certain enzymes (called SGOT and SGPT) in the blood. As the disease progresses, the concentration of these enzymes drops again, first to the pre-disease level and eventually to zero (when almost all of the liver cells have died). Monitoring the levels of these enzymes allows doctors to track the progress of a patient with this disease. If $C = f(t)$ is the concentration of the enzymes in the blood as a function of time,

    (a) sketch a possible graph of $C = f(t)$.

    (b) Mark on the graph the intervals where $f' > 0$ and where $f' < 0$.

    (c) What does $f'(t)$ represent, in practical terms?

18. The population of a herd of deer is given by

$$P(t) = 4000 + 400 \sin\left(\frac{\pi}{6}t\right) + 180 \sin\left(\frac{\pi}{3}t\right)$$

    where t is measured in months from the first of April.

    (a) Use a calculator or computer to sketch a graph showing how this population varies with time. Use the graph to answer the following:

    (b) When is the herd largest? How many deer are there at that time?

    (c) When is the herd smallest? How many deer are there then?

    (d) When is the herd growing the fastest? When is it shrinking the fastest?

    (e) How fast is the herd growing on April 1st?

19. A function defined for all x has the following properties:

    - f is increasing.

    - f is concave down.

    - $f(5) = 2$.

    - $f'(5) = 1/2$.

    (a) Sketch a possible graph for f.

    (b) How many zeros does f have?

    (c) What can you say about where the zeros are located?

    (d) What is $\lim_{x \to -\infty} f(x)$?

    (e) Is it possible that $f'(1) = 1$?

    (f) Is it possible that $f'(1) = 1/4$?

20. Roughly sketch the shape of the graph of a quadratic polynomial, f, if it is known that:

    - $(1, 3)$ is on the graph of f, and
    - $f'(0) = 3$, $f'(2) = 1$, $f'(3) = 0$.

21. Roughly sketch the shape of the graph of a cubic polynomial, f, if it is known that:

    - $(0, 3)$ is on the graph of f, and
    - $f'(0) = 4$, $f'(1) = 0$, $f'(2) = -\frac{4}{3}$, $f'(4) = 4$.

22. (a) Sketch the graph of a quintic (i.e., fifth degree) polynomial function whose slope equals 2 in exactly three different places. [You need not say what the particular function is—just make sure your graph doesn't do anything the graph of a quintic polynomial function couldn't do.] On your graph, locate and label the places where the slope of the curve is equal to 2, sketching in little tangential line segments.

    (b) Sketch the graph of a quintic polynomial function whose slope equals 2 in exactly two different places. [Again, don't worry about what the particular function is—just don't graph anything that does things the graph of a quintic couldn't do.] As in the above problem, locate and label the places where the slope of the curve is equal to 2.

    (c) Can the graph of a quintic polynomial function have only *one* place where the slope is equal to 2? If so, demonstrate by providing a particular example of a quintic polynomial function whose graph exhibits this characteristic. If not, explain precisely how you know.

Chapter 3

KEY CONCEPT: THE DEFINITE INTEGRAL

We started Chapter 2 by calculating the velocity from the distance traveled. This led us to the notion of derivative, or rate of change, of a function. Now we will consider the reverse problem: given the velocity, how can we calculate the distance traveled? This reverse problem will lead us to our second key concept, the *definite integral,* which computes the total change in a function from its rate of change. We will then discover that the definite integral can be used to calculate not only distance but also other quantities, such as the area under a curve and the average value of a function.

We end this chapter with the Fundamental Theorem of Calculus, which tells us that we can use a definite integral to get information about a function from its derivative. In this sense, derivatives and definite integrals work in opposite directions.

3.1 How Do We Measure Distance Traveled?

If the velocity is a constant, we can find the distance using the formula

$$\text{distance} = (\text{velocity}) \times (\text{time}).$$

In this section we will see how to estimate the distance when the velocity is not a constant.

A Thought Experiment

Suppose a car is moving with increasing velocity. First, let's suppose we are given the velocity every second:

time (sec)	0	1	2	3	4	5
velocity (ft/sec)	20	30	38	44	48	50

How far has the car moved? Since we don't know how fast the car is moving at every moment, we can't figure the distance out exactly, but we can make an estimate. Since the velocity is increasing, the car goes at least 20 feet during the first second. Likewise, it goes at least 30 feet during the next second, and at least 38 feet during the third second. In the fourth second, it goes at least 44 feet, and in the last second, at least 48 feet. During the five second period, it goes at least

$$20 + 30 + 38 + 44 + 48 = 180 \text{ feet}.$$

Thus, 180 feet is an underestimate of the total distance moved during the five seconds.

To get an overestimate, we can reason this way: In the first second the car moved no more than 30 feet; in the next second it moved no more than 38 feet; in the third second no more than 44 feet, and so on. Therefore, altogether it moved no more than

$$30 + 38 + 44 + 48 + 50 = 210 \text{ feet}.$$

Thus the total distance moved is between 180 and 210 feet:

$$180 \text{ feet} \leq \text{total distance traveled} \leq 210 \text{ feet}.$$

There is a difference of 30 feet between our upper and lower estimates.

What if we wanted a more accurate estimate? Then we should ask for more frequent velocity measurements, say every 0.5 seconds:

time (sec)	0	0.5	1.0	1.5	2.0	2.5	3.0	3.5	4.0	4.5	5.0
velocity (ft/sec)	20	26	30	35	38	42	44	46	48	49	50

As before, we get a lower estimate for each half second by using the velocity at the beginning of that half second. During the first half second, the velocity is at least 20 ft/sec, and so the car travels at least $(20)(0.5) = 10$ feet. During the next half second, the car moves at least $(26)(0.5) = 13$ feet, and so on. So now we can say

$$
\begin{aligned}
\text{Lower Estimate} \ = \ & (20)(0.5) + (26)(0.5) + (30)(0.5) + (35)(0.5) + (38)(0.5) + \\
& (42)(0.5) + (44)(0.5) + (46)(0.5) + (48)(0.5) + (49)(0.5) \\
= \ & 189 \text{ feet.}
\end{aligned}
$$

Notice that this is higher than our old lower estimate of 180 feet. We get a new upper estimate by considering the velocity at the end of each half second. During the first half second, the velocity is no more than 26 ft/sec, and so the car moves at most $(26)(0.5) = 13$ feet; in the next half second it moves at most $(30)(0.5) = 15$ feet, and so on. Thus

$$
\begin{aligned}
\text{Upper Estimate} \ = \ & (26)(0.5) + (30)(0.5) + (35)(0.5) + (38)(0.5) + (42)(0.5) \\
& + (44)(0.5) + (46)(0.5) + (48)(0.5) + (49)(0.5) + (50)(0.5) \\
= \ & 204 \text{ feet.}
\end{aligned}
$$

This is lower than our old upper estimate of 210 feet. Now we know that

$$189 \text{ feet} \leq \text{total distance traveled} \leq 204 \text{ feet.}$$

Notice that the difference between our new upper and lower estimates is now 15 feet, half what it was before. By halving our interval of measurement, we have halved the distance between the upper and lower estimates.

Visualizing Distance on the Velocity Graph

We can represent both upper and lower estimates on a graph of the velocity, as well as see how changing the interval of measurement changes the accuracy of our estimates.

The velocity can be represented by a smooth curve going through each of the given points, as in Figure 3.1 The area under the first dark rectangle is $(20)(1) = 20$, the lower estimate of the distance moved in the first second. The area of the second dark rectangle is $(30)(1) = 30$, the lower estimate for the distance moved in the next second. Therefore, the total area of the dark rectangles represents the lower estimate for the total distance moved during the five seconds.

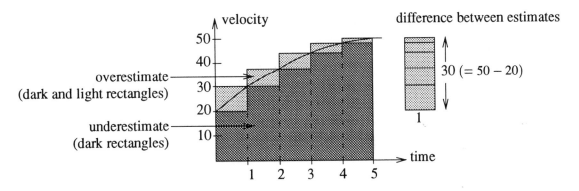

Figure 3.1: Velocity Measured Each Second

If the dark and light rectangles are considered together, the first area is $= (30)(1) = 30$, the upper estimate for the distance moved in the first second. Continuing this argument shows that the upper estimate for the total distance is represented by the area of the dark and light rectangles together. Therefore, the area of the light rectangles alone represents the difference between the two estimates.

To calculate the difference between the two estimates, imagine the light rectangles all pushed to the right and stacked on top of each other: this gives a rectangle of width 1 and height 30. Notice that the height 30 is just the difference between the initial and final values of the velocity ($30 = 50 - 20$) and the width, 1, is the time interval between velocity measurements.

The data for the velocities measured each half second is in Figure 3.2. The area of the dark rectangles again represents the lower estimate, and the dark and light rectangles together represent the upper estimate. As before, the difference between the two estimates is represented by the area of the light rectangles. This difference can be calculated by stacking the light rectangles vertically, giving a rectangle of the same height as before but of half the width. Its area is therefore half what it was before. Again, the height of this stack is $50 - 20 = 30$ and its width is the time interval 0.5.

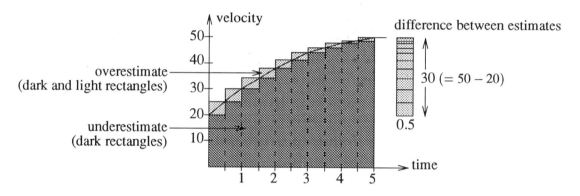

Figure 3.2: Velocity Measured Each Half-Second

❑ **Example 1**   What would be the difference between the upper and lower estimates if the velocity were given every tenth of a second? Every thousandth of a second?

Solution. Every tenth of a second: difference $= (50 - 20)(0.1) = 3$ feet.
Every thousandth of a second: difference $= (50 - 20)(\frac{1}{1000}) = 0.03$ feet.   ❑

❑ **Example 2**   How frequently must the velocity be recorded in order to estimate the total distance traveled to within $1/10$ foot?

Solution. The difference between the velocity at the beginning and end of the observation period is $50 - 20 = 30$. If the time between the measurements is h, then the difference between the upper and lower estimates is $(30)h$. We want

$$(30)h < 0.1 \quad \text{giving} \quad h < \frac{0.1}{30} \approx 0.0033.$$

So the measurements must be made less than 0.0033 seconds apart.   ❑

Determining Distance Precisely

In this section we will obtain a precise expression for the distance traveled. We express the exact distance traveled as a limit of estimates in much the same way as we expressed velocity as a limit of average velocities.

Suppose we want to know the distance traveled by a moving object over the time interval $a \leq t \leq b$. Let the velocity at time t be given by the function $v = f(t)$. Suppose we take measurements of $f(t)$ at equally spaced times $t_0, t_1, t_2, \ldots, t_n$ with time $t_0 = a$ and time $t_n = b$. The time interval between any two consecutive measurements is

$$\Delta t = \frac{b - a}{n}$$

where Δt (read 'delta t') means the change, or increment, in t.

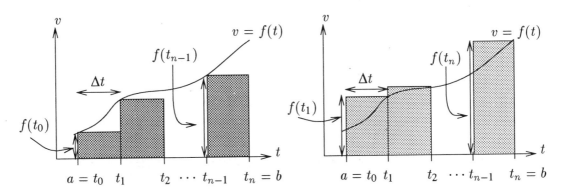

Figure 3.3: Left-Hand Sums   **Figure 3.4**: Right-Hand Sums

During the first time interval, the velocity can be approximated by $f(t_0)$, so the distance traveled is approximately

$$f(t_0)\Delta t.$$

During the second time interval, the velocity is about $f(t_1)$, so the distance traveled is about

$$f(t_1)\Delta t.$$

Continuing in this way and adding up all the estimates, we get an estimate for the total distance. In the last interval, the velocity is approximately $f(t_{n-1})$, so the last term is $f(t_{n-1})\Delta t$:

$$\begin{pmatrix} \text{Total distance traveled} \\ \text{between } a \text{ and } b \end{pmatrix} \approx f(t_0)\Delta t + f(t_1)\Delta t + f(t_2)\Delta t + \cdots + f(t_{n-1})\Delta t.$$

This is called a *left-hand sum* because we used the velocity from the left end of each time interval. It is represented by the sum of the areas of the rectangles in Figure 3.3.

We can also calculate a *right-hand sum* by using the velocity at the right end of each time interval. In that case, the estimate for the first interval is $f(t_1)\Delta t$, for the second interval it is $f(t_2)\Delta t$, and so on. The estimate for the last interval is now $f(t_n)\Delta t$ so

$$\left(\begin{array}{c} \text{Total distance traveled} \\ \text{between } a \text{ and } b \end{array}\right) \approx f(t_1)\Delta t + f(t_2)\Delta t + f(t_3)\Delta t + \cdots + f(t_n)\Delta t.$$

The right-hand sum is represented by the area of the rectangles in Figure 3.4.

If f is an increasing function, the left-hand sum will be an underestimate of the total distance traveled, since for each time interval we use the velocity at the start of that interval to compute the distance traveled, whereas in fact the velocity continues to increase after that measurement. We get an overestimate by using the velocity at the right end of each time interval. If f is decreasing, as in Figure 3.5, then the roles of the two sums are reversed: the left-hand sum is an overestimate and the right-hand sum is an underestimate.

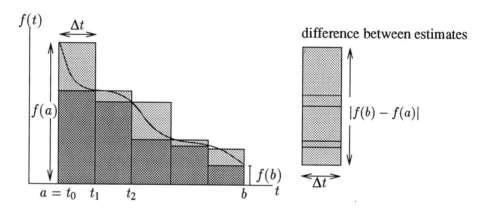

Figure 3.5: Left and Right Sums if f is Decreasing

For either increasing or decreasing functions, the exact value of the distance traveled lies somewhere between the two estimates. Thus the accuracy of our estimate depends on how close these two sums are. For a function which is increasing throughout or decreasing throughout the interval $[a, b]$:

$$\left|\begin{array}{c} \text{Difference between} \\ \text{upper and lower estimates} \end{array}\right| = \left|\begin{array}{c} \text{Difference between} \\ f(a) \text{ and } f(b) \end{array}\right| \times \Delta t = |f(b) - f(a)| \cdot \Delta t.$$

(Absolute values are used to make the difference nonnegative.) By making the measurements close enough together, we can make Δt as small as we like, and therefore we can make the difference between our lower and upper estimates as small as we like.

To find exactly the total distance traveled between a and b, we take the limit of the sums, as n, the number of subdivisions of the interval $[a, b]$, goes to infinity. The sum of the areas of the rectangles approaches the area under the curve between $t = a$ and $t = b$. So we conclude that

$$\begin{pmatrix} \text{Total distance traveled} \\ \text{between } t = a \text{ and } t = b \end{pmatrix} = \lim_{n \to \infty} (\text{left-hand sum})$$

$$= \lim_{n \to \infty} \left[f(t_0)\Delta t + f(t_1)\Delta t + \cdots + f(t_{n-1})\Delta t \right]$$

$$= \left(\text{Area under curve } f(t) \text{ from } t = a \text{ to } t = b \right)$$

and

$$\begin{pmatrix} \text{Total distance traveled} \\ \text{between } t = a \text{ and } t = b \end{pmatrix} = \lim_{n \to \infty} (\text{right-hand sum})$$

$$= \lim_{n \to \infty} \left[f(t_1)\Delta t + f(t_2)\Delta t + \cdots + f(t_n)\Delta t \right]$$

$$= (\text{Area under curve } f(t) \text{ between } t = a \text{ and } t = b).$$

Thus, if n is large enough, both the left-hand and the right-hand sums are accurate estimates of the distance traveled. This method of calculating the distance by taking the limit of a sum works even if the velocity is not increasing throughout or decreasing throughout the interval.

❖ Exercises for Section 3.1

1. A car comes to a stop five seconds after the driver slams on the brakes. While the brakes are on, the following velocities are recorded:

Time since brakes applied (sec)	0	1	2	3	4	5
Velocity (ft/sec)	88	60	40	25	10	0

(a) Give a lower and upper estimate of the distance the car traveled after the brakes were applied.

(b) On a sketch of velocity against time, show the lower and upper estimates and the difference between them.

2. Roger decides to run a marathon. Roger's friend Jeff rides behind him on a bicycle and clocks his pace every fifteen minutes. Roger starts out strong, but after an hour and a half is so exhausted that he has to stop. The data Jeff collected is summarized below:

Time spent running (mins)	0	15	30	45	60	75	90
Speed (mph)	12	11	10	10	8	7	0

(a) Assuming that Roger's speed is constantly decreasing, give upper and lower estimates for the distance Roger ran during the first half hour.

(b) Give upper and lower estimates for the distance Roger ran in total.

(c) How often would Jeff have needed to measure Roger's pace in order to find lower and upper estimates within 0.1 miles of the actual distance that he ran?

3. Natural gas can be produced from coal at a gasworks. The pollution from the gas is removed by scrubbers, which become less and less efficient as time goes on. Measurements made at the start of each month showing the rate at which pollutants are leaving the gasworks are as follows:

Time (months)	0	1	2	3	4	5	6
Rate pollution leaving (tons/month)	5	7	8	10	13	16	20

(a) Make an overestimate and an underestimate of the total amount of pollution that left the gasworks during the first month.

(b) Make an overestimate and an underestimate of the total amount of pollution that left the gasworks during the first six months.

(c) How often would measurements have to be made in order to find overestimates and underestimates which differ by less than 1 ton from the exact amount of pollution that left the gasworks in the first six months?

4. For the picture in Figure 3.6, use the grid and estimate the area of the region bounded by the curve, the horizontal axis and the lines $x = \pm 3$. Get an upper and a lower estimate that are within 4 square units of one another. Explain what you are doing.

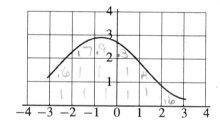

Figure 3.6: Graph for Problem 4

5. (a) For the picture in Fig. 3.7 estimate the shaded area with an error of at most 0.1.

(b) How can you approximate the shaded area to any desired degree of accuracy?

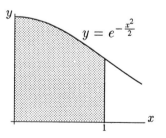

$y = e^{-\frac{x^2}{2}}$

Figure 3.7: Find the Area for Problem 5

6. Figure 3.8 shows the graph of the velocity, v, of an object (in m/sec). Estimate the total distance the object traveled between $t = 0$ and $t = 6$.

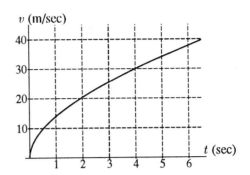

Figure 3.8: Velocity Graph for Problem 6

7. Your velocity is given by

$$v(t) = \sin(t^2), \text{ for } 0 \le t \le 1.1$$

Estimate the distance traveled during this time (to one decimal place).

8. For $0 \le t \le 1$, a bug is crawling at a velocity, v, determined by the formula

$$v = \frac{1}{1+t},$$

where t is in hours and v is in meters/hour. Estimate the distance that the bug crawls during this hour. Your estimate should be off by no more than 0.1 meter.

9. You jump out of an airplane. Before your parachute opens you fall faster and faster, but your acceleration decreases as you fall because of air resistance. The table below gives your acceleration, a (in m/sec^2), after t seconds.

t	0	1	2	3	4	5
a	9.81	8.03	6.53	5.38	4.41	3.61

(a) Give upper and lower estimates of your speed at $t = 5$.

(b) Get a new estimate by taking the average of your upper and lower estimates. What does the concavity of the graph of acceleration tell you about your new estimate?

10. An object has zero initial velocity and an acceleration of 32 ft/sec^2. Find a formula for its velocity as a function of time. Then use left and right sums to find upper and lower bounds on the distance that the object travels in 4 seconds. How can the precise distance be found? [Hint: Find the area under a curve.]

3.2 The Definite Integral

In Section 3.1 we saw how distance traveled and area under a curve can both be approximated by long sums and expressed exactly as the limit of a sum. This process of forming a sum and taking the limit has many interpretations. This section will be devoted to looking at this process in more detail. In particular, we will show how these sums can be defined for any function f, whether or not it represents a velocity. This is similar to what we did for the derivative: first we introduced the limit of difference quotients as the solution to the velocity problem, and then we studied the limit in its own right.

Suppose we have a function $f(t)$ which is continuous for $a \leq t \leq b$, except perhaps at a few points, and bounded everywhere. We divide the interval from $a \leq t \leq b$ into n equal subdivisions, and call the width of an individual subdivision Δt. Thus

$$\Delta t = \frac{b - a}{n}.$$

We will let $t_0, t_1, t_2, \ldots, t_n$ be endpoints of the subdivisions, as in Figure 3.9. As before, we construct two sums:

$$\text{Left-hand sum} = f(t_0)\Delta t + f(t_1)\Delta t + \cdots + f(t_{n-1})\Delta t$$

and

$$\text{Right-hand sum} = f(t_1)\Delta t + f(t_2)\Delta t + \cdots + f(t_n)\Delta t.$$

These sums may be represented graphically as the areas in Figure 3.9, provided $f(t) \geq 0$ and $a < b$. In the picture on the left, the first rectangle has width Δt and height $f(t_0)$, since the top of its left edge just reaches the graph, and hence it has area $f(t_0)\Delta t$; the second has width Δt and height $f(t_1)$, and hence has area $f(t_1)\Delta t$, and so on. The sum of all these areas is the left-hand sum. The right-hand sum, shown in the right-hand picture, is constructed in the same way, except that each rectangle touches the graph on its right edge instead of its left.

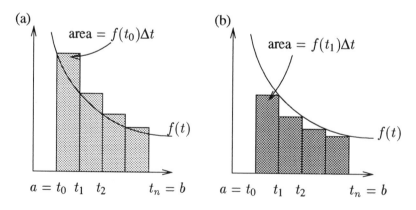

Figure 3.9: (a) Left-Hand Sums. (b) Right-Hand Sums

Writing Left and Right Sums Using Sigma Notation

Both the left-hand and right-hand sum can be written more compactly using *sigma*, or summation, notation. The symbol $\sum$ is a capital sigma, or Greek letter 'S'. We write

$$\text{Right-hand sum} = \sum_{i=1}^{n} f(t_i)\Delta t = f(t_1)\Delta t + f(t_2)\Delta t + \cdots + f(t_n)\Delta t$$

where the $\sum$ tells us to add terms of the form $f(t_i)\Delta t$. The "$i = 1$" at the base of the sigma sign tells us to start at $i = 1$, and the "n" at the top tells us to stop at $i = n$.

In the left-hand sum we start at $i = 0$ and stop at $i = n - 1$, so we write

$$\text{Left-hand sum} = \sum_{i=0}^{n-1} f(t_i)\Delta t = f(t_0)\Delta t + f(t_1)\Delta t + \cdots + f(t_{n-1})\Delta t.$$

Taking the Limit to Define the Definite Integral

In the previous section we took the limit of these sums as n went to infinity, and we do the same here. Whenever the integrand is continuous for $a \leq x \leq b$, and in fact for any function you are ever likely to meet, the limits of the left- and right-hand sums will exist and be equal. In that case, we define the *definite integral* to be the limit of these sums. When $f(t) \geq 0$ and $a < b$, the definite integral represents the area between the graph of $f(t)$ and the t-axis, from $t = a$ to $t = b$. The definite integral is well approximated by a left- or right-hand sum if n is large enough.

The **definite integral** of f from a to b, written

$$\int_a^b f(t)\,dt,$$

is the limit of the left-hand or right-hand sums with n subdivisions as n gets arbitrarily large. In other words:

$$\int_a^b f(t)\,dt = \lim_{n \to \infty} (\text{left-hand sum}) = \lim_{n \to \infty} \left(\sum_{i=0}^{n-1} f(t_i)\Delta t \right)$$

and

$$\int_a^b f(t)\,dt = \lim_{n \to \infty} (\text{right-hand sum}) = \lim_{n \to \infty} \left(\sum_{i=1}^{n} f(t_i)\Delta t \right).$$

Each of these sums is called a *Riemann sum*; f is called the *integrand*; a and b are called the *limits of integration*.

The "$\int$" notation comes from an old-fashioned "S", which stands for "sum" in the same way that $\sum$ does. The "dt" in the integral comes from the factor Δt. Notice that the limits on the $\sum$ sign are 0 and $n - 1$, or 1 and n, whereas the limits on the $\int$ sign are a and b.

☐ **Example 1** Find the left-hand and right-hand sums with $n = 2$ and $n = 10$ for $\int_1^2 \frac{1}{t}\,dt$. Compare the values of these sums with the integral.

Solution. Here $a = 1$ and $b = 2$, so for $n = 2$, $\Delta t = (2 - 1)/2 = 0.5$. Therefore $t_0 = 1$, $t_1 = 1.5$ and $t_2 = 2$. See Figure 3.10. Thus

$$
\begin{aligned}
\text{Left-hand sum} \;&=\; f(1)\Delta t + f(1.5)\Delta t \\
&=\; 1(0.5) + \frac{1}{1.5}(0.5) \\
&\approx\; 0.8333,
\end{aligned}
$$

and

$$
\begin{aligned}
\text{Right-hand sum} \;&=\; f(1.5)\Delta t + f(2)\Delta t \\
&=\; \frac{1}{1.5}(0.5) + \frac{1}{2}(0.5) \\
&\approx\; 0.5833.
\end{aligned}
$$

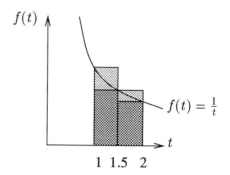

Figure 3.10: Approximation of $\int_1^2 \frac{1}{t}\,dt$ with $n = 2$.

From Figure 3.10 you can see that the left-hand sum is bigger than the area under the curve and the right-hand sum is smaller, so the area under the curve $f(t) = \frac{1}{t}$ from $t = 1$ to $t = 2$ is between 0.5833 and 0.8333.
Thus

$$
0.5833 < \int_1^2 \frac{1}{t}\,dt < 0.8333.
$$

When $n = 10$, $\Delta t = 0.1$ (see Figure 3.11), so

$$
\begin{aligned}
\text{Left-hand sum} \;&=\; f(1)\Delta t + f(1.1)\Delta t + \cdots + f(1.9)\Delta t \\
&=\; \left(1 + \frac{1}{1.1} + \cdots + \frac{1}{1.9}\right)0.1 \\
&\approx\; 0.7188,
\end{aligned}
$$

$$
\begin{aligned}
\text{Right-hand sum} \;&=\; f(1.1)\Delta t + f(1.2)\Delta t + \cdots + f(2)\Delta t \\
&=\; \left(\frac{1}{1.1} + \frac{1}{1.2} + \cdots + \frac{1}{2}\right)0.1 \\
&\approx\; 0.6688.
\end{aligned}
$$

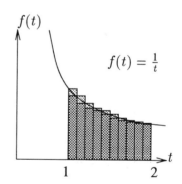

Figure 3.11: Approximation of $\int_1^2 \frac{1}{t}\,dt$ with $n = 10$.

From Figure 3.11 you can see that the left-hand sum is larger than the area under the curve, and the right-hand sum smaller, so

$$
0.6688 < \int_1^2 \frac{1}{t}\,dt < 0.7188.
$$

Notice that the left- and right-hand sums trap the true value of the integral between them. As the subdivisions become finer, the left- and right-hand sums get closer together. ☐

Convergence of Left and Right Sums

A function which is either increasing throughout an interval or decreasing throughout that interval is said to be *monotonic* there. It is usually easy to tell whether or not a function is monotonic just by looking at its graph.

When f is monotonic

The left- and right-hand sums trap the exact value of the integral between them. Let us continue the example

$$\int_1^2 \frac{1}{t}\,dt.$$

The left- and right-hand sums for $n = 2$, 10, 50, and 250 are in Table 3.1. Because the function $f(t) = \frac{1}{t}$ is decreasing, the left-hand sums converge down on the integral from above, and the right-hand sums converge up from below. From the last row of the table we can deduce that

$$0.6921 < \int_1^2 \frac{1}{t}\,dt < 0.6941$$

n	Left-hand sum	Right-hand sum
2	0.8333	0.5833
10	0.7188	0.6688
50	0.6982	0.6882
250	0.6941	0.6921

Table 3.1: Left- and right-hand sums for $\int_1^2 \frac{1}{t}\,dt$.

so $\int_1^2 \frac{1}{t}\,dt \approx 0.69$ to two decimal places. To approximate definite integrals accurately, we evaluate left- and right-hand sums with a large number of subdivisions, using a computer or calculator.

What if f is not monotonic?

Then the definite integral is not always bracketed between the left- and right-hand sums, but in practice you can still see the convergence quite easily. For example, Table 3.2 gives sums for the integral $\int_0^{2.5} \sin(t^2)\,dt$.

Although $\sin(t^2)$ is certainly not monotonic, by the time we get to $n = 250$, it is pretty clear that $\int_0^{2.5} \sin(t^2)\,dt \approx 0.43$ to two decimal places. As you can see, the sums wander around a bit at the beginning, and it's not certain that they won't wander later on. Nonetheless, it is a good rule of thumb that when the left- and right-hand sums begin to agree with each other, you are getting close to the true value of the integral.

n	Left-hand sum	Right-hand sum
2	1.2500	1.2085
10	0.4614	0.4531
50	0.4324	0.4307
250	0.4307	0.4304
1000	0.4306	0.4305

Table 3.2: Left- and right-hand sums for $\int_0^{2.5} \sin(t^2)\,dt$.

In this case we do not have a guarantee of the accuracy, as we do for monotonic functions. Notice that 0.43 does not lie between 1.2500 and 1.2850, the left- and right-hand sums for $n = 2$, nor even between 0.4614 and 0.4531, the two sums for $n = 10$. This shows that if the integrand is not monotonic, the left- and right-hand sums may both be larger (or both be smaller) than the integral. (See Problems 18-19, page 195, to see how this can happen.)

If f is not monotonic, it is often possible to construct upper and lower estimates for an integral by first splitting the interval $[a, b]$ into subintervals on which f is monotonic.

Error in Approximating a Definite Integral by Riemann Sums

For any approximation, it is useful to have an idea of how good it is. We define the *error* in an approximation to be the magnitude of the difference between the approximate and the true values. (Notice that error doesn't mean mistake here.) Since the true value of the integral is between the upper estimate and the lower estimate, the error must be less than the difference between these estimates. So, for a monotonic function, f, on the interval $[a, b]$:

$$\text{Error} \leq \left| \begin{array}{c} \text{Difference between} \\ \text{upper and lower estimates} \end{array} \right| = |f(b) - f(a)| \cdot \Delta t,$$

where $\Delta t = \frac{b-a}{n}$. Thus, we can estimate the error if the function is monotonic.

   If our approximation is taken to be the average of the lower and upper estimates (as in Example 2 below) then our bound for the error is halved.

☐ **Example 2** Compute $\int_1^3 t^2 \, dt$ to one decimal place.

Solution. Since $f(t) = t^2$ is an increasing function on the interval from 1 to 3, the left-hand sum will underestimate the integral and the right-hand sum will overestimate it, as shown in Figure 3.12. How large should n be? The problem asks us to find a value for the integral to one decimal place, i.e. find a value which we are certain is within 0.05 of the true value. (Note: This is *not* telling us that Δt is 0.05.) One way is to use a computer or calculator to calculate left- and right-hand sums for various values of n.
For $n = 100$, the left-hand sum = 8.5868 and the right-hand sum = 8.7468, so

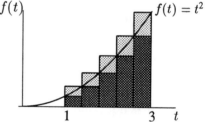

Figure 3.12: Approximation of $\int_1^3 t^2 \, dt$

$$8.5868 < \int_1^3 t^2 \, dt < 8.7468.$$

The left- and right-hand sums differ by 0.16, and averaging them gives

$$\int_1^3 t^2 \, dt \approx 8.6668,$$

which differs from the true value by at most $\frac{0.16}{2} = 0.08$. However, we want more accuracy, so we need a larger value of n. Taking $n = 5000$ makes the left-hand sum = 8.6650 (rounding down) and the right-hand sum = 8.6683 (rounding up) so

$$8.6650 < \int_1^3 t^2 \, dt < 8.6683.$$

Averaging the left and right sums gives

$$\int_1^3 t^2 \, dt \approx 8.6667.$$

   Since the left and right sums differ by 0.0033, the average must be within 0.00165 of the true value. (In fact, much smaller values of n will do. See Problem 17, page 195.) ☐

❖ Exercises for Section 3.2

For Problems 1–6 draw up a table of left- and right-hand sums with 2, 10, 50, and 250 subdivisions. Observe the limit to which your sums are tending as the number of subdivisions gets larger, and estimate the value of the definite integral.

1. $\int_0^1 x^3 \, dx$

2. $\int_0^{\pi/2} \cos x \, dx$

3. $\int_0^1 e^{t^2} \, dt$

4. $\int_1^2 x^x \, dx$

5. $\int_2^3 \sin(t^2) \, dt$

6. $\int_{0.2}^3 \sin(1/x) \, dx$

In problems 7–16, compute the definite integrals to one decimal place. In each case, say how many subdivisions you are using. [Note that, except for Problem 16, each function is monotonic over the given interval.]

7. $\int_0^5 x^2 \, dx$

8. $\int_1^4 \frac{1}{\sqrt{1+x^2}} \, dx$

9. $\int_1^{1.5} \sin x \, dx$

10. $\int_0^{\frac{\pi}{4}} \frac{d\theta}{\cos \theta}$

11. $\int_1^5 (\ln x)^2 \, dx$

12. $\int_{1.1}^{1.7} e^t \ln t \, dt$

13. $\int_1^2 2^x \, dx$

14. $\int_1^2 (1.03)^t \, dt$

15. $\int_{-2}^{-1} \cos^3 y \, dy$

16. $\int_{-3}^3 e^{-t^2} \, dt$
[Hint: The integrand is an even function.]

17. Consider the integral $\int_1^3 t^2 \, dt$. Use the error estimate at the end of this section to calculate the smallest number of subdivisions that will ensure that the average of the upper and lower estimates for this integral is accurate to one decimal place; in other words, the lower and upper estimates should differ by less than 0.1.

18. Using a picture, list in increasing order the following quantities: the integral $\int_0^{4\pi} (2+\cos x) \, dx$, the left-hand and right-hand sums with $n = 2$ subdivisions.

19. Sketch the graph of a function f (you do not need to give a formula for f) on an interval $[a, b]$ with the property that with $n = 2$ subdivisions:

$$\int_a^b f(x) \, dx < \text{left-hand sum} < \text{right-hand sum}.$$

3.3 Interpretations of the Definite Integral

We have seen how the definite integral of a velocity function can be interpreted as total distance traveled and as an area under a curve. In this section we will give some other interpretations of the definite integral.

The Definite Integral as Total Change

Remember that the rate of change of position is velocity, and that the integral of velocity gives the total change in position (or distance moved). We will generalize this result and show that the integral of the rate of change of any quantity will give the total change in that quantity.

Suppose $r(t)$ is the rate of change of some quantity Q, with respect to time, and that we are interested in the total change in Q between $t = a$ and $t = b$. As before, we divide the interval $a \le t \le b$ into n subdivisions, each of length Δt. For each small interval, we will calculate the change in Q, written ΔQ, and then add all these up. For each subinterval we assume the rate of change of Q is approximately constant so that we can say

$$\Delta Q \approx \text{(rate of change of } Q) \times \text{(time)}.$$

For the first subinterval from t_0 to t_1, the rate of change of Q is approximately $r(t_0)$, so

$$\Delta Q \approx r(t_0)\,\Delta t.$$

Similarly, for the second interval

$$\Delta Q \approx r(t_1)\,\Delta t.$$

Summing over all the subintervals,

$$\text{Total Change in } Q = \sum \Delta Q \approx \sum_{i=0}^{n-1} r(t_i)\,\Delta t.$$

Thus we have approximated the change in Q as a left-hand sum. By a similar argument, we can use a right-hand sum:

$$\text{Total Change in } Q = \sum \Delta Q \approx \sum_{i=1}^{n} r(t_i)\,\Delta t.$$

In addition, the change in Q between $t = a$ and $t = b$ can be written as $Q(b) - Q(a)$. Thus, taking the limit as n goes to infinity, we find that

---

If $r(t) =$ rate of change of Q with respect to time, then

$$Q(b) - Q(a) = \left(\begin{array}{c} \text{Total Change in } Q \\ \text{between } t = a \text{ and } t = b \end{array} \right) = \int_a^b r(t)\,dt.$$

☐ **Example 1**   After t hours, a population of bacteria is growing at a rate of 2^t million bacteria per hour. Estimate the total increase in the bacteria population during the first hour.

Solution. Since the rate the population is growing is $r(t) = 2^t$, we have

$$\text{Change in Population} = \int_0^1 2^t \, dt.$$

Since the rate is a monotonic function, the integral will be trapped between the left- and right-hand sums. For $n = 100$,

$$\text{Left sum} = 1.4377 \; < \; \int_0^1 2^t \, dt < 1.4477 = \text{Right sum}$$

so

$$\text{Change in Population} \; = \; \int_0^1 2^t \, dt \approx 1.4 \text{ million bacteria}.$$

☐

The Definite Integral as an Area

If $f(x)$ **is positive:** we can interpret each term $f(x_0)\Delta x, f(x_1)\Delta x, \ldots,$ in a left or right Riemann sum as the area of a rectangle (provided $\Delta x > 0$). As the width Δx of the rectangles approaches zero, the rectangles fit the curve of the graph more exactly, and the sum of their areas gets closer and closer to the area under the curve, shaded in Figure 3.13. Thus we know that:

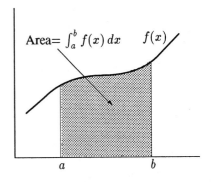

Area$= \int_a^b f(x) \, dx$ $f(x)$

> If $f(x) \geq 0$ and $a < b$:
>
> $$\left(\begin{array}{c} \text{Area under Graph of } f \\ \text{between } a \text{ and } b \end{array}\right) = \int_a^b f(x) \, dx.$$

Figure 3.13: The definite integral $\int_a^b f(x) \, dx$.

☐ **Example 2** Consider the integral $\int_{-1}^1 \sqrt{1 - x^2} \, dx$.

   (a) Interpret the integral as an area and find its exact value.

   (b) Estimate the integral using a left-hand sum with 100 subdivisions. Compare your answer to the exact value.

Solution.

(a) The integral is the area under the graph of $y = \sqrt{1 - x^2}$ between -1 and 1, which is a semicircle of radius 1 and area $\dfrac{\pi}{2}$ (see Figure 3.14).

(b) With $n = 100$, left-hand sum $= 1.569$. For comparison, $\dfrac{\pi}{2} = 1.5707\ldots$

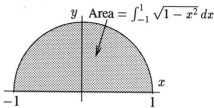

Figure 3.14: Area Interpretation of $\int_{-1}^{1} \sqrt{1 - x^2}\, dx$.

❑

What if $f(x)$ is not positive? We have assumed in drawing Figure 3.13 that the graph lies above the x-axis. If the graph lies below, then each $f(x)\Delta x$ is negative, and the area gets counted negatively. In that case, the definite integral is the negative of the area.

❑ **Example 3** What is the relation between the definite integral $\displaystyle\int_{-1}^{1}(x^2 - 1)\, dx$, and the area between the parabola $y = x^2 - 1$ and the x-axis?

Solution. The parabola lies below the axis between $x = -1$ and $x = 1$. (See Figure 3.15). So,

$$\int_{-1}^{1}(x^2 - 1)\, dx = -\text{Area} \approx -1.33.$$

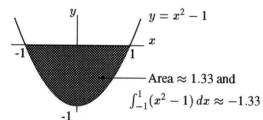

Figure 3.15: Area Interpretation of $\int_{-1}^{1}(x^2 - 1)\, dx$.

❑

If f is negative in some places and positive in others, areas below the axis get counted negatively in the definite integral and areas above get counted positively.

❑ **Example 4** Interpret the definite integral $\displaystyle\int_{0}^{2.5} \sin(x^2)\, dx$ in terms of areas.

Solution. The integral is the area of the first hump, A_1, minus the area of the first dip, A_2, in Figure 3.16. Approximating the integral with $n = 100$ (see Table 3.2, page 193) shows

$$\int_{0}^{2.5} \sin(x^2)\, dx \approx 0.43.$$

Now the graph of $y = \sin(x^2)$ crosses the x-axis where $x^2 = \pi$, i.e. at $x = \sqrt{\pi} \approx 1.8$. The next

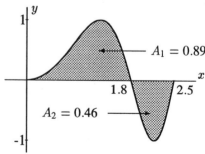

Figure 3.16: Graph of $y = \sin(x^2)$

crossing is at $x = \sqrt{2\pi} \approx 2.5$. Breaking the integral into two parts and calculating each one separately gives, after rounding,

$$\int_0^{1.8} \sin(x^2)\, dx \approx 0.89 \quad \text{and} \quad \int_{1.8}^{2.5} \sin(x^2)\, dx \approx -0.46$$

so $A_1 = 0.89$ and $A_2 = 0.46$. Then

$$\int_0^{2.5} \sin(x^2)\, dx = A_1 - A_2 \approx 0.89 - 0.46 = 0.43, \text{ as expected.}$$

❑

The Definite Integral as an Average

Let's look again at the right-hand sum with n subdivisions for a function f on the interval from a to b:

$$f(x_1)\Delta x + f(x_2)\Delta x + \cdots + f(x_n)\Delta x.$$

Since $\Delta x = (b - a)/n$, we can rewrite this as

$$(b - a) \cdot \left(\frac{f(x_1) + f(x_2) + \ldots + f(x_n)}{n} \right).$$

The second factor here is just the average of the n numbers $f(x_1), f(x_2), \ldots, f(x_n)$. As we take larger and larger values of n, this average takes in more and more values of f, distributed evenly over the interval from a to b. So as n gets arbitrarily large, the limit of this average may be thought of as an average of *all* the values of f on the interval. Thus we can say

$$\int_a^b f(x)\, dx = \lim_{n \to \infty} [f(x_1)\Delta x + f(x_2)\Delta x + \cdots + f(x_n)\Delta x]$$

$$= (b - a) \lim_{n \to \infty} \left(\frac{f(x_1) + f(x_2) + \ldots + f(x_n)}{n} \right), \quad \text{since} \quad \Delta x = \frac{b - a}{n}$$

$$= (b - a) \cdot \left(\begin{array}{c} \text{Average Value of } f \\ \text{from } a \text{ to } b \end{array} \right).$$

Writing I for the integral and A for the average value, we have shown that $I = (b - a)A$. Thus, if we interpret the integral as area under the graph of f, then we can think of the average value of f as the height of the rectangle on the same base which has the same area. (See Figure 3.17.) Also

$$\left(\begin{array}{c} \text{Average Value of } f \\ \text{from } a \text{ to } b \end{array} \right) = \frac{1}{b - a} \int_a^b f(x)\, dx.$$

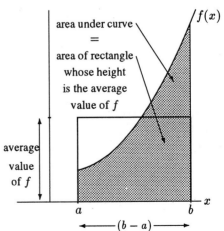

Figure 3.17: Area and Average Value

Using the Notation for the Definite Integral

Just as with the derivative, the notation for the definite integral is helpful in interpreting what it means. In trying to understand the derivative, we saw how dx could be thought of as an infinitesimal increment in x. ("Infinitesimal" means very, very small.) If we think of the interval $a \le x \le b$ as made up of infinitely many such increments, then we can think of

$$\int_a^b f(x)\, dx$$

as the sum of all the products $f(x)\, dx$. Often the key to interpreting the definite integral is understanding the $f(x)\, dx$ part. For example, if $v = f(t)$ is the velocity of a moving particle at time t, then $f(t)\, dt$ may be thought of informally as the distance traveled by the particle during the infinitesimally small time interval dt, since it is the velocity multiplied by the time. (It is often helpful to think of the units: if, for example, $f(t)$ is in ft/sec and dt is in secs, then $f(t)\, dt$ is in ft/sec $\times$ sec = ft.) The integral may then be thought of as the sum of all these infinitesimal distances, giving the total distance. Thus

$$\int_a^b f(t)\, dt$$

is the net change in the position of the particle between $t = a$ and $t = b$.

☐ **Example 5**   Oil is leaking out of a ruptured tanker at a rate of $v = f(t)$ gallons per minute. Write a definite integral expressing the total quantity of oil which leaks out of the tanker in the first hour.

Solution. Informally, we can imagine an infinitesimal time interval dt minutes. During this interval, oil is leaking out at an approximately constant rate of $f(t)$ gallons/minute. Thus, during this interval

$$\text{Amount of oil leaked} = (\text{rate}) \times (\text{time}) = f(t)\, dt$$

and the units of $f(t)\, dt$ are gallons/minute $\times$ minutes = gallons. The total amount of oil leaked is obtained by adding all these amounts between $t = 0$ and $t = 60$. (An hour is 60 minutes.) The sum of all these infinitesimal amounts is the integral

$$\left(\begin{array}{c} \text{Total amount of} \\ \text{oil leaked, in gallons} \end{array} \right) = \int_0^{60} f(t)\, dt.$$

☐

❖ Exercises for Section 3.3

1. (a) Sketch a graph of $f(x) = \sin(x^2)$ and mark on it the points where $x = \sqrt{\pi}, \sqrt{2\pi}, \sqrt{3\pi}, \sqrt{4\pi}$.

    (b) Use your graph and the area interpretation of the definite integral to decide which of the four numbers:

    $$\int_0^{\sqrt{n\pi}} \sin(x^2)\, dx, \quad n = 1, 2, 3, 4$$

    is largest. Which is smallest? How many of the numbers are positive?

2. Water is leaking out of a tank at a rate of $R(t)$ gallons/hr, where t is measured in hours.

   (a) Write a definite integral that expresses the total amount of water that leaks out in the first two hours.

   (b) Figure 3.18 is a graph of $R(t)$. On a sketch, shade in the region whose area represents the total amount of water that leaks out in the first two hours.

   (c) Give an upper and lower estimate of the total amount of water that leaks out in the first two hours.

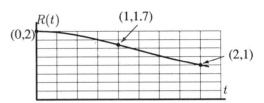

Figure 3.18: Rate of leakage in gal/hr.

(for Problem 2)

3. The rate at which the world's oil is being consumed is continuously increasing. Suppose the rate (in billions of barrels per year) is given by the function $r = f(t)$, where t is measured in years and $t = 0$ is the start of 1990.

   (a) Write a definite integral which represents the total quantity of oil used between the start of 1990 and the start of 1995.

   (b) Suppose $r = 32e^{0.05t}$. Using a left-hand sum with five subdivisions, find an approximate value for the total quantity of oil used between the start of 1990 and the start of 1995.

   (c) Interpret each of the five terms in the sum from part (b) in terms of oil consumption.

4. As this country's relatively rich coal deposits are depleted, a larger fraction of the country's coal will come from strip-mining, and it will become necessary to strip-mine larger and larger areas for each ton of coal. The graph in Figure 3.19 shows an estimate of the number of square miles/million tons that will be defaced during strip-mining as a function of the number of million tons already removed.

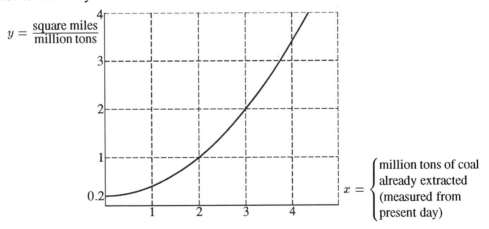

Figure 3.19: Strip-mining Estimates

(for Problem 4)

(a) Estimate the total number of square miles defaced in extracting the next 4 million tons of coal obtained in this country by strip-mining. Draw 4 rectangles under the curve and compute their area.

(b) Re-estimate the number of square miles defaced using rectangles above the curve.

(c) Use your answers to (a) and (b) to get a better estimate of the actual number of square miles defaced.

5. The number of hours, H, of daylight in Madrid, as a function of date is approximated by the formula

$$H = 12 + 2.4 \sin (0.0172(t - 80)),$$

where t is the number of days since the start of the year. Find the average number of hours of daylight in Madrid:

(a) in January

(b) in June

(c) over a whole year

(d) Comment on the relative magnitudes of your answers. Why are they reasonable?

6. The gravitational force F, on a rocket of mass m kg at a distance r km from the surface of the earth is given by

$$F = \frac{GMm}{(6400 + r)^2},$$

where M is the mass of the earth ($\approx 6 \times 10^{24}$ kg), and G is called the gravitational constant, whose value is about 6.7×10^{-17} if force is measured in newtons. What is the average force on a rocket of mass 12,000 kg between sea level and 1000 km above the surface of the earth?

7. For the function f graphed in Figure 3.20:

(a) Suppose you know $\int_0^2 f(x)\,dx$. What is $\int_{-2}^2 f(x)\,dx$?

(b) Suppose you know $\int_0^5 f(x)\,dx$ and $\int_2^5 f(x)\,dx$. What is $\int_0^2 f(x)\,dx$?

(c) Suppose you know $\int_{-2}^5 f(x)\,dx$ and $\int_{-2}^2 f(x)\,dx$. What is $\int_0^5 f(x)\,dx$?

8. For the function f graphed in Figure 3.20:

(a) Suppose you know $\int_{-2}^2 f(x)\,dx$ and $\int_0^5 f(x)\,dx$. What is $\int_2^5 f(x)\,dx$?

(b) Suppose you know $\int_{-2}^5 f(x)\,dx$ and $\int_{-2}^0 f(x)\,dx$. What is $\int_2^5 f(x)\,dx$?

(c) Suppose you know $\int_2^5 f(x)\,dx$ and $\int_{-2}^5 f(x)\,dx$. What is $\int_0^2 f(x)\,dx$.

9. For the function f in Figure 3.20, write an expression involving one or more definite integrals that denote(s):

(a) The average value of f for $0 \le x \le 5$.

(b) The average value of $|f|$ for $0 \le x \le 5$.

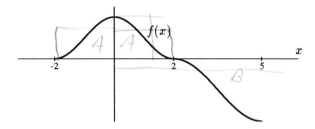

Figure 3.20: Use symmetry to express certain integrals of f in terms of others.
(for Problems 7–10)

10. For the function f in Figure 3.20, consider the average value of f over the following intervals:

    I: $0 \leq x \leq 1$ II: $0 \leq x \leq 2$ III: $0 \leq x \leq 5$ IV: $-2 \leq x \leq 2$

    (a) For which interval is the average value of f least?
    (b) For which interval is the average value of f greatest?
    (c) For which pair of intervals are the average values equal?

11. If you jump out of an airplane and your parachute fails to open, your downward velocity t seconds after the jump is approximated by

    $$v(t) = \frac{g}{k}(1 - e^{-kt}) \quad \text{where} \quad g = 9.8 \, \frac{\text{m}}{\text{sec}^2} \quad \text{and} \quad k = 0.2 \, \text{sec}^{-1}.$$

    (a) Write an expression for the distance you fall in T seconds.
    (b) If you jump from 5000 meters above the ground, for approximately how many seconds do you fall before hitting the ground?

3.4   The Fundamental Theorem of Calculus

In Chapter 2, we saw that a derivative could be thought of as a rate of change. Thus

$$F'(t) = \text{rate of change of } F(t) \text{ with respect to } t.$$

On the other hand, in this chapter we have seen that the definite integral of a rate of change of some quantity is the total change in that quantity. Thus

$$\begin{aligned} \int_a^b F'(t) \, dt &= \int_a^b (\text{rate of change of } F(t)) \, dt \\ &= (\text{Total change in } F(t) \text{ between } a \text{ and } b) \\ &= F(b) - F(a). \end{aligned}$$

We have just derived an important result:

The Fundamental Theorem of Calculus.
If $f = F'$, then

$$\int_a^b f(t) \, dt = F(b) - F(a).$$

The Fundamental Theorem is often used when the function f is known, and we want to learn about the unknown function F. The theorem enables us to reconstruct the function F from knowledge about its derivative $F' = f$.

❒ **Example 1**   Suppose you are given that $F'(t) = t \cos t$ and $F(0) = 2$. Find the values of $F(b)$ at the points $b = 0, 0.1, 0.2, \ldots, 1.0$.

Solution. We apply the Fundamental Theorem with $f(t) = t \cos t$ and $a = 0$ to get values for $F(b)$. Since

$$F(b) - F(0) = \int_0^b F'(t) \, dt = \int_0^b t \cos t \, dt$$

and $F(0) = 2$, we have

$$F(b) = 2 + \int_0^b t \cos t \, dt.$$

Using Riemann sums to estimate the definite integral $\int_0^b t \cos t \, dt$ for each of the values $b = 0, 0.1$, $0.2, \ldots, 1.0$ gives the following approximate values for F:

b	0	0.1	0.2	0.3	0.4	0.5	0.6	0.7	0.8	0.9	1.0
$F(b)$	2.000	2.005	2.020	2.044	2.077	2.117	2.164	2.216	2.271	2.327	2.382

Notice from the table that the function $F(b)$ is increasing between $b = 0$ and $b = 1$. This could have been predicted without the use of the Fundamental Theorem from the fact that the derivative, $t \cos t$, of $F(t)$ is positive for t between 0 and 1.   ❒

❒ **Example 2**   The graph of the derivative F' of some function F is given in Figure 3.21. If you are told that $F(20) = 150$, determine the maximum value attained by F.

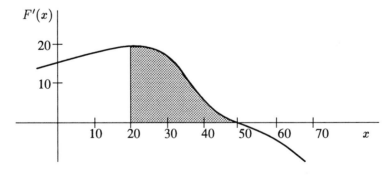

Figure 3.21: Graph of the Derivative F' of some function F

Solution. Let's begin by getting a rough idea of how F behaves. We know that $F(x)$ increases for $x < 50$ because the derivative of F is positive for $x < 50$. Similarly, $F(x)$ decreases for $x > 50$ because $F'(x)$ is negative for $x > 50$. Therefore the graph of F rises until the point at which $x = 50$, then it begins to fall. Evidently the highest point on the graph of F is at $x = 50$, and so the maximum value attained by F is $F(50)$. To evaluate $F(50)$, we use the Fundamental Theorem:

$$F(50) - F(20) = \int_{20}^{50} F'(x)\, dx,$$

which gives

$$F(50) = F(20) + \int_{20}^{50} F'(x)\, dx = 150 + \int_{20}^{50} F'(x)\, dx.$$

The definite integral equals the area of the shaded region under the graph of F', which we can guess is roughly 300. Therefore, the greatest value attained by F is $F(50) \approx 150 + 300 = 450$. ❏

❑ **Example 3**   Verify the Fundamental Theorem using $F(x) = 3x + 1$ and assuming $b > a > 0$.

Solution. The slope of the graph of $F(x) = 3x + 1$ is 3, so $F'(x) = 3$ everywhere. Now we want to show that

$$\int_a^b f(x)\, dx = F(b) - F(a)$$

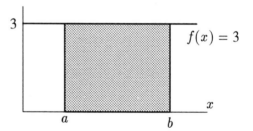

Figure 3.22: Area representing $\int_a^b 3\, dx$

where $f(x) = F'(x) = 3$.

Realize that $\int_a^b f(x)\, dx$ is the area under the graph of f between $x = a$ and $x = b$, shown in Figure 3.22. Since this region is rectangular with a width of $(b - a)$ and a height of 3:

$$\int_a^b f(x)\, dx = \int_a^b 3\, dx = 3(b - a).$$

On the other hand, we have

$$F(b) - F(a) = (3b + 1) - (3a + 1) = 3b - 3a = 3(b - a) = \int_a^b f(x)\, dx.$$

❏

☐ **Example 4** Biological activity in a pond is reflected in the rate at which carbon dioxide (CO_2) is added to or withdrawn from the water. Plants take CO_2 out of the water during the day for photosynthesis and put CO_2 into the water at night. Animals put CO_2 into the water all the time as they breathe. Biologists are interested in how the net rate at which CO_2 enters or leaves a pond varies during the day. Figure 3.23 shows this rate[1] as a function of time of day. The rate is measured in millimoles (mM) of CO_2 per liter of water per hour; time is measured in hours past dawn. At dawn, there were 2.600 mM of CO_2 per liter of water.

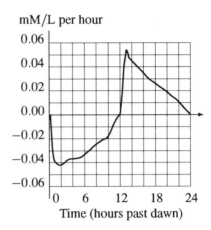

mM/L per hour

Figure 3.23: Rate CO_2 is entering/leaving water

(a) What can be concluded from the fact that the rate is negative during the day and positive at night?

(b) Some scientists have suggested that plants and animals respire (breathe) at a constant rate at night, and that plants photosynthesize at a constant rate during the day. Does Figure 3.23 support this view?

(c) When was the CO_2 content of the water at its lowest? How low did it go?

(d) How much CO_2 was released into the water during the twelve hours of darkness? Compare this quantity with the amount of CO_2 withdrawn from the water during the twelve hours of daylight. How can you tell by looking at the graph whether the CO_2 in the pond is in equilibrium?

(e) Estimate the CO_2 content of the water at 3-hour intervals throughout the day. Use your estimates to plot a graph of CO_2 content throughout the day.

Solution.

(a) CO_2 is being taken out of the water during the day, and returned at night. The pond must therefore contain some plants. (The data is in fact from pond water containing both plants and animals.)

(b) The graph in Figure 3.23 shows that the CO_2 content changes at a greater rate for the first 6 hours of daylight, $0 < t < 6$, than it does for the final 6 hours of daylight, $6 < t < 12$. It turns out that plants photosynthesize more vigorously in the morning than in the afternoon. Similarly, CO_2 content changes more rapidly in the first half of the night, $12 < t < 18$, than in the 6 hours just before dawn, $18 < t < 24$. The reason seems to be that at night plants quickly use up most of the sugar that they synthesized during the day, and then their respiration rate is inhibited.

---

[1]Data from *The Pattern of Photosynthesis and Respiration in Laboratory Microsystems*, by R.J. Beyers (Mem. 1st. Ital. Idrobiol. 1965.)

(c) We are now being asked about the total quantity of CO_2 in the pond, rater than the rate at which it is changing. We will let $f(t)$ denote the CO_2 content of the pond water (in mM/L) at t hours past dawn. Then Figure 3.23 is a graph of the derivative $f'(t)$. There are 2.600 mM/L of CO_2 in the water at dawn, so $f(0) = 2.600$.

The CO_2 content $f(t)$ decreases during the 12 hours of daylight, $0 < t < 12$, when $f'(t) < 0$, and then $f(t)$ increases for the next 12 hours. Thus $f(t)$ is at a minimum when $t = 12$, at dusk. By the Fundamental Theorem,

$$f(12) = f(0) + \int_0^{12} f'(t)\, dt = 2.600 + \int_0^{12} f'(t)\, dt.$$

We must approximate the definite integral by a Riemann sum. Use the graph in Figure 3.23 to make a brief table of values of the function $f'(t)$:

t	$f'(t)$	t	$f'(t)$	t	$f'(t)$	t	$f'(t)$	t	$f'(t)$	t	$f'(t)$
0	0.000	4	−0.039	8	−0.026	12	0.000	16	0.035	20	0.020
1	−0.042	5	−0.038	9	−0.023	13	0.054	17	0.030	21	0.015
2	−0.044	6	−0.035	10	−0.020	14	0.045	18	0.027	22	0.012
3	−0.041	7	−0.030	11	−0.008	15	0.040	19	0.023	23	0.005

The left Riemann sum with $n = 12$ terms, corresponding to $\Delta t = 1$, gives:

$$\int_0^{12} f'(t)\, dt \approx (0.000)(1) + (-0.042)(1) + (-0.044)(1) + \cdots + (-0.008)(1) = -0.346.$$

Therefore, at 12 hours past dawn, the CO_2 content of the pond water reaches its lowest level, which is approximately

$$2.600 - 0.346 = 2.254 \text{ mM per liter of water.}$$

(d) The increase in CO_2 during the 12 hours of darkness equals

$$f(24) - f(12) = \int_{12}^{24} f'(t)\, dt.$$

Using Riemann sums to estimate this integral, we find that about 0.306 mM/L of CO_2 was released into the pond during the night. In part (c) we calculated that about 0.346 mM/L of CO_2 was absorbed from the pond during the day. If the pond was in equilibrium, we would expect the daytime absorption to equal the nighttime release. These quantities are sufficiently close (0.346 and 0.306) that the difference is probably due to measurement error.

If the pond is in equilibrium, the area between the rate curve in Figure 3.23 and the axis for $0 \le t \le 12$ will equal the area between the rate curve and the axis for $12 \le t \le 24$. (The axis is the horizontal line at 0.00.) In this experiment the areas do look approximately equal.

(e) We must evaluate

$$f(b) = f(0) + \int_0^b f'(t)\,dt = 2.600 + \int_0^b f'(t)\,dt$$

for the values $b = 0, 3, 6, 9, 12, 15, 18, 21, 24$. Left Riemann sums with $\Delta t = 1$ give the following values:

b	0	3	6	9	12	15	18	21	24
$f(b)$	2.600	2.514	2.396	2.305	2.254	2.353	2.458	2.528	2.560

A graph of the CO_2 content against time is in Figure 3.24.

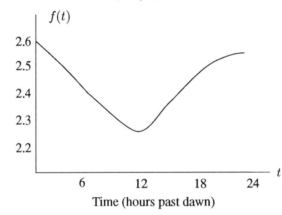

Carbon dioxide content (mM/L)

Figure 3.24: Carbon dioxide content in pond water throughout a day.

❏

❖ Exercises for Section 3.4

1. If the graph of f is as in the figure to the right and $F' = f$ and $F(0) = 0$, find $F(b)$ for $b = 1, 2, 3, 4, 5, 6$.

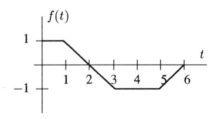

For Problems 2–4, suppose $F(0) = 0$ and $F'(\theta) = f(\theta) = \sin(\theta^2)$ for $0 \le \theta \le 2.5$.

2. Approximate $F(b)$ for $b = 0$, 0.5, 1, 1.5, 2, 2.5.

3. Using a graph of F', decide where F is increasing and where F is decreasing.

4. Does F have a maximum value for $0 \le \theta \le 2.5$? If so, at what value of θ does it occur, and approximately what is that maximum value?

5. If $F(t) = t(\ln t) - t$, it can be shown that $f(t) = F'(t) = \ln t$. Find $\displaystyle\int_{10}^{12} \ln t \, dt$ two ways:

    (a) Using left and right sums. (Approximate to one decimal place accuracy.)

    (b) Using the Fundamental Theorem of Calculus.

6. (a) Let $F(x) = \sin x$. Use graphs of $\sin x$ and $\cos x$ to explain why it is reasonable that $F'(x) = \cos x$.

    (b) Now find $\displaystyle\int_0^{\frac{\pi}{2}} \cos x \, dx$ two ways:

    i. Using left and right sums. (Approximate to one decimal place accuracy.)
    ii. Using the Fundamental Theorem of Calculus.

7. Verify the Fundamental Theorem of Calculus using $f(x) = 2$ and $F(x) = 2x$ by doing the following:

    (a) Finding $F'(x)$ and showing $F' = f$.

    (b) Interpreting $\displaystyle\int_a^b f(x) \, dx$ as an area, and finding that area using geometry.

    (c) Confirming that $\displaystyle\int_a^b f(x) \, dx = F(b) - F(a)$.

8. Same question as in Problem 7, but with $f(x) = 2x$ and $F(x) = x^2$.

For Problems 9–12, mark the following quantities on a copy of the graph of f in Figure 3.25.

9. A length representing $f(b) - f(a)$.

10. A slope representing $\dfrac{f(b) - f(a)}{b - a}$.

11. An area representing $F(b) - F(a)$, where $F' = f$.

12. A length roughly approximating $\dfrac{F(b) - F(a)}{b - a}$, where $F' = f$.

For Problems 13–15, assuming $F' = f$, mark the following quantities on a copy of the graph of F in Figure 3.26.

13. A slope representing $f(a)$.

14. A length representing $\displaystyle\int_a^b f(x)\, dx$.

15. A slope representing $\dfrac{1}{b - a}\displaystyle\int_a^b f(x)\, dx$.

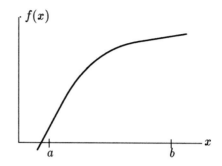

Figure 3.25: Mark these quantities on the graph
(for Problems 9–12)

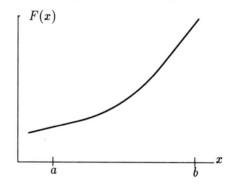

Figure 3.26: Mark these quantities on the graph
(for Problems 13–15)

16. A bicyclist is pedaling along a straight road with velocity, v, given in the figure to the right. Suppose the cyclist starts 5 miles from a lake, and that positive velocities take her away from the lake, negative towards. When is the cyclist furthest from the lake and how far away is she then?

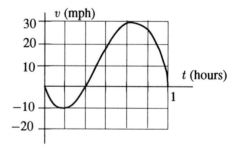

17. In Section 2.1, page 124, we defined the average velocity of a particle over the interval $a < t < b$ as the change in position during this interval divided by $(b - a)$. If the velocity of the particle is $v = f(t)$, the average velocity is also given by $\dfrac{1}{b - a}\displaystyle\int_a^b f(t)\, dt$. Show why these two ways of calculating the average velocity always lead to the same answer.

3.5   Further Notes on the Limit

When we defined the definite integral, we introduced a limit of the type

$$\lim_{n \to \infty}$$

This limit is different from the limit

$$\lim_{h \to 0}$$

that we used in the definition of the derivative. What is the difference? If we take the limit of a function as $h \to 0$, we are looking for a value approached by the function as $h = 0.1, 0.01, 0.001,$ $0.0001, \ldots$ (as well as $-0.1, -0.01, -0.001, -0.0001, \ldots$). In the case of the limit as $n \to \infty$, we're interested in the behavior of the function for values of n such as $n = 10^6, 10^{10}, 10^{100}, 10^{1000},$ $\ldots$ (or, for that matter, $n = 3 \times 10^{7700}$, $n = 6.9 \times 10^{9628}$). In other words, we want to know what happens to the function as n becomes arbitrarily large.

One technical difference between taking the limit as $h \to 0$ and $n \to \infty$ is that you needn't examine the limit at ∞ from "both sides"; it makes no sense to think of integers "larger than ∞", but we certainly can look at numbers approaching 0 (or any finite number) from either side.

More precisely, we say that the limit of $f(n)$ as $n \to \infty$ is L if we can make the value of f as close to L as we like by making n larger and larger. Here's an example:

$$\lim_{n \to \infty} \frac{1}{n}.$$

It's not hard to see that the value of the limit is $L = 0$: you can make $\frac{1}{n}$ as close to 0 as you like by taking a sufficiently large n. For example, to make $\frac{1}{n}$ smaller than $\frac{1}{1000}$, take $n > 1000$. To make $\frac{1}{n}$ smaller than 0.0000000001 take $n > 10^{10}$. And so on.

Here's another example:

$$\lim_{n \to \infty} n^2.$$

As n increases in size, the function n^2 doesn't approach any particular number—it too increases in size. In this case, we'd have to say that n^2 has no limit as $n \to \infty$, or that n^2 increases without bound.

Examples like $\lim_{n \to \infty} n^2$ rarely arise when using the limit to calculate a definite integral. Usually you can compute the value of the definite integral as accurately as you wish by taking sufficiently large values for n. Roughly, the reason that the sum

$$\sum_{i=1}^{n} f(x_i) \, \Delta x$$

doesn't increase without bound as n gets larger even though you're adding together more and more terms, is that the terms are getting smaller and smaller.

The limit is important because it allows us to avoid the philosophical problems of infinite sums. You might, at first, want to define the definite integral as an infinite sum of infinitesimally small terms. But what would it mean to add up infinitely many numbers? The Riemann sum definition of the definite integral as the limit of a finite sum, with the number of terms tending to infinity, avoids this problem.

❖ Exercises for Section 3.5

For Problems 1-3, use a graph to find the limit.

1. $\lim\limits_{t \to \infty} te^{-t}$

2. $\lim\limits_{t \to \infty} t^3 e^{-t}$

3. $\lim\limits_{x \to \infty} \dfrac{\sin x}{x}$.

4. (a) Using a calculator or computer, plot the graph of

$$y = \left(1 + \frac{1}{x}\right)^x \quad \text{for } x > 0.$$

(b) By extending the domain of your graph, investigate the value of y as x becomes very large; in other words, find $\lim\limits_{x \to \infty} \left(1 + \dfrac{1}{x}\right)^x$.

(c) You should recognize your answer to (b). What is it?

5. (a) What is wrong with the following "proof"?

$$\lim\limits_{n \to \infty} \left(1 + \frac{1}{n}\right) = 1$$

so

$$\lim\limits_{n \to \infty} \left(1 + \frac{1}{n}\right)^n = \lim\limits_{n \to \infty} 1^n = 1.$$

(b) Explain why, if you take n large enough (for example, $n = 10^{20}$ or more), your calculator thinks that

$$\lim\limits_{n \to \infty} \left(1 + \frac{1}{n}\right)^n = 1.$$

6. The Fibonacci sequence has been famous for many years. It is defined as follows: $F_0 = 0$, $F_1 = 1$, $F_2 = F_1 + F_0$, $F_3 = F_2 + F_1, ..., F_{n+1} = F_n + F_{n-1}, ...$ Work out and write in a column the Fibonacci numbers through F_{20}. Then, to the right of F_n, write the value of F_n/F_{n+1} evaluated to seven decimal places. As you work your way down the column, try to guess the value of each entry before you find it with your calculator. Work out the values of $(F_{n+1}/F_{n+2}) - (F_n/F_{n+1})$ as fractions for the first few values of n. What do you see? Is that consistent with what you noticed as you were writing the decimal values? Use algebra to see if you can prove that what you have noticed will indeed go on as n goes to infinity. (Fibonacci was the pseudonym of Leonardo of Pisa, a mathematician who died about 1240.)

3.6   Miscellaneous Exercises for Chapter 3

For Problems 1–6, find the integrals to one decimal place. In each case say how many subdivisions you used.

1. $\displaystyle\int_0^{10} 2^{-x}\,dx$

2. $\displaystyle\int_0^{\frac{\pi}{2}} \sin(2\theta^2 + 1)\,d\theta$

3. $\displaystyle\int_0^1 \sqrt{1 + \cos^2 t}\,dt$

4. $\displaystyle\int_{-1}^1 \frac{x^2 + 1}{x^2 - 4}\,dx$

5. $\displaystyle\int_2^3 -\frac{1}{(r+1)^2}\,dr$

6. $\displaystyle\int_0^1 \arcsin z\,dz$

7. Statisticians often use values of the function

$$F(b) = \int_0^b e^{-x^2}\,dx.$$

   (a) What is $F(0)$?

   (b) Does the value of F increase or decrease as b increases?

   (c) Estimate $F(1)$, $F(2)$, $F(3)$.

8. If $F(x) = e^{x^2}$, it can be shown that $f(x) = F'(x) = 2xe^{x^2}$.

   Find $\displaystyle\int_0^1 2xe^{x^2}\,dx$ two ways:

   (a) Using left and right sums. (Approximate to one decimal place.)

   (b) Using the Fundamental Theorem of Calculus.

9. If $F(t) = \frac{1}{2}\sin^2 t$, it can be shown that $f(t) = F'(t) = \sin t \cos t$.

   Find $\displaystyle\int_{0.2}^{0.4} \sin t \cos t\,dt$ two ways:

   (a) Using left and right sums. (Approximate to one decimal place.)

   (b) Using the Fundamental Theorem of Calculus.

10. A car accelerates smoothly from 0 to 60 mi/hr in 10 seconds.

    (a) How often would you have to measure its velocity to be able to determine within one foot the total distance the car travels in the 10 second period?

    (b) Suppose the car's velocity as a function of time is given in Figure 3.27. Estimate how far the car travels during the 10 second period. Include in your estimate a statement of the possible error.

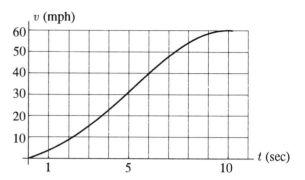

Figure 3.27: Velocity of accelerating car.
(for Problem 10)

11. The Environmental Protection Agency was recently asked to investigate a spill of radioactive iodine. Measurements showed the ambient radiation levels at the site to be four times the maximum acceptable limit, so the EPA ordered an evacuation of the surrounding neighborhood.

    It is known that the level of radiation from an iodine source decreases according to the formula

    $$R(t) = R_0 e^{-0.004t}$$

    where R = the radiation level (in millirems/hour) at time t, R_0 is the initial radiation level (at $t = 0$), and t is the time measured in hours.

    (a) How long will it take for the site to reach a acceptable level of radiation?

    (b) How much total radiation (in millirems) will have been given off by that time, assuming the maximum acceptable limit is 0.6 millirems/hour?

12. Each of the graphs in Figure 3.28 represents the velocity v of a particle moving along the x-axis for time $0 \leq t \leq 5$. The vertical scales of all graphs are the same. During this time interval, which particle(s)

    (a) has a constant acceleration?

    (b) ends up to the left of where it started?

    (c) ends up the furthest from its starting point

    (d) experiences the greatest instantaneous acceleration?

    (e) has the greatest average velocity?

(f) has the greatest average acceleration?

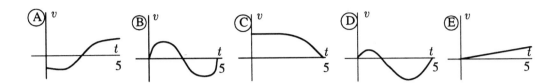

Figure 3.28: Velocity Graphs

13. Assume the population, P, of Mexico in millions, is given by

$$P = 67.38(1.026)^t,$$

where t is in years since 1980.

(a) What was the average population of Mexico between 1980 and 1990?

(b) What is the average of the population of Mexico in 1980 and the population in 1990?

(c) Explain, in terms of the concavity of the graph of P (see Figure 1.16 on page 20), why your answer to (b) is larger or smaller than your answer to (a).

14. How long is one arch of the sine curve (i.e., that portion of the graph of $y = \sin x$ from $x = 0$ to $x = \pi$)? Use a calculator and common sense to get an approximate answer. (You are not expected to use a formula for arclength.)

15. The Montgolfier brothers (Joseph and Etienne) were 18th Century pioneers in the field of hot-air ballooning. Had they had the appropriate instruments, they might have left us a record of one of their early experiments, like that shown in Figure 3.29. The graph is of their vertical velocity, v, with upwards as positive.

(a) Over what intervals was the acceleration positive? Negative? Zero?

(b) What was the greatest altitude achieved, and at what time?

(c) At what time was the magnitude of the acceleration greatest? What was the direction of their acceleration at that time?

(d) What might have happened during this flight to explain the answer to (c)?

(e) This particular flight ended on top of a hill. How do you know that it did, and what was the height of the hill above the starting point?

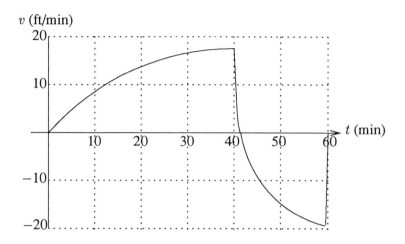

Figure 3.29: Flying with the Montgolfiers

(for Problem 15)

16. A mouse moves back and forth in a tunnel, attracted to bits of Camembert cheese alternatively introduced to and removed from the ends (right and left) of the tunnel. The graph of the mouse's velocity is given in Figure 3.30. Assuming that the mouse starts $(t = 0)$ at the center of the tunnel, use the graph to estimate the time(s) at which:

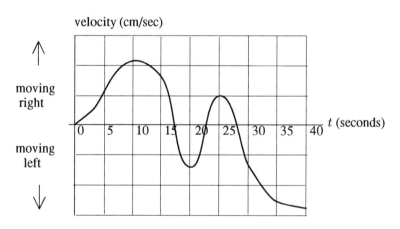

Figure 3.30: The famous Camembert Cheese Experiment

(for Problem 16)

(a) The mouse changes direction.

(b) The mouse is moving most rapidly to the right; to the left.

(c) The mouse is farthest to the right of center; farthest to the left.

(d) The mouse's speed (i.e. the magnitude of its velocity) is decreasing.

(e) The mouse is at the center of the tunnel.

17. The graph of some function f is given in Figure 3.31. List, from *smallest* to *largest*.

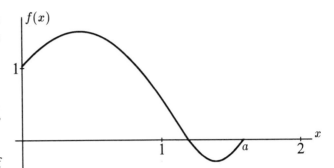

Figure 3.31: $y = f(x)$

(for Problem 17)

(a) $f'(1)$.

(b) The average value of $f(x)$, $0 \leq x \leq a$.

(c) The average value of the rate of change in $f(x)$, for $0 \leq x \leq a$.

(d) $\displaystyle\int_0^a f(x)\,dx$.

Chapter 4

SHORT-CUTS TO DIFFERENTIATION

In this chapter we make a systematic study of the derivatives of functions given by formulas. The chapter contains two kinds of results: the derivatives of specific functions, such as exponentials and the sine and cosine, and general rules, such as the product, quotient and chain rules, which allow us to differentiate combinations of functions.

4.1   Formulas for Derivative Functions

In Chapter 2, we defined the derivative

$$f'(x) = \lim_{h \to 0} \frac{f(x+h) - f(x)}{h}$$

and saw how the derivative represented a slope and a rate of change. We also saw how to find the derivative of a function given by a graph (by estimating the slope of the tangent at each point) and how to estimate the derivative of a function given by a table (by finding the average rate of change of the function between data values).

Useful Notation: We will write $\frac{d}{dx}(x^3)$, for example, to mean the derivative of x^3. Similarly, $\frac{d}{d\theta}\left(\sin(\theta^2)\right)$ denotes the derivative of $\sin(\theta^2)$ with θ as the variable.

Derivative of a Constant Function

The graph of a constant function $f(x) = c$ is a horizontal line, with slope 0 everywhere. Therefore, its derivative is 0 everywhere. See Figure 4.1.

$$\boxed{\text{If } f(x) = c, \text{then } f'(x) = 0}$$

For example: $\frac{d}{dx}(5) = 0$, $\frac{d}{dx}(\pi) = 0$.

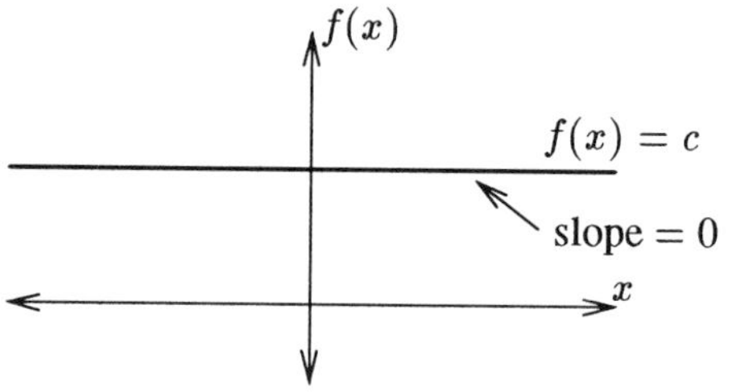

Figure 4.1: A Constant Function

Derivative of a Linear Function

We already know that the slope of a straight line is constant. This tells us that the derivative of a linear function is constant.

$$\boxed{\text{If } f(x) = b + mx, \text{then } f'(x) = \text{slope} = m.}$$

For example: $\frac{d}{dx}(5 - \frac{3}{2}x) = -\frac{3}{2}$.

Alternatively, we can use the definition of the derivative:

$$\begin{aligned} f'(x) &= \lim_{h \to 0} \frac{f(x+h) - f(x)}{h} \\ &= \lim_{h \to 0} \frac{(b + m(x+h)) - (b + mx)}{h} = \lim_{h \to 0} \frac{mh}{h} = \lim_{h \to 0} m \quad (\text{since } h \neq 0) \\ &= m. \end{aligned}$$

To find the second derivative of a linear function, we must differentiate m, which is a constant. Thus, if f is linear,

$$f''(x) = 0.$$

So as we would expect for a straight line, f is neither concave up nor concave down.

Derivative of a Constant times a Function

In Figure 4.2, you see the graph of $y = f(x)$ and of three multiples: $y = 3f(x)$, $y = \frac{1}{2}f(x)$, and $y = -2f(x)$. What is the relationship between the derivatives of these functions? In other words, for a particular x value, how are the slopes of these graphs related?

Multiplying by a constant "stretches" or "shrinks" the graph (and "flips" it over the x-axis if the constant is negative). The zeros of the function remain the same and the "peaks" and "valleys" occur at the same x values. What does change is the slope of the curve at each point. If the graph has been stretched, the "rises" have all been increased by the same factor, while the "runs" remain

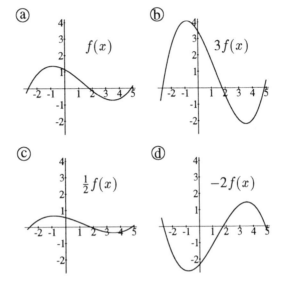

Figure 4.2: A Function and its Multiples

the same. Thus, the slopes are all steeper by the same factor. If the graph has been shrunk, the slopes are all smaller by the same factor. If the graph has been flipped over the x-axis, the slopes will all have their signs reversed. In other words, if a function is multiplied by a constant, c, so is its derivative:

$$\frac{d}{dx}(cf(x)) = cf'(x).$$

This result can be obtained algebraically, too:

$$\frac{d}{dx}(cf(x)) = \lim_{h \to 0} \frac{cf(x+h) - cf(x)}{h} = \lim_{h \to 0} c\,\frac{f(x+h) - f(x)}{h}$$

$$= c \lim_{h \to 0} \frac{f(x+h) - f(x)}{h} = cf'(x).$$

You may wonder why c can be taken across the limit sign: the reason is that c is a constant. If a function's value gets close to a certain number, c times that function gets close to c times that number.

Derivatives of Sums and Differences

Suppose we have two functions $f(x)$ and $g(x)$ with values in Table 4.1. Values of the sum $f(x)+g(x)$ are in the same table.

You can see that when the increments of $f(x)$ and the increments of $g(x)$ are added together they give the increments of $f(x) + g(x)$. For example, as x increases from 0 to 1, $f(x)$ increases by 10 and $g(x)$ increases by 0.2, while $f(x) + g(x)$ increases by $110.2 - 100 = 10.2$. Similarly, as x increases from 5 to 6, $f(x)$ increases by 60 and $g(x)$ by 0.2, while $f(x) + g(x)$ increases by $311.2 - 251.0 = 60.2$.

x	$f(x)$	$g(x)$	$f(x) + g(x)$
0	100	0	100
1	110	0.2	110.2
2	130	0.4	130.4
3	160	0.6	160.6
4	200	0.8	200.8
5	250	1.0	251.0
6	310	1.2	311.2
7	380	1.4	381.4

Table 4.1: Sum of Functions

From this example, you can see that the rate at which $f(x) + g(x)$ is increasing is the sum of the rates at which $f(x)$ and $g(x)$ are increasing. Stating this with derivatives:

$$\boxed{\frac{d}{dx}(f(x) + g(x)) = f'(x) + g'(x).}$$

Similarly for the difference:

$$\boxed{\frac{d}{dx}(f(x) - g(x)) = f'(x) - g'(x).}$$

Let's justify the sum rule more precisely using the definition of the derivative:

$$\frac{d}{dx}(f(x) + g(x)) = \lim_{h \to 0} \frac{(f(x + h) + g(x + h)) - (f(x) + g(x))}{h}$$

$$= \lim_{h \to 0} \left[\underbrace{\frac{f(x + h) - f(x)}{h}}_{\text{limit of this is } f'(x)} + \underbrace{\frac{g(x + h) - g(x)}{h}}_{\text{limit of this is } g'(x)} \right]$$

$$= f'(x) + g'(x).$$

You may have noticed that we have used the fact that the limit of a sum is the sum of the limits—a fact which is true, although we have not proved it. (Can you see why it is true?)

❖ Exercises for Section 4.1

1. Let $f(x) = -3x + 2$ and $g(x) = 2x + 1$.

   (a) If $k(x) = f(x) + g(x)$, find a formula for $k(x)$ and verify the sum rule by comparing $k'(x)$ with $[f'(x) + g'(x)]$.

   (b) If $j(x) = f(x) - g(x)$, determine a formula for $j(x)$ and compare $j'(x)$ with $[f'(x) - g'(x)]$.

2. Let $r(t) = 2t - 4$. If $s(t) = 3r(t)$, verify the constant rule by showing that $s'(t) = 3[r'(t)]$.

3. (a) If $f(x) = 5x - 3$, $g(x) = -2x + 1$, find the derivative of $f(g(x))$.

    (b) Based on your answer to part (a), make a conjecture about the derivative of the composition of two linear functions. Explain why what you say is true for any two linear functions.

4. If $r(x) = f(x) + 2g(x) + 3$, and $f'(x) = g(x)$ and $g'(x) = r(x)$, express

    (a) $r'(x)$ in terms of $f(x)$ and $g(x)$.

    (b) $f'(x)$ in terms of $f(x)$ and $r(x)$.

5. State in your own words why the limit of a sum is the sum of the limits.

4.2 Powers and Polynomials

Positive Integral Powers of x

First, let's look at what the graphs can tell us about the derivative of $f(x) = x^n$, with n a positive integer. We'll start with $n = 2$ and $n = 3$. The graphs of $f(x) = x^2$ and $g(x) = x^3$ are in Figure 4.3.

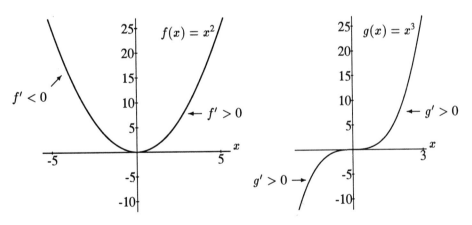

Figure 4.3: Graphs of $f(x) = x^2$ and $g(x) = x^3$

Derivative of x^2 and x^3

The shape of the graph of $f(x) = x^2$ in Figure 4.3 shows that its derivative is negative when $x < 0$, zero when $x = 0$, and positive for $x > 0$. The graph of $g(x) = x^3$, on the other hand, suggests that its derivative is positive when $x < 0$, is zero when $x = 0$, and becomes positive again when $x > 0$.

We'll calculate the derivative of $f(x) = x^2$ using the definition:

$$
\begin{aligned}
f'(x) &= \lim_{h \to 0} \frac{f(x+h) - f(x)}{h} = \lim_{h \to 0} \frac{(x+h)^2 - x^2}{h} \\
&= \lim_{h \to 0} \frac{x^2 + 2xh + h^2 - x^2}{h} = \lim_{h \to 0} \frac{2xh + h^2}{h} \\
&= \lim_{h \to 0} \frac{h(2x + h)}{h}.
\end{aligned}
$$

To take the limit, we look at what happens when h is close to 0, but we do not let $h = 0$. Therefore we can divide by h and say

$$
f'(x) = \lim_{h \to 0} \frac{h(2x+h)}{h} = \lim_{h \to 0} (2x + h) = 2x
$$

because as h gets close to zero, $2x + h$ gets close to $2x$. So

$$
f'(x) = \frac{d}{dx}(x^2) = 2x.
$$

Now $f'(x) = 2x$ has the behavior we expected: it is negative for $x < 0$, zero when $x = 0$, and positive for $x > 0$. The graph of f' is in Figure 4.4, along with the graph of g'.

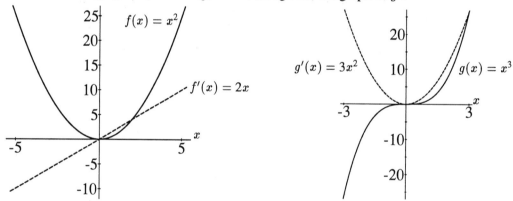

Figure 4.4: Derivatives of $f(x) = x^2$ and $g(x) = x^3$

☐ **Example 1** Find the derivative of $g(x) = x^3$.

Solution.

$$
\begin{aligned}
g'(x) &= \lim_{h \to 0} \frac{g(x+h) - g(x)}{h} = \lim_{h \to 0} \frac{(x+h)^3 - x^3}{h} \\
\text{multiplying out} \longrightarrow \quad &= \lim_{h \to 0} \frac{x^3 + 3x^2h + 3xh^2 + h^3 - x^3}{h} \\
&= \lim_{h \to 0} \frac{3x^2h + 3xh^2 + h^3}{h} \\
\text{dividing by } h \longrightarrow \quad &= \lim_{h \to 0} (3x^2 + 3xh + h^2) = 3x^2
\end{aligned}
$$

$$\nwarrow \text{looking at what happens as } h \to 0$$

so
$$g'(x) = \frac{d}{dx}(x^3) = 3x^2.$$

Again, $g'(x) = 3x^2$ is zero when $x = 0$, but positive everywhere else. $\square$

Further Examples

Similar calculations will show you that

$$\frac{d}{dx}(x^4) = 4x^3, \quad \frac{d}{dx}(x^5) = 5x^4,$$

and so on. These results might well lead you to conjecture that, for a positive integer n,

$$\boxed{\frac{d}{dx}(x^n) = nx^{n-1}}$$

and indeed this turns out to be true.

Justification of $\dfrac{d}{dx}(x^n) = nx^{n-1}$, for n a Positive Integer

To calculate the derivatives of x^2 and x^3, we had to multiply out $(x+h)^2$ and $(x+h)^3$. Here we must expand $(x+h)^n$. Let's look back at the previous expansions:

$$(x+h)^2 = x^2 + 2xh + h^2, \qquad (x+h)^3 = x^3 + 3x^2h + 3xh^2 + h^3,$$

and multiply out a few more examples:

$$
\begin{aligned}
(x+h)^4 &= x^4 + 4x^3h + 6x^2h^2 + 4xh^3 + h^4 \\
(x+h)^5 &= x^5 + 5x^4h + \underbrace{10x^3h^2 + 10x^2h^3 + 5xh^4 + h^5}_{\text{terms involving } h^2 \text{ and higher powers of } h}
\end{aligned}
$$

In general, we can say

$$(x+h)^n = x^n + nx^{n-1}h + \underbrace{\cdots + h^n}_{\text{terms involving } h^2 \text{ and higher powers of } h}.$$

Now to find the derivative

$$
\begin{aligned}
\frac{d}{dx}(x^n) &= \lim_{h\to 0}\frac{(x+h)^n - x^n}{h} \\
&= \lim_{h\to 0}\frac{(x^n + nx^{n-1}h + \cdots + h^n) - x^n}{h} \\
&= \lim_{h\to 0}\frac{nx^{n-1}h + \overbrace{\cdots + h^n}^{\text{terms involving } h^2 \text{ and higher powers of of } h}}{h} \\
&= \lim_{h\to 0}\frac{h(nx^{n-1} + \cdots + h^{n-1})}{h}.
\end{aligned}
$$

When you factor out h from terms involving h^2 and higher powers of h, each term will still have an h in it. Performing the division, we get:

$$\frac{d}{dx}(x^n) = \lim_{h \to 0} \big(nx^{n-1} + \overbrace{\cdots + h^{n-1}}^{\text{terms involving } h \text{ and higher powers of } h} \big).$$

But, as $h \to 0$, all terms involving an h will go to 0, so

$$\frac{d}{dx}(x^n) = \lim_{h \to 0} \big(nx^{n-1} + \underbrace{\cdots + h^{n-1}}_{\text{these terms go to 0}} \big) = nx^{n-1}.$$

□ **Example 2**   Differentiate $5x^2 - 7x^3$.

Solution.

$$
\begin{aligned}
\frac{d}{dx}(5x^2 - 7x^3) &= \frac{d}{dx}(5x^2) - \frac{d}{dx}(7x^3) \quad \text{(derivative of difference)} \\
&= 5\frac{d}{dx}(x^2) - 7\frac{d}{dx}(x^3) \quad \text{(derivative of multiple)} \\
&= 5(2x) - 7(3x^2) \\
&= 10x - 21x^2.
\end{aligned}
$$

□

Negative and Fractional Powers

We can calculate the derivatives of functions such as $g(x) = x^{-1}$ and $k(x) = x^{-2}$ using the definition of the derivative.

□ **Example 3**   Find $\dfrac{d}{dx}(x^{-2})$.

Solution.

$$
\begin{aligned}
\frac{d}{dx}\left(x^{-2}\right) &= \frac{d}{dx}\left(\frac{1}{x^2}\right) = \lim_{h \to 0}\left(\frac{\frac{1}{(x+h)^2} - \frac{1}{x^2}}{h}\right) \\
&= \lim_{h \to 0}\frac{1}{h}\left[\frac{x^2 - (x+h)^2}{(x+h)^2 x^2}\right] \quad \text{combining fractions over common denominator} \\
&= \lim_{h \to 0}\frac{1}{h}\left[\frac{x^2 - (x^2 + 2xh + h^2)}{(x+h)^2 x^2}\right] \quad \text{multiplying out} \\
&= \lim_{h \to 0}\frac{-2xh - h^2}{h(x+h)^2 x^2} \quad \text{simplifying numerator} \\
&= \lim_{h \to 0}\frac{-2x - h}{(x+h)^2 x^2} \quad \text{dividing by } h, \text{ top and bottom}
\end{aligned}
$$

$$= \frac{-2x}{x^4} \qquad \text{letting } h \to 0$$

$$= -\frac{2}{x^3} = -2x^{-3}.$$

So

$$\frac{d}{dx}(x^{-2}) = -2x^{-3}.$$

The graphs of $k(x) = x^{-2}$ and $k'(x) = -2x^{-3}$ are shown in Figure 4.5. Does the graph of k' have the features you'd expect? ❏

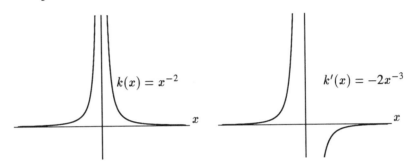

Figure 4.5: $k(x) = x^{-2}$ and its Derivative

This example suggests that the rule

$$\frac{d}{dx}(x^n) = nx^{n-1}$$

may hold for n a negative integer also, and indeed it does. It can be proved using a method similar to that for positive integral powers. In fact, this rule holds for any constant real number n, though we cannot yet easily justify it.

❏ **Example 4** Differentiate (a) $\frac{1}{x^3}$, (b) $x^{1/2}$, (c) $\frac{1}{\sqrt[3]{x}}$.

Solution.

(a) $\dfrac{d}{dx}\left(\dfrac{1}{x^3}\right) = \dfrac{d}{dx}(x^{-3}) = -3x^{-3-1} = -3x^{-4} = -\dfrac{3}{x^4}.$

(b) $\dfrac{d}{dx}\left(x^{1/2}\right) = \dfrac{1}{2}x^{(1/2)-1} = \dfrac{1}{2}x^{-1/2} = \dfrac{1}{2\sqrt{x}}.$

(c) $\dfrac{d}{dx}\left(\dfrac{1}{\sqrt[3]{x}}\right) = \dfrac{d}{dx}\left(x^{-1/3}\right) = -\dfrac{1}{3}x^{(-1/3)-1} = -\dfrac{1}{3}x^{-4/3} = -\dfrac{1}{3x^{4/3}}.$

❏

Derivatives of Polynomials

We already know how to differentiate powers, constant multiples, and sums. For example

$$\frac{d}{dx}(3x^5) = 3\frac{d}{dx}(x^5) = 3 \cdot 5x^4 = 15x^4$$

and

$$\frac{d}{dx}(x^5 + x^3) = \frac{d}{dx}(x^5) + \frac{d}{dx}(x^3) = 5x^4 + 3x^2.$$

Using these rules together, we can differentiate any polynomial.

☐ **Example 5**   Find the derivatives of   (a) $5x^2 + 3x + 2$,   (b) $\sqrt{3}x^7 - \dfrac{x^5}{5} + \pi$.

Solution. (a)

$$\begin{aligned}
\frac{d}{dx}(5x^2 + 3x + 2) &= 5\frac{d}{dx}(x^2) + 3\frac{d}{dx}(x) + \frac{d}{dx}(2) \\
&= 5 \cdot 2x + 3 \cdot 1 + 0 \qquad \text{since the derivative of a constant is zero} \\
&= 10x + 3.
\end{aligned}$$

(b)

$$\begin{aligned}
\frac{d}{dx}\left(\sqrt{3}x^7 - \frac{x^5}{5} + \pi\right) &= \sqrt{3}\frac{d}{dx}(x^7) - \frac{1}{5}\frac{d}{dx}(x^5) + \frac{d\pi}{dx} \\
&= \sqrt{3} \cdot 7x^6 - \frac{1}{5} \cdot 5x^4 + 0 \qquad \text{since } \pi \text{ is a constant} \\
&= 7\sqrt{3}x^6 - x^4.
\end{aligned}$$

☐

We can also use these rules to differentiate expressions which are not polynomials.

☐ **Example 6**   Differentiate   (a) $5\sqrt{x} - \dfrac{10}{x^2} + \dfrac{1}{2\sqrt{x}}$,   (b) $0.1x^3 + 2x^{\sqrt{2}}$.

Solution. (a)

$$\begin{aligned}
\frac{d}{dx}\left(5\sqrt{x} - \frac{10}{x^2} + \frac{1}{2\sqrt{x}}\right) &= \frac{d}{dx}\left(5x^{1/2} - 10x^{-2} + \frac{1}{2}x^{-1/2}\right) \\
&= 5 \cdot \frac{1}{2}x^{-1/2} - 10(-2)x^{-3} + \frac{1}{2}(-\frac{1}{2})x^{-3/2} \\
&= \frac{5}{2\sqrt{x}} + \frac{20}{x^3} - \frac{1}{4x^{3/2}},
\end{aligned}$$

(b)

$$\frac{d}{dx}(0.1x^3 + 2x^{\sqrt{2}}) = 0.1\frac{d}{dx}(x^3) + 2\frac{d}{dx}(x^{\sqrt{2}}) = 0.3x^2 + 2\sqrt{2}x^{\sqrt{2}-1}.$$

☐

Further Examples

☐ **Example 7** Find and interpret the second derivatives of

$$\text{(a) } f(x) = x^2, \qquad \text{(b) } g(x) = x^3, \qquad \text{(c) } k(x) = x^{\frac{1}{2}}.$$

Solution.

(a) For $f(x) = x^2$, $f'(x) = 2x$, so $f''(x) = \frac{d}{dx}(2x) = 2$. Since f'' is always positive, f is concave up, as expected for a parabola opening upwards. See Figure 4.6.

(b) For $g(x) = x^3$, $g'(x) = 3x^2$, so $g''(x) = \frac{d}{dx}(3x^2) = 3\frac{d}{dx}(x^2) = 3 \cdot 2x = 6x$. This is positive for $x > 0$ and negative for $x < 0$, which means x^3 is concave up for $x > 0$ and concave down for $x < 0$. See Figure 4.7.

(c) For $k(x) = x^{\frac{1}{2}}$, $k'(x) = \frac{1}{2}x^{\frac{1}{2}-1} = \frac{1}{2}x^{-\frac{1}{2}}$, so $k''(x) = \frac{d}{dx}\left(\frac{1}{2}x^{-\frac{1}{2}}\right) = \frac{1}{2} \cdot (-\frac{1}{2})x^{-\frac{1}{2}-1} = -\frac{1}{4}x^{-\frac{3}{2}}$. Now k' and k'' only make sense on the domain of k, i.e. $x \geq 0$. In fact, k' and k'' are undefined when $x = 0$, since $x^{-\frac{1}{2}}$ and $x^{-\frac{3}{2}}$ are not defined for $x = 0$. We have $k''(x)$ is negative for $x > 0$, so k is concave down. See Figure 4.8.

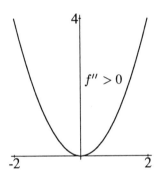

Figure 4.6: $f(x) = x^2$, $f''(x) = 2$

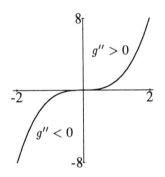

Figure 4.7: $g(x) = x^3$, $g''(x) = 6x$

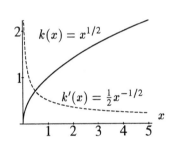

Figure 4.8: $k(x) = x^{1/2}$ and its derivative, for Examples 4(b) and 7(c)

☐ **Example 8** If the position of a body is given by

$$s = -4.9t^2 + 5t + 6,$$

find the velocity and acceleration of the body at time t.

Solution. The velocity, v, is the derivative of the position:

$$v = \frac{ds}{dt} = \frac{d}{dt}(-4.9t^2 + 5t + 6) = -9.8t + 5$$

and the acceleration, a, is the derivative of the velocity:

$$a = \frac{dv}{dt} = \frac{d}{dt}(-9.8t + 5) = -9.8.$$

☐ **Example 9** Figure 4.9 shows the graph of a cubic polynomial. Both graphically and algebraically, describe the range of values taken on by the derivative of this cubic.

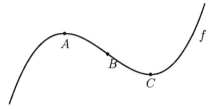

Figure 4.9: The Cubic of Example 9   **Figure 4.10**: Derivative of the Cubic of Example 9

Solution. Graphical approach: Suppose you move along the curve from left to right. To the left of A, the slope is positive; it starts very positive and decreases until the curve reaches A, where the slope is 0. Between A and C, the slope is negative. Between A and B, the slope is decreasing, getting more negative; it is most negative at B. Between B and C the slope is negative but increasing; at C the slope is zero. From C to the right, the slope is positive and increasing. The graph of the derivative function is shown in Figure 4.10.

Algebraic approach: f is a cubic that goes to $+\infty$ as $x \to +\infty$ so

$$f(x) = ax^3 + bx^2 + cx + d$$

with $a > 0$. Hence,

$$f'(x) = 3ax^2 + 2bx + c$$

whose graph is a parabola opening upwards, as in Figure 4.10.   ☐

❖ Exercises for Section 4.2

For Problems 1–16, find the derivative of the function given.

1. $y = x^{12}$

2. $y = x^{-12}$

3. $y = x^{4/3}$

4. $y = x^{3/4}$

5. $y = x^{-3/4}$

6. $f(x) = \dfrac{1}{x^4}$

7. $f(x) = \sqrt[4]{x}$

8. $f(x) = x^e$

9. $y = 4x^{3/2} - 5x^{1/2}$

10. $y = 6x^3 + 4x^2 - 2x$

11. $y = -3x^4 - 4x^3 - 6x + 2$

12. $y = 3t^5 - 5\sqrt{t} + \dfrac{7}{t}$

13. $y = 3t^2 + \dfrac{12}{\sqrt{t}} - \dfrac{1}{t^2}$

14. $y = z^2 + \dfrac{1}{2z}$

15. $y = \dfrac{x^2 + 1}{x}$

16. $y = \dfrac{\theta - 1}{\sqrt{\theta}}$

17. Find the functions in Problems 1–9 whose derivatives do not exist at $x = 0$.

For Problems 18–27, determine if the derivative rules from this section apply. If they do, find the derivative. If they don't apply, indicate why.

18. $y = \sqrt{x}$

19. $y = (x + 3)^{\frac{1}{2}}$

20. $y = 3x^2 + 4$

21. $y = \dfrac{1}{3z^2} + \dfrac{1}{4}$

22. $y = \dfrac{1}{3\sqrt{x}} + \dfrac{1}{4}$

23. $y = \dfrac{1}{3x^2 + 4}$

24. $y = 3^t$

25. $y = x^3$

26. $y = x^\pi$

27. $y = \pi^x$

28. If $f(t) = 2t^3 - 4t^2 + 3t - 1$, find $f'(t)$ and $f''(t)$.

29. (a) Find the *eighth* derivative of
$$f(x) = x^7 + 5x^5 - 4x^3 + 6x - 7.$$
   Think ahead! (The nth derivative is the result of differentiating n times.)

   (b) Find the seventh derivative of $f(x)$.

30. Find the equation of the line tangent to the graph of f at $(1, 1)$, where f is given by $f(x) = 2x^3 - 2x^2 + 1$.

31. The graph of the equation $y = x^3 - 9x^2 - 16x + 1$ has a slope equal to 5 at exactly two points. Find the coordinates of the points.

32. If $f(x) = 13 - 8x + \sqrt{2}x^2$ and $f'(r) = 4$, find r.

33. For what points in the domain of the function $f(x) = x^4 - 4x^3$ is the function both decreasing and concave up?

34. If $f(x) = 4x^3 + 6x^2 - 23x + 7$, find the intervals on which $f'(x) \geq 1$.

35. Given $p(x) = x^n - x$, find the intervals over which p is a decreasing function when:

   (a) $n = 2$ (b) $n = \frac{1}{2}$ (c) $n = -1$

36. Using a graph to help you, find the equations of all lines through the origin tangent to the parabola
$$y = x^2 - 2x + 4.$$
   Sketch your solutions.

37. Is there any value of n which makes $y = x^n$ a solution to the equation $13x\dfrac{dy}{dx} = y$? If so, what value?

38. Using the definition of derivative, justify the formula
$$\frac{d}{dx}(x^n) = nx^{n-1}.$$

   (a) for $n = -1$; $n = -3$.

   (b) for any negative integer n.

39. The period, T, of a pendulum is given in terms of its length, l, by

$$T = 2\pi\sqrt{\frac{l}{g}},$$

where g is the acceleration due to gravity (a constant).

(a) Find $\dfrac{dT}{dl}$.

(b) What is the sign of $\dfrac{dT}{dl}$? What does this tell you about the period of pendulums?

40. A ball is dropped from the top of the Empire State building to the ground below. The height, y, of the ball above the ground (in feet) is given as a function of time, t, in seconds by

$$y = 1250 - 16t^2.$$

(a) Find the velocity of the ball at time t. What is the sign of the velocity? Why is this to be expected?

(b) Show that the acceleration of the ball is a constant. What is the value and sign of this constant?

(c) When does the ball hit the ground, and how fast is it going at that time? Give your answer in feet/sec and in mph. (1 ft/sec = 15/22 mph).

41. (a) Use the formula for the area of a circle of radius r, $A = \pi r^2$, to find $\frac{dA}{dr}$.

(b) The result from (a) should look familiar. What does $\frac{dA}{dr}$ represent geometrically? Draw a picture.

(c) Use the difference quotient to explain the observation you made in (b).

42. What is the formula for $V(r)$, the volume of a sphere of radius r? Find $\frac{dV}{dr}$. What is the geometrical meaning of $\frac{dV}{dr}$?

43. We know that the graph of a tangent line lies close to the graph of the function near the point of tangency. However, usually, the further we go from the point of tangency, the greater the distance between the graph of the function and the tangent line. Let's investigate this for $f(x) = \frac{1}{x}$. Find the value of f at $x = 2$. Find the tangent to the curve at $x = 1$, and use it to approximate the value of f at $x = 2$. Now find the tangent to the curve at $x = 100$ and use it to approximate the value of f at $x = 2$. Which tangent line lies closer to the curve at $x = 2$? Is there something wrong with our basic idea? Explain.

4.3   The Exponential Function

What would you expect the graph of the derivative of the exponential function $f(x) = a^x$ to look like? The graph of this function is shown in Figure 4.11. The function increases slowly when $x < 0$ and more rapidly for $x > 0$, so the values of f' are small for $x < 0$ and larger for $x > 0$. Since the function is increasing for all real values of x, the graph of the derivative must lie above the x-axis. After some reflection, you may decide that the graph of f' must resemble the graph of f itself.

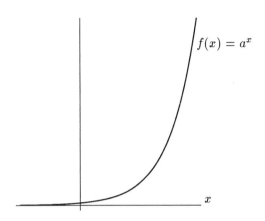

Figure 4.11: $f(x) = a^x$, with $a > 1$

The Derivatives of 2^x and 3^x

In Chapter 2, we saw that the derivative of $f(x) = 2^x$ at $x = 0$ is given by

$$f'(0) = \lim_{h \to 0} \frac{2^h - 2^0}{h} = \lim_{h \to 0} \frac{2^h - 1}{h} \approx 0.6931,$$

where we found the limit by evaluating $\frac{2^h - 1}{h}$ for small values of h. Similarly, we can calculate the derivative at $x = 1$ and $x = 2$:

$$f'(1) = \lim_{h \to 0} \frac{2^{1+h} - 2^1}{h} \approx 1.3863$$

$$f'(2) = \lim_{h \to 0} \frac{2^{2+h} - 2^2}{h} \approx 2.7726.$$

Can you see the relationship between these values of the derivative? If you notice that $1.3863 \approx 2(0.6931)$ and $2.7726 \approx 4(0.6931)$, you see

$$
\begin{aligned}
f'(0) &\approx & 0.6931 = 0.6931 \cdot 2^0 \\
f'(1) &\approx & 1.3863 \approx 0.6931 \cdot 2^1 \\
f'(2) &\approx & 2.7726 \approx 0.6931 \cdot 2^2
\end{aligned}
$$

Can you see a pattern? It looks as though $f'(x) \approx 0.6931 \cdot 2^x$. Let's look at the derivative function to see why this happens:

$$
\begin{aligned}
f'(x) &= \lim_{h \to 0} \left(\frac{2^{x+h} - 2^x}{h} \right) = \lim_{h \to 0} \left(\frac{2^x 2^h - 2^x}{h} \right) = \lim_{h \to 0} 2^x \left(\frac{2^h - 1}{h} \right) \\
&= 2^x \lim_{h \to 0} \left(\frac{2^h - 1}{h} \right) \qquad \text{\small since } x \text{ and } 2^x \text{ are fixed during this calculation}
\end{aligned}
$$

We have already found $\lim\limits_{h \to 0} \dfrac{2^h - 1}{h}$; its value is about 0.6931, so we have shown that

$$\boxed{\dfrac{d}{dx}(2^x) = f'(x) \approx (0.6931)2^x}$$

The graphs of $f(x) = 2^x$ and $f'(x) \approx (0.6931)2^x$ are shown in Figure 4.12. Notice that they do indeed resemble one another.

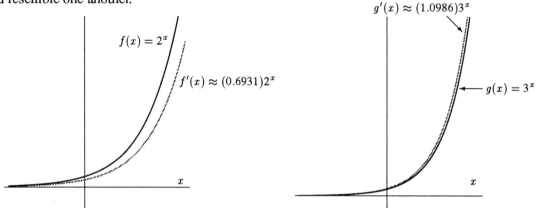

Figure 4.12: Graphs of $f(x) = 2^x$ and $f'(x) \approx (0.6931)2^x$

Figure 4.13: Graphs of $g(x) = 3^x$ and $g'(x) \approx (1.0986)3^x$

☐ **Example 1**   Find the derivative of $g(x) = 3^x$ and plot g and g' on the same axes.

Solution. As before

$$g'(x) = \lim_{h \to 0} \frac{3^{x+h} - 3^x}{h} = \lim_{h \to 0} \frac{3^x 3^h - 3^x}{h} = \lim_{h \to 0} 3^x \left(\frac{3^h - 1}{h} \right).$$

Using a calculator gives

$$\lim_{h \to 0} \left(\frac{3^h - 1}{h} \right) \approx 1.0986,$$

so

$$g'(x) \approx (1.0986)3^x.$$

The graphs are in Figure 4.13.   ❑

Notice that for both $f(x) = 2^x$ and $g(x) = 3^x$, *the derivative is proportional to the original function.* For $f(x) = 2^x$, $f'(x) \approx (0.6931)2^x$, so the constant of proportionality is less than one, and the graph of the derivative is below the graph of the original function. For $g(x) = 3^x$, $g'(x) \approx (1.0986)3^x$ and the constant is greater than one, so the graph of the derivative is above that of the original function.

Note on Round-Off Error and Limits

If you try to evaluate $\frac{2^h-1}{h}$ on a calculator by taking smaller and smaller values of h, the values of $\frac{2^h-1}{h}$ will at first get closer to 0.6931. However, they will eventually move away from 0.6931 again because of the *round-off error* (i.e., errors introduced by the fact that the calculator can only hold a certain number of digits).

As you try smaller and smaller values of h, how do you know when to stop? Unfortunately there is no fixed rule. A calculator can only *suggest* the value of a limit, but can never confirm that this value is correct. In this case, you can be pretty sure that the limit is about 0.6931 because the values of $\frac{2^h-1}{h}$ hover around 0.6931 for a while. But to be absolutely sure this is correct, you would have to find the limit by theoretical means.

The Derivative of a^x and the Definition of e

The calculation of the derivative of $f(x) = a^x$, for $a > 0$, works just the same as before and leads to

$$f'(x) = \lim_{h \to 0} \frac{a^{x+h} - a^x}{h} = a^x \lim_{h \to 0} \frac{a^h - 1}{h}.$$

The quantity $\lim_{h \to 0} \frac{a^h - 1}{h}$ doesn't depend on x, and so, for any particular a, is a constant. Therefore the derivative is again proportional to the original function, with constant of proportionality

$$\lim_{h \to 0} \frac{a^h - 1}{h}.$$

We can't use a calculator to evaluate this limit without knowing the value of a. However, when $a = 2$, we know that the limit is less than 1, and the derivative is smaller than the original function. When $a = 3$, the limit is more than 1, and the derivative is greater than the original function. Is there an in-between case, when derivative and function are exactly equal? In other words:

"Is there a value of a making $\frac{d}{dx}(a^x) = a^x$?"

If so, we have found a function with the remarkable property that it is equal to its own derivative. So let's look for such an a. This means we want to find an a such that

$$\lim_{h \to 0} \frac{a^h - 1}{h} = 1$$

or, for small h,

$$\frac{a^h - 1}{h} \approx 1.$$

Solving for a suggests that we can calculate a as follows:

$$a^h - 1 \approx h \qquad \text{or} \qquad a^h \approx 1 + h \qquad \text{so} \qquad a \approx (1 + h)^{1/h}.$$

Taking small values of h, as in Table 4.2, you can see $a \approx e = 2.718\ldots$, the number introduced in Chapter 1. Thus we see that for small h:

$$e \approx (1 + h)^{1/h} \approx 2.718.$$

h	$(1 + h)^{\frac{1}{h}}$
0.001	2.7169239
0.0001	2.7181459
0.00001	2.7182682

Table 4.2: Values of $(1 + h)^{1/h}$

In fact, it can be shown that

$$e = \lim_{h \to 0} (1 + h)^{1/h} = 2.718\ldots \quad \text{and} \quad \lim_{h \to 0} \frac{e^h - 1}{h} = 1$$

which means that e^x is its own derivative:

$$\boxed{\frac{d}{dx}(e^x) = e^x}$$

It turns out that the constants involved in the derivatives of 2^x and 3^x are natural logarithms. In fact, since $0.6931 \approx \ln 2$ and $1.0986 \approx \ln 3$, this suggests that:

$$\frac{d}{dx}(2^x) = (\ln 2)2^x \quad \text{and} \quad \frac{d}{dx}(3^x) = (\ln 3)3^x.$$

Later we will show that, in general,

$$\boxed{\frac{d}{dx}(a^x) = (\ln a)a^x}$$

Figure 4.14 shows the graph of the derivative of 2^x lies below the graph of the function; the derivative of 3^x lies above the function. With $e \approx 2.718$, the function e^x and its derivative are identical.

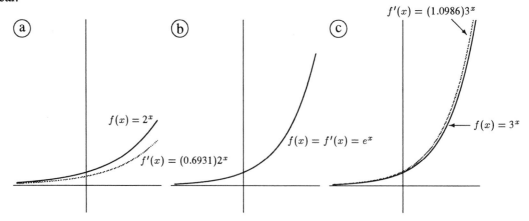

Figure 4.14: The functions 2^x, e^x, 3^x and their derivatives.

❏ **Example 2**   Differentiate $2 \cdot 3^x + 5e^x$.

Solution.
$$\frac{d}{dx}(2 \cdot 3^x + 5e^x) = \frac{d}{dx}(2 \cdot 3^x) + \frac{d}{dx}(5e^x) = 2\frac{d}{dx}(3^x) + 5\frac{d}{dx}(e^x)$$
$$\approx 2(1.0986)3^x + 5e^x = (2.1972)3^x + 5e^x.$$

❏

Numerical Demonstration of $\dfrac{d}{dx}(2^x) = (0.693)2^x$

❏ **Example 3**   Construct a table of values of 2^x for $x = 1.0, 1.1, \ldots, 1.5$. Use difference quotients with $h = 0.001$ to estimate its derivative at these points. Compare the values of 2^x and its derivative. What do you notice?

Solution. While constructing the table, keep all the digits your calculator holds (usually 8 to 12) throughout the calculation. The results are displayed to three decimal places in Table 4.3.

x	2^x	Difference Quotient $= \frac{2^{x+h}-2^x}{h}$	Difference Quotient $\frac{}{2^x}$
1.0	2.000	1.387	0.693
1.1	2.144	1.486	0.693
1.2	2.297	1.593	0.693
1.3	2.462	1.707	0.693
1.4	2.639	1.830	0.693
1.5	2.828	1.961	0.693

Table 4.3: Comparing values of 2^x and Difference Quotients (with $h = 0.001$)

The difference quotients in the third column approximate the derivative of 2^x. The fourth column contains the ratio $\frac{\text{Difference Quotient}}{2^x}$. The remarkable thing about the exponential function is that *all* the entries in the last column are 0.693 (to three decimal places). This suggests that the derivative of 2^x is proportional to the values of 2^x with constant of proportionality 0.693.

❏

We have seen that the derivative of a^x is proportional to a^x. Many quantities have rates of change which are proportional to themselves; for example, the pressure in the atmosphere as a function of altitude and the growth of a population both have this property. The fact that the constant of proportionality is 1 when $a = e$ makes e^x a particularly useful function.

❖ Exercises for Section 4.3

Find the derivatives of the functions in Problems 1–14.

1. $y = 5t^2 + 4e^t$

2. $f(x) = 2e^x + x^2$

3. $f(x) = 2^x + 2 \cdot 3^x$

4. $y = 4 \cdot 10^x - x^3$

5. $y = 3x - 2 \cdot 4^x$

6. $y = \dfrac{3^x}{3} + \dfrac{33}{\sqrt{x}}$

7. $f(x) = e^2 + x^e$

8. $f(x) = e^{1+x}$

9. $f(t) = e^{t+2}$

10. $y = e^{\theta - 1}$

11. $z = (\ln 4)e^x$

12. $z = (\ln 4)4^x$

13. $f(z) = (\ln 3)z^2 + (\ln 4)e^z$

14. $f(t) = (\ln 3)^t$

Which of the functions in Problems 15–24 can be differentiated using the rules we have developed so far? Perform the differentiation if you can; otherwise, indicate why the rules so far do not apply.

15. $y = x^2 + 2^x$

16. $y = \sqrt{x} - (\frac{1}{2})^x$

17. $y = x^2 \cdot 2^x$

18. $y = \dfrac{2^x}{x}$

19. $y = e^{x+5}$

20. $y = e^{5x}$

21. $f(s) = 5^s e^s$

22. $y = 4^{(x^2)}$

23. $f(z) = (\sqrt{4})^z$

24. $f(\theta) = 4^{\sqrt{\theta}}$

25. (a) Find the slope of the graph of $f(x) = 1 - e^x$ at the point where it crosses the x-axis.

   (b) Find the equation of the tangent line to the curve at this point.

   (c) Find the equation of the line which is perpendicular to the tangent line at this point. (This line is known as the *normal* line.)

26. Find the value of c in Figure 4.15, where the line l tangent to the graph of $y = 2^x$ at $(0, 1)$ intersects the x-axis.

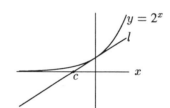

Figure 4.15: Where does the tangent line intersect the x-axis?

(for Problem 26)

27. With an inflation rate of 5%, prices are described by

$$P = P_0(1.05)^t$$

where P_0 is the price in dollars when $t = 0$ and t is time in years. Suppose $P_0 = 1$. How fast (in cents/year) is the price of the good rising when $t = 10$?

28. Since January 1, 1960, the population of Slim Chance has been described by the formula $P = 35{,}000(0.98)^t$ where P is the population of the city t years after 1960. At what rate was the population of the city changing on Jan. 1, 1983?

29. The *Global 2000 Report* gave the world's population, P, as 4.1 billion in 1975 and growing at 2% annually.

   (a) Give a formula for P in terms of time, t, measured in years since 1975.

   (b) Find $\dfrac{dP}{dt}$, $\left.\dfrac{dP}{dt}\right|_{t=0}$, $\left.\dfrac{dP}{dt}\right|_{t=15}$. What do each of these represent in practical terms?

30. Find the quadratic function $g(x) = ax^2 + bx + c$ which best fits the function $f(x) = e^x$ at $x = 0$, in the sense that

$$g(0) = f(0), \quad \text{and} \quad g'(0) = f'(0), \quad \text{and} \quad g''(0) = f''(0).$$

Using a computer or calculator, sketch graphs of f and g on the same axes. What do you notice?

31. Using the tangent line to e^x at $x = 0$, show that

$$e^x \geq 1 + x$$

for all values of x. A sketch will be helpful.

32. Find all solutions of the equation

$$2^x = 2x.$$

How do you know that you found all solutions?

33. Construct a table of values for 3^x, $x = 0, 0.1, 0.2, \ldots, 0.5$ like the one in Example 3 on page 237. Use difference quotients with $h = 0.001$ to estimate the derivative. Compute the ratio of the derivative and the function. What do you notice?

4.4 The Product and Quotient Rules

You now know how to find derivatives of powers and exponentials, and of sums and multiples of functions. This section will show you how to find the derivatives of products and quotients.

The Product Rule

Suppose we know the derivatives of $f(x)$ and $g(x)$ and want to calculate the derivative of the product, $f(x)g(x)$. The derivative of the product is calculated by taking the limit

$$\frac{d(f(x)g(x))}{dx} = \lim_{h \to 0} \frac{f(x+h)g(x+h) - f(x)g(x)}{h}.$$

To picture the quantity $f(x+h)g(x+h) - f(x)g(x)$, imagine a rectangle with sides $f(x+h)$ and $g(x+h)$. See Figure 4.16. Let's write Δf, meaning change in f, for $f(x+h) - f(x)$ and Δg for $g(x+h) - g(x)$. Then we have

$$f(x+h) = f(x) + \Delta f \quad \text{and} \quad g(x+h) = g(x) + \Delta g.$$

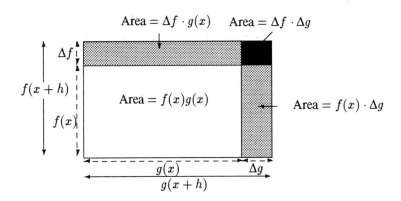

Figure 4.16: Illustration for the Product Rule (with Δf, Δg positive)

Then

$$f(x+h)g(x+h) - f(x)g(x) = \text{Area of Whole Rectangle} - \text{Unshaded Area}$$
$$= \text{Area of the Three Shaded Rectangles}$$
$$= \Delta f \cdot g(x) + f(x) \cdot \Delta g + \Delta f \cdot \Delta g$$

Now divide by h:

$$\frac{f(x+h)g(x+h) - f(x)g(x)}{h} = \frac{\Delta f}{h} \cdot g(x) + f(x) \cdot \frac{\Delta g}{h} + \frac{\Delta f \cdot \Delta g}{h}$$

To evaluate the limit as $h \to 0$, let's examine the three terms on the right separately. First, $\frac{\Delta f}{h} = \frac{f(x+h) - f(x)}{h} \to f'(x)$, so

$$\frac{\Delta f}{h} \cdot g(x) \to f'(x)g(x).$$

Second, $\frac{\Delta g}{h} = \frac{g(x+h) - g(x)}{h} \to g'(x)$, so

$$f(x) \cdot \frac{\Delta g}{h} \to f(x)g'(x).$$

Finally, write the third term as $\frac{\Delta f}{h} \cdot \frac{\Delta g}{h} h$. Then, as $h \to 0$,

$$\frac{\Delta f}{h} \cdot \frac{\Delta g}{h} \cdot h \to f'(x) \cdot g'(x) \cdot 0 = 0.$$

We conclude

$$\lim_{h \to 0} \frac{f(x+h)g(x+h) - f(x)g(x)}{h} = f'(x)g(x) + f(x)g'(x).$$

Thus we have

The Product Rule:

$$(fg)' = f'g + fg'.$$

In words:

The derivative of a product is the derivative of the first factor multiplied by the second, plus the first factor multiplied by the derivative of the second.

□ **Example 1** Differentiate (a) $x^2 e^x$, (b) $(3x^2 + 5x)e^x$, (c) $\dfrac{e^x}{x^2}$.

Solution.

(a) $\dfrac{d(x^2 e^x)}{dx} = \left(\dfrac{d(x^2)}{dx}\right) e^x + x^2 \dfrac{d(e^x)}{dx} = 2x e^x + x^2 e^x = (2x + x^2)e^x,$

(b)
$$\dfrac{d((3x^2 + 5x)e^x)}{dx} = \left(\dfrac{d(3x^2 + 5x)}{dx}\right) e^x + (3x^2 + 5x)\dfrac{d}{dx}e^x$$
$$= (6x + 5)e^x + (3x^2 + 5x)e^x = (3x^2 + 11x + 5)e^x,$$

(c) First we must write $\dfrac{e^x}{x^2}$ as the product $x^{-2}e^x$:

$$\dfrac{d}{dx}\left(\dfrac{e^x}{x^2}\right) = \dfrac{d(x^{-2}e^x)}{dx} = \left(\dfrac{d(x^{-2})}{dx}\right) e^x + x^{-2}\dfrac{d(e^x)}{dx}$$
$$= -2x^{-3}e^x + x^{-2}e^x = (-2x^{-3} + x^{-2})e^x.$$

□

The Quotient Rule

Suppose we want to differentiate a function of the form $Q(x) = \dfrac{f(x)}{g(x)}$. We'll find a formula for Q' in terms of f' and g'.

Since $f(x) = Q(x)g(x)$, we can use the product rule:

$$f'(x) = Q'(x)g(x) + Q(x)g'(x)$$
$$= Q'(x)g(x) + \dfrac{f(x)}{g(x)}g'(x).$$

Solving for $Q'(x)$:

$$Q'(x) = \dfrac{f'(x) - \frac{f(x)}{g(x)}g'(x)}{g(x)}.$$

Multiplying top and bottom by $g(x)$ to simplify gives

$$\frac{d}{dx}\left(\frac{f(x)}{g(x)}\right) = \frac{f'(x)g(x) - f(x)g'(x)}{(g(x))^2}$$

So we have

The Quotient Rule:

$$\left(\frac{f}{g}\right)' = \frac{f'g - fg'}{g^2}.$$

☐ **Example 2** Differentiate (a) $\dfrac{5x^2}{x^3 + 1}$, (b) $\dfrac{1}{2 + e^x}$, (c) $\dfrac{e^x}{x^2}$.

Solution.

(a)
$$\frac{d}{dx}\left(\frac{5x^2}{x^3 + 1}\right) = \frac{\left(\frac{d}{dx}(5x^2)\right)(x^3 + 1) - 5x^2\frac{d}{dx}(x^3 + 1)}{(x^3 + 1)^2}$$

$$= \frac{10x(x^3 + 1) - 5x^2(3x^2)}{(x^3 + 1)^2}$$

$$= \frac{-5x^4 + 10x}{(x^3 + 1)^2}.$$

(b)
$$\frac{d}{dx}\left(\frac{1}{1 + e^x}\right) = \frac{\left(\frac{d}{dx}(1)\right)(1 + e^x) - 1\frac{d}{dx}(1 + e^x)}{(1 + e^x)^2}$$

$$= \frac{0(1 + e^x) - 1(0 + e^x)}{(1 + e^x)^2}$$

$$= \frac{-e^x}{(1 + e^x)^2}.$$

(c) This is the same as part (c) of Example 1, but this time we will do it by the quotient rule.

$$\frac{d}{dx}\left(\frac{e^x}{x^2}\right) = \frac{\left(\frac{d(e^x)}{dx}\right)\cdot x^2 - e^x\left(\frac{d(x^2)}{dx}\right)}{(x^2)^2} = \frac{e^x x^2 - e^x 2x}{x^4}$$

$$= \frac{x^2 - 2x}{x^4}e^x = \frac{x - 2}{x^3}e^x.$$

This is in fact the same answer as before although it looks different. Can you show that it is the same?

☐

❖ Exercises for Section 4.4

1. If $f(x) = x^2(x^3 + 5)$, find $f'(x)$ two ways: using the product rule and by multiplying out. Do you get the same result? Should you?

2. If $f(x) = 2^x \cdot 3^x$, find $f'(x)$ two ways: using the product rule and by multiplying out. Do you get the same result?

For Problems 3–16, find the derivative. It may be to your advantage in some cases to simplify first.

3. $f(x) = xe^x$

4. $f(x) = \dfrac{x}{e^x}$

5. $y = x \cdot 2^x$

6. $y = \sqrt{x} \cdot 2^x$

7. $f(x) = (x^2 - \sqrt{x})3^x$

8. $w = (t^3 + 5t)(t^2 - 7t + 2)$

9. $z = (s^2 - \sqrt{s})(s^2 + \sqrt{s})$

10. $y = (t^3 - 7t^2 + 1)e^t$

11. $z = \dfrac{t^2 + 5t + 2}{t + 3}$

12. $w = \dfrac{y^3 - 6y^2 + 7y}{y}$

13. $f(x) = \dfrac{x^2 + 3}{x}$

14. $y = \dfrac{\sqrt{t}}{t^2 + 1}$

15. $f(x) = \dfrac{1 + x}{2 + 3x + 4x^2}$

16. $f(z) = \dfrac{3z^2}{5z^2 + 7z}$

17. If $f(x) = (3x + 8)(2x - 5)$, find $f'(x)$ and $f''(x)$.

18. Differentiate $f(t) = e^{-t}$ by writing it as $f(t) = \dfrac{1}{e^t}$.

19. Differentiate $f(x) = e^{2x}$ by writing it as $f(x) = e^x \cdot e^x$.

20. Differentiate $f(x) = e^{3x}$ by writing it as $f(x) = e^x \cdot e^{2x}$ and using the result of Problem 19.

21. Based on your answers to Problems 19 and 20, guess the derivative of e^{4x}.

22. (a) Differentiate $y = \dfrac{e^x}{x}$, $\quad y = \dfrac{e^x}{x^2}$, $\quad y = \dfrac{e^x}{x^3}$.

(b) What do you anticipate the derivative of $y = \dfrac{e^x}{x^n}$ will be?

23. Find the equation of the tangent line at $x = 1$ to $y = f(x)$ where $f(x) = \dfrac{3x^2}{5x^2 + 7x}$.

24. Given: $\left\{ \begin{array}{ll} H(3) = 1 & F(3) = 5 \\ H'(3) = 3 & F'(3) = 7 \end{array} \right\}$ find: $\left\{ \begin{array}{ll} \text{(a)} & G'(3) \quad \text{if } G(z) = F(z) \cdot H(z) \\ \text{(b)} & G'(3) \quad \text{if } G(w) = \frac{F(w)}{H(w)} \end{array} \right.$

25. (a) Suppose
$$f(x) = \frac{x}{x^2 - 1}.$$
For what values of x does this formula define f? Find a formula for $f'(x)$. Explain why $f(1.01)$ is so large, and why $f'(1.01)$ is even larger in magnitude.

(b) Now suppose
$$g(x) = \frac{x^2 + 3x - 4}{x^2 - 1}.$$
For what values of x does this formula define g? Find a formula for $g'(x)$. Explain why $g(1.01)$ and $g'(1.01)$ don't seem to be very large.

26. Using the product rule and the fact that $\dfrac{d(x)}{dx} = 1$, show that $\dfrac{d(x^2)}{dx} = 2x$ and $\dfrac{d(x^3)}{dx} = 3x^2$.

27. Find $f'(x)$ for the following functions without multiplying out by using the product rule.

(a) $f(x) = (x - 1)(x - 2)$.

(b) $f(x) = (x - 1)(x - 2)(x - 3)$.

(c) $f(x) = (x - 1)(x - 2)(x - 3)(x - 4)$.

28. Use the answer from Problem 27 to find $f'(x)$ for the following function.
$$f(x) = (x - r_1)(x - r_2)(x - r_3) \cdots (x - r_n)$$
where $r_1, r_2, \ldots, r_n$ are any real numbers.

29. The function
$$f(x) = e^x$$
has the properties
$$f'(x) = f(x) \text{ and } f(0) = 1.$$
Show that $f(x)$ is the only function with these properties.
[Hint: Assume $g'(x) = g(x)$, and $g(0) = 1$, for some function $g(x)$. Define $h(x) = \frac{g(x)}{e^x}$, and compute $h'(x)$. Use the fact that if $k'(x) = 0$ for all x, then $k(x)$ is a constant function.]

30. The quantity, q, of a certain skateboard sold depends on the selling price, p, so we write $q = f(p)$. You are given that $f(140) = 15{,}000$ and $f'(140) = -100$.

(a) What does $f(140) = 15{,}000$ and $f'(140) = -100$ tell you about the sale of the skateboards?

(b) The total revenue, R, earned by the sale of skateboards is given by $R = pq$. Find $\dfrac{dR}{dp}\Big|_{p=140}$.

(c) What is the sign of $\dfrac{dR}{dp}\Big|_{p=140}$? If the skateboards are currently selling for $140, how should the price be changed to increase revenues?

31. Let $f(v)$ be the gas consumption (in liter/km) of a car going at velocity v (in km/hr). In other words, $f(v)$ tells you how many liters of gas the car uses to go one kilometer if it is going at velocity v. You are told that

$$f(80) = 0.05 \text{ and } f'(80) = 0.0005.$$

(a) Let $g(v)$ be the distance the same car goes on one liter of gas at velocity v. What is the relationship between $f(v)$ and $g(v)$? Find $g(80)$ and $g'(80)$.

(b) Let $h(v)$ be the gas consumption in liters per hour. In other words, $h(v)$ tells you how many liters of gas the car uses in one hour if it is going at velocity v. What is the relationship between $h(v)$ and $f(v)$? Find $h(80)$ and $h'(80)$.

(c) How would you explain the practical meaning of the values of these functions and their derivatives to a driver who knows no calculus?

4.5 The Chain Rule

Composite functions like $\sin(x^2)$ or e^{3t} occur frequently in practice; in this section we will see how to differentiate such functions.

Composition of Linear Functions

Let's start by seeing what happens if we compose two linear functions. Suppose

$$z = g(x) = b_1 + m_1 x$$

and

$$y = f(z) = b_2 + m_2 z.$$

Then we can compose f and g:

$$
\begin{aligned}
y = f(g(x)) &= b_2 + m_2(b_1 + m_1 x) \\
&= b_2 + m_2 b_1 + m_2 m_1 x.
\end{aligned}
$$

Therefore the slope of the composite function is $m_2 m_1$, or the product of the slopes of the original lines. Since

$$m_2 = \frac{dy}{dz} \quad \text{and} \quad m_1 = \frac{dz}{dx}$$

in this case we have

$$\frac{dy}{dx} = m_1 m_2 = \frac{dy}{dz} \cdot \frac{dz}{dx}$$

The General Case

Since locally most functions look linear, we would expect the general case to be similar to that for lines. Suppose $f(g(x))$ is a composite function, with f being the outside function and g being the inside. Let us write

$$z = g(x) \quad \text{and} \quad y = f(z), \quad \text{so} \quad y = f(g(x)).$$

Then a small change in x, called Δx, generates a small change in z, called Δz. In turn, Δz generates a small change in y called Δy. In the limit as Δx, Δy, and Δz become infinitesimally small, they are denoted by dx, dy, and dz. Provided Δx and Δz are not zero, we can say:

$$\frac{\Delta y}{\Delta x} = \frac{\Delta y}{\Delta z} \cdot \frac{\Delta z}{\Delta x}.$$

This suggests that in the limit as $\Delta x, \Delta y$, and Δz become infinitesimally small

$$\frac{dy}{dx} = \frac{dy}{dz} \cdot \frac{dz}{dx}.$$

Since $\dfrac{dy}{dz} = f'(z)$ and $\dfrac{dz}{dx} = g'(x)$, this suggests that

$$\frac{d}{dx} f(g(x)) = f'(z) \cdot g'(x)$$

or, substituting $z = g(x)$, we have

The Chain Rule:

$$\frac{d}{dx} f(g(x)) = f'(g(x)) \cdot g'(x).$$

The derivative of a composite function is the product of the derivatives of the outside and inside functions. The derivative of the outside function must be evaluated at the inside function.

A rigorous justification of the chain rule is given in more advanced texts.

☐ **Example 1** Differentiate

(a) $(x^2 + 1)^{100}$,   (b) $\sqrt{3x^2 + 5x - 2}$,   (c) $\dfrac{1}{x^2 + x^4}$,   (d) $\sqrt{e^x + 1}$,   (e) e^{x^2}.

Solution.

(a) Here $z = g(x) = x^2 + 1$ is the inside function; $f(z) = z^{100}$ is the outside function. Now $g'(x) = 2x$ and $f'(z) = 100z^{99}$, so

$$\frac{d((x^2 + 1)^{100})}{dx} = 100z^{99} \cdot 2x = 100(x^2 + 1)^{99} \cdot 2x = 200x(x^2 + 1)^{99}.$$

(b) Here $z = g(x) = 3x^2 + 5x - 2$ and $f(z) = \sqrt{z}$, so $g'(x) = 6x + 5$ and $f'(z) = \frac{1}{2\sqrt{z}}$. Hence

$$\frac{d(\sqrt{3x^2 + 5x - 2})}{dx} = \frac{1}{2\sqrt{z}} \cdot (6x + 5) = \frac{1}{2\sqrt{3x^2 + 5x - 2}} \cdot (6x + 5).$$

(c) Let $z = g(x) = x^2 + x^4$ and $f(z) = \frac{1}{z}$, so $g'(x) = 2x + 4x^3$ and $f'(z) = -z^{-2} = -\frac{1}{z^2}$. Then

$$\frac{d}{dx}\left(\frac{1}{x^2 + x^4}\right) = -\frac{1}{z^2}(2x + 4x^3) = -\frac{2x + 4x^3}{(x^2 + x^4)^2}.$$

We could have done this problem using the quotient rule. Try it and see that you get the same answer!

(d) Let $z = g(x) = e^x + 1$ and $f(z) = \sqrt{z}$, so $g'(x) = e^x$ and $f'(z) = \frac{1}{2\sqrt{z}}$. We get

$$\frac{d(\sqrt{e^x + 1})}{dx} = \frac{1}{2\sqrt{z}}e^x = \frac{e^x}{2\sqrt{e^x + 1}}.$$

(e) To evaluate e^{x^2} you must first evaluate x^2 and then take e to that power. So $z = g(x) = x^2$ and $f(z) = e^z$, so $g'(x) = 2x$, and $f'(z) = e^z$, giving

$$\frac{d(e^{x^2})}{dx} = e^z \cdot 2x = e^{x^2} \cdot 2x = 2xe^{x^2}.$$

□

□ **Example 2** Find the derivative of e^{2x} by the chain rule and by the product rule.

Solution. chain rule: Let the inside function be $z = g(x) = 2x$ and the outside function be $f(z) = e^z$ so

$$\frac{d(e^{2x})}{dx} = f'(g(x)) \cdot g'(x) = e^{2x} \cdot 2 = 2e^{2x}.$$

Product rule: Let $e^{2x} = e^x \cdot e^x$ so

$$\frac{d(e^{2x})}{dx} = \frac{d(e^x e^x)}{dx} = \left(\frac{d(e^x)}{dx}\right)e^x + e^x\left(\frac{d(e^x)}{dx}\right) = e^x \cdot e^x + e^x \cdot e^x = 2e^{2x}.$$

□

Using the Product and Chain Rules to Differentiate a Quotient

If you prefer, you can differentiate a quotient by the product and chain rules, instead of by the quotient rule. The resulting formulas may look different, but they will be equivalent.

❏ **Example 3** Find $k'(x)$ if

$$k(x) = \frac{x}{(x^2 + 1)^2}.$$

Solution. One way is to use the quotient rule:

$$
\begin{aligned}
k'(x) &= \frac{1 \cdot (x^2 + 1)^2 - x \cdot 2(x^2 + 1) \cdot 2x}{[(x^2 + 1)^2]^2} \\
&= \frac{(x^2 + 1)[x^2 + 1 - 4x^2]}{(x^2 + 1)^4} \qquad \text{factoring } (x^2 + 1) \text{ out from top} \\
&= \frac{1 - 3x^2}{(x^2 + 1)^3}. \qquad \text{cancelling } (x^2 + 1)
\end{aligned}
$$

Alternatively you can write the original function as a product

$$k(x) = x \frac{1}{(x^2 + 1)^2} = x \cdot (x^2 + 1)^{-2}$$

and use the product rule:

$$k'(x) = 1 \cdot (x^2 + 1)^{-2} + x \cdot \frac{d}{dx}\left((x^2 + 1)^{-2}\right).$$

Now use the chain rule to differentiate $(x^2 + 1)^{-2}$. Let $z = x^2 + 1$, $f(z) = z^{-2}$, so

$$\frac{d}{dx}\left((x^2 + 1)^{-2}\right) = -2z^{-3} \cdot 2x = -2(x^2 + 1)^{-3} \cdot 2x = \frac{-4x}{(x^2 + 1)^3}.$$

Therefore

$$k'(x) = \frac{1}{(x^2 + 1)^2} + x \cdot \frac{-4x}{(x^2 + 1)^3} = \frac{1}{(x^2 + 1)^2} - \frac{4x^2}{(x^2 + 1)^3}.$$

Notice that if the product and chain rules are used, no cancelling is necessary. If you put these two fractions over a common denominator, you will get the same answer as from the quotient rule. ❏

❖ Exercises for Section 4.5

For Problems 1–23, find the derivative.

1. $f(x) = (x + 1)^{99}$

2. $f(x) = \sqrt{1 - x^2}$

3. $w = (t^2 + 1)^{100}$

4. $w = (t^3 + 1)^{100}$

5. $w = (\sqrt{t} + 1)^{100}$

6. $f(t) = e^{3t}$

7. $y = e^{\frac{3}{2}w}$

8. $y = e^{-4t}$

9. $y = \sqrt{s^3 + 1}$

10. $w = e^{\sqrt{s}}$

11. $y = te^{-t^2}$

12. $f(z) = \sqrt{z}e^{-z}$

13. $f(z) = \dfrac{\sqrt{z}}{e^z}$

14. $z = 2^{5t-3}$

15. $f(z) = \dfrac{1}{(e^z + 1)^2}$

16. $f(\theta) = \dfrac{1}{1 + e^{-\theta}}$

17. $f(x) = 6e^{5x} + e^{-x^2}$

18. $f(w) = (5w^2 + 3)e^{w^2}$

19. $w = (t^2 + 3t)(1 - e^{-2t})$

20. $f(y) = \sqrt{10^{(5-y)}}$

21. $f(x) = e^{-(x-1)^2}$

22. $f(y) = e^{e^{(y^2)}}$

23. $f(t) = 2 \cdot e^{-2e^{2t}}$

24. Find the equation of the tangent line at $x = 1$ to $y = f(x)$ where $f(x)$ is the function in Problem 17.

25. For what values of x is the graph of $y = e^{-x^2}$ concave down?

26. Given:
$$\begin{cases} F(2) = 1 & G(4) = 2 \\ F(4) = 3 & G(3) = 4 \\ F'(2) = 5 & G'(4) = 6 \\ F'(4) = 7 & G'(3) = 8 \end{cases}$$
find:
$$\begin{cases} \text{(a)} & H(4) & \text{if } H(x) = F(G(x)) \\ \text{(b)} & H'(4) & \text{if } H(x) = F(G(x)) \\ \text{(c)} & H(4) & \text{if } H(x) = G(F(x)) \\ \text{(d)} & H'(4) & \text{if } H(x) = G(F(x)) \\ \text{(e)} & H'(4) & \text{if } H(x) = \frac{F(x)}{G(x)} \end{cases}$$

27. Suppose $f(x) = (2x + 1)^{10}(3x - 1)^7$. Find a formula for $f'(x)$. Then decide on a reasonable way to simplify your result, and find a formula for $f''(x)$.

28. Is $x = \sqrt[3]{2t + 5}$ a solution to the equation $3x^2 \dfrac{dx}{dt} = 2$? Why or why not?

29. One gram of radioactive Carbon-14 decays according to the formula

$$Q = e^{-0.000121t},$$

where Q is the number of grams of Carbon-14 remaining after t years.

(a) Find the rate at which Carbon-14 is decaying (in grams/year).

(b) Sketch the rate you found in (a) against time.

30. The temperature, H, in degrees Fahrenheit (°F), of a can of soda that is put into a refrigerator to cool is given as a function of time, t, in hours by

$$H = 40 + 30e^{-2t}.$$

(a) Find the rate at which the temperature of the soda is changing (in °F/hour).

(b) What is the sign of $\dfrac{dH}{dt}$? Why?

(c) When, for $t \geq 0$, is the magnitude of $\dfrac{dH}{dt}$ largest? In terms of the can of soda, why is this?

31. The Theory of Relativity predicts that an object whose mass is m_0 when it is at rest will appear heavier when moving at speeds near the speed of light. When the object is moving at speed v, its mass m is given by

$$m = \frac{m_0}{\sqrt{1 - \frac{v^2}{c^2}}}$$

where c is the speed of light.

(a) Find $\dfrac{dm}{dv}$.

(b) In terms of physics, what does $\dfrac{dm}{dv}$ tell you?

4.6   The Trigonometric Functions

Derivatives of the Sine and Cosine

Since the sine and cosine functions are periodic, their derivatives must be periodic also. (Why?) Let's look at the graph of $f(x) = \sin x$ in Figure 4.17 and estimate the derivative function graphically.

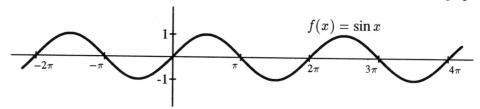

Figure 4.17: $f(x) = \sin x$

First you might ask yourself where the derivative is zero. (At $x = \pm\pi/2, \pm 3\pi/2, \pm 5\pi/2$, etc.) Then ask yourself where the derivative is positive and where it is negative. (Positive for $-\frac{\pi}{2} < x < \frac{\pi}{2}$; negative for $\frac{\pi}{2} < x < \frac{3\pi}{2}$, etc.) Since the largest positive slopes are at $x = 0, 2\pi$, etc., while the largest negative slopes are at $x = \pi, 3\pi$, etc., you get something like the graph in Figure 4.18.

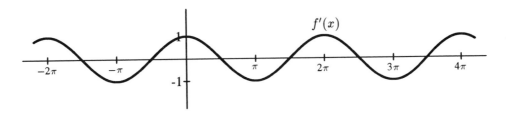

Figure 4.18: Derivative of $f(x) = \sin x$

The graph of the derivative certainly looks suspiciously like the graph of the cosine function and this might lead you to conjecture, quite correctly, that the derivative of the sine is the cosine.

Of course, we cannot be sure, just from the graphs, that the derivative of the sine really is the cosine. Even if the graph of the derivative of the sine looks exactly like the cosine, we can't be sure it isn't the graph of something else. However, we'll assume for now that the derivative of the sine *is* the cosine, and confirm the result at the end of the section.

One thing we can do now is to check that the derivative function in Figure 4.18 has amplitude 1 (as it ought to if it is the cosine). That means we have to show that the derivative of $f(x) = \sin x$ is 1 when $x = 0$. The next example shows us this is true when x is in radians.

☐ **Example 1** Using a calculator, estimate the derivative of $f(x) = \sin x$ at $x = 0$. Make sure your calculator is set in radians.

Solution. Since $f(x) = \sin x$,

$$f'(0) = \lim_{h \to 0} \frac{\sin(0 + h) - \sin 0}{h} = \lim_{h \to 0} \frac{\sin h}{h}.$$

Table 4.4 suggests that this limit is 1, so

$$f'(0) = \lim_{h \to 0} \frac{\sin h}{h} = 1.$$

h(radians)	$\frac{\sin h}{h}$
±0.1	0.99833
±0.01	0.99998
±0.001	1.00000
±0.0001	1.00000

Table 4.4: $\dfrac{\sin h}{h}$ for decreasing values of h

☐

Warning: It is important to notice that in the previous example h was in *radians*; any conclusions we have drawn about the derivative of $\sin x$ are valid *only* when x is in radians.

☐ **Example 2** Starting with the graph of the cosine function, sketch a graph of its derivative.

Solution. The graph of $g(x) = \cos x$ is in Figure 4.19(a). Its derivative is 0 at $x = 0, \pm\pi, \pm2\pi$, etc.; it is positive for $-\pi < x < 0$, $\pi < x < 2\pi$, etc., and negative for $0 < x < \pi$, $2\pi < x < 3\pi$, etc. The derivative is in Figure 4.19(b).

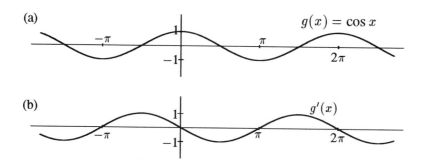

Figure 4.19: $g(x) = \cos x$ and its Derivative, $g'(x)$

As we did with the sine, we'll use the graphs to make a conjecture. The derivative of the cosine in Figure 4.19(b) looks exactly like the graph of sine, except reflected in the x-axis. But how can we be sure that the derivative is $-\sin x$?

☐ **Example 3** Use the relation $\dfrac{d}{dx}(\sin x) = \cos x$ to show that $\dfrac{d}{dx}(\cos x) = -\sin x$.

Solution. Since the cosine function is the sine function shifted to the left by $\frac{\pi}{2}$, we would expect the derivative of the cosine to be the derivative of the sine, shifted to the left. Since

$$\cos x = \sin\left(x + \frac{\pi}{2}\right),$$

using the chain rule gives

$$\frac{d}{dx}(\cos x) = \frac{d}{dx}\left(\sin\left(x + \frac{\pi}{2}\right)\right) = \cos\left(x + \frac{\pi}{2}\right).$$

But $\cos(x + \frac{\pi}{2})$ is the cosine shifted to the left by $\frac{\pi}{2}$, which gives a sine curve reflected in the x-axis. Thus

$$\frac{d}{dx}(\cos x) = \cos\left(x + \frac{\pi}{2}\right) = -\sin x.$$

In summary, for x in radians,

$$\frac{d}{dx}(\sin x) = \cos x \quad \text{and} \quad \frac{d}{dx}(\cos x) = -\sin x.$$

Now let's look at these formulas numerically.

☐ **Example 4** Construct a table of values for $\sin x$ for $x = 0, 0.1, 0.2, \ldots, 0.6$ (in radians). Use difference quotients with $h = 0.001$ to estimate the derivative at these points and compare with the values of $\cos x$.

Solution. The point of this example is to notice that the entries in the second column of the table to the right agree almost exactly with those in the third column. The fact that the difference quotients approximate $\cos x$ so closely suggests that $\frac{d}{dx}(\sin x) = \cos x$.

x	Difference Quotient $= \frac{\sin(x+h)-\sin x}{h}$	$\cos x$
0	1.000	1.000
0.1	0.995	0.995
0.2	0.980	0.980
0.3	0.955	0.955
0.4	0.921	0.921
0.5	0.877	0.878
0.6	0.825	0.825

$\square$

☐ **Example 5** Differentiate (a) $2\sin 3\theta$, (b) $\cos^2 x$, (c) $\cos(x^2)$, (d) $e^{-\sin t}$.

Solution. Use the chain rule:

(a) $\dfrac{d(2\sin 3\theta)}{d\theta} = 2\dfrac{d(\sin 3\theta)}{d\theta} = 2(\cos 3\theta)\dfrac{d(3\theta)}{d\theta} = 2(\cos 3\theta)3 = 6\cos 3\theta$

(b) $\dfrac{d(\cos^2 x)}{dx} = \dfrac{d\left((\cos x)^2\right)}{dx} = 2(\cos x)\cdot\dfrac{d(\cos x)}{dx} = 2(\cos x)(-\sin x) = -2\cos x \sin x$

(c) $\dfrac{d}{dx}(\cos(x^2)) = -\sin(x^2)\cdot\dfrac{d}{dx}(x^2) = -2x\sin(x^2)$

(d) $\dfrac{d(e^{-\sin t})}{dt} = e^{-\sin t}\dfrac{d(-\sin t)}{dt} = -(\cos t)e^{-\sin t}$

$\square$

Derivative of the Tangent

Since $\tan x = \dfrac{\sin x}{\cos x}$, we can differentiate $\tan x$ using the quotient rule:

$$\frac{d(\tan x)}{dx} = \frac{(\sin x)'(\cos x) - (\sin x)(\cos x)'}{\cos^2 x} = \frac{\cos^2 x + \sin^2 x}{\cos^2 x} = \frac{1}{\cos^2 x}$$

so, for x in radians,

$$\boxed{\frac{d}{dx}(\tan x) = \frac{1}{\cos^2 x}}$$

The graphs of $f(x) = \tan x$ and $f'(x) = \frac{1}{\cos^2 x}$ are in Figure 4.20. Does it seem reasonable to you that f' is always positive? Are the asymptotes of f' where you'd expect?

☐ **Example 6** Differentiate (a) $2\tan 3t$, (b) $\tan(1-\theta)$, (c) $\dfrac{1+\tan t}{1-\tan t}$.

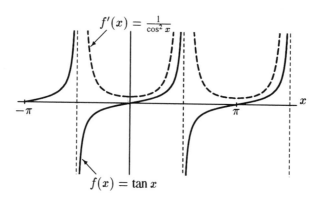

Figure 4.20: Tan x and its Derivative

Solution.

(a) Use the chain rule: $\dfrac{d}{dt}(2 \tan 3t) = 2 \dfrac{1}{\cos^2 3t} \dfrac{d(3t)}{dt} = \dfrac{6}{\cos^2 3t}$

(b) Use the chain rule: $\dfrac{d(\tan(1-\theta))}{d\theta} = \dfrac{1}{\cos^2(1-\theta)} \cdot \dfrac{d(1-\theta)}{d\theta} = \dfrac{-1}{\cos^2(1-\theta)}$.

(c) Use the quotient rule:

$$
\begin{aligned}
\frac{d}{dt}\left(\frac{1 + \tan t}{1 - \tan t}\right) &= \frac{(\frac{d(1+\tan t)}{dt})(1 - \tan t) - (1 + \tan t)\frac{d(1-\tan t)}{dt}}{(1 - \tan t)^2} \\
&= \frac{\frac{1}{\cos^2 t}(1 - \tan t) - (1 + \tan t)\left(-\frac{1}{\cos^2 t}\right)}{(1 - \tan t)^2} \\
&= \frac{2}{\cos^2 t \cdot (1 - \tan t)^2}.
\end{aligned}
$$

❑

Informal Justification of $\dfrac{d}{dx}(\sin x) = \cos x$

To find $\dfrac{d}{dx}(\sin x)$, we need to estimate

$$\frac{\sin(x + h) - \sin x}{h}$$

In Figure 4.21, the quantity $\sin(x + h) - \sin x$ is represented by the length QA. The arc QP is of length h, so

$$\frac{\sin(x + h) - \sin x}{h} = \frac{QA}{\text{Arc } QP}$$

Now, if h is small, QAP is more or less a right triangle because the arc QP is almost a straight line. Furthermore, using geometry, you can show that angle $AQP \approx x$. Thus, for small h:

$$\frac{\sin(x+h) - \sin x}{h} = \frac{QA}{\text{Arc } QP} \approx \cos x.$$

As $h \to 0$, the approximation gets better, so

$$\frac{d}{dx}(\sin x) = \lim_{h \to 0} \frac{\sin(x+h) - \sin x}{h} = \cos x.$$

Other derivations of this result are given in Exercises 32 and 33 at the end of this section.

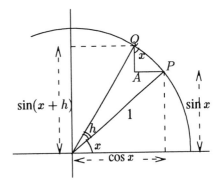

Figure 4.21: Unit Circle showing $\sin(x+h)$ and $\sin x$

❖ Exercises for Section 4.6

For Problems 1–20, find the derivative.

1. $f(x) = \sqrt{1 - \cos x}$

2. $f(x) = \cos(\sin x)$

3. $f(x) = \sin(3x)$

4. $z = \cos(4\theta)$

5. $w = \sin(e^t)$

6. $f(x) = x^2 \cos x$

7. $f(x) = e^{\cos x}$

8. $f(y) = e^{\sin y}$

9. $z = \theta e^{\cos \theta}$

10. $f(x) = 2x \sin(3x)$

11. $f(x) = \sin(2x) \cdot \sin(3x)$

12. $y = e^\theta \sin(2\theta)$

13. $f(x) = e^{-2x} \cdot \sin x$

14. $z = \sqrt{\sin t}$

15. $y = \sin^5 \theta$

16. $g(z) = \tan(e^z)$

17. $z = \tan(e^{-3\theta})$

18. $w = e^{-\sin \theta}$

19. $h(t) = t \cos t + \tan t$

20. $f(\theta) = \theta^2 \sin \theta + 2\theta \cos \theta - 2 \sin \theta$

21. Construct a table of values for $\cos x$, $x = 0, 0.1, 0.2, \ldots, 0.6$ as in Example 4, page 252. Estimate the derivative at these points, using $h = 0.001$, and compare with $-\sin x$.

22. A boat at anchor is bobbing up and down in the sea. The vertical distance, y, in feet, between the sea floor and the boat is given as a function of time, t, in minutes, by

$$y = 15 + \sin 2\pi t.$$

(a) Find the vertical velocity, v, of the boat at time t.

(b) Make a rough sketch of y and v against t.

23. On page 80 the depth, y, of water in Boston harbor was given by

$$y = 5 + 4.9 \cos\left(\frac{\pi}{6}t\right),$$

where t is the number of hours since midnight.

(a) Find $\dfrac{dy}{dt}$. What does $\dfrac{dy}{dt}$ represent, in terms of water level?

(b) For $0 \le t \le 24$, when is $\dfrac{dy}{dt}$ zero? (Figure 1.10 on page 81 may help.) Explain what it means (in terms of water level) for $\dfrac{dy}{dt}$ to be zero.

24. Consider the function $f(x) = e^{\sin x}$. Without plotting points, sketch the graph of this function by figuring out some of its properties. Is it periodic? Does it grow unboundedly like an exponential function? Does it ever achieve a negative value? Does it have any zeros? Does it have a maximum value? a minimum value? Check your answer using a graphing calculator or computer.

25. Repeat Problem 24 with $f(x) = \sin(e^x)$.

26. Find $\lim\limits_{h \to 0} \frac{\sin h}{h}$ when h is in degrees. Explain why this should be larger or smaller than the limit obtained with h in radians.

27. Suppose x is measured in degrees. Calculate the derivative of $y = \sin x$ when $x = 35°$.

28. If $w(t) = u(v(t))$ and $v(t) = \cos t$, and $u'(-1) = 2$, find $w'(\pi)$.

29. If $w(t)$ is given as in Problem 28, but $u'(-1)$ is not given, can you still find $w'(\pi)$? Explain why both algebraically and intuitively.

30. Consider the functions $k(p) = \tan(2p)$ and $r(p) = \tan(\frac{1}{2}p)$ for $-\pi < p < \pi$.

(a) Solve the equations $k'(p) = 1$ and $r'(p) = 1$.

(b) Interpret the results in part (a) graphically.

31. For $f(x) = \sin x$, find the equations of the tangent lines at $x = 0$ and at $x = \frac{\pi}{3}$. Use both tangent lines to approximate $\sin \frac{\pi}{6}$. Would you expect these results to be equally accurate, since they are taken equally far away from $x = \frac{\pi}{6}$ but on opposite sides? If the accuracy is different, can you account for the difference?

32. We will use the identities

$$\begin{aligned} \sin(a+b) &= \sin a \cos b + \sin b \cos a \\ \cos(a+b) &= \cos a \cos b - \sin a \sin b \end{aligned}$$

to calculate the derivatives of $\sin x$ and $\cos x$.

(a) Use the definition of the derivative to show that if $f(x) = \sin x$,

$$f'(x) = \sin x \lim_{h \to 0} \frac{\cos h - 1}{h} + \cos x \lim_{h \to 0} \frac{\sin h}{h}$$

(b) Evaluate the limits in part (a) with your calculator to conclude that $f'(x) = \cos x$.

(c) If $g(x) = \cos x$, use the definition of the derivative to show that $g'(x) = -\sin x$.

33. In this problem you will calculate the derivative of $\tan \theta$ rigorously (and without using the derivatives of $\sin \theta$ or $\cos \theta$). You will then use your result for $\tan \theta$ to calculate the derivatives of $\sin \theta$ and $\cos \theta$. Figure 4.22 shows $\tan \theta$ and $\Delta(\tan \theta)$, which is the change in $\tan \theta$, namely $\tan(\theta + \Delta\theta) - \tan \theta$.

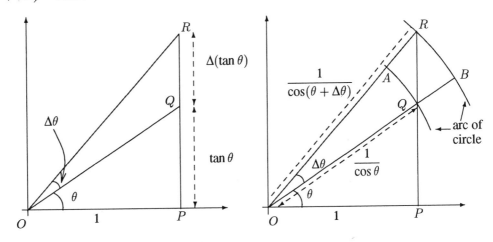

Figure 4.22: $\tan \theta$ and $\Delta(\tan \theta)$
(for Problem 33)

(a) By paying particular attention to how the two figures relate and using the fact that

$$\text{Area of Sector } OAQ \le \text{Area of Triangle } OQR \le \text{Area of Sector } OBR$$

explain why

$$\frac{\Delta\theta}{2\pi} \cdot \pi \left(\frac{1}{\cos\theta}\right)^2 \le \frac{1}{2} \cdot 1 \cdot \Delta(\tan\theta) \le \frac{\Delta\theta}{2\pi} \cdot \pi \left(\frac{1}{\cos(\theta + \Delta\theta)}\right)^2.$$

[Hint: A sector of circle with angle α at the center has area $\frac{\alpha}{2\pi}$ times the area of the whole circle.]

(b) Use part (a) to show as $\Delta\theta \to 0$ that

$$\frac{\Delta\tan\theta}{\Delta\theta} \to \left(\frac{1}{\cos\theta}\right)^2,$$

and hence that $\dfrac{d(\tan\theta)}{d\theta} = \left(\dfrac{1}{\cos\theta}\right)^2.$

(c) Derive the identity $(\tan\theta)^2 + 1 = (\frac{1}{\cos\theta})^2$. Then differentiate both sides of this identity with respect to θ, using the chain rule and the result of part (b) to show that $\dfrac{d}{d\theta}(\cos\theta) = -\sin\theta$.

(d) Differentiate both sides of the identity $(\sin\theta)^2 + (\cos\theta)^2 = 1$ and use the result of part (c) to show that $\dfrac{d}{d\theta}(\sin\theta) = \cos\theta.$

4.7 Applications of the Chain Rule

In this section we will use the chain rule to verify the derivatives of fractional powers and to find the derivatives of logarithms, exponentials, and the inverse trigonometric functions.

Derivative of $x^{1/2}$

Earlier we calculated the derivative of x^n, with n an integer, but we have been using the result for fractional values of n as well. We will now calculate the derivative of one of the powers we've so far taken on faith, $f(x) = x^{1/2}$. Since

$$(f(x))^2 = x,$$

the derivative of $(f(x))^2$ and the derivative of x must be equal, so

$$\frac{d}{dx}\left(f(x)\right)^2 = \frac{d}{dx}(x)$$

We can use the chain rule with $f(x)$ as the inside function to obtain:

$$2f(x) \cdot f'(x) = 1.$$

Solving for $f'(x)$ gives:

$$f'(x) = \frac{1}{2f(x)} = \frac{1}{2x^{1/2}}$$

so

$$\boxed{\frac{d(x^{1/2})}{dx} = \frac{1}{2x^{1/2}}}$$

☐ **Example 1** Differentiate (a) $x^{1/3}$, (b) $x^{1/n}$.

Solution.

(a) Since $(x^{1/3})^3 = x$, differentiating both sides gives

$$3(x^{1/3})^2 \cdot \frac{d}{dx}(x^{1/3}) = 1.$$

Thus $\frac{d}{dx}(x^{1/3}) = \frac{1}{3x^{2/3}} = \frac{1}{3}x^{-2/3}.$

(b) Since $(x^{1/n})^n = x$, differentiating both sides gives

$$n(x^{1/n})^{n-1} \cdot \frac{d}{dx}(x^{1/n}) = 1.$$

Thus $\frac{d}{dx}(x^{1/n}) = \frac{1}{nx^{(n-1)/n}} = \frac{1}{nx^{1-(1/n)}} = \frac{1}{n}x^{(1/n)-1}.$

$\square$

Derivative of $\ln x$

We'll again use the chain rule, and differentiate an identity which involves $\ln x$. Since $e^{\ln x} = x$, differentiating using the chain rule, we have:

$$\frac{d}{dx}(e^{\ln x}) = \frac{d}{dx}(x)$$

$$e^{\ln x} \cdot \frac{d}{dx}(\ln x) = 1.$$

Thus

$$\frac{d}{dx}(\ln x) = \frac{1}{e^{\ln x}} = \frac{1}{x},$$

so

$$\boxed{\frac{d(\ln x)}{dx} = \frac{1}{x}}$$

$\square$ **Example 2** Differentiate (a) $\ln(x^2 + 1)$, (b) $t^2 \ln t$, (c) $\sqrt{1 + \ln(1 - y)}$.

Solution.

(a) Using the chain rule:

$$\frac{d(\ln(x^2 + 1))}{dx} = \frac{1}{x^2 + 1}\frac{d(x^2 + 1)}{dx} = \frac{2x}{x^2 + 1}$$

(b) Using the product rule:

$$\frac{d(t^2 \ln t)}{dt} = \left(\frac{d(t^2)}{dt}\right)\ln t + t^2 \frac{d(\ln t)}{dt} = 2t \ln t + t^2 \cdot \frac{1}{t} = 2t \ln t + t$$

(c)

$$\frac{d\sqrt{1+\ln(1-y)}}{dy} = \frac{1}{2\sqrt{1+\ln(1-y)}} \cdot \frac{d(1+\ln(1-y))}{dy} \qquad \text{using chain rule}$$

$$= \frac{1}{2\sqrt{1+\ln(1-y)}} \cdot \frac{1}{1-y} \cdot \frac{d(1-y)}{dy} \qquad \text{using chain rule again}$$

$$= \frac{-1}{2(1-y)\sqrt{1+\ln(1-y)}}.$$

□

Derivative of a^x

Earlier we stated that the derivative of a^x is proportional to a^x, and that if $a = e$, the constant of proportionality is 1. Here we'll confirm this statement by calculating $\dfrac{d(a^x)}{dx}$ for a general constant a. The identity we will use is $\ln(a^x) = x \ln a$. Differentiating, using $\dfrac{d}{dx}(\ln x) = \dfrac{1}{x}$ and the chain rule on the left and remembering that $\ln a$ is a constant, we obtain:

$$\frac{1}{a^x} \cdot \frac{d}{dx}(a^x) = \ln a.$$

Solving gives

$$\boxed{\frac{d(a^x)}{dx} = (\ln a)a^x}$$

Derivatives of Inverse Trigonometric Functions

Recall that the inverse sine and tangent are defined as follows:

$\arcsin x =$ the number between $-\frac{\pi}{2}$ and $\frac{\pi}{2}$ (inclusive) whose sine is x
$\arctan x =$ the number between $-\frac{\pi}{2}$ and $\frac{\pi}{2}$ whose tangent is x

To find $\frac{d}{dx}(\arctan x)$ we use the identity $\tan(\arctan x) = x$. Differentiating using the chain rule gives

$$\frac{1}{\cos^2(\arctan x)} \cdot \frac{d}{dx}(\arctan x) = 1,$$

so

$$\frac{d}{dx}(\arctan x) = \cos^2(\arctan x).$$

Using the identity $1 + \tan^2 \theta = \dfrac{1}{\cos^2 \theta}$, we have

$$\cos^2(\arctan x) = \frac{1}{1+\tan^2(\arctan x)} = \frac{1}{1+x^2}.$$

Thus,

$$\frac{d}{dx}(\arctan x) = \frac{1}{1+x^2}$$

By a similar argument, one can show that

$$\frac{d}{dx}(\arcsin x) = \frac{1}{\sqrt{1-x^2}}$$

☐ **Example 3** Differentiate   (a) $\arctan(t^2)$ (b) $\arcsin(\tan\theta)$

Solution. Use the chain rule:

(a) $\dfrac{d}{dt}(\arctan(t^2)) = \dfrac{1}{1+(t^2)^2} \cdot \dfrac{d}{dt}(t^2) = \dfrac{2t}{1+t^4}$

(b) $\dfrac{d}{dt}(\arcsin(\tan\theta)) = \dfrac{1}{\sqrt{1-(\tan\theta)^2}} \cdot \dfrac{d}{d\theta}(\tan\theta) = \dfrac{1}{\sqrt{1-\tan^2\theta}} \cdot \dfrac{1}{\cos^2\theta}$

❑

❖ Exercises for Section 4.7

For Problems 1–15, find the derivative of the given function. In some cases, it may be to your advantage to simplify before differentiating.

1. $f(x) = \ln(1-x)$

2. $f(t) = \ln(t^2+1)$

3. $f(z) = \dfrac{1}{\ln z}$

4. $f(\theta) = \ln(\cos\theta)$

5. $f(x) = \ln(1-e^{-x})$

6. $f(\alpha) = \ln(\sin\alpha)$

7. $f(x) = \ln(e^x+1)$

8. $f(x) = \ln(e^{7x})$

9. $f(x) = e^{(\ln x)+1}$

10. $f(w) = \ln(\cos(w-1))$

11. $f(t) = \ln(e^{\ln t})$

12. $f(y) = \arcsin(y^2)$

13. $g(t) = \arctan(3t-4)$

14. $h(w) = w\arcsin w$

15. $g(\alpha) = \sin(\arcsin\alpha)$

16. Using the chain rule, find $\dfrac{d}{dx}(\arcsin x)$.

17. Using the chain rule, find $\dfrac{d}{dx}(\log x)$. (Recall that $\log x = \log_{10} x$.)

18. Imagine you are zooming in on the graph of each of the following functions near the origin:

$$y = x \qquad y = \sqrt{x} \qquad y = x^2 \qquad y = \sin x$$
$$y = x \sin x \qquad y = \tan x \qquad y = \sqrt{\frac{x}{x+1}} \qquad y = x^3$$
$$y = \ln(x+1) \qquad y = \tfrac{1}{2}\ln(x^2+1) \qquad y = 1 - \cos x \qquad y = \sqrt{2x - x^2}$$

Which of them look the same? Group together those functions which become indistinguishable, and give the equations of the lines they look like.

19. (a) Find the equation of the tangent line to $y = \ln x$ at $x = 1$.

(b) Use it to calculate approximate values for $\ln(1.1)$ and $\ln(2)$.

(c) Using a graph, explain whether the approximate values you have calculated are smaller or larger than the true values. Would the same result have held if you had used the tangent line to estimate $\ln(0.9)$ and $\ln(0.5)$? Why?

20. (a) Find the equation of the best quadratic approximation to $y = \ln x$ at $x = 1$. The best quadratic approximation will have the same first and second derivatives as $y = \ln x$ at $x = 1$.

(b) Use a computer or calculator to graph the approximation and $y = \ln x$ on the same set of axes. What do you notice?

(c) Use your quadratic approximation to calculate approximate values for $\ln(1.1)$ and $\ln(2)$.

21. (a) For $x > 0$, find the derivative of $f(x) = \arctan x + \arctan \frac{1}{x}$ and simplify.

(b) What does your result tell you about f?

22. For the amusement of the guests, some hotels have elevators on the outside of the building. Suppose such a hotel is 300 feet high and you are standing by a window 100 feet above the ground and 150 feet away from the hotel, and that the elevator descends at a constant speed of 30 ft/sec, starting at time $t = 0$. Let θ be the angle between the line of your horizon and your line of sight to the elevator. (See Figure 4.23.)

(a) Find the rate of change of θ with respect to the time t.

(b) If the rate of change of θ is a measure of how fast the elevator appears to you to be moving, at what height will the elevator be when it appears to be moving fastest?

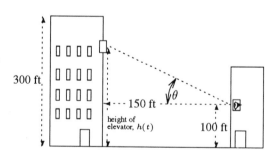

Figure 4.23: Descending Elevator
(for Problem 22)

4.8   Implicit Functions

In earlier chapters, most functions were written in the form $y = f(x)$; here y is said to be an *explicit* function of x. An equation such as

$$x^2 + y^2 = 4$$

is said to give y as an *implicit* function of x. Its graph is the circle in Figure 4.24. Since there are x values which correspond to two y values, y is not a function of x here. Solving gives

$$y = \pm\sqrt{4 - x^2}$$

where $y = \sqrt{4 - x^2}$ represents the top half of the circle and $y = -\sqrt{4 - x^2}$ represents the bottom half. On the top half and the bottom half of the circle, considered separately, y is a function of x.

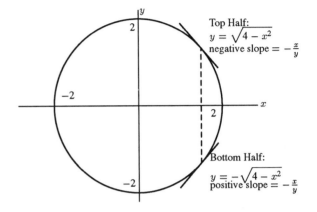

Top Half:
$$y = \sqrt{4 - x^2}$$
negative slope $= -\frac{x}{y}$

Bottom Half:
$$y = -\sqrt{4 - x^2}$$
positive slope $= -\frac{x}{y}$

Figure 4.24: $x^2 + y^2 = 4$

But let's consider the circle as a whole. The equation does represent a curve which has a tangent line at each point. The slope of this tangent can be found by differentiating the equation of the circle with respect to x:

$$\frac{d(x^2)}{dx} + \frac{d(y^2)}{dx} = \frac{d(4)}{dx}.$$

If you think of y as a function of x and use the chain rule, you get:

$$2x + 2y\frac{dy}{dx} = 0.$$

Solving gives

$$\frac{dy}{dx} = -\frac{x}{y}.$$

The derivative here depends on both x and y (instead of just on x). This is because for each x value (except for $x = \pm 2$) there are two y values, and the curve has a different slope at each one. Figure 4.24 shows that for x and y both positive, we are on the top right-half of the curve, and the slope should be negative (as the formula predicts). For x positive and y negative we are on the bottom right-half of the curve and the slope should be positive (as the formula predicts).

Differentiating the equation of the circle has given us the slope of the curve at all except two points, namely $(2, 0)$ and $(-2, 0)$, where the tangent is vertical. In general, an implicitly defined function does lead to a derivative whenever the expression for the derivative does not have a zero in the denominator.

☐ **Example 1**   Make a table of x and approximate y values for the equation $y^3 - xy = -6$ near $x = 7, y = 2$. Your table should include the x-values $6.8, 6.9, 7.0, 7.1, 7.2$.

Solution. We would like to solve for y in terms of x, but we cannot isolate y by factoring. There is a formula for solving cubics, somewhat like the quadratic formula, but it is too complicated to be useful here. Instead, first observe that $x = 7$, $y = 2$ does satisfy the equation. (Check this!)

Then find $\frac{dy}{dx}$ by implicit differentiation:

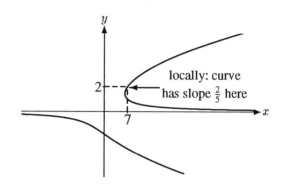

$$\frac{d}{dx}(y^3) - \frac{d}{dx}(xy) = -\frac{d}{dx}(6)$$

$$3y^2\frac{dy}{dx} - y - x\frac{dy}{dx} = 0$$

$$(3y^2 - x)\frac{dy}{dx} = y$$

$$\frac{dy}{dx} = \frac{y}{3y^2 - x}$$

Figure 4.25: Graph of $y^3 - xy = -6$

When $x = 7$ and $y = 2$, we have

$$\frac{dy}{dx} = \frac{2}{12 - 7} = \frac{2}{5}.$$

The equation of the tangent line at $(7, 2)$ is

$$y - 2 = \frac{2}{5}(x - 7)$$

or

$$y = 0.4x - 0.8.$$

Since the tangent lies very close to the curve near the point $(7, 2)$, we use the equation of the tangent line to calculate the approximate y values below:

x	6.8	6.9	7.0	7.1	7.2
approximate y	1.92	1.96	2.00	2.04	2.08

The point of this example is that although an equation like $y^3 - xy = -6$ may lead to a complicated curve which is difficult to deal with algebraically, it still looks like a straight line locally, just like the graph of any differentiable function. This means that the result we get for $\frac{dy}{dx}$ from implicit differentiation is very useful in analyzing the relationship between x and y, at least locally. In particular, we can use implicit differentiation to locate special points, such as where the tangent is horizontal or vertical.

❏ **Example 2**   Find all points where the tangent line to $y^3 - xy = -6$ is either horizontal or vertical.

Solution. From the previous example, $\dfrac{dy}{dx} = \dfrac{y}{3y^2 - x}$.

The tangent is horizontal when the numerator of $\dfrac{dy}{dx}$ equals 0, so $y = 0$. Since we also must satisfy $y^3 - xy = -6$, we get $0^3 - x \cdot 0 = -6$, which is impossible. We conclude that there are no points on the curve where the tangent line is horizontal.

The tangent is vertical when the denominator of $\dfrac{dy}{dx}$ is 0, giving $3y^2 - x = 0$. Thus, $x = 3y^2$ at any point with a vertical tangent line. Again, we must also satisfy $y^3 - xy = -6$, so

$$
\begin{aligned}
y^3 - (3y^2)y &= -6 \\
-2y^3 &= -6 \\
y &= \sqrt[3]{3} \approx 1.442
\end{aligned}
$$

We can then find x by substituting $y = \sqrt[3]{3}$ in $y^3 - xy = -6$. We get $3 - x(\sqrt[3]{3}) = -6$ so $x = \dfrac{9}{\sqrt[3]{3}} \approx 6.240$. $\qquad\qquad\qquad\qquad\qquad\qquad\qquad\qquad\qquad\qquad\qquad\qquad$ ☐

Using our expression for $\frac{dy}{dx}$ to find the points where the tangent is vertical or horizontal, as in the previous example, is a first step in obtaining an overall (global, not just local) picture of the curve $y^3 - xy = -6$. However, filling in the rest of the graph, even roughly, by using the sign of $\frac{dy}{dx}$ to tell us where the curve is increasing and where it is decreasing, can be difficult. Another method of getting the graph is described in Problem 15, on page 266.

❖ Exercises for Section 4.8

For problems 1 through 7, find $\frac{dy}{dx}$.

1. $x^2 + xy - y^3 = xy^2$
2. $x^2 + y^2 = \sqrt{7}$
3. $\sqrt{x} + \sqrt{y} = 25$

4. $\sin(xy) = 2x + 5$
5. $x \ln y + y^3 = \ln x$

6. $e^{\cos y} = x^3 \arctan y$
7. $\cos^2 y + \sin^2 y = y + 2$

For problems 8 through 10, find the equations the tangent lines to the following curves at the indicated points.

8. $x^{\frac{2}{3}} + y^{\frac{2}{3}} = a^{\frac{2}{3}}$ at $(a, 0)$
9. $y^2 = \dfrac{x^2}{xy - 4}$ at $(4, 2)$.
10. $xy^2 = 1$ at $(1, -1)$.

11. Assume y is a differentiable function of x and that $y + \sin y + x^2 = 9$. Use the fact that the point $x = 3, y = 0$ is on the curve to find $\frac{dy}{dx}$ at this point.

12. Show that the power rule for derivatives applies to rational powers of the form $y = x^{m/n}$ by raising both sides to the nth power and using implicit differentiation.

13. (a) Find the equations of the tangent lines to the circle $x^2 + y^2 = 25$ at the points where $x = 4$.

    (b) Find the equations of the normal lines to this circle at the same points. (The normal line to a curve at a point is perpendicular to the tangent there.)

    (c) At what point do the two normal lines intersect?

14. (a) Find the slope of the tangent line to the ellipse $\frac{x^2}{25} + \frac{y^2}{9} = 1$ at the point (x, y).

    (b) Are there any points where the slope is not defined?

15. Solve the equation $y^3 - xy = -6$ for x in terms of y and show how this enables you to sketch a graph of the equation by viewing x as a function of y. (Put the x-axis horizontally.) For which values of x are there three y-values? Two y-values? One y-value? Identify the points where the tangent line is vertical. Compare your answer to the result of Example 2, page 264.

16. Consider the equation $x^3 + y^3 - xy^2 = 5$.

    (a) Find $\frac{dy}{dx}$ by implicit differentiation.

    (b) Give a table of approximate y values near $x = 1, y = 2$ for $x = 0.96, 0.98, 1, 1.02, 1.04$.

    (c) Find the y-value for $x = 0.96$ by substituting $x = 0.96$ in the equation and solving for y using a computer or calculator. Compare with your answer in (b).

    (d) Find all points where the tangent line is horizontal or vertical.

17. Consider the curve $xe^{5y} = 3y$.

    (a) Find $\frac{dy}{dx}$.

    (b) Find the equation of the tangent line to the curve at $(0, 0)$.

    (c) If $x = 0.1$, estimate y using the tangent line.

4.9 Notes on the Tangent Line Approximation

Since the graph of a function and its tangent line have the same slope at the point of tangency (namely, the derivative at that point), the tangent line stays close to the graph of the function near that point. As we have seen, this means that we can approximate values of the function by values from the tangent line. This can be helpful because values on the line may be easier to calculate.

The Tangent Line Approximation

The slope of the tangent to the graph of $y = f(x)$ at $x = a$ is $f'(a)$, so the equation of the tangent line is

$$y - f(a) = f'(a)(x - a)$$

or

$$y = f(a) + f'(a)(x - a).$$

Now we approximate the values of f by the y values from the line, giving:

The Tangent Line Approximation: For values of x near a,

$$f(x) \approx f(a) + f'(a)(x - a)$$

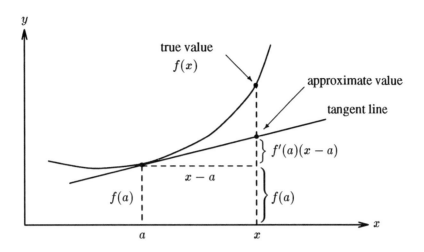

Figure 4.26: Local Linearization: Approximation by the Tangent Line

Consider Figure 4.26; since we are thinking of a as fixed, $f(a)$ and $f'(a)$ are constant, and the right-hand side of this "almost equality" is a linear function of x. We call the function

$$f(a) + f'(a)(x - a)$$

the *local linearization* of f near $x = a$.

☐ **Example 1**   What is the tangent line approximation for $\sin x$ near $x = 0$?

Solution. The local linearization of f near $x = 0$ is

$$f(x) \approx f(0) + f'(0)(x - 0).$$

When $f(x) = \sin x$, $f'(x) = \cos x$ so $f(0) = \sin 0 = 0$ and $f'(0) = \cos 0 = 1$. Our approximation is thus

$$\sin x \approx x.$$

This means that, near $x = 0$, the function $f(x) = \sin x$ can be approximated by the function $y = x$. Another way of saying this is that if you zoom in on the graphs of the functions $\sin x$ and x near the origin, you won't be able to tell them apart. See Figure 4.27.

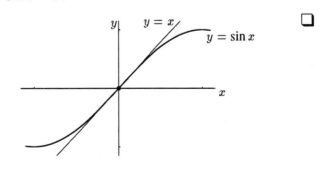

Figure 4.27: Tangent approximation to $y = \sin x$

❑ **Example 2** Use local linearity to find $\lim\limits_{h \to 0} \dfrac{\sin 3h}{h}$.

Solution. When h is small, $3h$ is small too, so Example 1 tells us that $\sin 3h \approx 3h$, so $\dfrac{\sin 3h}{h} \approx \dfrac{3h}{h} = 3$. The approximation gets better and better as h gets smaller, so

$$\lim_{h \to 0} \frac{\sin 3h}{h} = 3.$$

❑

❑ **Example 3** What is the local linearization of e^{kx} near $x = 0$?

Solution. With $f(x) = e^{kx}$, we see that the local linearization of f near $x = 0$,

$$f(x) \approx f(0) + f'(0)(x - 0),$$

becomes

$$e^{kx} \approx e^{k \cdot 0} + f'(0)x,$$

where $f'(0)$ is the derivative of e^{kx} at $x = 0$. By the chain rule, if $f(x) = e^{kx}$, then $f'(x) = ke^{kx}$, so $f'(0) = ke^{k \cdot 0} = k$. Thus our formula reduces to

$$e^{kx} \approx 1 + kx.$$

This is the local linearization of e^{kx} near $x = 0$. In other words, if you zoom in on the functions $f(x) = e^{kx}$ and $y = 1 + kx$ near the origin, you won't be able to tell them apart. ❑

❑ **Example 4** Linearize $\sqrt{1 + x}$ near $x = 0$.

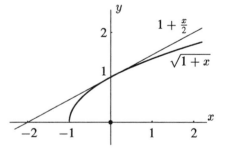

Solution. With $f(x) = \sqrt{1 + x}$, the chain rule gives $f'(x) = 1/(2\sqrt{1 + x})$, so $f(0) = 1$ and $f'(0) = 1/2$. Therefore the local linearization of f near $x = 0$,

$$f(x) \approx f(0) + f'(0)(x - 0),$$

becomes

$$\sqrt{1 + x} \approx 1 + \frac{x}{2}.$$

Figure 4.28: Tangent approximation to $\sqrt{1 + x}$

This means that, near $x = 0$, the function $\sqrt{1 + x}$ can be approximated by its tangent line $y = 1 + x/2$. See Figure 4.28. ❑

❑ **Example 5 Compound Interest.** One easy way to estimate how quickly money in a savings account grows is to use the "Rule of 70", which says that the time in years it takes money to double in an account bearing $i\%$ annual interest compounded yearly is approximately $70/i$ for small values of i. Find the local linearization of $\ln(1 + x)$ and use it to verify this rule.

Solution. Let $r = \frac{i}{100} = i\%$. (For example if $i = 5\%$, $r = 0.05$.) Then the balance $\$B$ after t years is given by

$$B = P(1 + r)^t,$$

where $\$P$ is the original deposit. If we are doubling our money, then $B = 2P$, so we wish to solve for t in the equation $2P = P(1 + r)^t$, which is equivalent to

$$2 = (1 + r)^t.$$

Taking natural logarithms of both sides and solving for t yields

$$\ln 2 = t \ln(1 + r)$$
$$t = \frac{\ln 2}{\ln(1 + r)}.$$

We now approximate $\ln(1 + r)$ near $r = 0$. Let $f(r) = \ln(1 + r)$. Then $f'(r) = 1/(1 + r)$. So $f(0) = 0$, and $f'(0) = 1$, so

$$f(r) \approx f(0) + f'(0)r$$

becomes

$$\ln(1 + r) \approx r.$$

Therefore

$$t = \frac{\ln 2}{\ln(1 + r)} \approx \frac{\ln 2}{r} = \frac{100 \ln 2}{i} \approx \frac{69.3}{i} \approx \frac{70}{i},$$

as claimed. We expect this approximation to hold for small values of r and i; it turns out that values of i up to 10 give good enough answers for most everyday purposes. ❑

This example illustrates a handy feature of local linearization; such estimates as the Rule of 70, or $\sin x \approx x$ for small x, are fairly accurate and very useful for quick calculations.

❖ Exercises for Section 4.9

1. Show that $e^{-x} \approx 1 - x$ near $x = 0$.

2. Show that $1 - x/2$ is the local linearization of $1/(\sqrt{1 + x})$ near $x = 0$.

3. What is the local linearization of $\dfrac{1}{x}$ near $x = 1$?

4. Using the local linearization of e^x, evaluate $\displaystyle\lim_{h \to 0} \frac{e^{kh} - 1}{h}$.

5. Interpret the local linearization $e^{rt} \approx 1 + rt$ in terms of compound interest if r is the nominal annual interest rate, and t is the time in years.

6. (a) Show that $1 + kx$ is the local linearization of $(1 + x)^k$ near $x = 0$.

   (b) Someone claims that the square root of 1.1 is about 1.05. Without using a calculator, do you think that this estimate is about right?

   (c) Is the actual number above or below 1.05?

7. (a) Use the identity $\cos x = 1 - 2\sin^2(\frac{x}{2})$ and the tangent line approximation to $\sin x$ near $x = 0$ to obtain

$$\cos x \approx 1 - \frac{x^2}{2}.$$

   (b) Using this approximation, evaluate

   i. $\displaystyle\lim_{h \to 0} \frac{\cos h - 1}{h}$

   ii. $\displaystyle\lim_{h \to 0} \frac{\cos h - 1}{h^2}$

8. Multiply the local linearization of e^x near $x = 0$ by itself to obtain an approximation for e^{2x}. Compare this with the actual local linearization of e^{2x}. Explain why they are consistent, and discuss which one is more accurate.

9. (a) Show that $1 - x$ is the local linearization of $\dfrac{1}{1 + x}$ near $x = 0$.

   (b) From your answer to part (a), show that near $x = 0$,

   $$\frac{1}{1 + x^2} \approx 1 - x^2.$$

   (c) Without differentiating, what do you think the derivative of $\dfrac{1}{1 + x^2}$ is at $x = 0$?

10. From the local linearizations of e^x and $\sin x$ near $x = 0$, write down the local linearization of the function $e^x \sin x$. From this result, write down the derivative of $e^x \sin x$ at $x = 0$. Using this technique, write down the derivative of $\dfrac{e^x \sin x}{1 + x}$ at $x = 0$.

11. Derive the product rule,

$$[f(x)g(x)]' = f'(x)g(x) + f(x)g'(x),$$

   using local linearization. [Hint: Use the definition of the derivative, and use local linearization for $f(x + h)$ and $g(x + h)$.]

12. Derive the chain rule using local linearization. [Hint: In other words, differentiate $f(g(x))$, using $g(x + h) \approx g(x) + g'(x)h$ and $f(z + k) \approx f(z) + f'(z)k$.]

4.10   Miscellaneous Exercises for Chapter 4

For Problems 1–21, find the derivative. Assume a and k are constants.

1. $f(x) = (3x^2 + \pi)(e^x - 4)$

2. $g(x) = 2x - \frac{1}{\sqrt[3]{x}} + 3^x - e$

3. $f(z) = \dfrac{z^2 + 1}{\sqrt{z}}$

4. $h(r) = \dfrac{r^2}{2r + 1}$

5. $g(t) = e^{(1+3t)^2}$

6. $f(t) = 2te^t - \dfrac{1}{\sqrt{t}}$

7. $h(x) = xe^{\tan x}$

8. $g(w) = \dfrac{1}{2^w + e^w}$

9. $f(y) = \ln \ln(2y^3)$

10. $f(x) = (2 - 4x - 3x^2)(6x^e - 3\pi)$

11. $r(\theta) = \sin((3\theta - \pi)^2)$

12. $s(\theta) = \sin^2(3\theta - \pi)$

13. $g(\theta) = \sqrt{a^2 - \sin^2 \theta}$

14. $g(x) = \dfrac{x^2 + \sqrt{x} + 1}{x^{\frac{3}{2}}}$

15. $w(\theta) = \dfrac{\theta}{\sin^2 \theta}$

16. $p(\theta) = \dfrac{\sin(5 - \theta)}{\theta^2}$

17. $h(t) = \ln\left(e^{-t} - t\right)$

18. $g(x) = x^k + k^x$

19. $s(x) = \arctan(2 - x)$

20. $r(\theta) = e^{(e^\theta + e^{-\theta})}$

21. $r(y) = \dfrac{y}{(\cos y) + a}$

22. Given: $r(2) = 4$, $s(2) = 1$, $s(4) = 2$ $r'(2) = -1$, $s'(2) = 3$, $s'(4) = 3$.

Compute the following derivatives or state what additional information you would need to be able to compute the derivative.

(a) $H'(2)$ if $H(x) = r(x) \cdot s(x)$

(b) $H'(2)$ if $H(x) = \sqrt{r(x)}$

(c) $H'(2)$ if $H(x) = r(s(x))$

(d) $H'(2)$ if $H(x) = s(r(x))$

23. Imagine you are zooming in on the graphs of the following functions near the origin:

$$y = \arcsin x \quad y = \sin x - \tan x \quad y = x - \sin x \quad y = \frac{x}{x+1}$$
$$y = \frac{\sin x}{1+\sin x} \quad y = \frac{x^2}{x^2+1} \quad y = \frac{1-\cos x}{\cos x} \quad y = \frac{x}{x^2+1}$$
$$y = \frac{\sin x}{x} - 1 \quad y = -x \ln x \quad y = e^x - 1 \quad y = x^{10} + \sqrt[10]{x}$$
$$y = \arctan x$$

Which of them look the same? Group together those functions which become indistinguishable, and give the equation of the line they look like. [Note: $\frac{\sin x}{x} - 1$ and $-x \ln x$ never quite make it to the origin.]

24. The graphs of $\sin x$ and $\cos x$ intersect once between 0 and $\frac{\pi}{2}$. What is the angle between the two curves at the point where they intersect? (You need to think about how the angle between two curves should be defined.)

25. The acceleration due to gravity, g, at a distance r from the center of the earth is given by

$$g = \frac{GM}{r^2},$$

    where M is the mass of the earth and G is a constant.

    (a) Find $\dfrac{dg}{dr}$.

    (b) What is the practical interpretation (in terms of acceleration) of $\dfrac{dg}{dr}$? Why would you expect it to be negative?

26. (Continuation of Problem 25.)
    You are told that $M = 6 \times 10^{24}$ and $G = 6.67 \times 10^{-20}$ where M is in kg and r in km.

    (a) What is the value of $\dfrac{dg}{dr}$ at the surface of the earth ($r = 6400$ km) ?

    (b) What does this tell you about whether or not it is reasonable to assume g is constant near the surface of the Earth?

27. In 1975, the population of Mexico was about 84 million and growing at 2.6% annually, while the population of the U.S. was about 250 million and growing at 0.7% annually. Which population was growing faster, if we measure growth rates in people/year? Explain your answer.

28. A museum has decided to sell one of its paintings and to invest the proceeds. The price the painting will fetch changes with time, and is denoted by $P(t)$, where t is the number of years since 1990. If the picture is sold between 1990 and 2010 (so $0 \le t \le 20$), and the money from the sale is invested in a bank account earning 5% annual interest compounded once a year, the balance, $B(t)$, in the account in the year 2010 is given by

$$B(t) = P(t)(1.05)^{20-t}.$$

    (a) Explain why $B(t)$ is given by this formula.

    (b) Show that the formula for $B(t)$ is equivalent to

$$B(t) = (1.05)^{20} \frac{P(t)}{(1.05)^t}.$$

    (c) Find $B'(10)$, given that $P(10) = 150,000$ and $P'(10) = 5000$.

29. If you invest $\$P$ in a bank account at an annual interest rate of $r\%$, after t years you will have $\$B$, where

$$B = P\left(1 + \frac{r}{100}\right)^t.$$

(a) Find $\dfrac{dB}{dt}$, assuming P and r are constant. In terms of money, what does $\dfrac{dB}{dt}$ represent?

(b) Find $\dfrac{dB}{dr}$, assuming P and t are constant. In terms of money, what does $\dfrac{dB}{dr}$ represent?

30. Suppose the distance, s, of a moving body from a fixed point is given as a function of time by $s = 20e^{t/2}$.

(a) Find the velocity, v, of the body as a function of t.

(b) Find v as a function of s, and hence show that s satisfies the differential equation $s' = \dfrac{1}{2}s.$

31. Air pressure at sea level is 30 inches of mercury. At an altitude of h feet above sea level, the air pressure, P, in inches of mercury is given by

$$P = 30e^{-3.23 \times 10^{-5}h}$$

(a) Sketch a rough graph of P against h.

(b) Find the equation of the tangent line to the graph at $h = 0$.

(c) A rough rule of thumb used by travelers is that air pressure drops about 1 inch for every 1000 feet increase in height above sea level. Write a formula for the air pressure given by this rule of thumb.

(d) What is the relation between your answers to (b) and (c)? Explain why the rule of thumb works.

(e) Are the predictions made by the rule of thumb too large or too small? Why?

32. (a) Using the chain rule, find g' for $g(x) = e^{n \ln x}$, where n is any real number.

(b) Using the fact that $x^n = e^{n \ln x}$, show you can use g' to verify the power rule, $\dfrac{d(x^n)}{dx} = nx^{n-1}$, for all real numbers, n.

33. (a) Assuming that $\dfrac{d}{dx}(\ln x) = \dfrac{1}{x}$, show that

$$\lim_{n \to \infty} \left(1 + \frac{1}{n}\right)^n = e$$

by writing $\left(1 + \frac{1}{n}\right)^n = e^{n \cdot \ln\left(1 + \frac{1}{n}\right)}$ and showing that $n \ln\left(1 + \frac{1}{n}\right)$ is the slope of the secant to $f(x) = \ln x$ between $x = 1$ and $x = \left(1 + \frac{1}{n}\right)$.

(b) What does the formula

$$\lim_{n \to \infty} \left(1 + \frac{1}{n}\right)^n = e$$

tell you about compound interest?

(c) Explain in terms of compound interest why it is reasonable to expect that

$$\left(1 + \frac{1}{n+1}\right)^{n+1} > \left(1 + \frac{1}{n}\right)^n$$

for all positive integers n.

(d) Demonstrate the formula

$$\left(1 + \frac{1}{n+1}\right)^{n+1} > \left(1 + \frac{1}{n}\right)^n$$

using again the hint above. Explain your reasoning with a sketch.

34. Given a number $a > 1$, the equation

$$a^x = 1 + x$$

has the solution $x = 0$ for all a. Are there any other solutions? How does your answer depend on the value of a? [Hint: Graph the functions on both sides of the equation.]

35. Suppose you put a yam in a hot oven, maintained at a constant temperature 200°C. Suppose that at time $t = 30$ minutes, the temperature T of the yam is 120° and is increasing at a (instantaneous) rate of 2°/min. Newton's law of cooling (or, in our case, warming) implies that the temperature at time t will be given by a formula of the form

$$T(t) = 200 - ae^{-bt}.$$

Find a and b.

36. A spherical cell is growing at a constant rate of $400\mu\text{m}^3/\text{day}$ ($1\mu\text{m} = 10^{-6}\text{m}$). At what rate is its radius increasing when the radius is $10\mu\text{m}$?

37. When the growth of a spherical cell depends on the flow of nutrients through the surface, it is reasonable to assume that the growth rate $\frac{dV}{dt}$ is *proportional* to the surface area, S. Assume that for a particular cell $\frac{dV}{dt} = \frac{1}{3} \cdot S$. At what rate is its radius r increasing?

38. Suppose the total number of people, N, who have contracted a disease by a time t days after its outbreak is given by

$$N = \frac{1,000,000}{1 + 5,000e^{-0.1t}}.$$

   (a) In the long run, how many people will have had the disease?

   (b) Is there any day on which more than a million people fall sick? Half a million? Quarter of a million? (Note: You do not have to try to find out on what days these things happen.)

39. (a) Provide a 3-dimensional analogue for the geometrical demonstration of the formula for the derivative of a product, given on page 240. In other words, find a formula for the derivative of $F(x) \cdot G(x) \cdot H(x)$ using Figure 4.29.

   (b) Verify your results by writing $F(x) \cdot G(x) \cdot H(x)$ as $(F(x) \cdot G(x)) \cdot H(x)$ and using the Product rule twice.

   (c) Generalize your result to n functions: what is the derivative of $f_1(x) \cdot f_2(x) \cdot f_3(x) \cdot \ldots \cdot f_n(x)$?

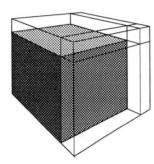

Figure 4.29: A graphical representation of the 3-dimensional product rule.

(for Problem 39)

Chapter 5

USING THE DERIVATIVE

The main purpose of this chapter is to use the first and second derivatives to analyze the qualitative behavior of functions. We would like to learn, for instance, where a function is increasing or decreasing, where its graph is concave up or concave down, and where it is largest or smallest. From Chapter 2 we recall that

- If $f' > 0$ on an interval, then f is increasing on that interval.
- If $f' < 0$ on an interval, then f is decreasing on that interval.
- If $f'' > 0$ on an interval, then the graph of f is concave up on that interval.
- If $f'' < 0$ on an interval, then the graph of f is concave down on that interval.

We can do more with these principles now than we could in Chapter 2 because we can now differentiate the standard functions.

5.1 Maxima and Minima

How to get a Helpful Graph

This section is about how to use the derivative of a function to understand its graph. When we graph a function on a computer or graphing calculator, we only see part of the picture, and what we see can be misleading. The derivative can often direct our attention to important features on the graph.

☐ **Example 1** Sketch a helpful graph of the function $f(x) = x^3 - 9x^2 - 48x + 52$.

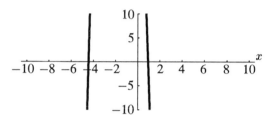

Solution. Since f is a cubic, we expect a graph that is roughly S-shaped. Plotting this function with $-10 \le x \le 10$, $-10 \le y \le 10$, gives the two nearly vertical lines in Figure 5.1. We know that there is more going on than this, but how do we tell where to look?

Figure 5.1: Unhelpful graph of $f(x) = x^3 - 9x^2 - 48x + 52$

We'll use the derivative to decide where the function is increasing and where it is decreasing. The derivative of f is

$$f'(x) = 3x^2 - 18x - 48$$

A good way to find the intervals where $f' > 0$ or $f' < 0$ is to find the points where $f'(x) = 0$, that is, where $3x^2 - 18x - 48 = 0$. The solutions of the quadratic are $x = -2$ and $x = 8$ and these points divide the x-axis (which is the domain of f) into the intervals $-\infty < x < -2$, $-2 < x < 8$ and $8 < x < \infty$. Since $f' = 0$ *only* at $x = -2$ and $x = 8$, and since f' is continuous, f' cannot change sign *within* any of these intervals.

How can we tell what happens within any interval? One way is simply to check the sign of f' at any point in each of the intervals, since f' will then have the same sign in the entire interval. For example, since -3 is in the interval $-\infty < x < -2$ and $f'(-3) = 33 > 0$, f' will be positive on all of $-\infty < x < -2$, which means that f is increasing on $-\infty < x < -2$. Similarly, since $f'(0) = -48$ and $f'(10) = 72$, we know that f decreases up to $x = 8$ and then increases for $x > 8$.

f	incr. ↗	-2	decr. ↘	8	incr. ↗	x
f'	$+$	0	$-$	0	$+$	

We find that $f(-2) = 104$ and $f(8) = -396$. Hence on the interval $-2 < x < 8$ the function decreases from a high of 104 to a low of -396. (Now we see why not much showed up in our first calculator graph.) One more point on the graph is easy to get: the y intercept, $f(0) = 52$. With just these three points we can get a much more helpful graph. By setting the domain and range in our calculator to $-10 \le x \le 20$ and $-400 \le y \le 400$, we get Figure 5.2. Where does the graph in Figure 5.1 fit into this one?

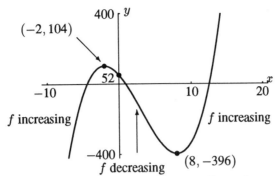

Figure 5.2: Helpful graph of $f(x) = x^3 - 9x^2 - 48x + 52$

Critical Points

We have been using the fact that if $f' > 0$ on an interval, the function f is increasing there; if $f' < 0$ on an interval, the function f is decreasing there. In the preceding example, the points $x = -2$ and $x = 8$, where the derivative switched between increasing and decreasing, played a key role. Points such as these where the derivative is zero are often important and lead to the following definition.

> For any function f, a point p in the domain of f where $f'(p) = 0$ or $f'(p)$ is undefined, is called a **critical point** of the function. In addition, the point $(p, f(p))$ on the graph of f is also called a critical point. Geometrically, if $f'(p) = 0$ then the line tangent to the graph of f at p is horizontal. A **critical value** of f is the value, $f(p)$, of the function at a critical point, p.

Notice that the words "critical point of f" can refer either to special points in the domain of f or to special points on the graph of f. You will know which meaning is intended from the context.

At a point where a function is not differentiable there is no horizontal tangent to the graph— there's either a vertical tangent or no tangent at all. For example, $x = 0$ is a critical point for the absolute value function $f(x) = |x|$. However, most of the functions we will work with will be differentiable everywhere, and therefore most of our critical points will be of the $f'(p) = 0$ variety.

What do the Critical Points Tell Us?

Determining the critical points is often the first step in understanding the qualitative behavior of a function. For a given function f on an interval $a < x < b$, the points where $f'(p) = 0$ (or $f'(p)$ is undefined) divide the interval into sub-intervals on which the sign of the derivative is constant, either positive or negative. Therefore, *between two successive critical points the graph of a function cannot change direction; it either goes up or down.*

A function may have any number of critical points or none at all. For example, any quadratic function has exactly one critical point (See Figure 5.3), while the cubic in Example 1 has two critical points. On the other hand the cubic $f(x) = x^3 + x + 1$ has no critical points at all since its derivative is $f'(x) = 3x^2 + 1$, which is always positive. (See Figure 5.4). The function $f(x) = \sin x$

has infinitely many critical points. They occur where $f'(x) = \cos x = 0$, namely at the points $\pm\frac{\pi}{2}, \pm\frac{3\pi}{2}, \pm\frac{5\pi}{2}, \ldots$. The critical values of $\sin x$ are 1 and -1. (See Figure 5.5).

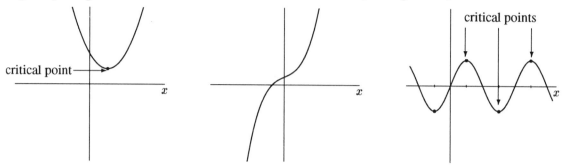

Figure 5.3: A Quadratic: One Critical Point

Figure 5.4: $f(x) = x^3 + x + 1$: No Critical Points

Figure 5.5: $f(x) = \sin x$: Infinitely Many Critical Points

Local Maxima and Minima

What happens to a function at a critical point? Suppose that $f'(p) = 0$. We know that the graph has a horizontal tangent there, but what happens *near* p? If f' has different signs on either side of p, then the graph changes direction at p, so the graph must look like one of the possibilities in Figure 5.6. In the first case we say that f has a local minimum[1] at p and in the second we say that f has a local

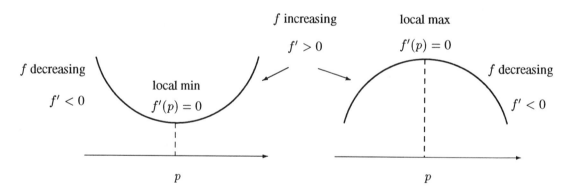

Figure 5.6: Changes in Direction: Local Maxima and Minima

maximum at p. We use the adjective "local" because we are concerned only with what happens near p.

> Suppose p is a critical point of a function f. Then
>
> - f is a **local minimum** at p if, near p, the values of f get no smaller than $f(p)$.
>
> - f is a **local maximum** at p if, near p, the values of f get no larger than $f(p)$.

---

[1]The word "minima" is the Latin plural of minimum.

How do we decide which critical points are local maxima and which are local minima?

As Figure 5.6 illustrates, we have the following criterion.

The First Derivative Test for Local Maxima and Minima
If p is a critical point, and if f' changes sign at p, then f has either a local minimum or a local maximum at p.

- if f' is negative to the left of p and positive to the right of p, then f has a local minimum at p.

- if f' is positive to the left of p and negative to the right of p, then f has a local maximum at p.

☐ **Example 2** (continuation of Example 1, page 278)
Identify the local maxima and minima of $f(x) = x^3 - 9x^2 - 48x + 52$.

Solution. The function $f(x) = x^3 - 9x^2 - 48x + 52$ has derivative $f'(x) = 3x^2 - 18x - 48 = 3(x^2 - 6x - 16) = 3(x - 8)(x + 2)$. To the left of $x = -2$, f' is positive. Between $x = -2$ and $x = 8$, f' is negative. To the right of $x = 8$, f' is positive. By the First Derivative Test, f has a local maximum at $x = -2$ and a local minimum at $x = 8$. See Figure 5.7 ☐

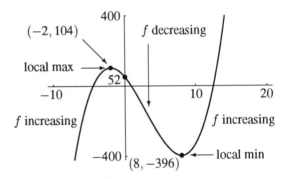

Figure 5.7: Helpful graph of $f(x) = x^3 - 9x^2 - 48x + 52$

Warning: There doesn't *have* to be a change in the sign of f' at a critical point. Consider $f(x) = x^3$ whose graph is in Figure 5.8. The derivative, $f'(x) = 3x^2$ is positive on both sides of $x = 0$, so the first derivative test tells us nothing. There is neither a local maximum nor a local minimum at $x = 0$. In other words, *a function doesn't have to have a local maximum or local minimum at every critical point.*

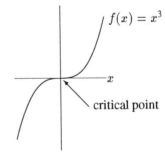

Figure 5.8: Critical Point but no change in direction.

Global Maxima and Minima

The local maxima and minima tell us where a function is locally largest or smallest. However we are often more interested in where the function is absolutely largest or smallest in a given domain. We say

> - f has a *global minimum* at p if all values of f are greater than or equal to $f(p)$.
>
> - f has a *global maximum* at p if all values of f are less than or equal to $f(p)$.

How do we find global maxima and minima?

This depends on what domain we are interested in. For a continuous function defined on a closed interval $a \le x \le b$ (i.e., an interval containing its endpoints), Figure 5.9 demonstrates that the global maximum or minimum occurs either at a local maximum or a local minimum or at one of the endpoints, $x = a$ or $x = b$.

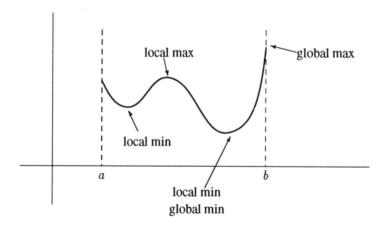

Figure 5.9: Global Maximum and Minimum on a Closed Interval $a \le x \le b$

> **To find the global maximum and minimum** of a continuous function on a closed interval $a \le x \le b$, compare values of the function at all the critical points in the interval and at the endpoints.

What if the function is defined on an open interval $a < x < b$ (i.e. an interval not including its endpoints), or on the entire real line? We say there is no global maximum in Figure 5.10 because the function has no actual largest value. The global minimum in Figure 5.10 coincides with the local minimum and is marked. Thus, Figure 5.10 and Figure 5.11 show there may not be a global maximum or minimum, on an open interval.

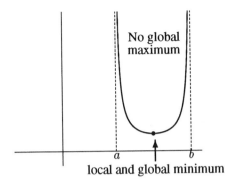

Figure 5.10: Global Maximum and Minimum on $a < x < b$

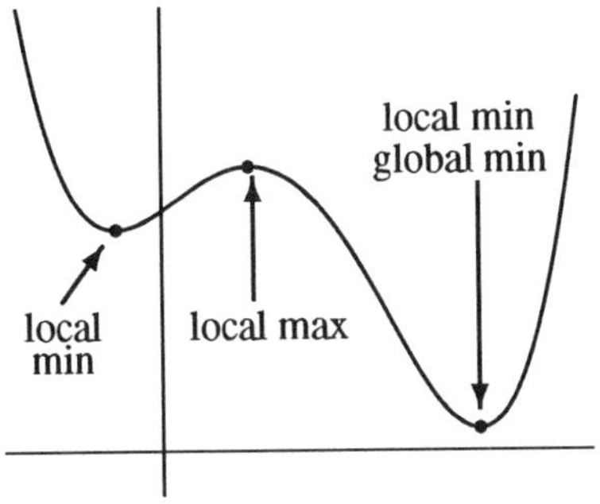

Figure 5.11: Global Maximum and Minimum when the domain is the real line.

> **To find the global maximum and minimum** of a continuous function on an open interval $a < x < b$ or on the entire real line, find the value of the function at all the critical points and sketch a graph.

Warning: Notice that local maxima or minima of a function f occur at critical points, where $f'(p) = 0$ (or $f'(p)$ undefined). Since global maxima or minima can occur at endpoints (where f' is not necessarily 0 or undefined), *not every global maximum or minimum is a local maximum or minimum.*

☐ **Example 3** Find the global maxima and minima of

$$f(x) = x^3 - 9x^2 - 48x + 52$$

on (a) $-5 \leq x \leq 12$ (b) $-5 \leq x \leq 14$ (c) $-5 \leq x < \infty$

Solution.

(a) We have calculated the critical points of this function previously using:

$$f'(x) = 3x^2 - 18x - 48 = 3(x + 2)(x - 8)$$

so $x = -2$ and $x = 8$ are critical points. Evaluating f at the critical points and the endpoints, we discover

$$\begin{aligned} f(-5) &= (-5)^3 - 9(-5)^2 - 48(-5) + 52 = -58 \\ f(-2) &= 104 \\ f(8) &= -396 \\ f(12) &= -92. \end{aligned}$$

So the global maximum on $[-5, 12]$ is 104 and occurs at $x = -2$, and the global minimum on $[-5, 12]$ is -396 and occurs at $x = 8$.

(b) For the interval $[-5, 14]$, we compare

$$f(-5) = -58, \quad f(-2) = 104, \quad f(8) = -396$$

with the value of the function at the new
endpoint:

$$f(14) = 360.$$

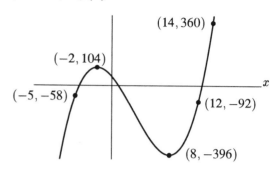

The global maximum is now 360 and occurs at $x = 14$, and the global minimum
is still -396 and occurs at $x = 8$. Notice
that since the function is increasing for
$x > 8$, changing the right hand end of the
interval from $x = 12$ to $x = 14$ alters
the global maximum but not the global
minimum. See Figure 5.12.

Figure 5.12: Graph of $f(x) = x^3 - 9x^2 - 48x + 52$
(For Example 3)

(c) Figure 5.12 shows that for $-5 \leq x < \infty$ there is no global maximum, because we can make
$f(x)$ as large as we please by choosing x large enough. The global minimum remains -396
at $x = 8$.

□ **Example 4** Consider the function $f(x) = \dfrac{1}{x(x-1)}$. Use a graph of this function to find the local
maxima and minima, and confirm your result analytically. What are the global maxima and minima
on $0 < x < 1$, on $-\infty < x < 0$, and on $1 < x < \infty$?

Solution. Using $-2 < x < 2$ and $-10 < y < 10$
gives a helpful graph; see Figure 5.13. The graph
shows that there are no local minima but that
there is a local maximum at about $x = \frac{1}{2}$. Explaining this result analytically means showing
that it is what we would expect using derivatives.

Now $f'(x) = -\dfrac{2x-1}{(x(x-1))^2}$, so $f'(x) = 0$
where

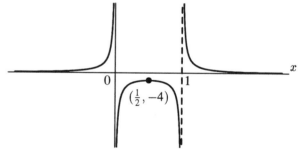

Figure 5.13: Graph of $f(x) = \frac{1}{x(x-1)}$

$2x - 1 = 0$; the only critical point in the domain
is $x = \frac{1}{2}$. Furthermore
$f'(x) > 0$ where $0 < x < \frac{1}{2}$ and $f'(x) < 0$ where $\frac{1}{2} < x < 1$, thus, by the First Derivative Test
the critical point $x = \frac{1}{2}$ is a local maximum. It is also a global maximum for f on $0 < x < 1$. The
function does not have a global minimum on $0 < x < 1$ since $f(x)$ tends to $-\infty$ both as x tends to
0 from the right and as x tends to 1 from the left.

For $-\infty < x < 0$ or $1 < x < \infty$, there are no critical points and therefore no local or global
maxima or minima. Although $\frac{1}{x(x-1)} \to 0$ both as $x \to \infty$ and as $x \to -\infty$, we don't say 0 is a
minimum (either local or global) because $\frac{1}{x(x-1)}$ never actually *equals* 0.

Notice that although $f' > 0$ everywhere that it is defined on $-\infty < x < \frac{1}{2}$, f is not increasing throughout this interval. The problem is that f and f' are not defined at $x = 0$, so we cannot conclude from $f' > 0$ that f is increasing throughout this interval. □

☐ **Example 5** The graph of $f(x) = \sin x + 2e^x$ is in Figure 5.14. Explain why there are no local maxima or minima for $x \geq 0$.

Solution. Local maxima and minima can occur only where

$$f'(x) = \cos x + 2e^x = 0.$$

But $\cos x$ is always between -1 and 1 and $2e^x \geq 2$ for $x \geq 0$, so $f'(x)$ cannot be 0 for any $x \geq 0$. Therefore there are no local maxima or minima there. However, $x = 0$ gives the global minimum for $x \geq 0$. The point $x = 0$ is not a local minimum because it is not a critical point.

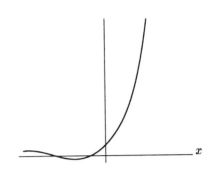

Figure 5.14: $f(x) = \sin x + 2e^x$

□

❖ Exercises for Section 5.1

1. Indicate all critical points on the graph of f to the right and determine which correspond to local maxima of f, which to local minima, and which to neither.

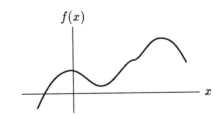

$f(x)$

2. Sketch graphs of two continuous functions f and g, each of which has exactly five critical points, the points $A–E$ on the graph to the right; also

   (a) $\lim\limits_{x \to -\infty} f(x) = \infty$
        $\lim\limits_{x \to \infty} f(x) = \infty$

   (b) $\lim\limits_{x \to -\infty} g(x) = -\infty$
        $\lim\limits_{x \to \infty} g(x) = 0$

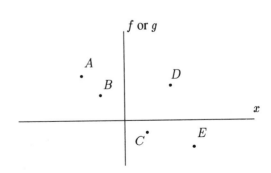

f or g

A

B

D

C

E

3. Indicate on the graph of the derivative function f' to the right the x-values that are critical points of the function f itself. At which critical points does f have local maxima, local minima, or neither?

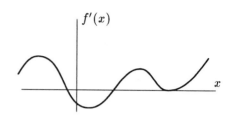

In each of Problems 4–9, find all critical points and then use the first derivative test to determine local maxima and minima. When you are finished, use a calculator or computer to sketch a graph of each function to check your work.

4. $f(x) = 2x^3 + 3x^2 - 36x + 5$

5. $f(x) = 3x^4 - 4x^3 + 6$

6. $(x) = (x^2 - 4)^7$

7. $f(x) = (x^3 - 8)^4$

8. $f(x) = \dfrac{x}{x^2 + 1}$

9. $f(x) = 2x^2 e^{5x} + 1$

10. Plot the graph of $f(x) = x^3 - e^x$ using a graphing calculator or computer to find all local and global maxima and minima for:

   (a) $-1 \le x \le 4$

   (b) $-3 \le x \le 2$

11. For $y = f(x) = x^{10} - 10x$, and $0 \le x \le 2$, find the value(s) of x for which

   (a) $f(x)$ has a local maximum or local minimum. Which is which?

   (b) $f(x)$ has a global maximum or global minimum. Which is which?

12. For $f(x) = x - \ln x$, and $0.1 \le x \le 2$, find the value(s) of x for which

   (a) $f(x)$ has a local maximum or local minimum. Which is which?

   (b) $f(x)$ has a global maximum or global minimum. Which is which?

13. For $f(x) = \sin^2 x - \cos x$, and $0 \le x \le \pi$, find, to two decimal places, the value(s) of x for which

   (a) $f(x)$ has a local maximum or local minimum. Which is which?

   (b) $f(x)$ has a global maximum or global minimum. Which is which?

14. (a) Water is flowing at a constant rate into a cylindrical container standing vertically. Sketch a graph showing the depth of water against time.

   (b) Water is flowing at a constant rate into a cone-shaped container standing on its point. Sketch a graph showing the depth of the water against time.

15.   (a) Sketch a graph of $y = 10^{-x}$.

   (b) If x is between 0 and 5, find bounds for 10^{-x}; that is, find values of A and B such that $A \leq 10^{-x} \leq B$. Your bounds, A and B, should be as close together as possible. Mark these values on the graph so that it is clear why these are the bounds.

16. Under what conditions on a, b and c, is the cubic

$$f(x) = x^3 + ax^2 + bx + c$$

increasing everywhere?

17. Choose the constants a and b in the function

$$f(x) = axe^{bx}$$

in such a way that $f(\frac{1}{3}) = 1$ and the function has a maximum at $x = \frac{1}{3}$.

18. Use the derivative formulas and algebra to find the intervals over which the function $f(x) = \dfrac{x + 50}{x^2 + 525}$ is increasing and is decreasing. It is possible to solve this problem on a graphing calculator or computer, but it is somewhat difficult; describe the difficulty.

19. Sketch the graph of
$$f(x) = 2x^3 - 9x^2 + 12x + 1.$$

Use the graph to decide how many solutions the following equations have:

(a) $f(x) = 10$ (b) $f(x) = 5$ (c) $f(x) = 0$ (d) $f(x) = 2e$

You need not find these solutions.

20.   (a) Draw a picture of a smooth curve whose slope starts out being slightly negative, then gradually becomes more negative, then gradually becomes less negative, then quickly becomes positive, and then gradually becomes less and less positive until it becomes zero, and remains zero as x goes to infinity.

   (b) Underneath the picture you drew in part (a), sketch a picture of the derivative of the curve you drew in (a). Make sure that the units on your x-axis are the same as in (a), and that they line up vertically with the units on the x-axis in (a). After sketching the derivative, compare your sketch to the instructions for (a).

   (c) Label the concavity of the various parts of the curve in (a).

21. *(Continuation of Problem 20)* Your graph in Problem 20(b) should have a local maximum (i.e., a place where the *slope* of the curve drawn in 20(a) has a local maximum). Label the local maximum on the curve in 20(b) and label the corresponding point "*A*" on the curve in 20(a). Draw the tangent line through *A* in 20(a). Next find a point on the curve slightly to the right of *A*. Sketch the tangent line through this point. Is the slope of this tangent line smaller or larger than the slope of the tangent line through *A*? Do the same thing for a point on the curve to the left of *A*. Is the slope smaller or larger? What is the concavity to the left of *A* and to the right of *A*? *A* is called an inflection point for the curve in Problem 20(a).

22. Do the same procedure as in problem 21 for a local *minimum* on the curve in problem 20(b). Use the letter "*B*" this time!

23. Use derivatives and the fact that $\sin 0 = 0$ to show that $\sin x \leq x$ for all $x \geq 0$.

24. Suppose f has a continuous derivative everywhere. From the values of $f'(\theta)$ in the table below, estimate the θ values with $1 < \theta < 2.1$ at which $f(\theta)$ has a local maximum or a minimum, and say which is which.

θ	1.0	1.1	1.2	1.3	1.4	1.5	1.6	1.7	1.8	1.9	2.0	2.1
$f'(\theta)$	2.37	0.31	−2.00	−3.45	−3.34	−1.70	0.76	2.80	3.61	2.76	0.69	−1.62

25. (a) On a computer or calculator, graph $f(\theta) = \theta - \sin\theta$. Can you tell whether the function has any zeros in the interval $0 \leq \theta \leq 1$?

    (b) Find f'. What does the sign of f' tell you about the zeros of f in the interval $0 \leq \theta \leq 1$?

26. Assume the function f is differentiable everywhere and has just one critical point, at $x = 3$. In (a) through (d) below, you are given additional conditions. In each case decide whether $x = 3$ is a local maximum, a local minimum, or neither. Explain your reasoning. Also sketch possible graphs for all four cases.

    (a) $f'(1) = 3$ and $f'(5) = -1$
    (b) $\lim_{x \to \infty} f(x) = \infty$ and $\lim_{x \to -\infty} f(x) = \infty$,
    (c) $f(1) = 1$, $f(2) = 2$, $f(4) = 4$, $f(5) = 5$
    (d) $f'(2) = -1$, $f(3) = 1$, $\lim_{x \to \infty} f(x) = 3$

27. (a) Show that
$$x > 2 \ln x$$
    for all $x > 0$. [Hint: Find the minimum of $f(x) = x - 2\ln x$.]

    (b) Use the above result to show that
$$e^x > x^2$$
    for all positive x.

    (c) Is
$$x > 3 \ln x$$
    for all positive x?

28. Show that for all values of x

$$x^4 - 4x > -4.$$

29. The rabbit population on a small Pacific island is approximated by

$$P(t) = \frac{2000}{1 + e^{(5.3-0.4t)}}$$

with t measured in years since 1774, when Captain James Cook left 10 rabbits on the island. Using a calculator or computer,

(a) graph P. Does the population level off?

(b) Estimate when the rabbit population grew most rapidly. How large was the population at that time?

(c) What natural causes could lead to the shape of the graph of P?

5.2   Concavity and Inflection Points

Recall how the sign of the second derivative determines the concavity of a function on an interval. If $f'' > 0$ on an interval then the graph of f is concave upon the interval, and if $f'' < 0$ then the graph is concave down on the interval. Geometrically, if a curve is concave up near some point then it lies above its tangent line at that point and if it is concave down it lies below its tangent line at that point, as in Figure 5.15.

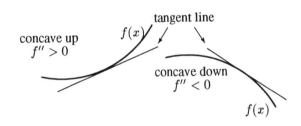

Figure 5.15: Concavity and the Tangent Line

Points of Inflection

Many functions change concavity; their graphs are concave up on some intervals and concave down on others.

A point p at which the graph of a function changes concavity is called an **inflection point** of f. The words "inflection point of f" can refer either to a point in the domain of f or to a point on the graph of f. The context of the problem will tell you which is meant.

How do you locate an inflection point?

At a point where the concavity changes, the second derivative changes sign. Thus on one side of an inflection point, the graph will be concave down ($f'' < 0$), and on the other side, the graph will be concave up ($f'' > 0$). At the inflection point, f'' will be zero or undefined. See Figure 5.16.

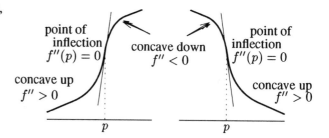

Figure 5.16: Change in Concavity, the graph crosses its tangent line

☐ **Example 1** Find the inflection points of $f(x) = x^3 - 9x^2 - 48x + 52$.

Solution.
Since $f(x) = x^3 - 9x^2 - 48x + 52$ is concave down at its maximum, $(-2, 104)$ and concave up at its minimum $(8, -396)$, it must have an inflection point in between. Since $f'(x) = 3x^2 - 18x - 48$, and $f''(x) = 6x - 18$, we see that $f''(x) < 0$ when $x < 3$ and $f''(x) > 0$ when $x > 3$ (and that $f''(3) = 0$). Hence $x = 3$ is an inflection point. This is where the graph changes concavity. See Figure 5.17. ☐

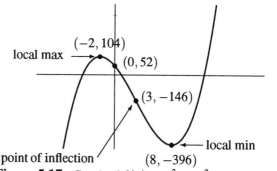

Figure 5.17: Graph of $f(x) = x^3 - 9x^2 - 48x + 52$, again

☐ **Example 2** Find the inflection points for $g(x) = xe^{-x}$ and sketch the graph.

Solution. We have $g'(x) = (1 - x)e^{-x}$ and $g''(x) = (x - 2)e^{-x}$. Hence $g'' < 0$ when $x < 2$ and $g'' > 0$ when $x > 2$, so $x = 2$ is an inflection point. The graph is sketched in Figure 5.18. ☐

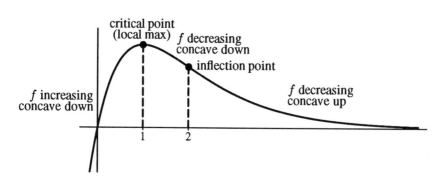

Figure 5.18: Graph of $g(x) = xe^{-x}$

Warning: *Not every point x where $f''(x) = 0$ (or f'' is undefined) is an inflection point* (just as not every point where $f' = 0$ is a local maximum or minimum). For instance $f(x) = x^4$ has $f''(x) = 12x^2$ so $f''(0) = 0$, but $f'' > 0$ when $x > 0$ and when $x < 0$, so there is *no* change in concavity at $x = 0$. See Figure 5.19.

$f'' = 0$ but no inflection point here

Inflection points can also be easily interpreted in terms of first derivatives. Recall that the graph of a function f is concave up where f' is increasing and is concave down where f' is decreasing. A change in concavity must therefore occur where f' changes from increasing to de-

Figure 5.19: Graph of $f(x) = x^4$

creasing or from decreasing to increasing; i.e., points where f' has a local maximum or a local minimum. The same conclusion can be arrived at another way. After all, if p is an inflection point, then $f''(p) = 0$ (or $f''(p)$ is undefined) and hence p is a critical point of the derivative function f'. Since f'' changes sign at p, an inflection point is a local maximum or minimum of f'.

Inflection Points and Local Maxima and Minima

> A function f has an inflection point at p
>
> - if f' has a local minimum or a local maximum at p
>
> - if f'' changes sign at p

☐ **Example 3**   Sketch the graph of $f(x) = x + \sin x$ and find the points where it is increasing most rapidly, and where it is increasing least rapidly.

Solution. We first examine the derivative, $f'(x) = 1 + \cos x$ which is shown in Figure 5.20. The critical points of f are the points where $f'(x) = 0$, so they occur at $x = \ldots, -3\pi, -\pi, \pi, 3\pi, \ldots$. At these points, the line tangent to the graph of f is horizontal. According to Figure 5.20, the derivative f' does not change sign at the critical

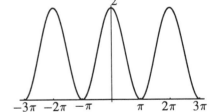

Figure 5.20: Graph of $f'(x) = 1 + \cos x$

points of f since $f'(x) \geq 0$ for all x, so f has no local minima and no local maxima. The function f simply increases between successive critical points.

The critical points of f are, however, inflection points of f. This is because f' has local minima at these points, as is easily seen from its graph. At these points f' changes from decreasing to increasing so the graph of f changes from concave down to concave up.

But there are other inflection points! We see that f' has local maxima at $x = \ldots - 2\pi, 0, 2\pi, \ldots$ where f' changes from increasing to decreasing. At these points, f changes from concave up to concave down.

Where is f increasing most rapidly? At the points $x = \ldots, -2\pi, 0, 2\pi, 4\pi, 6\pi, \ldots$, because these points are local maxima for f' and f' has the same value at each of them. Likewise f is increasing most slowly at the points $x = \ldots, -3\pi, -\pi, \pi, 3\pi, 5\pi, \ldots$, since these points are local minima for f'. The graph of f is pretty flat at $x = \ldots, -3\pi, -\pi, \pi, 3\pi, 5\pi, \ldots$ since both f' and f'' are zero at these points. See Figure 5.21. ☐

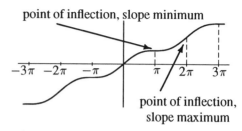

Figure 5.21: Graph of $f(x) = x + \sin x$

☐ **Example 4**   Suppose that water is being poured into the vase in Figure 5.22 at a constant rate. Graph $y = f(t)$, the depth of the water against time, t. Explain the concavity and mark inflection points.

Solution. At first the water level, y, rises quite slowly because the base of the vase is wide and so it takes a lot of water to make the depth increase much. However, as the vase narrows, the rate at which the water is rising increases. This means that initially y is increasing at an increasing rate, and the graph is concave up. The rate of increase in the water level is at a maximum when the water reaches the middle of the vase where the diameter is smallest; this is an inflection point. After that, the rate at which y increases starts to decrease again and so the graph is concave down. See Figure 5.23.

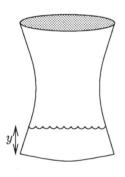

Figure 5.22: A vase

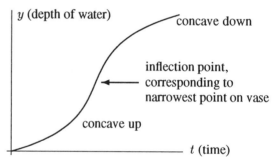

Figure 5.23: Graph of depth of water in the vase, y, against time, t

☐

☐ **Example 5**   If f is increasing and its graph is concave up, what can you conclude about $\lim\limits_{x \to \infty} f(x)$? What can you say about $\lim\limits_{x \to -\infty} g(x)$ if g is decreasing and its graph is concave down?

Solution. We must have $\lim\limits_{x \to \infty} f(x) = \infty$ and $\lim\limits_{x \to -\infty} g(x) = -\infty$. Can you draw a sketch to show why this must be true? ☐

We can also use the fact that concavity tells us whether the graph lies above or below the tangent line to get estimates for functions.

☐ **Example 6**   Explain why, for all values of x, $e^x \geq 1 + x$.

Solution. The graph of the function $f(x) = e^x$ is concave up everywhere and the equation of its tangent line at the point $(0, 1)$ is $y = x + 1$. Since the graph always lies above its tangent we have the inequality

$$e^x \geq 1 + x.$$

Problem 30 shows how this inequality can be used to compute e. $\qquad\square$

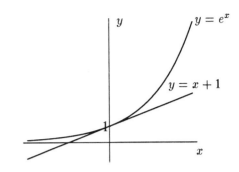

Figure 5.24: Graphs of $y = e^x$, $y = 1 + x$

The Second Derivative Test for Local Maxima and Minima

The second derivative gives us another way of testing whether a function has a local maximum or a local minimum at a critical point.

Suppose $f'(p) = 0$ so that p is a critical point of f and the graph of f has a horizontal tangent line at this point. If $f''(p) > 0$, the graph of f will be concave up at p. Thus f has a local minimum at p. Likewise, if $f''(p) < 0$, f will be concave down at p and f will have a local maximum at p. See Figure 5.25. We have what is called:

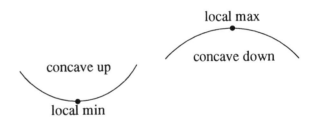

Figure 5.25: Local Maxima and Minima and Concavity

The Second Derivative Test for Local Maxima and Minima

- If $f'(p) = 0$ and $f''(p) > 0$ then f has a local minimum at p.

- If $f'(p) = 0$ and $f''(p) < 0$ then f has a local maximum at p.

Warning: We cannot draw any conclusion if both $f'(p) = 0$ and $f''(p) = 0$. For example, if $f(x) = x^3$ and $g(x) = x^4$, both $f'(0) = g'(0) = 0$ and $f''(0) = g''(0) = 0$. However, the point $x = 0$ is a minimum for g but is neither a maximum nor a minimum for f, as you can see from the graphs. When the second derivative test fails to give information about a critical point p because $f''(p) = 0$, the first derivative test can still be useful. For example, g' changes sign from $+$ to $-$ at 0, so we know g has a local minimum there.

Here are some examples of the use of the second derivative test.

☐ **Example 7** Classify as local maxima or local minima the critical points of

$$\text{(a) } f(x) = x^3 - 9x^2 - 48x + 52 \quad \text{(b) } g(x) = xe^{-x} \quad \text{(c) } h(x) = x + \frac{1}{x}$$

Solution.

(a) As we saw in Example 2 on page 281,

$$f'(x) = 3x^2 - 18x - 48$$

and the critical points of f are $x = -2$ and $x = 8$. We have

$$f''(x) = 6x - 18.$$

Thus

$$f''(8) = 6(8) - 18 = 30 > 0.$$

So f has a local minimum at $x = 8$. Since

$$f''(-2) = 6(-2) - 18 = -30 < 0,$$

f has a local maximum at $x = -2$.

(b) In Example 2 on page 290, we saw that for $g(x) = xe^{-x}$ we have

$$g'(x) = e^{-x} - xe^{-x} = (1 - x)e^{-x}.$$

Hence $x = 1$ is the only critical point. We see that g' changes from positive to negative at $x = 1$ since e^{-x} is always positive, so by the First Derivative Test g has a local maximum at $x = 1$. If we wish to use the Second Derivative Test, we compute

$$g''(x) = (x - 2)e^{-x}$$

and thus $g''(1) = (-1)e^{-1} < 0$, so again $x = 1$ gives a local maximum.

(c) For $h(x) = x + \frac{1}{x}$ we calculate

$$h'(x) = 1 - \frac{1}{x^2}$$

and so the critical points of h are at $x = \pm 1$. Now

$$h''(x) = \frac{2}{x^3}$$

so $h''(1) = 2 > 0$ and $x = 1$ gives a local minimum. On the other hand $h''(-1) = -2 < 0$ so $x = -1$ gives a local maximum.

☐

❖ Exercises for Section 5.2

1. Indicate on the graph of the derivative f' the x-values that are inflection points of the function f itself.

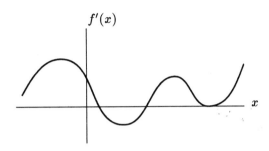

$f'(x)$

x

2. Indicate on the graph of the second derivative f'' the x-values that are inflection points of the function f itself.

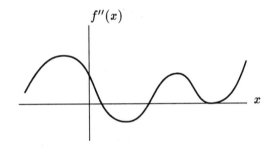

$f''(x)$

x

3. A certain function f has derivative $f'(x) = \cos(x^2) + 2x - 1$. Does f have a local maximum, a local minimum, or neither at its critical point $x = 0$?

Sketch the functions in Problems 4–7. Indicate clearly the coordinates of all maxima and minima and points of inflection.

4. $y = 0.5xe^{-10x}$

5. $y = 2 + 3\cos x$, for $0 \le x \le 6\pi$

6. $f(x) = 3x^5 - 5x^3$, for $|x| \le 1.5$

7. $f(x) = \sin x + \cos x$, for $0 \le x \le 6\pi$

8. For $f(x) = e^{\frac{x}{2}} - \ln(x^2 + 1)$, find the coordinates of all intercepts, maxima, minima and inflection points to two decimal places.

9. For $f(x) = x^{10} - 10x$, and $0 \le x \le 2$, find the value(s) of x for which $f(x)$ is increasing most rapidly; decreasing most rapidly.

10. For $f(x) = x - \ln x$ and $0.1 \le x \le 2$, find the value(s) of x for which $f(x)$ is increasing most rapidly; decreasing most rapidly.

11. For $f(x) = x - \cos x$, $0 \le x \le \pi$, find the value(s) of x for which

    (a) $f(x)$ is greatest; $f(x)$ is least.
    (b) $f(x)$ is increasing most rapidly; decreasing most rapidly.
    (c) The slopes of the lines tangent to the graph of f are increasing most rapidly.

12. For the function given in the graph to the right:

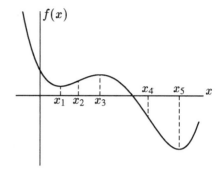

    (a) Sketch $f'(x)$.

    (b) Where does $f'(x)$ change its sign?

    (c) Where does $f'(x)$ have local maxima or minima?

13. Suppose that you know that a certain function f has $f(0) = 0$ and derivative given by

$$f'(x) = (\ln x)^2 - 2(\sin x)^4 \qquad \text{on the interval} \qquad 0 < x \le 7.5$$

    Use a calculator or computer to graph f' and its derivative. Clearly state where $f(x)$ is increasing and where it is decreasing, and where $f(x)$ is concave up and where it is concave down. Use this information to sketch a graph of $f(x)$.

14. Using your answer to Problem 12 as a guide, write a short paragraph (using complete sentences) which describes the relationship between the following features of a function f:

    (a) The local maxima and minima of f.
    (b) The points at which f changes concavity.
    (c) The sign changes of f'.
    (d) The local maxima and minima of f'.

15. Use concavity to show that $\ln x \le x - 1$.

16. Use the fact that $\ln x$ and e^x are inverse functions to show that the inequalities $e^x \ge 1 + x$ and $\ln x \le x - 1$ are equivalent.

17. For $f(x) = e^{\left(\frac{1}{x}\right)}$, with $x \ne 0$

    (a) describe the behavior of the function as $x \to \infty$, as $x \to -\infty$, as $x \to 0^+$, and as $x \to 0^-$ (i.e., as $x \to 0$ from the right and from the left).

    (b) For what values of x is the function increasing? decreasing? neither?

    (c) For what values of x is the graph of the function concave up? concave down? neither?

    (d) Use the above results to sketch a graph of f.

18. For $f(x) = \ln(x^3 + 1)$, with $x > -1$

    (a) describe the behavior of the function as $x \to \infty$ and as $x \to (-1)^+$ (i.e., as $x \to -1$ from the right).

    (b) For what values of x is the function increasing? decreasing? neither?

    (c) For what values of x is the the graph of the function concave up? concave down? neither?

    (d) Use the above results to sketch a graph of f.

19. Assume the polynomial f has exactly two local maxima and one local minimum.

    (a) Sketch a possible graph of f.

    (b) What is the largest number of zeros f could have?

    (c) What is the least number of zeros f could have?

    (d) What is the least number of inflection points f could have?

    (e) Is the degree of f even or odd? How can you tell?

    (f) What is the smallest degree f could have?

    (g) Find a possible formula for $f(x)$.

20. Sketch a smooth curve $f(x)$ for $0 \le x \le 10$ satisfying the following:

    - $|f''(x)| \le 0.5$ for all x.

    - $f'(x)$ takes on the values -2 and $+2$ somewhere.

    - f'' is not constant.

21. Let f be a function with positive values. Set $g = \dfrac{1}{f}$.

    (a) If f is increasing in an interval around x_0, what about g?

    (b) If f has a local maximum at x_1, what about g?

    (c) If f is concave down at x_2, what about g?

For Problems 22–27, sketch a possible graph of $y = f(x)$, using the given information about the derivative of f. Assume that the function is defined and continuous for all real x.

22.

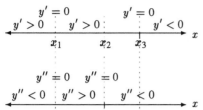

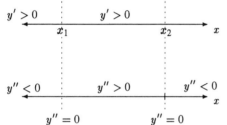

23.

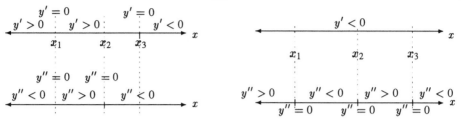

24.

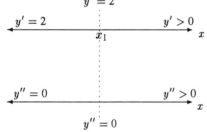

25.

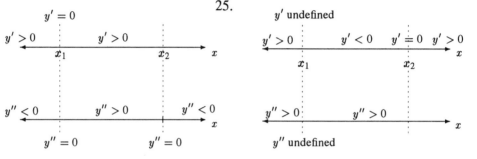

26.

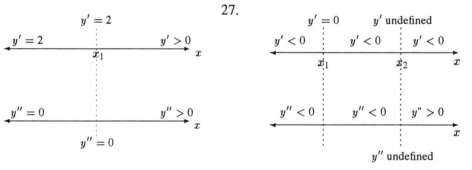

27.

28. If water is flowing at a constant rate (i.e. constant volume per unit time) into the Grecian urn in Figure 5.26, sketch a graph of the depth of the water against time. Mark on the graph the time at which the water reaches the widest point of the vase.

Figure 5.26: Urn for Problem 28

29. If water is flowing at a constant rate (i.e. constant volume per unit time) into the vase in Figure 5.27, sketch a graph of the depth of the water against time. Mark on the graph the time at which the water reaches the 'corner' of the vase.

Figure 5.27: Vase for Problem 29

30. Given the fact that $e^x \geq 1 + x$ for all x, let $x = \frac{1}{n}$ and show that for every positive integer n,

$$e > \left(1 + \frac{1}{n}\right)^n.$$

Consider $x = -\frac{1}{n+1}$ and show that

$$e < \left(1 + \frac{1}{n}\right)^{n+1}$$

for all positive integers n.

Use your calculator with some specific choices of n to prove that $2.7 < e < 2.8$. Be clear about what choice of n you are making. Suppose you want to calculate e to 10 decimal places this way. Figure out a specific value of n that will enable you to pinch e into an interval of length less than 10^{-10}. (Try this on your calculator. Did you get a reasonable answer? If not, why not?)

31. Suppose $g(t) = \dfrac{\ln t}{t}$ for $t > 0$.

    (a) Does g have a global maximum and minimum on $0 < t < \infty$? If so, where, and what is its value?

    (b) What does your answer to (a) tell you about the number of roots to the equation

    $$\frac{\ln x}{x} = \frac{\ln 5}{5}?$$

    [Note: There are many ways to investigate the number of solutions to this equation. We are asking you to draw a conclusion from your answer to (a).]

    (c) Find the solution(s).

5.3 Families of Curves: A Qualitative Study

Our study of calculus has given us experience in using the first and second derivatives to study the qualitative properties of a function. In this section we will study *families of functions*. Confronted with the problem of modeling some phenomenon, a crucial first step involves recognizing which families of functions might fit the available data. You can think of our work in this section as an introduction to some of the skills needed in mathematical modeling, without much discussion of the phenomena being modeled, or as a function watcher's guide to a few colorful and interesting species. We urge you to make your own guidebook.

We saw in Chapter 1 that knowledge of one function can provide knowledge of the graphs of many others. The shape of the graph of $y = x^2$ also tells us, indirectly, about the shape of the graphs of $y = x^2 + 2$, $y = (x + 2)^2$, $y = 2x^2$, and countless other functions. Using the sine function as another example, we say that all functions of the form $y = A \sin(Bx + C) + D$ form a *family*; their graphs are identical to that of $y = \sin x$ but for shifts and dilations determined by the values of A, B, C, and D. We can apply the tools of calculus to the analysis of such families.

Curves of the form $y = xe^{-bx}$

We will consider only $x \geq 0$. The graph of one member of the family, with $b = 1$, is shown in Figure 5.28. Such a graph represents a quantity that increases at first, then rapidly decays toward zero. For example, the number of bacteria in a person during the course of an infection, from onset to cure, might be described in this way. The question is how exactly does the shape of the curve change when b varies? For simplicity we will consider only curves with $b > 0$. Note that all curves of the family pass through the origin. The behavior of the curves as $x \to \infty$ is also pretty clear. Since $b > 0$ (by assumption), and e^{bx} grows much faster than x as $x \to \infty$, we must have

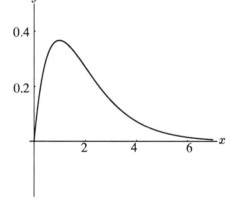

Figure 5.28: Graph of $y = xe^{-x}$

$$xe^{-bx} = \frac{x}{e^{bx}} \to 0 \quad \text{as} \quad x \to \infty.$$

Geometrically, this means that the x-axis is a horizontal asymptote.

Next we find and classify critical points and then we check concavity. We compute

$$\frac{dy}{dx} = x\left(-be^{-bx}\right) + e^{-bx} = (1 - bx)e^{-bx},$$

$$\frac{d^2y}{dx^2} = (1 - bx)\left(-be^{-bx}\right) + (-b)e^{-bx} = -b(2 - bx)e^{-bx}.$$

There is a critical point where $\frac{dy}{dx} = 0$, i.e., where $(1 - bx)e^{-bx} = 0$. Since e^{-bx} is never zero, the only critical point is $x = 1/b$. Looking at the expression for $\frac{dy}{dx}$, we can see that if $x < 1/b$ then $\frac{dy}{dx} > 0$, and if $x > 1/b$ then $\frac{dy}{dx} < 0$. hence there is a local maximum at $x = 1/b$. The maximum value is $y = \frac{1}{b}e^{-b(1/b)} = \frac{1}{be}$.

Examination of the second derivative shows that $\frac{d^2y}{dx^2} = 0$ only where $2 - bx = 0$. Solving for x gives $x = 2/b$. It is easy to see from the expression for $\frac{d^2y}{dx^2}$ that $\frac{d^2y}{dx^2}$ changes from negative to positive at $x = 2/b$, so the graph changes from concave down to concave up at $x = 2/b$, which is therefore an inflection point. At that point $y = \frac{2}{b}e^{-b(2/b)} = \frac{2}{be^2}$. See Figure 5.29.

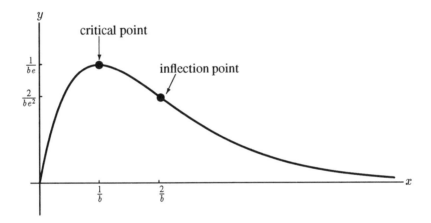

Figure 5.29: Graph of $y = xe^{-bx}$

The most obvious effect of varying b is to move the critical point (local maximum) of the curve. If b is large, then $1/b$ is small and so xe^{-bx} reaches a low maximum very quickly. If b is small, near zero, then $1/b$ is very large so the curve rises very high before the decay towards zero begins. The inflection point moves as b varies, too. It will always be twice as far from the y-axis as the critical point. See Figure 5.30.

Figure 5.31 shows several curves in the family on the same set of axes. Note also that all the curves have the same tangent line at the origin. This is because the derivative $\frac{dy}{dx}$ has the same value at $x = 0$ for every b, namely $\frac{dy}{dx} = 1$. (Note that the scales on the x and y axes are different.)

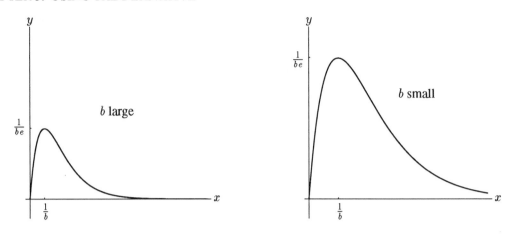

Figure 5.30: Graph of $y = xe^{-bx}$ with large b and small b

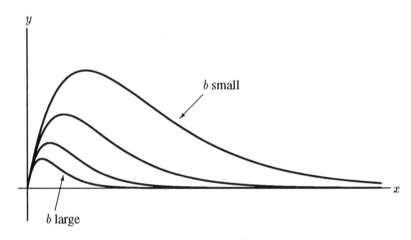

Figure 5.31: Graph of $y = xe^{-bx}$, with b varying.

Curves of the Form $y = a\left(1 - e^{-bx}\right)$

The graph of one member of the family with $a = 2$ and $b = 1$ is shown in Figure 5.32. Such a graph represents a quantity which is increasing but levelling off. For example, a body falling against air resistance speeds up and its velocity levels off as it approaches the terminal velocity. Similarly, if a pollutant pouring into a lake builds up towards a saturation level, its concentration may be described in this way. The graph might also indicate the temperature of a cooking yam.

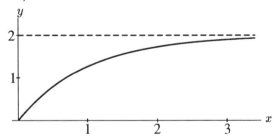

Figure 5.32: $y = 2(1 - e^{-x})$

How does the shape change if a and b vary? For simplicity, we will assume that a and b are positive. We are interested in how the shape of the graph changes when one parameter is fixed and the other is allowed to vary.

First, fix $b > 0$ and let a vary. As $x \to \infty$, $e^{-bx} \to 0$ and so $y = a(1 - e^{-bx})$ approaches $y = a$ from below; thus $y = a$ is a horizontal asymptote. Physically, it represents the terminal velocity of a falling object or the saturation pollutant level in a pond.

What happens with a fixed and b varying? How does the graph look when b is small and when b is large? Following the usual line of inquiry, we try to locate and classify critical points and check concavity. We compute

$$\frac{dy}{dx} = abe^{-bx}, \qquad \frac{d^2y}{dx^2} = -ab^2e^{-bx}$$

Since e^{-bx} is never zero there are no critical points. We see that $\dfrac{dy}{dx}$ is always positive and $\dfrac{d^2y}{dx^2}$ is always negative, and hence each curve is always increasing and always concave down.

How sharply does the curve rise and how soon does it get close to a? The answer to these questions depends on the size of b.

Since $\dfrac{dy}{dx} = abe^{-bx}$, the slope of the tangent to the curve at $x = 0$ is ab. How long does it take the curve to climb halfway up from $y = 0$ to $y = a$? Where $y = \frac{a}{2}$, we have

$$ae^{-bx} = \frac{a}{2}, \qquad \text{so} \qquad x = \frac{\ln 2}{b}.$$

If b is large then $\frac{\ln 2}{b}$ is small, so in a very short distance the curve is already half the way up to a. If b is small, then $\frac{\ln 2}{b}$ is large and you have to go a long way out to get up to $\frac{1}{2}a$. See Figure 5.33.

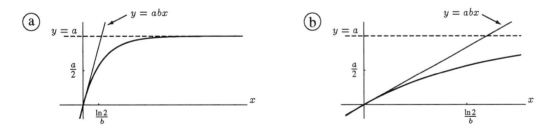

Figure 5.33: $y = a(1 - e^{-bx})$, with a fixed and (a) b large, (b) b small

Fixing a and letting b vary from small to large gives the family in Figure 5.34.

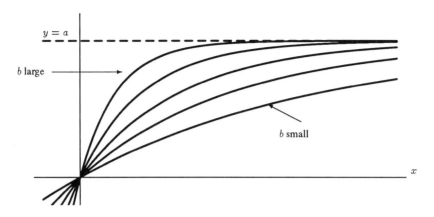

Figure 5.34: $y = a(1 - e^{-bx})$, with a fixed, b varying

Curves of the Form $y = ax^3 + bx^2 + cx + d$.

This is the family of cubics. From the specific examples we have seen earlier, we expect the graph to be S-shaped. Here we have four parameters, $a, b, c,$ and d, so the situation is somewhat complicated. The effect of the parameter d is easy to see. It is the y-intercept of the curve and changing d shifts the entire curve up or down without changing its shape. Since this is easy enough to account for, we set $d = 0$ and study the family $y = ax^3 + bx^2 + cx$; each of these curves passes through the origin.

One effect of the parameter a, called the leading coefficient, is also easy to observe. No matter what the size of a, provided $a \neq 0$, for large positive and negative values of x the term ax^3 dominates $bx^2 + cx$, and so for such values the curve looks like $y = ax^3$. If $a > 0$ then as $x \to +\infty$, y also tends to $+\infty$, and as $x \to -\infty$, $y \to -\infty$. However, if $a < 0$, then as $x \to +\infty$, y instead tends to $-\infty$, and as $x \to -\infty$, $y \to +\infty$. If $a = 0$, we really have a quadratic polynomial whose graph will be a parabola. Having noted this, let us now assume that $a > 0$ for the rest of our discussion.

For our family, $y = ax^3 + bx^2 + cx$, we have

$$\frac{dy}{dx} = 3ax^2 + 2bx + c, \qquad \frac{d^2y}{dx^2} = 6ax + 2b.$$

The formula for $\frac{dy}{dx}$ is a quadratic polynomial, so it may equal zero at exactly 0, 1, or 2 points. Hence a given cubic may have *no* critical points (e.g., $y = x^3 + x$), one critical point (e.g., $y = x^3$), or *two* critical points (e.g., $y = x^3 - x$).

But every cubic has exactly one inflection point. It occurs where $\frac{d^2y}{dx^2} = 0$, which is at

$$x = \frac{-b}{3a}.$$

The second derivative changes from negative to positive at this point, so the graph changes from concave down to concave up there. Note that the x-value of the inflection point does not depend on c.

Since the graph of a cubic looks like an S-shaped curve, it must look like one of the three possible cases in Figure 5.35. (Remember, we're assuming that $a > 0$.) The reasoning behind these sketches is as follows: in (a), there is no critical point so the curve must be always increasing. This is so because $\frac{dy}{dx}$ can never change sign and hence must be always positive, or always negative. Since $a > 0$ and $y \to +\infty$ as $x \to +\infty$, $\frac{dy}{dx}$ must be always positive. A similar argument goes for (b) but we allow the curve to flatten out at one point. It is not hard to show that in this case the critical point and the inflection point coincide. For (c) the curve must get from $y = -\infty$ to $y = +\infty$, as x goes from $-\infty$ to $+\infty$, changing concavity once and having two horizontal tangents. Disregarding the size of the critical values (how high the peak, how low the valley) the shape must be something like (c). Broadly speaking, the S "pivots" around the inflection point. In each case, the graph can be shown to be symmetric about the inflection point.

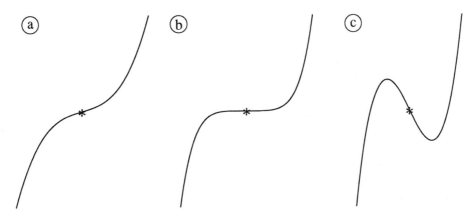

Figure 5.35: A cubic with (a) no critical points, (b) one critical point, (c) two critical points (The * marks the inflection point.)

Curves of the Form $y = e^{-\frac{(x-a)^2}{b}}$

This is a multiple of the *normal density* function, used in probability and statistics. In applications, it is common to scale the function by multiplying it by a positive constant, but we will just consider the two parameter family depending on a and b, and we will assume that $b > 0$. The graph of any member of the family is a bell-shaped curve as shown in Figure 5.36 (where we have $a = 1$ and $b = 2$).

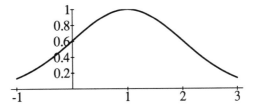

Figure 5.36: Graph of $y = e^{-\frac{(x-1)^2}{2}}$

As we saw in Chapter 1, the role of a is to move the curve to the right or left. Notice that for any curve in the family, y is always positive. Since $b > 0$, $y \to 0$ as $x \to \pm\infty$ for any choice of a.

Thus the x-axis is always a horizontal asymptote. To investigate the critical points and points of inflection, we calculate

$$\frac{dy}{dx} = -\frac{2(x-a)}{b}e^{-\frac{(x-a)^2}{b}}$$

and using the Product rule

$$\begin{aligned}\frac{d^2y}{dx^2} &= -\frac{2}{b}\left(\left(\frac{d}{dx}(x-a)\right)e^{-\frac{(x-a)^2}{b}} + (x-a)\frac{d}{dx}\left(e^{-\frac{(x-a)^2}{b}}\right)\right) \\ &= -\frac{2}{b}\left(e^{-\frac{(x-a)^2}{b}} + (x-a)\left(-\frac{2(x-a)}{b}e^{-\frac{(x-a)^2}{b}}\right)\right) \\ &= \frac{2}{b}e^{-\frac{(x-a)^2}{b}}\left(\frac{2(x-a)^2}{b} - 1\right).\end{aligned}$$

There is a critical point where

$$\frac{dy}{dx} = 0, \quad \text{i.e. where} \quad -\frac{2(x-a)}{b}e^{-\frac{(x-a)^2}{b}} = 0.$$

Since $e^{-\frac{(x-a)^2}{b}}$ is never zero, the only critical point is where $x = a$. At that point, $y = 1$. (Why?) Looking at the expression for $\frac{dy}{dx}$, we can see that if $x < a$, then $\frac{dy}{dx} > 0$, and if $x > a$, then $\frac{dy}{dx} < 0$. Hence there is a local maximum at $x = a$. Inflection points occur where the second derivative changes sign; thus we wish to find values of x for which

$$\frac{d^2y}{dx^2} = 0.$$

Since $e^{-\frac{(x-a)^2}{b}}$ is never zero, this only occurs where

$$\frac{2(x-a)^2}{b} - 1 = 0.$$

Solving for x gives

$$x = a \pm \sqrt{\frac{b}{2}}.$$

Looking at the expression for $\frac{d^2y}{dx^2}$, we see that $\frac{d^2y}{dx^2}$ is negative for $x = a$ and positive for very large x, both positive and negative. (Why?) Therefore the concavity changes at $x = a + \sqrt{\frac{b}{2}}$, and at $x = a - \sqrt{\frac{b}{2}}$, so we have inflection points there. These points are symmetrically located at a distance of $\sqrt{\frac{b}{2}}$ on either side of $x = a$. (See Figure 5.37.) At these points, $y = e^{-1/2} \approx 0.6$.

With this information we can already see the effect of the parameters. The parameter a determines the location of the center of the 'bell' and the parameter b determines how narrow or wide the bell is. (See Figure 5.38.) If b is small then the inflection points are close to a and the bell is sharply peaked near a, whereas if b is large the inflection points are further away from a and the bell is more spread out.

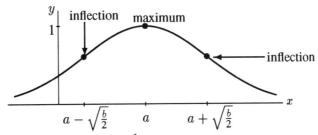

Figure 5.37: Graph of $y = e^{-\frac{(x-a)^2}{b}}$: Bell-Shaped curve with peak at $x = a$

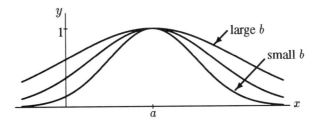

Figure 5.38: Graph of $y = e^{-\frac{(x-a)^2}{b}}$ with various values of b (a fixed)

Curves of the Form $y = e^{-ax}\sin bx$

As usual, we will assume that the parameters a and b are positive. The shape of $y = \sin bx$ is a sine curve with period $\frac{2\pi}{b}$. So if b is small, the period will be large and the graph of the function will be a 'stretched-out' sine curve. If b is large, the period is small and the graph oscillates rapidly. What is the effect of multiplying by e^{-ax}? The e^{-ax} represents the amplitude of the wave. Since e^{-ax} decreases as x increases, the amplitude of the wave continuously shrinks, or is *damped*. These curves represent *damped harmonic motion*, such as we see with a pendulum oscillating with friction.

To see the shape of $y = e^{-ax}\sin bx$ we draw $y = e^{-ax}$ and $y = -e^{-ax}$ (its reflection across the x-axis) and fit an ever shrinking version of $y = \sin bx$ between these two curves. See Figure 5.39.

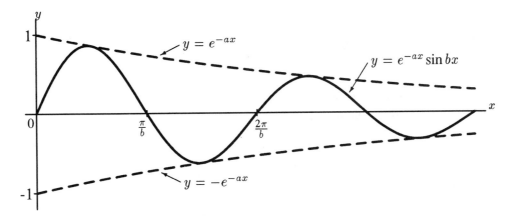

Figure 5.39: Graph of $y = e^{-ax}\sin bx$

As you can see in Figure 5.40(a), if a is large relative to b then the damping takes place very fast. If a is small relative to b, as in Figure 5.40(b), then the damping takes longer, so there may be many oscillations before we notice much damping. It is important to note however, that no matter the size of a or b the limiting value is zero, i.e. $\lim_{x \to \infty} e^{-ax} \sin bx = 0$; that is, the oscillations die out as $x \to \infty$.

(a) (b)

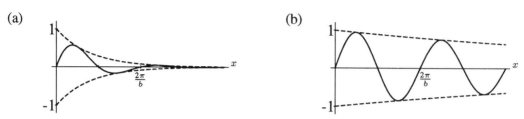

Figure 5.40: $y = e^{-ax} \sin bx$, with b fixed and (a) a large, (b) a small.

❖ Exercises for Section 5.3

1. Consider the function $p(x) = x^3 - ax$, where a is constant.

   (a) If $a < 0$, show that $p(x)$ is always increasing.

   (b) If $a > 0$, show that $p(x)$ has a local maximum and a local minimum.

   (c) Sketch and label typical graphs for the cases when $a < 0$ and when $a > 0$.

2. Consider the function $p(x) = x^3 - ax$, where a is constant and $a > 0$.

   (a) Find the local maxima and minima of p.

   (b) What effect does increasing the value of a have on the positions of the maxima and minima?

   (c) On the same axes, sketch and label the graphs of p for three different positive values of a.

3. What effect does increasing the value of a have on the graph of $f(x) = x^2 + 2ax$? Consider roots, maxima and minima, and both positive and negative values of a.

4. (a) Find all critical points of $f(x) = x^4 + ax^2 + b$.

   (b) Under what conditions on a and b will this function have exactly one critical point? What is the one critical point and is it a local maximum, a local minimum, or neither?

   (c) Under what conditions on a and b will this function have exactly three critical points? What are they and which are local maxima and which are local minima?

   (d) Is it ever possible for this function to have two critical points? No critical points? More than three critical points?

5. The number, N, of people who have heard a rumor spread by mass media is given by the following function of time, t;

$$N = a(1 - e^{-kt})$$

Suppose that there are 200,000 people in the population who all hear the rumor eventually. If 10% of them heard in the first day, find a and k, assuming t is measured in days.

6. The velocity, v, of a parachutist is given by

$$v = \frac{mg}{k}\left(1 - e^{-\frac{kt}{m}}\right)$$

where t is time, m is the mass of the parachutist, g is the acceleration due to gravity (a constant), and k is a positive constant. If t is measured in seconds and v is in feet per second, $g = 32 \text{ ft/sec}^2$.

   (a) What is the terminal velocity? That is, in terms of the constants in the problem, what is $\lim_{t \to \infty} v$?

   Suppose you are skydiving (so mg is your own weight) and that in the first 10 seconds you reach 80% of your terminal velocity.

   (b) Find k.

   (c) Find your terminal velocity in ft/sec and in mph. (Note: 1 mph = (22/15) ft/sec.)

7. Suppose the temperature, T, of a yam put into a hot oven maintained at 200°C is given as a function of time by

$$T = a(1 - e^{-kt}) + b$$

where T is in degrees Centigrade and t is in minutes.

   (a) If the yam starts at 20°C, find a and b.

   (b) If the temperature of the yam is initially increasing at 2°C/minute, find k.

8. Consider the family of curves

$$y = A(x + B)^2$$

   (a) Assuming $B = 0$, what is the effect of varying A on the graph? Consider

       i. $A > 0$ and $A < 0$,

       ii. $A > 1$ and $0 < A < 1$.

   (b) Assuming $A = 1$, what is the effect of varying B on the graph?

   (c) Write a couple of sentences describing the role played by A and B in determining the shape of the graph. Illustrate your answer by sketches.

9. Consider the family

$$y = \frac{A}{x + B}$$

(a) If $B = 0$, what is the effect of varying A on the graph?

(b) If $A = 1$, what is the effect of varying B?

(c) On one set of axes, graph the function for several values of A and B.

10. Consider the family of curves

$$y = ae^{-bx^2}.$$

Analyze the effect of varying a and b on the shape of the curve. Illustrate your answer with sketches.

11. Consider the 'surge function'

$$y = axe^{-bx}.$$

(a) Find the maxima, minima, and points of inflection.

(b) How does varying a and b affect the shape of the graph?

(c) On one set of axes, graph this function for several values of a and b.

12. Consider the family of curves

$$y = A\sin(Bx + C).$$

What is the effect of varying A, B, and C on the shape of the graph? For each parameter, sketch some graphs for various values (holding the others constant).

13. (a) Assuming $x \geq 0$ and $a > 0$, find the maximum and minimum values of

$$y = e^{-ax}\sin x$$

in terms of a.

(b) Describe in words how the position and values of these maxima and minima change if a increases. Give an intuitive justification for this.

14. The study of resonance leads to the family

$$y = \frac{1}{(1 - x^2)^2 + 2ax^2}, \qquad x \geq 0 \quad \text{and} \quad a > 0.$$

(a) Find and classify the critical points, then explain why the family is most interesting for $0 < a < 1$.

(b) Show that for a very near 0, the critical point is approximately $\left(1, \frac{1}{2a}\right)$.

(c) Show that $y < \frac{1}{(1 - x^2)^2}$ for $x > 0$.

(d) Sketch the curves for $a = 0.05, 0.10, 1.0$, and 3.0 on one set of axes, together with $y = \frac{1}{(1 - x^2)^2}$, the curve with $a = 0$.

(e) What is the geometric significance of the parameter a?

5.4 Economic Applications: Marginality

Management decisions within a particular firm or industry usually depend on the costs and revenues involved. In this section we will look at applications of the derivative to the cost and revenue functions.

> The *cost function*, $C(q)$, gives the total cost of producing a quantity q of some good.

What sort of function do you expect C to be? The more goods that are made, the higher the cost, so C is an increasing function. For most goods, such as cars or cases of soda, q can only be a non-negative integer, so the graph of C might look like that in Figure 5.41(a).

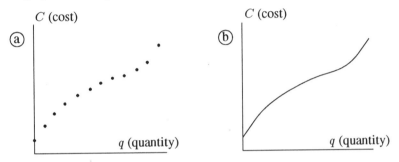

Figure 5.41: Cost: (a) positive integer values of q and (b) all positive values of q.

However, economists usually imagine the graph of C as the smooth curve drawn through these points, as in Figure 5.41(b). This is equivalent to assuming that $C(q)$ is defined for all non-negative values of q, not just for integers.

Cost functions usually have the general shape shown in Figure 5.41. The intercept on the C axis represents the *fixed costs*, which are incurred even if nothing is produced. (This includes, for instance, the machinery needed to begin production.) The cost function increases fast at first and then more slowly because producing larger quantities of a good is usually more efficient than small—this is called economy of scale. At still higher production levels, the cost function starts to increase faster again as resources become scarce, and sharp increases may occur when new factories have to be built. Thus, $C(q)$ starts out concave down and becomes concave up later on.

> The *revenue function*, $R(q)$, represents the total revenue received by the firm from selling a quantity q of some good.

If the price is a constant, p, then

$$\text{revenue} = (\text{price}) \cdot (\text{quantity}), \quad \text{so} \quad R = pq$$

and the graph of R is a straight line through the origin. See Figure 5.42(a). In practice, for large values of q, the market may become glutted causing the price to drop, giving R the shape in Figure 5.42(b).

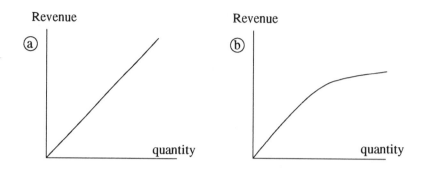

Figure 5.42: Revenue: (a) Constant Price and (b) Decreasing Price

Decisions are often made by considering the profit, usually written as π (to distinguish it from the price, p; this π has nothing to do with the area of a circle, and merely stands for the Greek equivalent of 'p') and defined by

$$\boxed{\text{Profit} = \text{Revenue} - \text{Cost} \quad \text{or} \quad \pi = R - C.}$$

☐ **Example 1** If $C(q)$ and $R(q)$ are given by the graphs in Figure 5.43, for what values of q does the firm make a profit?

Solution. The firm makes a profit whenever revenues are greater than costs, i.e. when $R > C$. The graph of R is above the graph of C approximately when $110 < q < 220$, so production between $q = 110$ and $q = 220$ will generate a profit. ☐

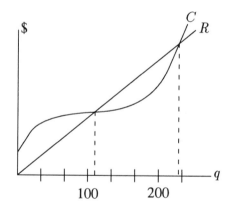

Figure 5.43: Costs and Revenues for Example 1

Marginal Analysis

Many economic decisions are based on an analysis of the costs and revenues 'at the margin' or at the edge. Let's look at this idea through an example.

Suppose you are running an airline and that you are trying to decide whether to put on an additional flight. How should you decide? We'll assume that the decision is to be made purely on financial grounds: if the flight will make money for the company, it should fly. Obviously you need to consider the costs and revenues involved. Since the choice is between adding this flight and leaving things the way they are, the crucial question is whether the *additional costs* incurred are greater or smaller than the *additional revenues* generated by the flight. These additional costs and revenues are called *marginal costs* and *marginal revenues*.

Suppose $C(q)$ is the function giving the total cost of running q flights. If the airline had originally planned to run 100 flights, its costs would be $C(100)$. With the additional flight, its costs would be $C(101)$. Therefore,

$$\text{Marginal Cost} = C(101) - C(100).$$

Now $C(101) - C(100) = \frac{C(101)-C(100)}{1}$, which is the average rate of change of cost between 100 and 101 flights. In Figure 5.44 the average rate of change is the slope of the secant line. If the cost function is not curving too fast near the point, the slope of the secant line will be close to the slope of the tangent line there. Therefore, the average rate of change is close to the instantaneous rate of change.

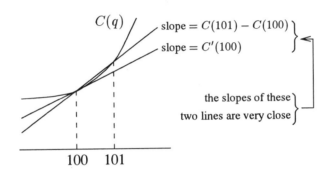

Figure 5.44: Marginal Cost: Slope of One of These Lines

Many economists choose to define marginal cost, MC, as the instantaneous rate of change of cost with quantity:

$$\boxed{\text{Marginal Cost} = MC = C'(q).}$$

Similarly if the revenue generated by q flights is $R(q)$, the additional revenue generated by increasing the number of flights from 100 to 101 is

$$\text{marginal revenue} = R(101) - R(100).$$

Now $R(101) - R(100)$ is the average rate of change of revenue between 100 and 101 flights. As before, the average rate of change is usually almost equal to the instantaneous rate of change, so economists often define

$$\boxed{\text{Marginal Revenue} = MR = R'(q).}$$

☐ **Example 2** If $C(q)$ and $R(q)$ for the airline are given in Figure 5.45, should the company put on the 101st flight?

Solution. The marginal revenue is the slope of the revenue line; it is a constant. The marginal cost is the slope of the tangent to the graph of C at the point 100. From Figure 5.45, you can see that the slope of the tangent at the point A is smaller than the slope at B so $MC < MR$. This means that the airline will make more in extra revenue than it will spend in extra costs if it runs another flight, so it should go ahead and run the 101st flight. ☐

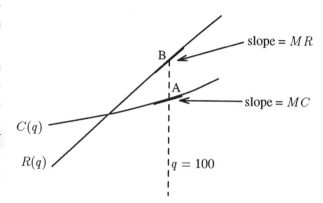

Figure 5.45: Cost and Revenue for Example 2

Since MC and MR are derivative functions, they can be estimated from the graphs of total cost and total revenue.

☐ **Example 3** If R and C are given by the graphs in Figure 5.46, sketch graphs of $MR = R'(q)$ and $MC = C'(q)$.

Solution. The revenue graph is a line through the origin, with equation

$$R = pq \qquad \text{where } p \text{ is a constant price}$$

so

$$MR = R'(q) = p.$$

The total cost is increasing, so the marginal cost is always positive.
For small q values, the total cost curve is concave down, so the marginal cost is decreasing. For larger q, say $q > 100$, the total cost is concave up and the marginal cost is increasing. Thus the marginal cost has a minimum at about $q = 100$. See Figure 5.47. ☐

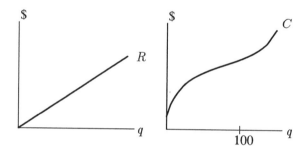

Figure 5.46: Total Revenue and Cost for Example 3

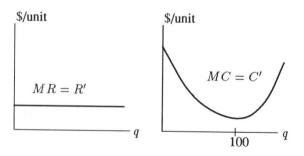

Figure 5.47: Marginal Revenue and Costs for Example 3

Maximizing Profit

Now let's look at an important problem: given functions for total revenue
and cost, how can we maximize total
profit? This is clearly a fundamental issue for any producer of goods. The next
example looks at this problem and ar-
rives at a result which is valid generally.

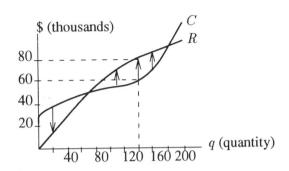

☐ **Example 4**   Find the maximum profit
if the total revenue and total cost are
given by the curves R and C, respectively, in Figure 5.48.

Figure 5.48: Maximum Profit

Solution. The profit is represented by the vertical difference between the curves, and is marked by
the vertical arrows on the graph. When revenue is below cost, the company is taking a loss; when
revenue is above cost, the company is making a profit. You can see that the profit is largest at about
$q = 140$, so this is the production level we're looking for.

You should observe that at $q = 140$ the tangents to the two curves are parallel. To the left of
$q = 140$, the revenue curve has a larger slope than the cost curve, and the profit increases with q.
To the right of $q = 140$, the slope of the revenue curve is less than the slope of the cost curve, and
the profit is decreasing. At the point where the slopes are equal, the profit has a local maximum.
Otherwise the profit could be increased by moving to the left or to the right. To be sure that the local
maximum is a global maximum, we need to check the endpoints. At $q = 0$ and $q = 200$, the profit
is negative, so the global maximum is indeed at $q = 140$.

Thus we have found the quantity (namely $q = 140$) which gives the maximum profit. If we
wanted the actual profit, we would have to estimate the vertical distance between the curves at
$q = 140$. This tells us that the maximum profit = $80,000 - 60,000 = $20,000.

Suppose for some reason we wanted to find the minimum profit (which is the same thing as the
maximum loss). For the curves in this example, this will occur at one of the endpoints, when $q = 0$
or $q = 200$.   ☐

Let's look at the general situation: to maximize or minimize profit over an interval, we can
optimize the profit function π where

$$\pi(q) = R(q) - C(q).$$

According to our analysis in Section 5.1, global maxima and minima can only occur either at critical
points or at the endpoints, if any, of the interval. To find critical points of π, look for zeros of the
derivative:

$$\pi'(q) = R'(q) - C'(q) = 0.$$

So

$$R'(q) = C'(q),$$

i.e. the slopes are equal. This is the same observation that we made in the previous example. In economic language, the maximum (or minimum) profit can occur when

Marginal Cost = Marginal Revenue.

Of course, maximal or minimal profit does not *have* to occur where $MR = MC$—there are always the endpoints to consider. However, this relationship is much more powerful than the answer to our one specific problem, because it is the condition that helps determine the maximum (or minimum) profit in general.

☐ **Example 5** Find the quantity q which will maximize profit if the total revenue and total cost (in dollars) are given by

$$R(q) = 5q - 0.003q^2$$

$$C(q) = 300 + 1.1q$$

where $0 \le q \le 1000$ units. What production level will give the minimum profit?

Solution. We can begin by looking for production levels that give Marginal Revenue = Marginal Cost:

$$MR = R'(q) = 5 - 0.006q$$

$$MC = C'(q) = 1.1.$$

So

$$5 - 0.006q = 1.1$$
$$q = \frac{3.9}{0.006} = 650 \text{ units.}$$

Does this represent a local maximum or minimum of π? We can tell by looking at what is going on at production levels of 649 units and 651 units. When $q = 649$ we have $MR = \$1.106$, which is greater than the (constant) marginal cost of 1.1. This means that producing one more unit will bring in more revenue than its cost, so profit will increase. When $q = 651$, $MR = \$1.094$, which is now *less* than MC, so it is not profitable to produce the 651st unit. We conclude that $q = 650$ is a local maximum for the profit function π. The profit earned by producing and selling this quantity is $\pi(650) = R(650) - C(650) = \$1982.50 - \$1015 = \$967.50$.

To check for global maxima we need to look at the endpoints. If $q = 0$, the only cost is $300 (the fixed costs) and there is no revenue, so $\pi(0) = -300$. At the upper limit of $q = 1000$, $R(1000) = \$2000$, $C(1000) = \$1400$ and so $\pi(1000) = \$600$. Therefore, the maximum profit is where $MR = MC$, and the minimum profit occurs when $q = 0$. ☐

❖ Exercises for Section 5.4

1. A manufacturer reports the total cost and total revenue functions shown in Figure 5.49. Sketch graphs, as a function of quantity, of:

   (a) Total profit,

   (b) Marginal cost,

   (c) Marginal revenue.

   Label the points q_1 and q_2 on your graphs.

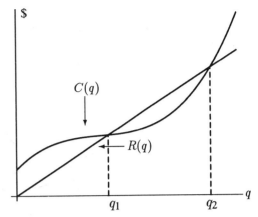

Figure 5.49: Cost and Revenue for a Good
(for Problem 1)

2. Let $C(q)$ be the total cost of producing a quantity q of a certain good. (See Figure 5.50.)

   (a) What is the meaning of $C(0)$?

   (b) Describe in words how the marginal cost changes as the quantity produced increases.

   (c) Explain the concavity of the graph (in terms of economics).

   (d) Explain the economic significance (in terms of marginal cost) of the point at which the concavity changes.

   (e) Do you think the graph of $C(q)$ looks like this for all types of goods?

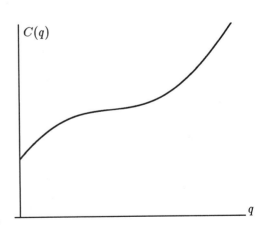

Figure 5.50: Cost of Producing a Good
(for Problem 2)

3. The *average cost* of manufacturing a quantity q of a good is defined to be

$$a(q) = \frac{C(q)}{q}.$$

You are given a graph of the average cost $a(q)$ in Figure 5.51

(a) Sketch a graph of the marginal cost $C'(q)$.

(b) Show that if

$$a(q) = b + mq$$

then

$$C'(q) = b + 2mq.$$

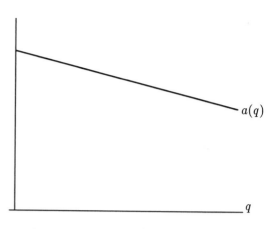

Figure 5.51: Average Cost of a Good

(for Problem 3)

4. $C(q)$ is the total cost of producing a quantity q. The average cost $a(q) = \frac{C(q)}{q}$ is given in Figure 5.52.

The following rule is used by some economists to determine the marginal cost $C'(q_0)$, for any q_0:

- Construct the tangent t_1 to $a(q)$ at q_0.

- Let t_2 be the line with the same y-intercept as t_1, but twice the slope of t_1.

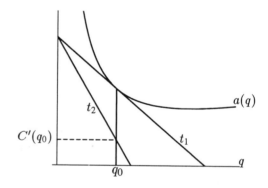

Figure 5.52: Using Average Cost to find Marginal Cost

(for Problem 4)

Then $C'(q_0)$ is as shown in Figure 5.52. Explain why this rule works.

5. Assume a manufacturer's cost of producing a good is given by the graph of $C(q)$ in Figure 5.53. Assume also that the manufacturer can sell the product for a price p each (regardless of the quantity sold), so that the total revenue from selling a quantity q is $R(q) = pq$. (See Figure 5.53.)

(a) The difference $\pi(q) = R(q) - C(q)$ is the total profit. For which quantity q_0 is the profit maximal? Mark your answer on a sketch of the graph.

(b) What is the relationship between p and $C'(q_0)$? (Note that p is the slope of the line $R(q) = pq$). Explain your result both graphically and analytically. ($\pi(q)$ has a maximum at $q = q_0$, so $\pi'(q_0) = 0$.) What does this mean in terms of economics?

(c) Graph $C'(q)$ and p (as a horizontal line) on the same axes. Mark q_0 on the q-axis.

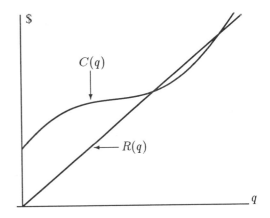

Figure 5.53: Cost and Revenue for a Good
(for Problem 5)

5.5 Optimization

The real world is full of problems where it is important to find the maximum or minimum value of some quantity. For example, engineers want to cut the strongest beam from a log of wood, scientists want to calculate which wavelength carries the maximum radiation at a given temperature, urban planners want to design traffic patterns to minimize delays, and financial forecasters want to predict when a stock market dive is bottoming out (meaning prices are at their lowest). All the techniques for finding such values make up the field called *optimization*. In this section and the next, you'll see how the derivative provides an efficient way of solving many optimization problems.

In this section, we'll look at some optimization problems where the function that we're trying to optimize is given; for example, by a formula or a graph. The section that follows this one will examine situations in which the function needs to be constructed, or modeled, using information provided by the problem.

☐ **Example 1** When an arrow is shot into the air, its range R is defined as the horizontal distance from the archer to the point at which the arrow hits the ground (see Figure 5.54). Clearly R depends on how fast the arrow is going and the initial angle that the arrow makes with the ground. If we fix the initial speed, some angles will yield larger Rs than others: shooting the arrow nearly vertically will give a (dangerously) small R. A nearly horizontal angle will also give a small R as the arrow hits the ground before it has had a chance to travel far horizontally.

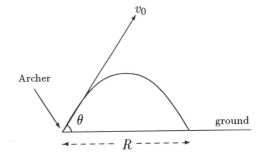

Figure 5.54: Arrow's Path

What initial angle will give a maximum for R? To answer this question requires knowing how

R depends on the angle, θ. If the ground is horizontal and we neglect air resistance, it can be shown that

$$R = \frac{v_0^2 \sin(2\theta)}{g}$$

where v_0 is the initial velocity of the arrow and g is the (constant) acceleration due to gravity. In this formula, θ is measured from the horizontal and $0 \le \theta \le \frac{\pi}{2}$.

Solution. We want to find the global maximum of R for $0 \le \theta \le \frac{\pi}{2}$. We first search for critical points of R in the interval $0 \le \theta \le \frac{\pi}{2}$:

$$\frac{dR}{d\theta} = \frac{v_0^2 \cdot 2\cos(2\theta)}{g}.$$

Setting $\dfrac{dR}{d\theta}$ equal to 0 we find

$$0 = \cos(2\theta) \quad \text{giving} \quad 2\theta = \pm\frac{\pi}{2}, \pm\frac{3\pi}{2}, \pm\frac{5\pi}{2}, \ldots$$

so $\frac{\pi}{4}$ is the only critical point in the interval $0 \le \theta \le \frac{\pi}{2}$. The critical value is $R\left(\frac{\pi}{4}\right) = \frac{v_0^2}{g}$. Now we must check the value of R at the endpoints $\theta = 0$ and $\theta = \frac{\pi}{2}$ (when the arrow is shot exactly horizontally and exactly vertically). Since $R = 0$ at each endpoint, the critical value $\theta = \frac{\pi}{4}$ is both a local and global maximum on $0 \le \theta \le \frac{\pi}{2}$. Therefore if you want the arrow to go furthest, you should shoot it at an angle of $\frac{\pi}{4}$ or $45°$. $\square$

☐ **Example 2** Some numbers are smaller than their squares (2 is smaller than 4), some are larger than their squares (1/3 is larger than 1/9) and some are equal to their squares (0 and 1). Any number x with $0 \le x \le 1$ will be less than or equal to its square. Among all such numbers, find the ones that differ from their squares the *most*, i.e. for which the function $s(x) = x - x^2$ is a maximum.

Solution. We have already observed that s is 0 at both of the endpoints of the interval, and that it is positive for any x inside the interval. Let's look for critical points of s:

$$
\begin{aligned}
s'(x) &= 1 - 2x \\
0 &= 1 - 2x \\
x &= 1/2.
\end{aligned}
$$

So there is only one critical point ($x = 1/2$) in the interval. We can see that $x = 1/2$ is a local maximum since $s' > 0$ when $x < 1/2$ and $s' < 0$ when $x > 1/2$. Now consider the endpoints, $x = 0$ and $x = 1$. Since s is 0 at both endpoints, $x = 1/2$ is the global maximum. The maximum difference is $s(1/2) = 1/2 - (1/2)^2 = 1/4$.

Alternatively, we could have recognized that s is just a parabola, and we know about parabolas: this one is concave down, with zeros at $x = 0, 1$, so it must have its maximum at $x = 1/2$. (Sometimes using calculus to solve a problem is overkill.) $\square$

Finding Upper and Lower Bounds

A problem which is closely related to finding maxima and minima is finding the *bounds* of a function. In Example 2 above, the difference between a number and its square ranges from 0 (when $x = 0, 1$) to $1/4$ (when $x = 1/2$). So, in Example 2,

$$0 \leq s(x) \leq \frac{1}{4}$$

and we say that the function s is bounded below by 0 and bounded above by $1/4$. Of course, we could also say that

$$-10 \leq s(x) \leq 170$$

so that s is also bounded below by -10 and above by 170. However, we consider the 0 and 1/4 to be 'better' bounds because they describe more accurately how the function s behaves.

The *best possible bounds* for a function, f, over an interval are numbers A and B such that, for all x in the interval,

$$A \leq f(x) \leq B$$

and where A and B are as close together as possible. A is called the *greatest lower bound* and B is the *least upper bound*.

☐ **Example 3** Find the best possible bounds for the function

$$f(x) = x + \sin x$$

over the interval $0 \leq x \leq 2\pi$.

Solution. Looking at the graph of f in Figure 5.55 will suggest that f is nondecreasing over the entire interval. You can confirm this by looking at the derivative:

$$f'(x) = 1 + \cos x$$

Since $\cos x \geq -1$, we have $f' \geq 0$ everywhere, so f never decreases. This means that a lower bound for f is 0 (its value at the left endpoint of the interval) and an upper bound is 2π (its value at the right endpoint). That is, if $0 \leq x \leq 2\pi$:

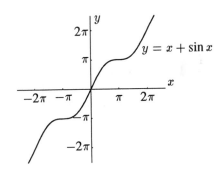

Figure 5.55: Graph of $y = x + \sin x$

$$0 \leq f(x) \leq 2\pi.$$

These are clearly the best bounds for the function f over the interval. ☐

☐ **Example 4** Find the best possible bounds for the oscillation represented for $t \geq 0$ seconds by the function

$$f(t) = e^{-t} \cos t.$$

Solution. What does the graph of f look like? You can think of it as a cosine curve with a decreasing amplitude of e^{-t}; in other words, it is a cosine curve squashed between the graphs of $y = e^{-t}$ and $y = -e^{-t}$, forming a wave with lower and lower crests and shallower and shallower troughs. See Figure 5.56.

From the graph you can see that for $t \geq 0$, the graph of f lies between the horizontal lines $y = -1$ and $y = 1$. This means that -1 and 1 are bounds:

$$-1 \leq e^{-t} \cos t \leq 1$$

The line $y = 1$ is the best possible upper bound because the graph does come up that high (at $t = 0$). However, we can find a better lower bound if we find the global minimum value of f for $t \geq 0$; this minimum occurs in the first trough between $t = \frac{\pi}{2}$ and $t = \frac{3\pi}{2}$ because later troughs are squashed closer to the t-axis. At that point $f'(t) = 0$. Using the Product rule

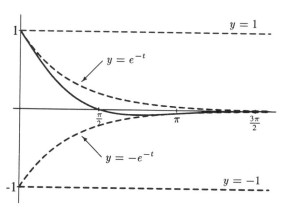

Figure 5.56: $f(t) = e^{-t} \cos t$ for $t \geq 0$

$$f'(t) = (-e^{-t}) \cos t + e^{-t}(-\sin t) = -e^{-t}(\cos t + \sin t) = 0$$

Since e^{-t} is never 0, we must have

$$\cos t + \sin t = 0$$

so $\tan t = -1$ giving $t = \dfrac{3\pi}{4}$

Thus the minimum occurs at $t = \frac{3\pi}{4}$. The value of that minimum is

$$f\left(\frac{3\pi}{4}\right) = e^{-\frac{3\pi}{4}} \cos\left(\frac{3\pi}{4}\right) \approx -0.067.$$

Rounding down so that the inequalities still hold for all $t \geq 0$ gives:

$$-0.07 \leq e^{-t} \cos t \leq 1.$$

Notice how much smaller in magnitude the lower bound is than the upper: 0.07 (less than $\frac{1}{10}$) versus 1. This is a reflection of how quickly the factor e^{-t} causes the oscillation to die out. $\square$

Minimizing Gas Consumption

The next example looks at a way of setting driving speeds to maximize fuel efficiency. We will assume some information about gas consumption, g (in gallons/hour), as a function of velocity, v (in mph). Notice that we are measuring gas consumption in gallons per *hour*, so g represents the rate at which gas is being consumed by the engine.

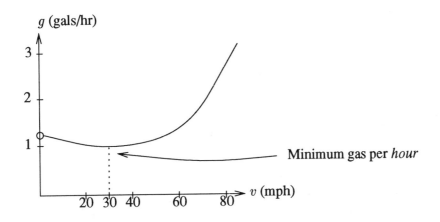

Figure 5.57: Gas Consumption versus Velocity

For low speeds, gas consumption drops as the speed increases because an engine runs ineffi-ciently at low speeds. At higher speeds, gas consumption increases again. If gas consumption at different speeds were measured experimentally, we might get a graph like that in Figure 5.57.

Looking at the graph, you can see the minimum gas per hour is used by going about 30 mph. But that is not necessarily what we want for fuel efficiency. If we go as slow as 30 mph, the journey might take so long that we'd use up less gas by going faster. In fact, if we assume that the distance we drive is fixed, then what we want to minimize is the gas consumption *per mile*, not the gas consumption per hour.

☐ **Example 5**   Using Figure 5.57, estimate the velocity which gives the minimum gas consumption, G, in gallons per mile.[2]

Solution. First we must find G, the average gas consumption per mile. Since

$$\frac{\Delta y}{\Delta x} = \frac{\text{gals/hour}}{\text{miles/hour}} = \frac{\text{gals}}{\text{mile}}$$

we can say that

$$G = \frac{g}{v} \text{ gals/mile.}$$

Thus we want to find the minimum value of $\frac{g}{v}$ when g and v are related by the graph in Figure 5.57. What does $\frac{g}{v}$ represent graphically?

[2]Adapted from *Calculus: The Analysis of Functions*, by Peter D. Taylor (Toronto: Wall & Emerson, Inc., 1992)

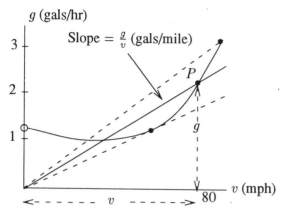

Figure 5.58: Graphical Representation of Gas Consumption per Mile, $G = \dfrac{g}{v}$

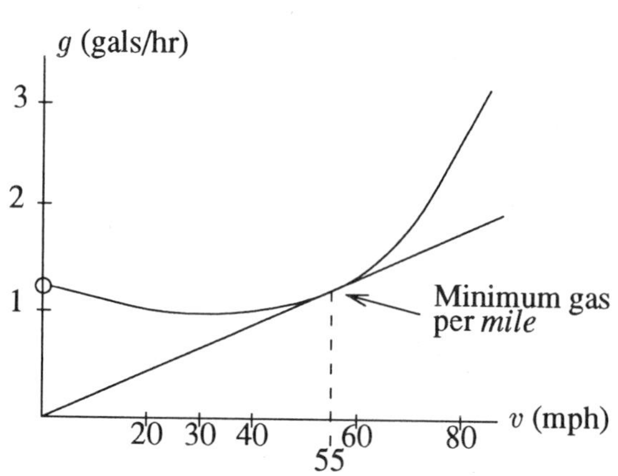

Figure 5.59: Velocity for Maximum Fuel Efficiency

Figure 5.58 shows that $\frac{g}{v}$ is the slope of the line from the origin to the point P. Where on the curve should P be to make the slope a minimum? From the possible positions of the line shown in Figure 5.58, you can see that the slope of the line is a minimum when the line is tangent to the curve. From Figure 5.59, you can see that the velocity at this point is about 55 mph. As we expected, this is larger than 30 mph. Thus to minimize gas consumption per mile, you should drive about 55 mph. ∎

❖ Exercises for Section 5.5

1. When you cough, your windpipe contracts. The velocity, v, with which air comes out depends on the radius, r, of your windpipe. If R is the normal (rest) radius of your windpipe, then for $r \le R$, the velocity is given by:

$$v = a(R - r)r^2$$

where a is a constant. What value of r maximizes the velocity?

2. The temperature change, T, in a patient generated by a dose, D, of a drug is given by

$$T = \left(\frac{C}{2} - \frac{D}{3}\right) D^2$$

where C is a positive constant.

(a) What dosage maximizes the temperature change?

(b) The sensitivity of the body, at dosage D, to the drug is defined as $\frac{dT}{dD}$. What dosage maximizes sensitivity?

3. A smokestack deposits soot on the ground with a concentration inversely proportional to the square of the distance from the stack. With two smokestacks 20 miles apart, the concentration of the combined deposits on the line joining them, at a distance x from one stack, is given by

$$S = \frac{k_1}{x^2} + \frac{k_2}{(20 - x)^2}$$

where k_1 and k_2 are constants which depend on the quantity of smoke each stack is emitting. If $k_1 = 7k_2$, find the point on the line joining the stacks where the concentration of the deposit is a minimum.

Find the best possible bounds for the functions in Problems 4–7, i.e., determine numbers A and B such that $A \le f(x) \le B$ for all numbers x in the domain specified. A and B should be as close together as possible. Many of these bounds will be irrational numbers that can be accurately expressed only symbolically (e.g., $\ln 3$). In such cases give both the symbolic value of the bound and a decimal representation to two decimal places. Round A down and B up so that your inequality $A \le f(x) \le B$ still holds.

4. e^{-x^2},   for $|x| \le 0.3$

5. $\ln(1 + x)$,   for $x \ge 0$

6. $\ln(1 + x^2)$,   for $-1 \le x \le 2$

7. $x^3 - 4x^2 + 4x$,   for $0 \le x \le 4$

8. (a) For which positive number x is $\sqrt[x]{x}$ largest? Justify your answer.

   (b) For which positive integer n is $\sqrt[n]{n}$ largest? Justify your answer.

   (c) Use your answer to (a) to decide which is larger:

$$3^{\frac{1}{3}} \text{ or } \pi^{\frac{1}{\pi}}.$$

[Hint: You may want to write $x = e^{\ln x}$.]

9. The *arithmetic mean* of two numbers a and b is defined as $\frac{a+b}{2}$; the *geometric mean* of two positive numbers a and b is defined as $\sqrt{ab}$.

   (a) For two positive numbers, which of the two means is larger? Justify your answer. [Hint: Define $f(x) = \frac{a+x}{2} - \sqrt{ax}$ for fixed a.]

   (b) For three positive numbers a, b, c, the arithmetic and geometric mean are $\frac{a+b+c}{3}$ and $\sqrt[3]{abc}$, respectively. Which of the two means of three numbers is larger?
   [Hint: Redefine $f(x)$ for fixed a and b.]

10. Given numbers a_1, a_2, a_3, define

$$f(x) = (x - a_1)^2 + (x - a_2)^2 + (x - a_3)^2.$$

Where is $f(x)$ minimal? Why is this result reasonable?

11. (a) For any numbers a_1, a_2, a_3, a_4, find the values of x that minimizes

$$f(x) = (x - a_1)^2 + (x - a_2)^2 + (x - a_3)^2 + (x - a_4)^2.$$

   (b) Use your answer to a) and to Problem 10 to guess the value of x that minimizes

$$f(x) = (x - a_1)^2 + (x - a_2)^2 + (x - a_3)^2 + \cdots + (x - a_n)^2.$$

12. When birds lay eggs, they do so in clutches of several at a time. When the eggs hatch, each clutch gives rise to a brood of baby birds. We want to determine the clutch size which maximizes the number of birds surviving to adulthood per brood. If the clutch is small, there are few baby birds in the brood; if the clutch is large, there are so many baby birds to feed that most die of starvation. The number of surviving birds per brood as a function of clutch size is shown in Figure 5.60[3].

   (a) Estimate the clutch size which maximizes the number of survivors per brood.

   (b) Suppose also that there is a biological cost to having a larger clutch: the female survival rate is reduced by large clutches. This cost is represented by the dotted line in Figure 5.60. If we take cost into account by assuming that the optimal clutch size in fact maximizes the vertical distance between the curves, what is the new optimal clutch size?

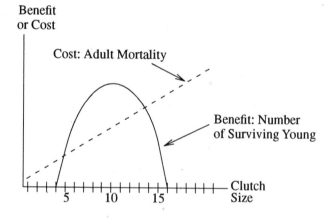

Figure 5.60: Clutch Size of Birds

(data from Oxford, England)

13. Let $f(v)$ be the energy consumption of a flying bird, measured in joules per second (a joule is a unit of energy), as a function of its speed v (in meters/sec). See Figure 5.61.

---

[3]Data from C. M. Perrins and D. Lack reported in an Introduction to Behavioral Ecology by J. R. Krebs and N. B. Davies

(a) Explain the shape of this graph (in terms of the way birds fly).

Now let $a(v)$ be the energy consumption of the same bird, measured in joules *per meter*.

(b) What is the relationship between $f(v)$ and $a(v)$?

(c) Where is $a(v)$ minimal?

(d) Should the bird try to minimize $f(v)$ or $a(v)$ when it is flying? Why?

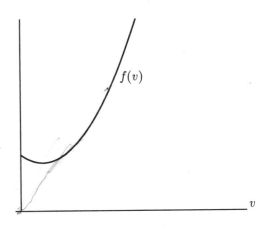

Figure 5.61: Energy Consumption of a Flying Bird

(for Problem 13)

14. If you invest x dollars in a certain project, your return is $R(x)$. Suppose you want to choose x to maximize your return per dollar invested, which is:

$$r(x) = \frac{R(x)}{x}.$$

(a) Suppose the graph of $R(x)$ has the form in Figure 5.62, with $R(0) = 0$. Illustrate on a copy of this graph that the optimum value of $r(x)$ is obtained at a point where the line from the origin to the point is tangent to the graph.

(b) Of course, it is also true that the maximum of $r(x)$ should occur at a point at which the slope of the $r(x)$ graph is zero. On the same set of axes as (a), draw a rough version of the graph of $r(x)$, corresponding to your R graph, and illustrate that the maximum occurs where the slope is 0.

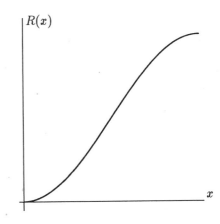

Figure 5.62: Revenue vs. Investment

(for Problem 14)

(c) Show, by taking the derivative of the above formula for $r(x)$, that the conditions in (a) and (b) are equivalent: the point where the line from the origin is tangent to the graph of R is the same point where the graph of r has zero slope.[4]

---

[4]From *Calculus: The Analysis of Functions*, by Peter D. Taylor (Toronto: Wall & Emerson, Inc., 1992)

15. A bird such as a starling feeds worms to its young. To collect worms, the bird flies to a site where worms are to be found, picks up several in its beak, and flies back to its nest. The *loading curve* in Figure 5.63[5] shows how the number of worms (the load) a starling collects depends on the time it has been searching for them. The curve is concave down because the bird can pick up worms more efficiently when its beak is empty; when its beak is already partly full, the bird becomes much less efficient. The traveling time (from nest to site and back) is represented by the distance PO in Figure 5.63. Suppose the bird wants to maximize the rate at which it brings worms to the nest, where

$$\text{Rate Worms Arrive at Nest} = \frac{\text{Load}}{(\text{Traveling Time} + \text{Searching Time})}$$

(a) Draw a line in Figure 5.63 whose slope is this rate.

(b) Using the graph, estimate the load which maximizes this rate.

(c) If the traveling time is increased, does the optimal load increase or decrease? Why?

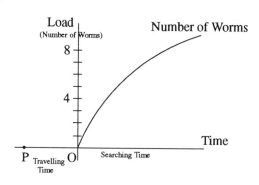

Figure 5.63: Bird's Loading Curve

5.6   More Optimization: Introduction to Modeling

Finding the global maxima and minima once the formula for the function is known need not be difficult. What is challenging is the translation of a given problem into a function and a domain over which it is to be optimized. This process of translation is called *modeling*. Often constructing a model given a description of a problem will depend on recognizing how physics, economics, geometry, or trigonometry can be applied. By looking at the examples that follow, you will get the flavor of some kinds of modeling.

A Problem in Physics

☐ **Example 1**   A grapefruit is tossed straight up with an initial velocity of 100 ft/sec. The moment that it leaves the thrower's hand it is 6 feet above the ground (you met this piece of fruit back in Chapter 2—remember?). How high does it go before returning to the ground?

---

[5]Alex Kacelnick(1984). Reported in Introduction to Behavioral Ecology, J. R. Krebs and N. B. Davis.

Solution. Here the quantity that we want to maximize is the height, y, of the grapefruit above the ground. We can make measurements to determine y at 1-second intervals and obtain a graph like that shown in Figure 5.64. You can see that the grapefruit goes above 150 feet and reaches its highest point after about 3 seconds.

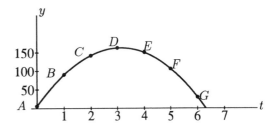

Figure 5.64: Height of the Grapefruit

We can get more exact results by constructing a model of the problem; in this case, a model will consist of a formula for the function $y = f(t)$ that gives the grapefruit's height in terms of elapsed time t. One such model, which can be obtained from physics, is

$$y = -16t^2 + 100t + 6.$$

In this formula, the 100 comes from the initial velocity and the 6 comes from the initial height. The 16 is related to the acceleration due to gravity, and is an approximate value. Other features of this formula, which make it a *model* of a flying grapefruit rather than an exact description, are that it assumes no air resistance or wind. If these factors were added to the model, it would be more accurate yet more complex. Often models are devised in stages, progressing from the simplest to the more detailed, as required by the desired accuracy of the solution. We'll be content with this simple model for the path of the grapefruit in this example.

Using the derivative we can find exactly when the grapefruit is at the highest point. We can think of this in two ways. By common sense, at the peak of the grapefruit's flight, the velocity, $\frac{dy}{dt}$, must be zero. Alternately, we are looking for a global maximum of y, so we look for critical points where $\frac{dy}{dt} = 0$. Thus we have

$$\frac{dy}{dt} = 100 - 32t = 0 \quad \text{and thus} \quad t = \frac{100}{32} = 3.125 \text{ secs.}$$

So far we have the time at which the height is a maximum; the maximum value of y is then

$$y = -16(3.125)^2 + 100(3.125) + 6 = \ 162.25 \text{ feet.}$$

❑

Problems in Geometry

❏ **Example 2**   What are the dimensions of an aluminum can that can hold 10 in^3 of juice that will use the least material (i.e. aluminum)? Assume that the can is cylindrical, and is capped on both ends.

Solution. It is *always* a good idea to think about a problem in general terms before trying to solve it. Since we're trying to use as little material as possible, why not make the can very small, e.g. the size of a peanut? We can't, since the can must hold 10 in^3. If we make the can short, to try to use less in the sides where the label is, we'll have to make it fat as well, so it can hold 10 in^3. You can see that saving aluminum in the sides might actually use more aluminum in the top and bottom in a short, fat can. See Figure 5.65(a).

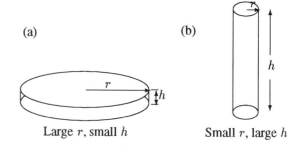

(a)   (b)

Large r, small h Small r, large h

Figure 5.65: Various Cylindrical-Shaped Cans

If we try to save material by making the top and bottom small, the can has to be somewhat tall, again to accommodate the 10 in^3 of juice. So any savings we get by using a small top and bottom might be outweighed by the tallness of the sides. See Figure 5.65(b). This is, in fact, true, as Table 5.1 shows. The table gives the amount of material used in the can for some choices of r and h. You can see that r and h change in opposite directions, and that more material is used at the extremes (very large or very small r and h) than in the middle. From the table you can see that the optimal radius for our can lies somewhere between $0.5 \leq r \leq 2.0$.

r	h	Material used, M (in^2)
0.1	318.31	200.06
0.5	12.73	41.57
1.0	3.18	26.28
2.0	0.80	35.13
3.0	0.35	63.22
10.0	0.03	630.32

Table 5.1: Amount of Material Used in the Can for Various Choices of r and h

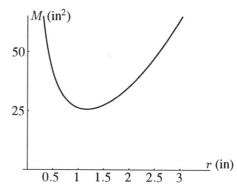

Figure 5.66: Total Material Used in Can, M, as a function of radius of ends, r

If we consider the material used as a function of the radius r, a graph of this function looks like Figure 5.66. The graph shows that the global minimum we want is at a critical point.

Both the table and the preceding graph were obtained from a model, which in this case is a formula for the material used in making the can. Finding such a formula will obviously rely on a knowledge of the geometry of a cylinder, in particular its area and volume. We'll label the height of the can h and the radius of the circular ends r. Now we construct the model:

$$
\begin{aligned}
M &= \text{material used in the can} \\
&= \text{material in the side} + \text{material in ends}
\end{aligned}
$$

where

$$\text{material in ends} \;=\; 2 \times \text{the area of a circle with radius } r$$
$$=\; 2 \cdot \pi r^2$$
$$\text{material in the side} \;=\; \text{the area of cylinder with height } h \text{ and radius } r$$
$$=\; 2\pi r h$$

As we discussed above, h is not independent of r, but changes in the opposite direction: if r grows, h shrinks and vice-versa. To find a precise relationship, we'll have to use the fact that the volume of the cylindrical can, $\pi r^2 h$, is equal to the constant 10 in^3, so

$$\pi r^2 h = 10 \quad \text{giving} \quad h = \frac{10}{\pi r^2}$$

So,

$$\text{material used in the side} = 2\pi r h = 2\pi r \frac{10}{\pi r^2} = \frac{20}{r}.$$

Therefore we obtain the formula for the total material used in a can of radius r:

$$M = 2\pi r^2 + \frac{20}{r}.$$

This function could be used to create Table 5.1 and Figure 5.66. Note that the domain of this function is $0 < r < \infty$, so that there are no endpoints to consider.

Now that our model is complete, we can use calculus to find the minimum of M. We look for the location of any critical points:

$$\frac{dM}{dr} = 4\pi r - \frac{20}{r^2} = 0 \quad \text{at a critical point,} \quad \text{so} \quad 4\pi r = \frac{20}{r^2}.$$

Therefore

$$\pi r^3 = 5 \quad \text{so} \quad r = \left(\frac{5}{\pi}\right)^{\frac{1}{3}} \approx 1.17 \text{ inches.}$$

This gives

$$h = \frac{10}{\pi r^2} \approx \frac{10}{\pi (1.17)^2} \approx 2.33 \text{ inches.}$$

and the material used, $f(1.17)$, is about 25.7 in^2. Figure 5.66 shows that M has one critical point which is the global minimum, so 25.7 in^2 must be that global minimum.   ❑

Practical Tips for Modeling Optimization Problems

In the last example, we used an approach that has broad application to many kinds of optimization problems. Here are some highlights:

1. Think about the problem: Be sure that you are clear about what quantity or function is to be optimized. Try different situations to get a feel for whether it even makes sense to try to optimize the quantity.

2. If possible, make several sketches showing how the elements that vary are related. Label your sketches clearly by assigning variables to quantities which change.

3. Try to obtain a formula for the function to be optimized in terms of the variables that you identified in the previous step. If necessary, eliminate from this formula all but one variable. Identify the domain over which this variable varies.

4. Now apply the techniques of Section 5.2 to obtain candidates for local maxima and minima, and evaluate the function at these points and at the endpoints, if any, to find the global maxima and minima.

The next example, another problem in geometry, illustrates this approach.

☐ **Example 3** Alaina wants to get to the bus stop as quickly as possible. The bus stop is across a grassy park, 2000 feet west and 600 feet north of her starting position. Alaina can walk west along the edge of the park on the sidewalk at a speed of 6 ft/sec. She can also travel through the grass in the park, but only at a rate of 4 ft/sec (the park is a favorite place to walk dogs, so she must move with care). What path will get her to the bus stop the fastest?

Solution. Thinking about the problem, you might assume that she should take a path that is the shortest distance. Unfortunately, the path that follows the shortest distance to the bus stop is entirely in the park, where her speed is slow (See Figure 5.67(a). That distance is $\sqrt{2000^2 + 600^2} \approx 2100$ feet, which will take her about 525 seconds to traverse. She could instead walk quickly the entire 2000 feet along the sidewalk, which would leave her just the 600 foot northward journey through the park (See Figure 5.67(b)). This method would take $2000/6 + 600/4 \approx 483$ seconds total walking time.

But can she do even better? Perhaps another combination of sidewalk and park will give a smaller travel time. For example, what is the travel time if she walks 1000 feet west along the sidewalk and the rest of the way through the park (See Figure 5.67(c))?. (Answer: about 454 seconds.)

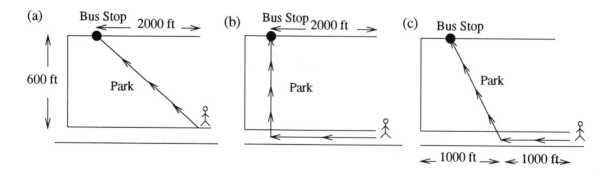

Figure 5.67: Three Possible Paths to the Bus Stop

Let's make a model for this problem. We can label the distance that she walks west along the sidewalk x and the distance she walks through the park y, as in Figure 5.68. Then the total time, t, will be

$$t = t_{\text{sidewalk}} + t_{\text{park}}.$$

Since

$$\text{time} = \text{distance/speed},$$

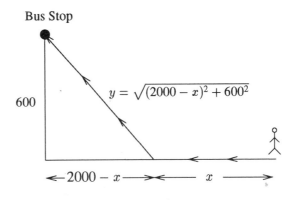

Figure 5.68: Modeling Time to Bus Stop

and she can walk 6 ft/sec on the sidewalk and 4 ft/sec in the park, we have that:

$$t = \frac{x}{6} + \frac{y}{4}.$$

Now, by the Pythagorean theorem, $y = \sqrt{(2000 - x)^2 + 600^2}$ feet through the park. The total time for both pieces is therefore

$$t = \frac{x}{6} + \frac{\sqrt{(2000 - x)^2 + 600^2}}{4}.$$

If you try searching for critical points of this function, you run into quite a bit of algebra. You may find it easier to graph the function on a calculator or computer, and estimate the critical point which turns out to be $x \approx 1463$ feet and gives the minimum total time of about 445 seconds. ❑

❑ **Example 4**   One hallway which is 4 feet wide meets another hallway which is 8 feet wide in a right angle. See Figure 5.69. What is the length of the longest ladder which can be carried horizontally around the corner?

Solution. We begin by thinking about the problem. We imagine the ladder being carried on its side and ignore its width. To allow the longest ladder possible we carry the ladder around the corner so that it just touches both walls (at A and C) and just touches the corner at B. Let's draw some lines that do this. See Figure 5.69. As you can see, the length of the line $\overline{ABC}$ decreases as the corner is turned, then increases again. The minimum length would be the *largest* length of a ladder that could make it around the corner. Certainly a smaller ladder would work (it wouldn't touch A, B and C simultaneously) but a larger one would not fit.

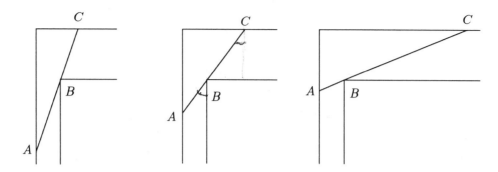

Figure 5.69: Various Ladders that Touch Both Walls and Corner

Thus the way to approach this problem of finding the largest ladder that can make it around the corner in the hallway is to find the *smallest* length of the line $\overline{ABC}$. That smallest length will be the length of the longest possible ladder. We will express the length, l, as a function of θ, the angle between the line and the wall of the narrow hall. See Figure 5.70.

$$l = \overline{AB} + \overline{BC}.$$

Now $\overline{AB} = \dfrac{4}{\sin \theta}$ and $\overline{BC} = \dfrac{8}{\cos \theta}$, so

$$l = \frac{4}{\sin \theta} + \frac{8}{\cos \theta}.$$

Here the domain of θ is $0 < \theta < \frac{\pi}{2}$. The next step is to look for a minimum of l by searching for critical points.

$$\frac{dl}{d\theta} = -\frac{4}{(\sin \theta)^2}(\cos \theta) - \frac{8}{(\cos \theta)^2}(-\sin \theta).$$

When $\dfrac{dl}{d\theta} = 0$,

$$-4\frac{\cos \theta}{(\sin \theta)^2} + 8\frac{\sin \theta}{(\cos \theta)^2} = 0$$

so

$$(\cos \theta)^3 = 2(\sin \theta)^3$$

giving

$$\frac{(\sin \theta)^3}{(\cos \theta)^3} = \frac{1}{2}$$

so

$$\tan \theta = \sqrt[3]{0.5} \approx 0.79$$

and

$$\theta \approx 0.67 \text{ radians.}$$

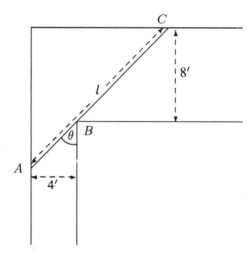

Figure 5.70: Ladder and Hallway

Thus $\theta \approx 0.67$ is a critical point, and our preliminary discussion of what happens for θ near 0 and θ near $\frac{\pi}{2}$ assures us that l will have a global minimum there. The minimum value of l is therefore

$$l = \frac{4}{\sin(0.67)} + \frac{8}{\cos(0.67)} \approx 16.65 \text{ feet,}$$

so the ladder can be at most 16.65 feet long, and still make it around the corner. ❑

❖ Exercises for Section 5.6

1. If you have 100 feet of fencing and want to enclose a rectangular area up against a long straight wall, what is the largest area you can enclose?

2. Of points on the parabola whose equation is $y = x^2$, which is nearest to $(1,0)$? Find the coordinates to two decimals. [Hint: Minimize the *square* of the distance—this avoids square roots.]

3. Find the point(s) on the ellipse

$$\frac{x^2}{9} + y^2 = 1$$

   (a) closest to the point $(2,0)$.

   (b) closest to the focus $(\sqrt{8},0)$.

   [Hint: Minimize the *square* of the distance—this avoids square roots.]

4. Sketch the graph of $f(x) = 2 - 2\sin x$. Express the distance from the origin to a point on the graph of f as a function of x. Find the point on the graph of f that is closest to the origin. Label this point on the graph, draw the line segment from the origin to this point, and state how long this minimum distance is.

5. The graphs of $y = \sqrt{x}$, $x = 9$, and $y = 0$ bound a region in the first quadrant. Find the dimensions of the rectangle of maximum *area* and the dimensions of the rectangle of maximum *perimeter* that can be inscribed in this region. (The sides of the rectangles should be parallel to the axes.)

6. The graphs of $y = \sqrt{x}$, $x = c$ $(c > 1)$, and $y = 0$ bound a region in the first quadrant. Show that the height of the rectangle of maximum *area* inscribed therein depends on the value of c but that the height of the rectangle of maximum *perimeter* is independent of c. (The sides of the rectangle should be parallel to the axes.)

7. The graphs of $y = x^n$ $(n > 0, n \neq 1)$, $x = c$ $(c > 1)$, and $y = 0$ bound a region in the first quadrant. Show that the height of the rectangle of maximum area inscribed therein depends on the values of n and c but that the height of the rectangle of maximum perimeter depends only on n. (The sides of the rectangle should be parallel to the axes.)

8. Refer to Figure 5.71.

   (a) Bueya spends her days as follows: She is at Washington University in the morning, at her job in East St. Louis in the afternoon, and late in the evening she has a beer in her favorite bar. She has lunch and dinner at home. Where on the road should she look for an apartment to minimize her daily traveling distance? (See Figure 5.71)

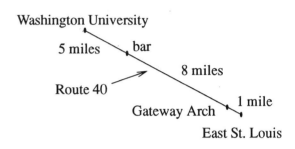

Figure 5.71: St. Louis

   (b) Her colleague Marie-Josée goes to a gym near the Gateway Arch before breakfast (which she eats at home), and she spends the rest of the day the same way as Bueya. Where on the road should she look for an apartment?

9. To get the best view of the Statue of Liberty in in the figure to the right you should be at the position where θ is a maximum. If the statue stands 92 meters high, including the pedestal which is 46 meters high, how far from the base should you be?

10. Assume that a pigeon is released from a boat (point B in Figure 5.72) floating on a lake. Because of falling air over the cool water, the energy required to fly one meter over the lake is twice the corresponding energy e required for flying over the bank ($e = 3$ joule/meter). To minimize the energy required to fly from B to the loft L the pigeon will head to a certain point P on the bank and then fly along the bank to L.

   (a) What is the optimal angle $\theta = $ angle BPA?

   (b) Does your answer change if $\overline{AL}, \overline{AB}$ and e have different numerical values?

   (c) Do you think a pigeon would actually fly that way?

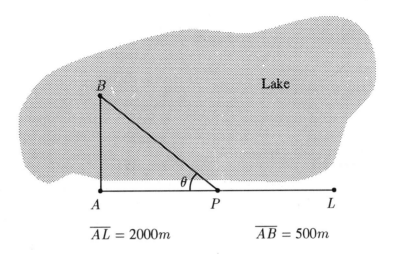

$$\overline{AL} = 2000m \qquad \overline{AB} = 500m$$

Figure 5.72: Pigeon Lake
(for Problem 10)

11. On the same side of a straight river are two towns and a pumping station, S, that supplies water to them. The pumping station is at the river's edge with pipes extending straight to the two towns. The distances are shown in the figure to the right. Where should the pumping station be located to minimize the total length of pipe?

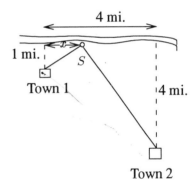

5.7   Newton's Method

Many problems in mathematics involve finding the root of an equation. For example, we might have to locate the zeros of a complicated polynomial, or determine the point of intersection of two curves. Since this problem is so common, mathematicians have developed a wide variety of methods for finding roots. When you learned algebra you probably spent a good deal of time learning how to solve certain types of equations—for example, quadratics. Here we will concentrate on numerical methods for approximating solutions which cannot be calculated any other way.

In Chapter 1, you have already seen one such method: bisection. It is very simple, yet effective. However, it has two major drawbacks. First, it cannot locate a root where the curve is tangent to, but does not cross, the x-axis. Second, it is relatively slow in the sense that it requires a considerable number of iterations to achieve a desired level of accuracy. While speed may not be very important in solving a single equation, you should realize that in practical problems, one may have to solve thousands of equations as a parameter changes. In such a case, any saving in the number of steps can be important.

Using Newton's Method

We now consider a powerful root-finding method developed by Newton. Suppose we have a function $y = f(x)$. The equation $f(x) = 0$ has a root at $x = r$, as shown in Figure 5.73. We begin with an initial estimate, x_0, for this root. (This can be a guess.) We will now obtain a better estimate x_1. To do this, construct the tangent line to the curve at the point $x = x_0$ and extend it until it crosses the x-axis, as shown in Figure 5.73. The point where it crosses the axis is usually much closer to r and we use that point as the next estimate, x_1. Having found x_1, we now repeat the process using x_1 instead of x_0. We construct another tangent line to the curve at $x = x_1$, extend it until it crosses the x-axis, and use that x-intercept as the next approximation, x_2, and so on. The resulting sequence of x-intercepts usually converges to the root r at an extremely rapid rate.

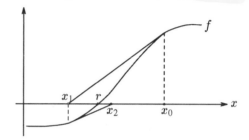

Figure 5.73: Newton's Method: Successive Approximations x_0, x_1, x_2, ... to the Root, r.

Let's see how this looks algebraically. We know that the slope of the tangent line at the initial estimate x_0 is $f'(x_0)$ and so the equation of the tangent line is

$$y - f(x_0) = f'(x_0)(x - x_0)$$

At the point where this tangent line crosses the x-axis, we have $y = 0$ and $x = x_1$, so that

$$0 - f(x_0) = f'(x_0)(x_1 - x_0)$$

Solving for x_1, we obtain

$$x_1 = x_0 - \frac{f(x_0)}{f'(x_0)}$$

provided that $f'(x_0)$ is not zero. We now repeat this argument and find the next approximation is

$$x_2 = x_1 - \frac{f(x_1)}{f'(x_1)}$$

and, in general, for any $n = 0, 1, 2, \ldots$,

$$\boxed{x_{n+1} = x_n - \frac{f(x_n)}{f'(x_n)}}$$

provided that $f'(x_n)$ is not zero. This is *Newton's Method*.

☐ **Example 1** Use Newton's Method to find the fifth root of 23. (By calculator, this is 1.872171231, correct to nine decimal places.)

Solution. To use Newton's Method, we have to find an equation of the form $f(x) = 0$ having $23^{\frac{1}{5}}$ as a root. Since $23^{\frac{1}{5}}$ is a root of $x^5 = 23$ or $x^5 - 23 = 0$, we take $f(x) = x^5 - 23$. The root of this equation is less than 2 (since $2^5 = 32$), so we will choose $x_0 = 2$ as our initial estimate. Now $f'(x) = 5x^4$, so we can set up Newton's Method as

$$x_{n+1} = x_n - \frac{(x_n^5 - 23)}{5x_n^4}.$$

In this case, we can simplify using a common denominator to obtain

$$x_{n+1} = \frac{4x_n^5 + 23}{5x_n^4}.$$

Therefore, starting with $x_0 = 2$, we find that $x_1 = 1.8875$. This leads to $x_2 = 1.872418193$ and $x_3 = 1.872171296$. These values are in Table 5.2. Since $f(1.87218231) > 0$ and $f(1.872171230) < 0$, the root lies between 1.872171230 and 1.872171231. Therefore, in just four iterations of Newton's Method, we have achieved nine decimal accuracy.

n	x_n	$f(x_n)$
0	2	9
1	1.8875	0.957130661
2	1.872418193	0.015173919
3	1.872171296	0.000004020
4	1.872171231	0.000000027

Table 5.2: Newton's Method: $x_0 = 2$

☐

As a general guideline for Newton's Method, once the first correct decimal place is found, each successive iteration approximately doubles the number of correct digits.

What happens if we select a very poor initial estimate? In the above example, suppose x_0 were 10 instead of 2. The results are in Table 5.3.

Notice that even with $x_0 = 10$, the sequence of values moves reasonably quickly toward the solution: we achieve seven decimal place accuracy by the 11th iteration.

n	x_n
0	10
1	8.000460000
2	6.401419079
3	5.123931891
4	4.105818871
5	3.300841811
6	2.679422313
7	2.232784753
8	1.971312452
9	1.881654220
10	1.872266333
11	1.872171240

Table 5.3: Newton's Method: $x_0 = 10$

☐ **Example 2** Find the first point of intersection of the curves given by $f(x) = \sin x$ and $g(x) = e^{-x}$.

Solution. The graphs in Figure 5.74 make it clear that there are an infinite number of points of intersection, all with $x > 0$. In order to find the first one, we consider the function

$$F(x) = f(x) - g(x) = \sin x - e^{-x}$$

whose derivative is $F'(x) = \cos x + e^{-x}$. The point we want must be fairly close to $x = 0$, so we start with $x_0 = 0$. The values in Table 5.4 are approximations to the root. Since $F(0.588532744) > 0$

and $F(0.588532743) < 0$, the root lies between 0.588532743 and 0.588532744. (Remember, your calculator must be set in radians.) ◻

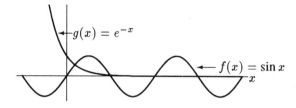

n	x_n
0	0
1	0.5
2	0.585643817
3	0.588529413
4	0.588532744
5	0.588532744

Figure 5.74: $f(x) = \sin x$ and $g(x) = e^{-x}$

Table 5.4: Successive Approximations to Root of $\sin x = e^{-x}$

◻ **Example 3** Find the next positive point of intersection of $f(x) = \sin x$ and $g(x) = e^{-x}$. You might have to try several different initial values for x_0.

When Does Newton's Method Fail?

In most practical situations, Newton's Method works well. Occasionally, however, the sequence x_0, x_1, x_2, ... fails to converge or fails to converge to the root you wanted. Sometimes, for example, the sequence can 'jump' from one root to another. This is particularly likely to happen if the magnitude of the derivative $f'(x_n)$ is small for some x_n. In this case, the tangent line is nearly horizontal and so x_{n+1} will be far from x_n. See Figure 5.75.

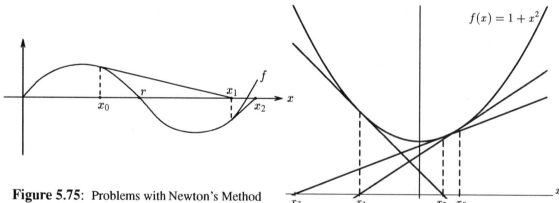

Figure 5.75: Problems with Newton's Method

Figure 5.76: $f(x) = 1 + x^2$

If the equation $f(x) = 0$ has *no* root, then the sequence will not converge. In fact the sequence obtained by applying Newton's Method to $f(x) = 1 + x^2$ is one of the best known examples of *chaotic behavior* and has attracted considerable research interest recently. See Figure 5.76.

❖ Exercises for Section 5.7

1. Suppose you want to find a solution of the equation

$$x^3 + 3x^2 + 3x - 6 = 0.$$

Consider $f(x) = x^3 + 3x^2 + 3x - 6$.

(a) Find $f'(x)$ and use it to show that $f(x)$ increases everywhere.

(b) How many roots does the original equation have?

(c) For each root, find an interval which contains it.

(d) Find each root to two decimal places, using Newton's method.

For Problems 2–4, use Newton's method to find the given quantities to two decimal places:

2. $\sqrt[3]{50}$ 3. $\sqrt[4]{100}$ 4. $10^{-\frac{1}{3}}$

For Problems 5–9, solve each equation and give each answer to two decimal places:

5. $\sin x = 1 - x$ 8. $(\ln x) + 1 = x$

6. $\cos x = x$ 9. $e^x \cos x = 1$, for $0 < x < 2\pi$.

7. $e^{-x} = \ln x$

10. Find, to two decimal places, all solutions of $\ln x = \dfrac{1}{x}$.

11. How many zeros do the functions below have? For each zero, find an upper and a lower bound which differ by no more than 0.1.
    (a) $f(x) = x^3 + x - 1$,   (b) $f(x) = \sin x - \frac{2}{3}x$,   (c) $f(x) = 10xe^{-x} - 1$.

12. Find the largest zero of

$$f(x) = x^3 + x - 1$$

to six decimal places, using Newton's method. How do you know your approximation is as good as you claim?

13. The problem of calculating the square root, $\sqrt{a}$, of any positive number a, is often done by applying Newton's Method to the function $f(x) = x^2 - a$. Apply the method to obtain an expression for x_{n+1} in terms of x_n. Use this to approximate $\sqrt{a}$ for $a = 2, 10, 1000$, and π, correct to four decimal places, starting at $x_0 = \frac{a}{2}$ in each case.

14. What is the smallest positive zero of the function $f(x) = \sin x$? Let us apply Newton's method, with initial guess $x_0 = 3$, and see how fast it converges to π.

   (a) Compute the first two approximations, x_1 and x_2; compare x_2 with π.

   (b) Newton's method works very well here. Explain why. To do this, you will have to outline the basic idea behind Newton's method.

   (c) Starting with 3 and 4 as original estimates, how many bisections would you need to get an estimate as good as x_2?

15. Newton's method can be very sensitive to your initial estimate, x_0. Using a Newton's method program on your calculator or computer, try the following experiment. With $f(x) = \sin x - \frac{2}{3}x$

   (a) Do Newton's method with the following initial estimates:

   $$x_0 = 0.904, \quad x_0 = 0.905, \quad x_0 = 0.906.$$

   (b) What happens?

5.8 Working Backwards: Antiderivatives

Chapter 2 began with the idea that velocity is the derivative of distance and that acceleration is the derivative of velocity. In this section we will analyze the motion of a body falling freely under the influence of gravity. This involves going from acceleration to velocity to position. The general idea of working backwards from the derivative of a function to the function itself turns out to be just as important as going the other way.

Uniform Motion

To set the stage, let's briefly consider a familiar problem. A body moving in a straight line with constant velocity is said to be in *uniform motion*. Imagine a car moving at 50 mph. How far does it go in a given time? The answer is given by

$$\text{Distance} = (\text{Rate}) \times (\text{Time})$$

or

$$s = 50t$$

where s is the distance of the car (in miles) from a fixed reference point and t is time in hours. Alternatively, we can describe the motion by writing the equation

$$\frac{ds}{dt} = 50.$$

We can view this as an equation, a *differential equation*, for the *function s*, and we say that the solution to this equation results from our answering the question: What function $s = f(t)$ do we differentiate to get 50? We know from our experience in taking derivatives that one answer is $s = 50t$ because:

$$\frac{d}{dt}(50t) = 50\frac{d}{dt}(t) = 50.$$

Solving the differential equation means working backwards from $\frac{ds}{dt}$ to s itself. Notice that the solution $s = 50t$ is the same as the one we obtained before by using distance $=$ (rate) $\times$ (time).

Is this the end of it? Even for this simple example there is something more to say. Notice that if C is a constant then

$$s = 50t + C$$

is also a possible expression for s because the derivative of C is zero.

The equation $s = 50t + C$ tells us that $s = C$ when $t = 0$. Thus, the constant C represents the initial distance of the car from the reference point. (Note that the car does not necessarily start at the reference point.) Call this initial distance s_0. Then the solution to the *initial value problem*

$$\frac{ds}{dt} = 50, \text{ and } s = s_0 \text{ when } t = 0$$

is

$$s = 50t + s_0.$$

Uniformly Accelerated Motion

Now we will consider a body moving with *constant acceleration* along a straight line, or *uniformly accelerated motion*. It has been known since Galileo's time that this is the kind of motion experienced by a body falling under the influence of gravity. (By the way, "falling" is meant to include "falling up", as when the body is tossed up into the air). Thus, if v is the upward velocity and t is the time

$$\frac{dv}{dt} = -g$$

where g is the constant *acceleration due to gravity*, whose value is determined by experiment. In today's most frequently used units its value is approximately

$$g = 32 \text{ ft/sec}^2, \text{ or } g = 9.81 \text{ m/sec}^2, \text{ or } 981 \text{ cm/sec}^2.$$

The negative sign comes from the fact that velocity is measured upward, while gravity acts downward.

☐ **Example 1** Find, as a function of time, the position and velocity of a stone dropped from a 100 foot high building. When does it hit the ground and how fast is it going at that time?

Solution. If we measure distance, s, in feet above the ground, and the velocity, v, in ft/sec upwards, then the acceleration due to gravity is 32 ft/sec^2 downwards, so

$$\frac{dv}{dt} = -32,$$

where t is measured in seconds from when the stone was dropped. From what we know about taking derivatives, we must have

$$v = -32t + C$$

where C is some constant. Since $v = C$ when $t = 0$, the constant C represents the initial velocity, v_0. The fact that the stone is "dropped" rather than "thrown" off the top of the building tells us that the initial velocity is zero, so $C = v_0 = 0$. Thus

$$v = -32t$$

But now we can write

$$v = \frac{ds}{dt} = -32t.$$

Finding s means finding a function whose derivative is $-32t$. One possible function is $-16t^2$ because

$$\frac{d}{dt}(-16t^2) = -16\frac{d}{dt}(t^2) = -16(2t) = -32t.$$

As before we could also have

$$s = -16t^2 + K$$

where K is another constant, because the derivative of K is zero.

Since the stone starts at the top of the building, $s = 100$ when $t = 0$. Substituting into $s = -16t^2 + K$ gives

$$100 = -16(0^2) + K$$

so

$$K = 100$$

and therefore

$$s = -16t^2 + 100.$$

Thus we have found both v and s as functions of time.

The stone hits the ground when $s = 0$, so we must solve

$$0 = -16t^2 + 100$$

giving $t^2 = 100/16$ or $t = \pm 10/4 = \pm 2.5$ secs. Since t must be positive, $t = 2.5$ secs. At that time, $v = -32(2.5) = -80$ ft/sec. (Note that the velocity is negative because we are considering up as the positive direction and down as the negative direction.)   ❑

❑ **Example 2**   An object is thrown vertically upwards with a speed of 10 m/sec from a height of 2 meters. Find the highest point it reaches and when it hits the ground.

Solution. We must first find the equation of motion. This time, the velocity is in m/sec, so we use $g = 9.8$ m/sec^2. Measuring distance in meters upward from the ground, we have

$$\frac{dv}{dt} = -9.8$$

As before, v is a function whose derivative is constant, so

$$v = -9.8t + C.$$

Since the initial velocity is 10 meters/sec upwards, we know that $v = 10$ when $t = 0$. Substituting gives

$$10 = -9.8(0) + C$$

so

$$C = 10.$$

Thus

$$v = -9.8t + 10.$$

To find s, we use

$$v = \frac{ds}{dt} = -9.8t + 10$$

and look for a function that has $-9.8t + 10$ as its derivative. Let's think about each term separately. Since

$$\frac{d}{dt}\left(-9.8\frac{t^2}{2}\right) = -9.8\frac{2t}{2} = -9.8t \quad \text{and} \quad \frac{d}{dt}(10t) = 10,$$

a possible formula is $s = -9.8\frac{t^2}{2} + 10t = -4.9t^2 + 10t$. As before, we can add any constant so

$$s = -4.9t^2 + 10t + K.$$

To find K, we use the fact that the object starts at a height of 2 meters, so $s = 2$ when $t = 0$. Substituting gives

$$2 = -4.9(0)^2 + 10(0) + K$$

so $K = 2$ and therefore

$$s = -4.9t^2 + 10t + 2$$

The object reaches its highest point when the velocity is 0, or

$$v = -9.8t + 10 = 0$$

which occurs when

$$t = \frac{10}{9.8} \approx 1.02 \text{ secs.}$$

When $t = 1.02$ secs,

$$s = -4.9(1.02)^2 + 10(1.02) + 2 \approx 7.10 \text{ meters.}$$

So the maximum height reached is 7.10 meters. The object reaches the ground when $s = 0$:

$$0 = -4.9t^2 + 10t + 2.$$

Solving this using the quadratic formula gives

$$t \approx -0.18 \quad \text{and} \quad t \approx 2.22 \text{ seconds.}$$

Since the time at which it hits the ground must be positive, $t \approx 2.22$ seconds.

Antiderivatives

We were able to solve the problem of uniformly accelerated motion by working backwards from the derivative of a function to the function itself. Reversing the process of differentiation is called finding an *antiderivative*. Given a function f we ask for a function F with $F' = f$. The function F is then called an *antiderivative* of f.

We can also say that we want to solve the *differential equation*

$$\frac{dy}{dx} = f(x).$$

Here f is a known function and we want to find a function $y = F(x)$ with $F'(x) = \frac{dy}{dx} = f(x)$; the function $y = F(x)$ is the "unknown" in the equation. For example:

An antiderivative of $f(x) = x^2$ is $F(x) = \frac{1}{3}x^3$ because $\frac{d}{dx}\left(\frac{1}{3}x^3\right) = \frac{1}{3} \cdot 3x^2 = x^2$.

An antiderivative of $g(x) = x^3$ is $G(x) = \frac{1}{4}x^4$ because $\frac{d}{dx}\left(\frac{1}{4}x^4\right) = \frac{1}{4} \cdot 4x^3 = x^3$.

An antiderivative of $h(x) = e^x$ is $H(x) = e^x$. (Why?)

We obtained these antiderivatives "by inspection"; that is our experience with differentiating gives us a chance of answering the reverse question: "Which function do I differentiate to get …?" Obviously, any specific result we have about derivatives gives us a corresponding result about antiderivatives. We can easily make a table of columns of functions and their derivatives and then use the table to solve

$$\frac{dy}{dx} = f(x)$$

whenever we happen to find $f(x)$ in the column of derivatives. What if we can't find a given $f(x)$ in our table? This is a question for Chapter 6. In this section we will look at two other important questions.

What is the solution to the differential equation $\dfrac{dy}{dx} = 0$?

Here we are asking for a function whose derivative is *always* zero. A constant function has this property – are there any other functions whose derivative is always zero? Common sense says no. A function whose derivative is zero everywhere on an interval must have a horizontal tangent line at every point of its graph, and the only way this can happen is if the graph itself is a horizontal line, i.e. if the function is constant. Put another way, if we think of the derivative as a velocity, and if the velocity is always zero, then the object is standing still. Whatever its initial position is, that's where it stays; the position function is constant. A rigorous proof of this result using the definition of the derivative is surprisingly subtle. We have:

If the derivative of a function is zero everywhere on an interval, then the function is constant in the interval: i.e.,

$$\text{if } F'(x) = 0 \text{ then } F(x) = C.$$

What is the most general antiderivative of f?

Suppose that we have two functions F and G with $F' = f$ and $G' = f$, that is F and G are both antiderivatives of the same function f. Do we necessarily have $F = G$? Not quite. Since $F' = G'$ we have $F' - G' = 0$ or $(F - G)' = 0$. But this means that we must have $F - G = $ constant, so $F - G = C$, or $F = G + C$ where C is a constant. Thus, any two antiderivatives of the same function differ by a constant:

> If F and G are both antiderivatives of a function f, then $F = G + C$ for some constant C. Equivalently, if $y = F(x)$ is solution to the differential equation
> $$\frac{dy}{dx} = f(x),$$
> then the most general solution is $y = F(x) + C$ where C is a constant.

We used this fact implicitly in our study of motion when we wrote that the general solution of $\frac{ds}{dt} = 50$ was $s = 50t + C$. We can also make sense of this from a geometric point of view. When we are asked to solve

$$\frac{dy}{dx} = f(x)$$

for the unknown function $y = F(x)$, what we are told about F is that its graph has slope f. But shifting the graph of F up or down, that is, looking at $F + C$, produces a curve with the same slope and hence a function satisfying the same differential equation. Hence "the" solution of a differential equation is actually not one function but a *family* of functions; not one curve but a *family* of 'parallel' curves.

☐ **Example 3** Find and graph the general solution of the differential equation

$$\frac{dy}{dx} = \sin x + 2.$$

Solution. We are asking for a function whose derivative is $\sin x + 2$. Since the derivative of a sum is the sum of the derivatives, we can try to find antiderivatives of $\sin x$ and 2 separately and then add the results. An antiderivative of $\sin x$ is $-\cos x$, since $\frac{d}{dx}(-\cos x) = \sin x$, and an antiderivative of 2 is $2x$. Thus a solution of the equation is $y = -\cos x + 2x$. The general solution is therefore $y = -\cos x + 2x + C$, where C is any constant. Figure 5.77 shows several curves in this family. ☐

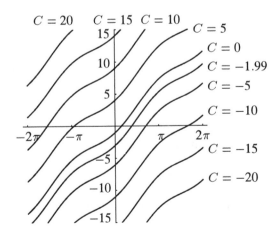

Figure 5.77: Solution Curves of $\frac{dy}{dx} = \sin x + 2$.

How can we pick out *one* antiderivative of a function f?

Picking out one antiderivative is equivalent to selecting a value of C, which, as we saw in the motion examples, involves knowing an extra piece of information, such as the initial velocity or initial position. In general, we can pick out one particular curve in the family of curves corresponding to solutions of $\frac{dy}{dx} = f(x)$ by specifying, *as an extra condition*, that we want the curve that passes through a given point (x_0, y_0). Provided that F is defined at x_0, there will be exactly one solution curve of $\frac{dy}{dx} = f(x)$ that passes through a given point (x_0, y_0). Equivalently, there will be exactly one antiderivative F of f for which $F(x_0) = y_0$; imposing an extra condition allows us to determine the constant C.

We say that the differential equation *plus* an extra condition

$$\frac{dy}{dx} = f(x), \quad y(x_0) = y_0$$

constitutes an *initial value problem*. An initial value problem has a unique solution.

Notation: $y(x_0) = y_0$ is shorthand for $y = y_0$ when $x = x_0$. Although it involves using the same letter for the variable y and the function f, this notation is frequently used for differential equations.

❑ **Example 4** Find the solution of the initial value problem

$$\frac{dy}{dx} = \sin x + 2, \quad y(3) = 5.$$

Solution. We have already seen that the general solution to the differential equation is $y = -\cos x + 2x + C$. The initial condition will allow us to determine the constant C. Substituting $y(3) = 5$, gives

$$5 = y(3) = -\cos 3 + 2 \cdot 3 + C$$

so C is given approximately by:

$$C = 5 + \cos 3 - 6 = -1.99$$

Thus the (unique) solution is

$$y = -\cos x + 2x - 1.99.$$

Figure 5.77 shows this particular solution (marked with $C = -1.99$). ❑

❑ **Example 5** Suppose f' is given in Figure 5.78. Sketch a graph of f in the cases when $f(0) = 0$ and $f(0) = 1$.

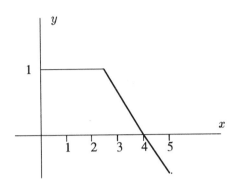

Figure 5.78: Graph of f' for Example 5

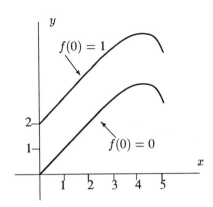

Figure 5.79: Graph of f for Solutions to Example 5

Solution. For $0 \leq x \leq 2$, f has a constant slope of 1 and so is a straight line. For $2 \leq x \leq 4$, f is increasing but more slowly; it has a maximum at $x = 4$ and decreases thereafter. See Figure 5.79.

The solutions for $f(0) = 0$ and $f(0) = 1$ start at different points on the vertical axis, but have the same shape. ◻

Example 6 Figure 5.80 shows a graph of a function f. Sketch the family of antiderivatives, F, of f.

Solution. Let's first sketch any one antiderivative F of f. We get all the others by shifting the graph of F up or down by a constant. The function f gives us the slope of F. Looking at the graph of f we see that F has a horizontal tangent at $x = -2$ and $x = 0$ because $f = 0$ there. Since $f(-2) = 0$ and f is decreasing at $x = -2$, F has a local maximum there; f is increasing at $x = 0$, so F has a local minimum there. Furthermore F has inflection points at $x = -1$ and $x = 1$ since f is at its minimum and maximum there. See Figure 5.81. The x-axis is a horizontal asymptote for f, because $f \to 0$ as $x \to \infty$; this means that the graph of F levels off as $x \to \infty$. The antiderivatives of f will all have these properties, since they are all of the form $F(x) + C$ for some constant C.

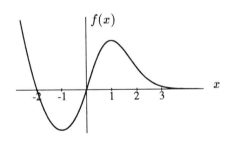

Figure 5.80: Graph of Slope Function, f.
(For Example 6)

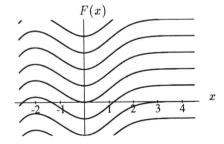

Figure 5.81: Graph of Antiderivatives, F.
(For Example 6)

◻

❖ Exercises for Section 5.8

For Problems 1–22, find an antiderivative.

1. $f(x) = 5$

2. $f(x) = 5x$

3. $f(x) = x^2$

4. $g(t) = t^2 + t$

5. $h(t) = \cos t$

6. $g(z) = \sqrt{z}$, for $z > 0$.

7. $h(z) = \dfrac{1}{z}$, for $z > 0$

8. $r(t) = \dfrac{1}{t^2}$, for $t > 0$

9. $g(z) = \dfrac{1}{z^3}$, for $z > 0$

10. $f(z) = e^z$

11. $g(t) = \sin t$

12. $f(t) = 2t^2 + 3t^3 + 4t^4$

13. $p(t) = t^3 - \dfrac{t^2}{2} - t$

14. $q(y) = y^4 + \dfrac{1}{y}$, for $y > 0$

15. $f(x) = 6\sqrt{x} - \dfrac{1}{x^2} + \dfrac{10}{x}$

16. $p(\theta) = 2\sin(2\theta)$

17. $r(t) = e^t + 5e^{5t}$

18. $q(t) = (t+1)^2$

19. $f(x) = 5^x$

20. $h(t) = 2t\cos(t^2)$

21. $f(t) = t\cos(t^2)$

22. $f(x) = xe^{x^2}$

For each of the functions in Problems 23–28, find a function F such that $F' = f$ and $F(0) = 2$.

23. $f(x) = 3$

24. $f(x) = e^x$

25. $f(x) = x^2$

26. $f(x) = \cos x$

27. $f(x) = \sin x$

28. $f(x) = (x-1)^2$

For each function in Problems 29–32, sketch two functions F where $F' = f$. In one case, let $F(0) = 0$, and in the other, let $F(0) = 1$.

29.

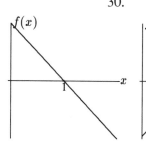

30.

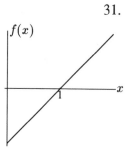

31.

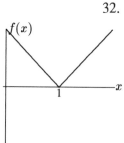

32.

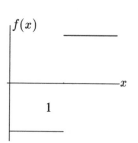

33. The vertical velocity of a cork bobbing up and down on the waves in the sea is given by Figure 5.82. Upwards is considered positive. Describe the motion of the cork at each of the labeled points. At which point(s), if any, is the acceleration zero? Sketch a graph of the height of the cork above the sea floor as a function of time.

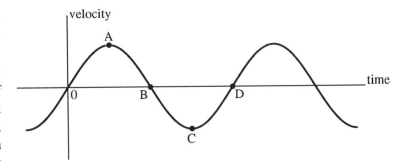

Figure 5.82: Velocity of a Bobbing Cork

34. If a car goes from 0 to 80 mph in 6 seconds with constant acceleration, what is the acceleration?

35. A car going 80 ft/sec (about 55 mph) brakes to a stop in 5 seconds. Assume the deceleration is constant.

    (a) Graph the velocity against time, t, for $0 \leq t \leq 5$ seconds.

    (b) Represent, as an area on the graph, the total distance traveled from the time the brakes are applied until the car comes to a stop.

    (c) Find this area and hence the distance traveled.

    (d) Now find the total distance traveled using antidifferentiation.

36. A 727 jet needs to be going 200 mph to take off. If it can accelerate from 0 to 200 mph in 30 seconds, how long must the runway be? (Assume constant acceleration.)

37. Suppose a car going at 30 ft/sec decelerates at a constant rate of 5 ft/sec^2.

    (a) Draw up a table showing the velocity of the car every $\frac{1}{2}$ second. When does the car come to rest?

    (b) Using your table, find left and right sums which estimate the total distance traveled before the car comes to rest. Which is an overestimate and which is an underestimate?

    (c) Now find a formula for the velocity of the car as a function of time, and hence find the total distance traveled by antidifferentiation. What is the relationship between your answer to (c) and your estimates in (b)?

38. A ball that is dropped from a window hits the ground in 5 seconds. How high is the window? (Use $g = -32$ ft/sec^2.)

39. On the moon, the acceleration due to gravity is about 1.6 m/sec^2 (compared to $g \approx 9.8$ m/sec^2 on Earth). If you drop a rock on the moon (with initial velocity 0), find formulas for:

    (a) its velocity $v(t)$ at time t.

    (b) The distance $s(t)$ it falls in time t.

40. (a) Imagine throwing a rock straight up in the air. What should its initial velocity be if the rock reaches a maximum height of 100 feet? ($g = 32$ ft/sec^2)

    (b) Now imagine being transplanted to the moon and throwing a moon rock vertically upwards with the same velocity as in part (a). How high will it go? (On the moon, $g = 5.2$ ft/sec^2.)

41. The graph of f'' is given in Figure 5.83. Draw graphs of f and f', assuming both contain the origin, and use them to decide at which of the labeled x values

    (a) $f(x)$ is greatest

    (b) $f(x)$ is least

    (c) $f'(x)$ is greatest

    (d) $f'(x)$ is least

    (e) $f''(x)$ is greatest

    (f) $f''(x)$ is least

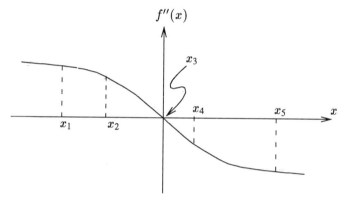

Figure 5.83: Note that this is a graph of f'' against x

(for Problem 41)

5.9 Notes on Equations of Motion: Why acceleration?

The problem of a body moving freely under the influence of gravity near the surface of the earth intrigued mathematicians and philosophers from Greek times onward and was finally solved by Galileo and Newton. The questions to be answered were: How do the velocity and the position of the body vary with time? We define s to be the position, or height, of the body above a fixed point (often ground level); v then is the velocity of the body measured upwards. We assume that the acceleration of the body is a constant, $-g$ (the negative sign means that the acceleration is downward). Thus

$$\text{acceleration} = \frac{dv}{dt} = -g.$$

Thus velocity is the antiderivative of $-g$:

$$v = -gt + C.$$

If the initial velocity is v_0, then $C = v_0$, so

$$v = -gt + v_0.$$

How about the position? We know that

$$\frac{ds}{dt} = v = -gt + v_0.$$

Therefore we can find s by antidifferentiating again, giving:

$$s = -\frac{gt^2}{2} + v_0 t + C.$$

If the initial position is s_0, then we must have

$$s = -\frac{gt^2}{2} + v_0 t + s_0.$$

Our derivation of the formulas for the velocity and the position of the body took little effort. It hides an almost 2000 year struggle to understand the mechanics of falling bodies, from Aristotle's *Physics* to Galileo's *Dialogues Concerning Two New Sciences*. Before sketching Galileo's solution of the problem we want to make a few more observations on uniform and uniformly accelerated motion.

Though it is an over simplification of his ideas, we can say that Aristotle's conception of motion was primarily in terms of *change of position*. This seems entirely reasonable; it is what we commonly observe, and this view dominated discussions of motion for centuries. But it misses a subtlety that was brought to light by Descartes, Galileo and, with a different emphasis, by Newton. That subtlety is now usually referred to as the *principle of inertia*.

This holds that a body traveling undisturbed at constant velocity in a straight line will continue in this motion indefinitely. Stated another way, one cannot distinguish in any absolute sense, i.e. by performing an experiment, between "being at rest" and "moving with constant velocity in a straight line". If you are reading this book in a closed room and have no external reference points, there is no experiment that will tell you, one way or the other, whether you are "at rest" or whether you, the room, and everything in it are moving with constant velocity in a straight line. Therefore, as Newton saw, an understanding of motion should be based on *change of velocity* rather than change of position. Since acceleration is the rate of change of velocity, it is acceleration that must play a central role in the description of motion.

How does acceleration come about? How does the velocity change? Through the action of *forces*. With Newton there is a new emphasis on the importance of forces. Newton's laws of motion do not say what a force *is*, they say how it *acts*. His first law is the principle of inertia, which says what happens in the *absence* of a force – nothing. His second law says that a force acts to produce a change in velocity, i.e. an acceleration. It states that $F = ma$ where m is the mass of the object, F is the net force and a is the acceleration produced by this force.

Let's return to Galileo for a moment. We owe to him the demonstration that a body falling under the influence of gravity does so with constant acceleration. Furthermore, assuming we can neglect air resistance, this constant acceleration is independent of the mass of the body. This last fact was the outcome of Galileo's famous observation that a heavy and light ball dropped off the leaning tower of Pisa hit the ground at the same time. Whether or not he actually performed this experiment, Galileo presented a very clear thought experiment in the *Dialogues* to prove the same point. (This point was again counter to Aristotle's more common sense notion that the heavier ball would reach the ground first.) Mathematically, Galileo showed that the mass of the object does *not* appear as a variable in the equation of motion. Thus the same constant acceleration equation applies to all bodies falling under the influence of gravity.

Nearly a hundred years after Galileo's experiment, Newton formulated his laws of motion and gravity, which led to differential equations describing the motion of a falling body. In Newton's system, Galileo's observation that the acceleration is independent of the mass is reflected in the fact that the mass of a falling object cancels out of the equation of motion.

According to Newton, acceleration is caused by force, and in the case of falling bodies, that force is the force of gravity. Newton's law of gravity says that the gravitational force between two bodies is attractive and given by

$$F = \frac{GmM}{r^2}$$

where G is the gravitational constant, m and M are the masses of the two bodies, and r is their separation. This is the famous "inverse square law". For a falling body, we take M to be the mass of the earth and r to be the distance from the body to the center of the Earth. So actually r changes as the body falls, but for anything we can easily observe (say a ball dropped from the tower of Pisa) it won't change significantly over the course of the motion. Hence, as an approximation, it is reasonable to assume that

$$\frac{GM}{r^2} = \text{constant.}$$

Call this constant g; then the gravitational force, F, is acting downwards, so $F = -mg$. However, according to Newton's second law,

$$F = ma = m\frac{d^2s}{dt^2}.$$

Hence

$$m\frac{d^2s}{dt^2} = -mg,$$

or

$$\frac{d^2s}{dt^2} = -g.$$

Thus Newton derived Galileo's experimental fact that as an approximation bodies falling under the influence of gravity do so with a constant acceleration which is independent of the mass of the body.

❖ Exercises for Section 5.9

1. Galileo was the first person to show that the distance traveled by a body falling from rest is proportional to the square of the time it has traveled, and independent of the mass of the body. Derive this result from the fact that the acceleration due to gravity is a constant.

2. While attempting to understand the motion of bodies under gravity, Galileo stated that:

> The time in which any space is traversed by a body starting at rest and uniformly accelerated is equal to the time in which that same space would be traversed by the same body moving at a uniform speed whose value is the mean of the highest velocity and the velocity just before acceleration began.

(a) Write Galileo's statement in symbols, defining all the symbols you use.

(b) Verify Galileo's statement for a body dropped off a 100 foot building accelerating from rest under gravity until it hits the ground below. (Use $g = 32$ ft/sec^2.)

(c) Show why Galileo's statement is true in general.

3. In his *Dialogues Concerning Two New Sciences*, Galileo wrote:

> The distances traversed during equal intervals of time by a body falling from rest stand to one another in the same ratio as the odd numbers beginning with unity.

Assume, as is now known, that $s = -\frac{1}{2}gt^2$, where s is the total distance traveled in time t and g is the acceleration due to gravity.

(a) How far does a falling body travel in the first second (between $t = 0$ and $t = 1$)? During the second second (between $t = 1$ and $t = 2$)? The third second? The fourth second?

(b) What do your answers tell you about the truth of Galileo's statement?

5.10   Miscellaneous Exercises for Chapter 5

For each of the functions in Problems 1–4 do the following:

- Find f' and f''.
- Find the critical points of f.
- Evaluate f at the critical points and at the endpoints.
- Find any inflection points.
- Sketch f. Indicate clearly where f is increasing or decreasing, and its concavity.
- Identify the local and global maxima and minima of f.

1. $f(x) = x^3 - 3x^2$ $(-1 \le x \le 3)$

2. $f(x) = x + \sin x$ $(0 \le x \le 2\pi)$

3. $f(x) = e^{-x} \sin x$ $(0 \le x \le 2\pi)$

4. $f(x) = x^{-\frac{2}{3}} + x^{\frac{1}{3}}$ $(1.2 \le x \le 3.5)$

For each of the functions in Problems 5–7, find the limits as x tends to $+\infty$ and $-\infty$ and then proceed as in Problems 1–4. (i.e., find f', etc.)

5. $f(x) = 2x^3 - 9x^2 + 12x + 1$

6. $f(x) = \dfrac{4x^2}{x^2 + 1}$

7. $f(x) = xe^{-x}$

8. Sketch a graph of $e^{-\frac{x^2}{2}}$, marking local maxima and minima and points of inflection.

For each of the functions in Problems 9–14, use derivatives to find and identify local maxima and minima and points of inflection and sketch a graph. Confirm your answers using a calculator or computer.

9. $f(x) = x^3 + 3x^2 - 9x - 15$

10. $f(x) = x^5 - 15x^3 + 10$

11. $f(x) = x - 2\ln x \qquad$ for $x > 0$

12. $f(x) = x^2 e^{5x}$

13. $f(x) = e^{-x^2}$

14. $f(x) = \dfrac{x^2}{x^2 + 1}$

Find bounds for the functions in Problems 15–16:

15. $e^{-x}\sin x, \quad$ for $x \geq 0$.

16. $x\sin x, \quad$ for $0 \leq x \leq 2\pi$.

For the graphs in Problems 17–20 decide:

(a) Over what intervals is f increasing? decreasing?

(b) Does f have maxima or minima? If so, which, and where?

17.

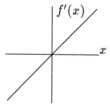

18.

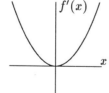

19.
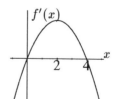

20.
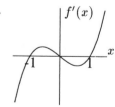

21. Consider the vase in Figure 5.84. Assume the vase is filled with water at a constant rate (i.e. constant volume per unit time).

    (a) Graph $y = f(t)$, the depth of the water, against time, t. Pay attention to the concavity. Mark interesting points on the graph.

    (b) Where does $y = f(t)$ grow fastest, slowest? Estimate the ratio between these two growth rates.

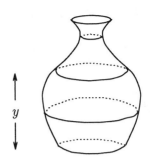

Figure 5.84: Vase for Problem 21

22. Sketch several members of the family

$$y = x^3 - ax^2$$

on the same coordinate plane. Show that the critical points lie on the curve $y = -\frac{1}{2}x^3$.

Find antiderivatives for the functions in Problems 23–32

23. $f(x) = x + x^5 + x^{-5}$

24. $f(x) = x^6 - \frac{1}{7x^6}$

25. $g(t) = 5 + \cos t$

26. $h(r) = 2\sqrt{r} + \frac{1}{2\sqrt{r}}$

27. $g(x) = (x + 1)^3$

28. $h(t) = \frac{(t - 1)^2}{t^2}$

29. $p(y) = \frac{1}{y} + y + 1$

30. $f(z) = e^z + 3$

31. $g(\theta) = \sin \theta - 2 \cos \theta$

32. $p(r) = 2\pi r$

33. A cat, walking along the window ledge of a New York apartment, knocks off a flower pot which falls to the street 200 feet below. How fast is the flower pot traveling when it hits the street? (Give your answer in ft/sec and in mph, given that 1 ft/sec = 15/22 mph.)

34. Of all the cars surveyed in the April 1991 issue of *Car and Driver*, the Acura NSX going at 70 mph stops in the shortest distance: 157 feet. Find the acceleration, assuming it to be constant.

35. An object is thrown vertically upwards with a velocity of 160 ft/sec.

    (a) Sketch a graph of the velocity of the object (with upward as positive) against time.

    (b) Mark on the graph the points at which the object reaches its highest point and when it hits the ground.

    (c) Express the maximum height reached by the object as an area on the graph, and hence find this height.

    (d) Now express the velocity of the object as a function of time and find the greatest height by antidifferentiation.

36. An object is thrown vertically upwards with a velocity of 80 ft/sec.

    (a) Make a table showing the velocity every second.

    (b) When does the object reach its highest point? When does the object hit the ground?

    (c) Using your table, write left and right sums which under and overestimate the total distance traveled to the highest point.

    (d) Use antidifferentiation to find the greatest height reached.

37. Quantum mechanics predicts that the force, F, between two gas molecules separated by a distance r is given by

$$F = -\frac{A}{r^7} + \frac{B}{r^{13}}$$

for some positive constants A and B. A sketch of F against r is in Figure 5.85. The potential energy, V, of the gas molecules satisfies $F = -\dfrac{dV}{dr}$ and $V \to 0$ as $r \to \infty$.

    (a) Find a formula for V.

    (b) Sketch a graph of V against r. Mark r_0 on it.

    (c) What does r_0 represent, both in terms of the force, and in terms of potential energy of the gas molecules?

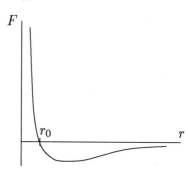

Figure 5.85: Force between Two Gas Molecules

38. For $a > 0$, the line

$$a(a^2 + 1)y = a - x$$

forms a triangle in the first quadrant with the x- and y-axis.

    (a) Find, in terms of a, the x- and y-intercepts of the line.

    (b) Find the area of the triangle, as a function of a.

    (c) Find the value of a making the area maximal.

    (d) What is this greatest area?

    (e) If you want the triangle to have area $\frac{1}{5}$, what choices do you have for a?

39. Let $C(q)$ be the total cost of producing a quantity q of a certain good. The *average cost* is $a(q) = \frac{C(q)}{q}$.

    (a) Interpret $a(q)$ graphically, as the slope of a line in Figure 5.86.

    (b) Find on the graph the quantity q_0 where $a(q)$ is minimal.

    (c) What is the relationship between $a(q_0)$ and the marginal cost, $C'(q_0)$? Explain your result both graphically and analytically (using the fact that $a'(q_0) = 0$, since $a(q)$ has a minimum at q_0). What does this result mean, in terms of economics?

    (d) Graph $C'(q)$ and $a(q)$ on the same axes. Mark q_0 on the q-axis.

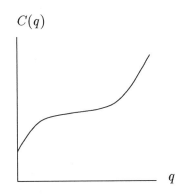

$C(q)$

q

Figure 5.86: Cost of Producing a Good
(for Problem 39)

40. Consider a large tank of water, with temperature $W(t)$. The ambient temperature $A(t)$ (i.e., the temperature of the surrounding air) is given by the graph in Figure 5.87.

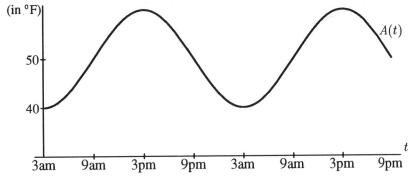

Figure 5.87: Temperature of Air Surrounding a Tank of Water
(for Problem 40)

The temperature of the water is affected by the temperature of the surrounding air.

    (a) How does the temperature of the water change if the water is colder than the surrounding air? What if the water is warmer?

    (b) Using your answer in (a), sketch a possible graph for $W(t)$, on the same axes as $A(t)$.

    (c) Explain the relationship between the extrema of $W(t)$ and the points where the two graphs intersect.

    (d) What is the relationship between the rate at which the temperature of the water changes and the difference $A(t) - W(t)$?

(e) What is the relationship between the inflection points of $W(t)$ and the points where $A(t) - W(t)$ has an extremum?

(f) Assume the tank is refilled with cold water (35° F) at 3 am. Sketch a possible graph for $W(t)$. Pay attention to the concavity.

41. Any body radiates energy at various wavelengths. The power of the radiation (per meter^2 of surface) and the distribution of the radiation amongst the wavelengths vary with temperature. (To read more about this, look up the Stefan-Boltzmann law and Wien's Displacement law in a physics text.)

The function in Figure 5.88 represents the intensity of the radiation of a black body at a temperature $T = 3000°$ K, as a function of the wavelength. The intensity

of the radiation is highest in the infrared range, i.e., at wavelengths longer than that of visible light (0.4–0.7μm). Max Planck's radiation law, announced to the Berlin Physical Society on October 19, 1900, states that

$$r(\lambda) = \frac{a}{\lambda^5 (e^{\frac{b}{\lambda}} - 1)},$$

where a and b are empirical constants, chosen to best fit the experimental data.

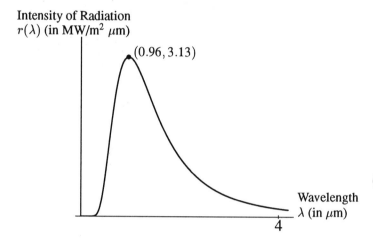

Figure 5.88: Intensity of Energy Radiated from a Heated Body (for Problem 41)

Find a and b so that the formula fits the graph.

(Later in 1900 Planck was able to derive his radiation law entirely from theory. He found that $a = 2\pi c^2 h$ and $b = \frac{hc}{Tk}$ where c = speed of light, h = Planck's constant, and k = Boltzmann's constant.)

42. Populations with a growth limit have been modeled with the logistic family

$$y = \frac{A}{1 + Be^{-Cx}} \qquad -\infty < x < \infty \quad \text{and} \quad A, B, C > 0.$$

(a) Sketch a graph of $g(x) = \frac{A}{1 + e^{-Cx}}$. What is the significance of the parameter A?

(b) Evaluate $g(-x) + g(x)$ and say what the sum means geometrically.

(c) What happens to the graph of g if A is kept constant and C is increased?

(d) Show that the curve $y = \frac{A}{1 + Be^{-Cx}}$ is a horizontal shift of the graph of g.

Chapter 6

THE INTEGRAL

We have seen that the derivative has many useful interpretations, and the same is true of the definite integral. Up until now, we have approximated definite integrals by left- and right-hand sums. In this chapter we will learn more efficient numerical methods as well as how to find definite integrals exactly, using the Fundamental Theorem of Calculus.

6.1   The Definite Integral Revisited

In Chapter 3 we defined the definite integral to allow us to reconstruct the distance traveled from velocity. If f is bounded and continuous (except perhaps at a few points) on the interval $a \le x \le b$, we divide the interval into n equal subdivisions of length $\Delta x = \frac{b-a}{n}$ using the points $a = x_0, x_1,$ $\ldots, x_n = b$, and then we form the Riemann sums:

$$\text{Left-hand sum} = f(x_0)\Delta x + f(x_1)\Delta x + \cdots + f(x_{n-1})\Delta x$$

$$= \sum_{i=0}^{n-1} f(x_i)\Delta x.$$

$$\text{Right-hand sum} = f(x_1)\Delta x + f(x_2)\Delta x + \cdots + f(x_n)\Delta x$$

$$= \sum_{i=1}^{n} f(x_i)\Delta x.$$

We expect that as n goes to infinity and the subdivisions become finer both sums will approach the same number. This common limit is defined to be the *definite integral*:

$$\int_a^b f(x)\,dx = \lim_{n \to \infty} (\text{left-hand sum}) = \lim_{n \to \infty} \sum_{i=0}^{n-1} f(x_i)\Delta x$$

$$= \lim_{n \to \infty} (\text{right-hand sum}) = \lim_{n \to \infty} \sum_{i=1}^{n} f(x_i)\Delta x.$$

The function f is called the *integrand*, the numbers a and b are the *limits of integration*. One reason for using both left- and right- hand sums is that when f is monotonic from a to b (i.e. when it is either increasing throughout or decreasing throughout the interval) the two sums trap the integral between them.

Interpretation of the Definite Integral

The three principal interpretations of the integral $\int_a^b f(x)\,dx$, which were discussed in Chapter 3, are:

- The **area** between the graph of f and the x-axis from a to b, assuming $f \ge 0$ and $a < b$. See Figure 6.1(a).

- If f is the rate of change of a quantity Q, then the integral can be interpreted as the **total change** in Q:

$$\left(\begin{array}{c} \text{Total Change in } Q \text{ between} \\ x = a \text{ and } x = b \end{array} \right) = \int_a^b f(x)\,dx.$$

- $(b - a)$ times the **average value** of f over the interval $[a, b]$, or

$$\left(\begin{array}{c} \text{Average value of } f \\ \text{from } a \text{ to } b \end{array} \right) = \frac{1}{b - a} \int_a^b f(x)\, dx.$$

The average value of f is the height of the rectangle with base $(b - a)$ and whose area equals the area under the graph of $f(x)$ between $x = a$ and $x = b$. See Figure 6.1(b).

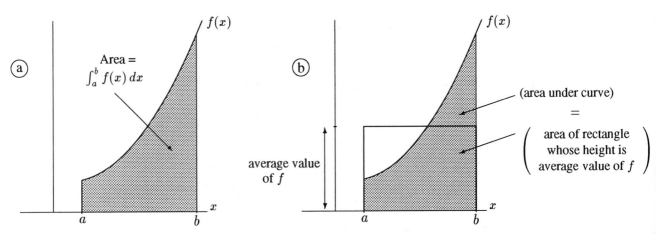

Figure 6.1: Area and Average Value

What if $f = F'$?

There is one special situation where $\int_a^b f(x)\, dx$ is very easy to compute, and that is when f is the derivative of another function F. We call F the antiderivative of f, and we have:

$$f(x) = F'(x).$$

Then we can interpret $f(x)$ as the instantaneous rate of change of the quantity $F(x)$, and $\int_a^b f(x)\, dx$ as the total change in $F(x)$ from time a to b:

$$\left(\begin{array}{c} \text{Total change in } F(x) \\ \text{between } a \text{ and } b \end{array} \right) = \int_a^b f(x)\, dx.$$

Since the change in $F(x)$ can also be expressed as $F(b) - F(a)$, we have the result seen at the end of Chapter 3:

The Fundamental Theorem of Calculus. If f is the derivative of F, that is, $f = F'$, then

$$\int_a^b f(x)\, dx = F(b) - F(a).$$

This theorem has tremendous practical importance: it says that *if* we know F, we don't need to calculate the left and right sums to find a definite integral, but instead we can simply subtract $F(a)$ from $F(b)$.

Notation: Why the dx?

The notation for the definite integral may seem a little overloaded. Since the only things you need to compute $\int_a^b f(x)\,dx$ are the lower and upper limits, a and b, and the function f, why include the dx? It turns out to be useful because

- the dx identifies the independent variable of the function being integrated. For example, in $\int_a^b cx\,dx$, the dx reminds us that x is the variable, and c is a constant.

- the dx reminds us of the structure of the Riemann sum, i.e. $\int_a^b f(x)\,dx$ reminds us of $\sum f(x)\,\Delta x$.

Notice that, for $a \leq b$, and $f \geq 0$, both

$$\int_a^b f(x)\,dx \quad \text{and} \quad \int_a^b f(t)\,dt$$

represent the area under the graph of f from a to b. Since the area under the graph of f is the same whether the variable is called x or t, the value of both integrals will be the same. Since the name of the variable (whether it is called x or t) doesn't affect the value of the integral, the x or t is said to be a *dummy variable*.

❖ Exercises for Section 6.1

1. Estimate $\int_1^2 x^2\,dx$ using left- and right-hand sums with 4 subdivisions. How far from the true value of the integral could your estimate be?

2. If the graph of f is in Figure 6.2:

   (a) What is $\int_1^6 f(x)\,dx$?

   (b) What is the average value of f on $[1,6]$?

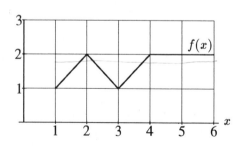

Figure 6.2: Graph of $f(x)$

(for Problem 2)

3. What is the average value of $f(x) = 4x + 7$ on $[1, 3]$?

4. Given $f(x) = x$, find intervals $[a, b]$ satisfying:

   (a) $\int_a^b f(x)\, dx$ is less than the average value of f on the interval $[a, b]$.

   (b) $\int_a^b f(x)\, dx$ is equal to the average value of f on the interval $[a, b]$.

   (c) $\int_a^b f(x)\, dx$ is greater than the average value of f on the interval $[a, b]$.

5. Use the Fundamental Theorem of Calculus to find $\displaystyle\int_0^2 (3x^2 + 1)\, dx$.

6. If the velocity of a particle at time t is given by $v(t) = \cos t$, use the Fundamental Theorem of Calculus to find the total distance traveled by the particle between $t = 0$ and $t = \frac{\pi}{2}$.

7. Without computing the integral, decide if

$$\int_0^{2\pi} e^{-x} \sin x \, dx$$

is positive or negative, and explain your decision. [Hint: Sketch $e^{-x} \sin x$.]

8. A new sales agent finds that as she gains experience, she increases the number of large appliances she sells each month. In the first month she sells only 7, but each month she sells two more than the month before, so that the number she sells in month t is $2t + 5$.

   (a) Find the average number of large appliances she sells per month over the first year arithmetically, by calculating the number of appliances sold each month and then taking the average over twelve months.

   (b) Now find the average by integration as though the sales function applied for all values of t (instead of just for integers).

   (c) How well do the two results compare?

   (d) Thinking of the answer you found in (a) as the true answer, and the integral answer as an approximation, why might anyone want to use the integral answer instead of the true answer?

   (e) Draw a picture showing both answers as the area of some region. Mark on your picture a region representing the error in the integral answer.

9. The graph of $f(x) = \dfrac{1}{x+1}$ is in Figure 6.3. For $x \geq 0$, define a new function $F(x)$ as follows: $F(x)$ is the area of the region bounded by the x-axis, the y-axis, the graph of $f(x)$, and the vertical line at x.

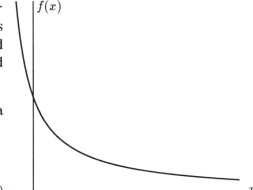

Figure 6.3: $f(x) = \dfrac{1}{x+1}$

(for Problem 9)

(a) For positive x and h, sketch the area represented by $F(x+h) - F(x)$.

(b) Using your sketch, show that

$$h \cdot f(x) > F(x+h) - F(x) > h \cdot f(x+h).$$

(c) Using (b), find $F'(x)$ as the limit of the difference quotient.

10. Whether a resource is distributed evenly among members of a population is often an important political or economic question. How can we measure this? How can we decide if the distribution of wealth in this country is becoming more or less equitable over time? How can we measure which country has the most equitable income distribution? This problem describes a way of doing this.

Suppose the resource is distributed evenly. Then any 20% of the population will have 20% of the resource. Similarly, any 30% will have 30% of the resource and so on. If, however, the resource is not distributed evenly, the poorest $P\%$ of the population (in terms of this resource) will not have $P\%$ of the goods. Suppose $F(x)$ represents the fraction of the resources owned by the poorest fraction x of the population. Thus $F(0.4) = 0.1$ means that the poorest 40% of the population owns 10% of the resource.

(a) What would F be if the resource was distributed evenly?

(b) What must be true of any such F? What must $F(0)$ and $F(1)$ equal? Is F increasing or decreasing? Concave up or concave down?

(c) Gini's index of inequality, G, is one way to measure how evenly the resource is distributed. It is defined by

$$G = 2 \int_0^1 (x - F(x)) \, dx.$$

Show graphically what G represents.

6.2 Properties of the Definite Integral

This section contains properties of the definite integral that will enable us to say something about the value of an integral without having to actually evaluate it. This is analogous to being able to say that $\sqrt{24}$ is less that $\sqrt{37}$, without knowing the exact value of either.

Facts About Limits of Integration. *Let a, b, and c be any numbers, then*

1. $\displaystyle\int_b^a f(x)\,dx = -\int_a^b f(x)\,dx.$

2. $\displaystyle\int_a^c f(x)\,dx + \int_c^b f(x)\,dx = \int_a^b f(x)\,dx.$

In words,

1. *the integral from b to a is the negative of the integral from a to b.*

2. *the integral from a to c plus the integral from c to b is the integral from a to b.*

By interpreting the integrals as areas, we can justify these results for $f \geq 0$. In fact, they are true for all functions for which the integrals make sense.

Why is $\int_b^a f(x)\,dx = -\int_a^b f(x)\,dx$?

Both integrals are approximated by sums of the form $\sum f(x_i)\Delta x$, where the x_i's are the same in each case, the only difference in the sums for $\int_b^a f(x)\,dx$ and $\int_a^b f(x)\,dx$ is that in the first, $\Delta x = \frac{(a-b)}{n} = -\frac{(b-a)}{n}$ and in the second $\Delta x = \frac{(b-a)}{n}$. Since everything else about the sums is the same, we must have $\int_b^a f(x)\,dx = -\int_a^b f(x)\,dx$.

Why is $\int_a^c f(x)\,dx + \int_c^b f(x)\,dx = \int_a^b f(x)\,dx$?

Suppose $a < c < b$. Figure 6.4(a) suggests that $\int_a^c f(x)\,dx + \int_c^b f(x)\,dx = \int_a^b f(x)\,dx$ since the area under f from a to c plus the area under f from c to b together comprise the whole area under f from a to b.

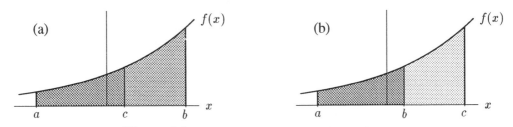

Figure 6.4: Additivity of the definite integral.

Actually, this property holds for all numbers a, b, and c, not just those satisfying $a < c < b$. See Figure 6.4(b). For example, the area under f from 3 to 6 is equal to the area from 3 to 8 *minus* the area from 6 to 8, so

$$\int_3^6 f(x)\,dx = \int_3^8 f(x)\,dx - \int_6^8 f(x)\,dx = \int_3^8 f(x)\,dx + \int_8^6 f(x)\,dx.$$

☐ **Example 1** Suppose that you know that

$$\int_0^{\frac{\pi}{2}} \cos x \, dx = 1 \text{ and } \int_0^{\frac{\pi}{4}} \cos x \, dx = \frac{1}{\sqrt{2}}.$$

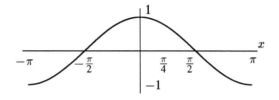

See Figure 6.5. What can you say about the
following integrals?

Figure 6.5: Graph of $\cos x$.

(a) $\int_{\frac{\pi}{4}}^{\frac{\pi}{2}} \cos x \, dx$ (b) $\int_{-\frac{\pi}{2}}^{\frac{\pi}{2}} \cos x \, dx$ (c) $\int_{\pi}^{\frac{\pi}{2}} \cos x \, dx$ (d) $\int_0^{\pi} \cos x \, dx$ (e) $\int_0^{\frac{\pi}{3}} \cos x \, dx$

Solution.

(a) Since $\int_0^{\frac{\pi}{2}} \cos x \, dx = \int_0^{\frac{\pi}{4}} \cos x \, dx + \int_{\frac{\pi}{4}}^{\frac{\pi}{2}} \cos x \, dx$ by the additivity property, we get
$1 = \frac{1}{\sqrt{2}} + \int_{\frac{\pi}{4}}^{\frac{\pi}{2}} \cos x \, dx$, so $\int_{\frac{\pi}{4}}^{\frac{\pi}{2}} \cos x \, dx = 1 - \frac{1}{\sqrt{2}}$.

(b) $\int_{-\frac{\pi}{2}}^{\frac{\pi}{2}} \cos x \, dx = \int_{-\frac{\pi}{2}}^{0} \cos x \, dx + \int_0^{\frac{\pi}{2}} \cos x \, dx$. From the symmetry of the cosine function
about the y-axis, it is clear that $\int_{-\frac{\pi}{2}}^{0} \cos x \, dx = \int_0^{\frac{\pi}{2}} \cos x \, dx$. So $\int_{-\frac{\pi}{2}}^{\frac{\pi}{2}} \cos x \, dx = 1 + 1 = 2$.

(c) By symmetry, $\int_{\frac{\pi}{2}}^{\pi} \cos x \, dx = -\int_0^{\frac{\pi}{2}} \cos x \, dx = -1$.
Therefore, $\int_{\pi}^{\frac{\pi}{2}} \cos x \, dx = -\int_{\frac{\pi}{2}}^{\pi} \cos x \, dx = -(-1) = 1$.

(d) $\int_0^{\pi} \cos x \, dx = \int_0^{\frac{\pi}{2}} \cos x \, dx + \int_{\frac{\pi}{2}}^{\pi} \cos x \, dx$. Again, from the symmetry of the cosine function,
we know that the shape of the curve between $\frac{\pi}{2}$ and π is the same as the shape between 0 and
$\frac{\pi}{2}$, except that it is below the x-axis. The value for $\int_{\frac{\pi}{2}}^{\pi} \cos x \, dx$ will thus be the negative of
the value of $\int_0^{\frac{\pi}{2}} \cos x \, dx$, so $\int_0^{\pi} \cos x \, dx = 1 + (-1) = 0$.

(e) From the graph of $\cos x$, we can see that the area between 0 and $\frac{\pi}{3}$ is greater than the area
between 0 and $\frac{\pi}{4}$ and less than the area between 0 and $\frac{\pi}{2}$. So $\frac{1}{\sqrt{2}} < \int_0^{\frac{\pi}{3}} \cos x \, dx < 1$.

☐

Facts about Sums and Constant Multiples of the Integrand.   *Let f and g be functions and let c be a constant.*

1. $\displaystyle\int_a^b \left(f(x) \pm g(x)\right) dx = \int_a^b f(x)\, dx \pm \int_a^b g(x)\, dx.$

2. $\displaystyle\int_a^b cf(x)\, dx = c\int_a^b f(x)\, dx.$

In words,

1. *the integral of the sum (or difference) of two functions is the sum (or difference) of their integrals,*

2. *the integral of a constant times a function is that constant times the integral of the function.*

Why do these properties hold?

Both can be understood by thinking of the definition of the definite integral as a limit of a sum of areas of rectangles.

For property (1), let's suppose that f and g are positive on the interval $[a, b]$, so that the area under $f(x) + g(x)$ is approximated by the sum of the areas of rectangles like the one shaded in Figure 6.6. The area of this rectangle is

$$(f(x_i) + g(x_i))\Delta x = f(x_i)\Delta x + g(x_i)\Delta x.$$

Since $f(x_i)\Delta x$ is the area of a rectangle under the graph of f, and $g(x_i)\Delta x$ is the area of a rectangle under the graph of g, the area under $f(x) + g(x)$ is the sum of the areas under $f(x)$ and $g(x)$.

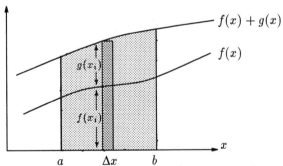

Figure 6.6: Area $= \int_a^b \left(f(x) + g(x)\right)\, dx = \int_a^b f(x)\, dx + \int_a^b g(x)\, dx$

For property (2), notice that multiplying a function by c stretches or flattens it in the vertical direction by a factor of c, and hence multiplies the area by c.

☐ **Example 2**   Evaluate the following definite integral: $\displaystyle\int_0^2 (1 + 3x)\, dx.$

Solution. We can break this integral up as follows:

$$\int_0^2 (1 + 3x)\, dx = \int_0^2 1\, dx + \int_0^2 3x\, dx = \int_0^2 1\, dx + 3\int_0^2 x\, dx.$$

This expresses our original integral in terms of two simpler integrals. From the area interpretation of the integral we see that

$$\int_0^2 1\, dx = 2,$$

since it represents the area under the horizontal line $y = 1$ between $x = 0$ and $x = 2$ (see Figure 6.7(a)). Similarly,

$$\int_0^2 x\, dx = \frac{1}{2} \cdot 2 \cdot 2 = 2$$

because it is the area of the triangle in Figure 6.7(b). Therefore

$$\int_0^2 (1 + 3x)\, dx = \int_0^2 1\, dx + 3\int_0^2 x\, dx = 2 + 3(2) = 8.$$

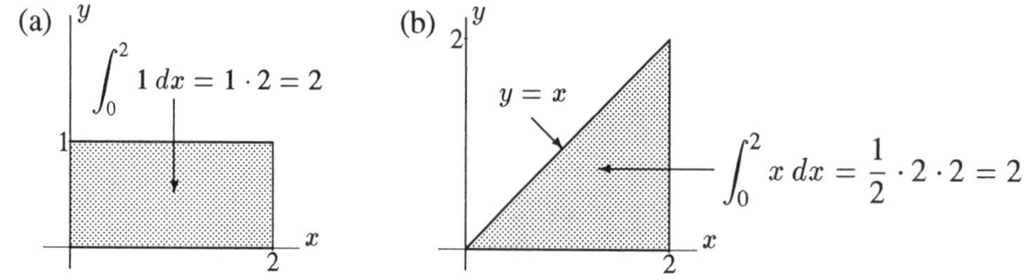

(a)

$$\int_0^2 1\, dx = 1 \cdot 2 = 2$$

(b)

$$y = x$$

$$\int_0^2 x\, dx = \frac{1}{2} \cdot 2 \cdot 2 = 2$$

Figure 6.7: $\int_0^2 1\, dx + 3\int_0^2 x\, dx$ (for Example 2)

How to Compare Integrals

Although we need a computer or calculator to evaluate $\int_0^{\sqrt{\pi}} \sin(x^2)\, dx$ accurately, we can say immediately that it is less than $\sqrt{\pi}$, since, for all x,

$$\sin(x^2) \le 1$$

the area under $y = \sin(x^2)$ is less than the rectangular are under the line $y = 1$ between $x = 0$ and $x = \sqrt{\pi}$ (see Figure 6.8). The rectangle has area $\sqrt{\pi} \cdot 1 = \sqrt{\pi}$, so

$$\int_0^{\sqrt{\pi}} \sin(x^2)\, dx \le \sqrt{\pi}$$

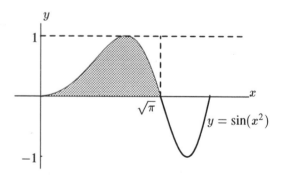

Figure 6.8: Graph showing that $\int_0^{\sqrt{\pi}} \sin(x^2)\, dx < \sqrt{\pi}$.

More generally, we have the following fact:

Comparing Definite Integrals.   If $f(x) \leq g(x)$ for $a \leq x \leq b$, then

$$\int_a^b f(x)\, dx \leq \int_a^b g(x)\, dx.$$

This result says that, under these assumptions, the area under the graph of f is smaller than the area under the graph of g. See Figure 6.9.

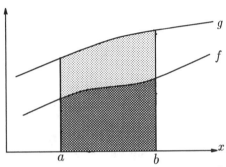

Figure 6.9: If $f(x) \leq g(x)$ then $\int_a^b f(x)\, dx \leq \int_a^b g(x)\, dx$.

☐ **Example 3**   Find an upper estimate for the number $\int_0^1 \sin x\, dx$?

Solution. As Figure 6.10 suggests[1], for all $x \geq 0$,

$$\sin x \leq x.$$

By comparison

$$\int_0^1 \sin x\, dx \leq \int_0^1 x\, dx.$$

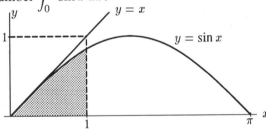

Figure 6.10: Graph showing $\int_0^1 \sin x\, dx \leq \int_0^1 x\, dx$.

The second integral is the area of the right triangle bounded by the line $y = x$, the x-axis, and the line $x = 1$ (see Figure 6.10). Hence $\int_0^1 x\, dx = 1/2$ and we get the estimate

$$\int_0^1 \sin x\, dx \leq \frac{1}{2}.$$

☐

☐ **Example 4**   Find a lower estimate for $\int_0^1 \sin x\, dx$.

Solution. From Figure 6.11, you can see that $\sin x$ is concave downward for $0 \leq x \leq 1$, so

---
[1]This inequality is justified in Problem 23 in Section 5.1.

$$\int_0^1 \sin x \, dx \;=\; \text{area under curve}$$

$$> \;\; \text{area of shaded triangle}$$

$$= \;\; \frac{1}{2} \cdot 1 \cdot \sin 1$$

$$\approx \;\; 0.4207$$

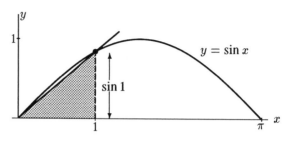

Figure 6.11: Graph showing $\frac{(\sin 1)}{2} \le \int_0^1 \sin x \, dx$.

Putting the results of the last two examples together, we see

$$0.4207 \le \int_0^1 \sin x \, dx \le 0.5.$$

For comparison, we compute the true value of the integral using left- and right-hand sums:

$$\int_0^1 \sin x \, dx \approx 0.4597.$$

❖ Exercises for Section 6.2

1. Suppose you know $\int_1^3 3x^2 \, dx = 26$ and $\int_1^3 2x \, dx = 8$. What is $\int_1^3 (x^2 - x) \, dx$?

2. (a) Devise a formula, using geometry, for

$$\int_a^b 1 \, dx$$

   in terms of a and b. Assume $a \le b$.

   (b) Use the result of part (a) to find: i. $\int_2^5 1 \, dx$ ii. $\int_{-3}^8 1 \, dx$ iii. $\int_1^3 23 \, dx$

3. (a) Devise a formula, using geometry, for

$$\int_a^b x \, dx$$

   in terms of a and b. Assume that $0 \le a \le b$.

   (b) Use the result of part (a) to find: i. $\int_2^5 x \, dx$ ii. $\int_{-3}^8 x \, dx$ iii. $\int_1^3 5x \, dx$.

4. Find $\int_{-1}^{1} |x| \, dx$ geometrically.

5. The function for the *standard normal distribution*, which is often used in statistics, has the formula

$$\frac{1}{\sqrt{2\pi}} e^{-\frac{x^2}{2}}$$

and the graph in Figure 6.12. Statistics books usually contain tables showing only the area under the curve from 0 to b, for different values of b. Table 6.1 is part of such a table.

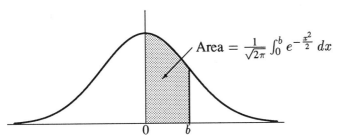

Area $= \frac{1}{\sqrt{2\pi}} \int_0^b e^{-\frac{x^2}{2}} \, dx$

b	$\frac{1}{\sqrt{2\pi}} \int_0^b e^{-\frac{x^2}{2}} \, dx$
1	0.3413
2	0.4772
3	0.4987
4	0.5000

Figure 6.12: The Normal Curve

Table 6.1: Values of $\frac{1}{\sqrt{2\pi}} \int_0^b e^{-\frac{x^2}{2}} \, dx$

Use the information given in the table and the symmetry of the standard normal curve about the y-axis to find

a) $\frac{1}{\sqrt{2\pi}} \int_1^3 e^{-\frac{x^2}{2}} \, dx$.

b) $\frac{1}{\sqrt{2\pi}} \int_{-2}^3 e^{-\frac{x^2}{2}} \, dx$.

6. Use the property $\int_b^a f(x) \, dx = -\int_a^b f(x) \, dx$ to show that $\int_a^a f(x) \, dx = 0$.

7. Is $\int_{-1}^1 e^{x^2} \, dx$ positive, negative or zero? Explain.

8. Explain why $0 < \int_0^1 e^{x^2} \, dx < 3$.

9. Using left and right sums, verify the assertion made at the end of Section 6.2 that

$$\int_0^1 \sin x \, dx \approx 0.4597.$$

Be sure that your calculator is set in radians.

10. Let $F(x) = \int_0^x 2t \, dt$. Then $F(x)$ is the area under the line $y = 2t$ from the origin to x.

    (a) Construct a table showing the values of F for $x = 0, 1, 2, 3, 4, 5$.
    (b) Is F increasing or decreasing when $x > 0$? Concave up or down? Explain.
    (c) When $t < 0$, the line $y = 2t$ is below the t-axis (the horizontal axis). Explain why $F(-1)$ is positive.

6.3 Antiderivatives and the Fundamental Theorem

Why is the Fundamental Theorem Useful?

Knowing the Fundamental Theorem of Calculus, which tells us that when $F' = f$,

$$\int_a^b f(x)\, dx = F(b) - F(a),$$

gives us a whole new way of calculating definite integrals. To find $\int_a^b f(x)\, dx$, we first try to find F, and then calculate $F(b) - F(a)$. This method of computing definite integrals has an important advantage over using left- and right-hand sums: it gives an exact answer quickly. However, the method only works in situations where you can find $F(x)$. This is not always easy; for example, none of the functions we have encountered so far is an antiderivative of $\sin(x^2)$. Nonetheless, because the method is so helpful when it does work, we will spend the next few sections trying to make it work as often as we can.

☐ **Example 1**   Compute $\int_1^3 2x\, dx$ using the Fundamental Theorem.

Solution. We need a function with derivative $2x$. Since

$$\frac{d}{dx} x^2 = 2x,$$

$F(x) = x^2$ is an antiderivative of $f(x) = 2x$. Thus

$$\int_1^3 2x\, dx = F(3) - F(1) = 3^2 - 1^2 = 8.$$

❑

Notation: The Indefinite Integral

If you have found an antiderivative for a function, you can always find another one by adding a constant. For example, x^2 and $x^2 + 1$ are both antiderivatives of $2x$. In fact, $x^2 + C$ is an antiderivative for any constant C. This is because the derivative of a constant is zero:

$$\frac{d}{dx}(x^2 + C) = 2x + 0 = 2x.$$

Which antiderivative should you use to compute a definite integral? It makes no difference because the constant cancels out when you subtract $F(a)$ from $F(b)$. For example, if we had used $x^2 + C$ in computing $\int_1^3 2x\, dx$, we would have arrived at the same answer as before:

$$\int_1^3 2x\, dx = F(3) - F(1) = (3^2 + C) - (1^2 + C) = 8.$$

The general antiderivative is written $F(x) + C$. Because of the connection with the definite integral, we introduce a notation for this general antiderivative that looks like the definite integral without the limits, and is called the *indefinite integral*:

$$\int f(x)\,dx = F(x) + C.$$

It is important to understand the difference between

$$\int_a^b f(x)\,dx \qquad \text{and} \qquad \int f(x)\,dx.$$

The first is a number, the second is actually a *family* of functions. They have such similar notations because the second is helpful in computing the first. Because the notation is similar, the word 'integration' is frequently used for the process of finding the antiderivative as well as of finding the definite integral. The context usually makes clear which is intended.

We will also introduce a shorthand notation for $F(b) - F(a)$: we will write it as

$$\left. F(x) \right|_a^b$$

For example:

$$\int_1^3 2x\,dx = x^2 \Big|_1^3 = 3^2 - 1^2 = 8.$$

Visualizing the Antiderivative

The role of the constant C can be visualized on a graph of F. For different values of C, the graphs of $F(x) + C$ are all the same shape, but shifted vertically.

☐ **Example 2** For the function f given in Figure 6.13, sketch a graph of three antiderivative functions F, where $F' = f$ and (a) $F(0) = 0$, (b) $F(0) = 1$, (c) $F(0) = 2$.

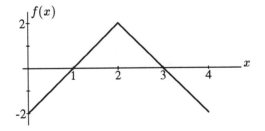

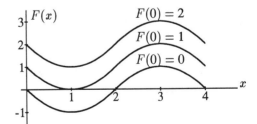

Figure 6.13: Function, f, for Example 2.   **Figure 6.14**: Antiderivatives for Example 2: $F' = f$

Solution. As we saw at the end of Chapter 5, we are looking for a function F whose slope at any point is equal to the value of f there. Where f is above the x-axis, F is climbing; where f is below the x-axis, F is falling. If f is increasing, F is concave up; if f is decreasing, F is concave down.

To draw a graph of F, start at the point on the vertical axis specified by the initial condition ($F(0) = 0$ or $F(0) = 1$ or $F(0) = 2$) and move across the plane with a slope given by the value of f. See Figure 6.14. Different initial conditions lead to different graphs for F, but for a given x value they all have the same slope (because the value of $F'(x) = f(x)$ is the same for each). Thus the different F curves are obtained from one another by a vertical shift. $\square$

Finding Antiderivatives

Finding antiderivatives of functions is like taking square roots of numbers: if you pick a number at random, like 7 or 493, you will have trouble saying what its square root is without a calculator. But if you happen to pick a number like 25 or 64, which you know is a square, then you can find its square root exactly. Similarly, if you happen to pick a function which you recognize as a derivative, then you can find its antiderivative easily.

We saw this when we noticed that $2x$ was the derivative of x^2; this told us that x^2 was an antiderivative of $2x$. If we divide by 2, then we find that

$$\text{an antiderivative of } x \text{ is } \frac{x^2}{2}.$$

To check this statement, take the derivative of $x^2/2$:

$$\frac{d}{dx}\left(\frac{x^2}{2}\right) = \frac{1}{2} \cdot \frac{d}{dx}x^2 = \frac{1}{2} \cdot 2x = x.$$

What about an antiderivative of x^2? The derivative of x^3 is $3x^2$, so the derivative of $\frac{x^3}{3}$ is $\frac{3x^2}{3} = x^2$. Thus,

$$\text{an antiderivative of } x^2 \text{ is } \frac{x^3}{3}.$$

Can you see the pattern? It looks like

$$\text{an antiderivative of } x^n \text{ is } \frac{x^{n+1}}{n+1}.$$

(We assume $n \neq -1$, or we would have $\frac{x^0}{0}$, which doesn't make sense.) It is easy to check this formula by differentiation:

$$\frac{d}{dx}\left(\frac{x^{n+1}}{n+1}\right) = \frac{(n+1)x^n}{n+1} = x^n.$$

In indefinite integral notation, we have shown that

$$\int x^n\, dx = \frac{x^{n+1}}{n+1} + C, \quad n \neq -1.$$

So what about when $n = -1$? In other words, what is the antiderivative of $1/x$? Fortunately, we know a function whose derivative is $1/x$, namely the natural logarithm. Thus, since $\frac{d}{dx}(\ln x) = \frac{1}{x}$,

$$\int \frac{1}{x}\, dx = \ln x + C, \quad \text{for } x > 0.$$

If $x < 0$, then $\ln x$ is not defined, so it can't be an antiderivative of $1/x$. In this case, we can use $\ln(-x)$:

$$\frac{d}{dx}\ln(-x) = (-1)\frac{1}{-x} = \frac{1}{x}$$

so

$$\int \frac{1}{x}\, dx = \ln(-x) + C, \quad \text{for } x < 0.$$

So $\ln x$ is an antiderivative of $1/x$ if $x > 0$, and $\ln(-x)$ is an antiderivative of $1/x$ if $x < 0$. Since $|x| = x$ when $x > 0$ and $|x| = -x$ when $x < 0$:

$$\text{an antiderivative of } \frac{1}{x} \text{ is } \ln|x|.$$

Therefore

$$\boxed{\int \frac{1}{x}\, dx = \ln|x| + C.}$$

Since the exponential function is its own derivative, it is also its own antiderivative; thus

$$\boxed{\int e^x\, dx = e^x + C.}$$

Also, antiderivatives of the sine and cosine are easy to guess: since

$$\frac{d}{dx}\sin x = \cos x \qquad \text{and} \qquad \frac{d}{dx}\cos x = -\sin x,$$

we get

$$\boxed{\int \cos x\, dx = \sin x + C} \qquad \text{and} \qquad \boxed{\int \sin x\, dx = -\cos x + C.}$$

□ **Example 3** Find $\displaystyle\int (3x + x^2)\, dx$.

Solution. We know that $x^2/2$ is an antiderivative of x and that $x^3/3$ is an antiderivative of x^2. Putting these together we get

$$\int (3x + x^2)\, dx = 3\left(\frac{x^2}{2}\right) + \frac{x^3}{3} + C.$$

You should always check your antiderivatives by differentiation—it's easy to do. Here

$$\frac{d}{dx}\left(\frac{3}{2}x^2 + \frac{x^3}{3} + C\right) = \frac{3}{2} \cdot 2x + \frac{3x^2}{3} = 3x + x^2.$$

❑

The previous example illustrates that the sum and constant multiplication rules of differentiation work in reverse:

Facts about Antiderivatives: Sums and Constant Multiples.

1. *An antiderivative of the sum (or difference) of two functions is the sum (or difference) of their antiderivatives.*

2. *An antiderivative of a constant times a function is the constant times an antiderivative of the function.*

In indefinite integral notation, this says that

1. $\displaystyle\int \left(f(x) \pm g(x)\right)\, dx = \int f(x)\, dx \pm \int g(x)\, dx$, *and*

2. $\displaystyle\int cf(x)\, dx = c \int f(x)\, dx$.

❑ **Example 4** Find $\displaystyle\int (\sin x + 3 \cos x)\, dx$.

Solution.

$$\int (\sin x + 3 \cos x)\, dx = \int \sin x\, dx + 3 \int \cos x\, dx = -\cos x + 3 \sin x + C.$$

Check by differentiating:

$$\frac{d}{dx}(-\cos x + 3 \sin x + C) = \sin x + 3 \cos x$$

❑

❑ **Example 5** Find $\displaystyle\int_0^1 \sin x\, dx$.

Solution. The indefinite integral is

$$\int \sin x\, dx = -\cos x + C,$$

so

$$\int_0^1 \sin x\, dx = -\cos x \Big|_0^1 = (-\cos 1) - (-\cos 0) \approx -0.5403 + 1 = 0.4597.$$

❑

❖ Exercises for Section 6.3

For Problems 1–9, find an antiderivative $F(x)$ with $F'(x) = f(x)$ and $F(0) = 0$. Is there only one possible solution in each case?

1. $f(x) = 3$

2. $f(x) = 2x$

3. $f(x) = -7x$

4. $f(x) = \dfrac{1}{4}x$

5. $f(x) = x^2$

6. $f(x) = \sqrt{x}$

7. $f(x) = 2 + 4x + 5x^2$

8. $f(x) = 5 - \dfrac{x^6}{6} - x^7$

9. $f(x) = \sin x$

Find an antiderivative of each function in Problems 10–15.

10. $f(x) = \dfrac{1}{\sqrt{x}}$

11. $f(x) = x^3 + 5\sqrt{x} - \dfrac{2}{x^2}$

12. $g(x) = x^{3/2} + x^{-3/2}$

13. $k(t) = 7 - \dfrac{t^8}{8} + \dfrac{1}{t}$

14. $h(y) = \dfrac{y^2 + 1}{y}$

15. $F(\theta) = \cos \theta + \sin \theta$

16. Find an antiderivative $F(x)$ of $f(x) = e^x$ that satisfies $F(1) = 0$.

17. Show that $\dfrac{x^2 \sin x}{2}$ is not an antiderivative of $x \cos x$.

18. (a) Show that $\dfrac{\frac{x^2}{2} + 6x}{\frac{x^3}{3}}$ is not an antiderivative of $\dfrac{x + 6}{x^2}$.

    (b) Find an antiderivative of $\dfrac{x + 6}{x^2}$.

19. Show that $\arctan 2y$ is an antiderivative of $\dfrac{2}{1 + 4y^2}$.

Find the indefinite integrals in Problems 20–38.

20. $\int (5x + 7)\, dx$

21. $\int \left(\dfrac{3}{t} - \dfrac{2}{t^2} \right) dt$

22. $\int \left(3\cos\psi + 3\sqrt{\psi} \right) d\psi$

23. $\int \left(x^{\frac{3}{2}} + \dfrac{\sqrt{x}}{5} - \dfrac{2}{x} \right) dx$

24. $\int \left(\dfrac{x+1}{x} \right) dx$

25. $\int \left(\dfrac{x^2 + x + 1}{x} \right) dx$

26. $\int (e^x + 5)\, dx$

27. $\int (3\cos x - 7\sin x)\, dx$

28. $\int \left(x + \dfrac{2}{x} + \pi\sin x \right) dx$

29. $\int (2e^x - 8\cos x)\, dx$

30. $\int \dfrac{1}{\cos^2 x}\, dx$

31. $\int 2^x\, dx$ [Hint: What is $\frac{d}{dx}(2^x)$?]

32. $\int (x+1)^2\, dx$

33. $\int (x+1)^3\, dx$

34. $\int (x+1)^9\, dx$

35. $\int \dfrac{1}{x+1}\, dx$

36. $\int \dfrac{1}{2x-1}\, dx$ [Hint: What is $\frac{d}{dx}\ln|2x-1|$?]

37. $\int (e^{5+x} + e^{5x})\, dx$ [Hint: What is $\frac{d}{dx}(e^{5x})$?]

38. $\int (\cos 2x - 2\sin x)\, dx$ [Hint: What is $\frac{d}{dx}(\sin 2x)$?]

39. Find an antiderivative of   (a) e^{2t} [Hint: What is $\frac{d}{dt}e^{2t}$?]   (b) $\dfrac{1}{e^{3\theta}}$

40. Consider the function $f(x) = x^3$. Sketch the graph of f and three of its antiderivatives F, where $F(0) = 0$, $F(0) = 1$, $F(0) = -1$.

For the functions in Problems 41–42, graph a function $F(x)$ such that $F'(x) = f(x)$ and $F(0) = 0$.

41.

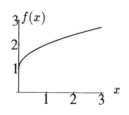

42.

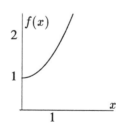

For each function in Problems 43–45, sketch two functions F where $F'(x) = f(x)$. In one case, let $F(0) = 0$, and in the other, let $F(0) = 1$. In each case, mark x_1, x_2, and x_3 on the x-axis of your graph.

43.

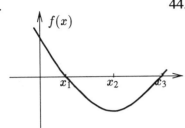

44.

45.

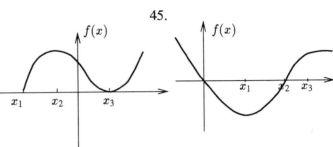

Using the Fundamental Theorem, evaluate the definite integrals in Problems 46–56 symbolically (as in $\ln(3\pi)$) and numerically ($\ln(3\pi) \approx 2.243$):

46. $\displaystyle\int_2^5 (x^3 - \pi x^2)\, dx$

52. $\displaystyle\int_0^1 2e^x\, dx$

47. $\displaystyle\int_0^1 \sin\theta\, d\theta$

53. $\displaystyle\int_0^{\frac{\pi}{4}} \frac{1}{\cos^2 x}\, dx$

48. $\displaystyle\int_1^2 \frac{1 + y^2}{y}\, dy$

54. $\displaystyle\int_0^1 \frac{1}{x^{(19/20)}}\, dx$

49. $\displaystyle\int_0^2 \left(\frac{x^3}{3} + 2x\right) dx$

55. $\displaystyle\int_{-1}^1 2^x\, dx$ [Hint: What is $\frac{d}{dx}2^x$?]

50. $\displaystyle\int_0^{\pi/4} (\sin t + \cos t)\, dt$

56. $\displaystyle\int_{\frac{\pi}{6}}^0 (\sin x + \cos 2x)\, dx$

[Hint: What is $\frac{d}{dx}(\sin 2x)$?]

51. $\displaystyle\int_{-3}^{-1} \frac{2}{r^3}\, dr$

57. (a) Find the following derivatives:

   i. $\frac{d}{dx}\sin(5x)$ ii. $\frac{d}{dx}\cos(x^2)$ iii. $\frac{d}{dx}e^{\sin x}$

   iv. $\frac{d}{dx}\sin(\cos x)$ v. $\frac{d}{dx}(\ln(\cos x))$ vi. $\frac{d}{dx}\ln(\cos(x^4))$

   (b) Use the above to find the following indefinite integrals:

   i. $\int \cos(5x)\, dx$ ii. $\int x\sin(x^2)\, dx$ iii. $\int e^{\sin x}\cos x\, dx$

   iv. $\int \sin x\cos(\cos x)\, dx$ v. $\int \tan x\, dx$ vi. $\int x^3\tan(x^4)\, dx$

   (c) We shall formalize the technique used in these problems in Section 6.4. Can you make any generalizations now about what is happening in these integrals?

58. (a) Find derivatives for the following functions.

    i. $\ln(1+x) - \ln(2+x)$ ii. $\ln(\cos x)$ iii. e^{x^2} iv. $(1+x^2)^{15}$

    (b) Using your answers to part (a), find antiderivatives for the following functions.

    i. $\tan x$ (Recall how to express tan in terms of sin and cos.)

    ii. $xe^{x^2/2}$ iii. $x(1+x^2)^{14}$ iv. $\dfrac{1}{(1+x)(2+x)}$ v. $x(1+x^2)^{12}$ vi. $\dfrac{1}{(1+x)(3+x)}$

59. The average value of the function $v(x) = \dfrac{6}{x^2}$ on the interval $[1, c]$ is equal to 1. Find the value of c.

60. (a) What is the average value of $f(t) = \sin t$ over $0 \le t \le 2\pi$? Why is this a reasonable answer?

    (b) Find the average value of $f(t) = \sin t$ over $0 \le t \le \pi$.

61. Sketch the parabola $y = x(x - \pi)$ and the curve $y = \sin x$, showing their points of intersection. Find the area between the two graphs.

62. Consider the costs of drilling an oil well. There are two types of costs: *fixed costs* (independent of drilling) and *marginal costs* (the increment in costs that comes from drilling one more meter). Together, they make up the total cost, C. The marginal costs depend on the depth at which you are drilling; drilling becomes more expensive, per meter, as you dig deeper into the earth. Suppose the fixed costs are 1,000,000 Riyal (the Riyal is the unit of currency of Saudi Arabia), and the marginal costs are

$$C'(x) = 4{,}000 + 10x$$

in Riyal/meter, where x is the depth in meters. Find the total cost of drilling a well x meters deep.

63. One of the earliest pollution problems brought to the attention of the Environmental Protection Agency (EPA) was the case of the Sioux Lake in eastern South Dakota. For years, a small paper plant located nearby had been discharging waste containing carbon tetrachloride (CCl_4) into the waters of the lake. At the time the EPA learned of the situation, the chemical was entering at a rate of 16 cubic yards per year.

The Agency immediately ordered the installation of filters designed to slow (and eventually stop) the flow of CCl_4 from the mill. Implementation of this program took exactly three years, during which the flow of pollutant was steady at 16 cubic yards/year. Once the filters were installed, the flow declined. From the time the filters were installed until the time the flow stopped, the rate of flow was well approximated by

$$\text{Rate (in cubic yards/year)} = t^2 - 14t + 49$$

where t is time measured in years since the EPA learned of the situation.

    (a) Draw a graph showing the rate of CCl_4 flow into the lake as a function of time, beginning at the time the EPA first learned of the situation.

(b) How many years elapsed between the time the EPA learned of the situation and the time the pollution flow stopped entirely?

(c) How much CCl$_4$ entered the waters during the time shown in the graph in part (a)?

6.4 Integration by Substitution—Part I

In Chapter 4, we learned rules which enabled you to differentiate any function obtained by combining constants, powers of x, $\sin x$, $\cos x$, a^x, and $\ln x$, using addition, multiplication, division, or composition of functions. These are called *elementary* functions. Thus to differentiate

$$f(x) = (\cos x)\sqrt{x^3 + 1},$$

you recognize $f(x)$ as a product of $\cos x$ and $\sqrt{x^3 + 1}$ and apply the product rule; to differentiate $\sqrt{x^3 + 1}$, you use the chain rule.

Are there similar rules for antidifferentiation? For example, to antidifferentiate a sum you antidifferentiate each term separately, so there is a sum rule. But what about a product rule? Or a chain rule? It turns out there are no such rules for antidifferentiation. However, in the next few sections, we'll introduce two methods of antidifferentiation: the substitution method and integration by parts, which sometimes help to reverse the chain and product rules, respectively.

You'll find that there's a great difference between looking for derivatives and looking for antiderivatives. Every elementary function has an elementary derivative found by using the differentiation rules. However, most elementary functions do not have an elementary antiderivative. Some examples are $\sqrt{x^3 + 1}$, $(\sin x)/x$, and e^{-x^2}. These are not strange monster functions, either, but rather simple functions that can and do arise naturally. On the other hand, there are many functions that do have elementary antiderivatives.

The Guess-And-Check Method

A good strategy for finding simple antiderivatives is to *guess* an answer (using your knowledge of differentiation rules), then *check* your answer by differentiating it. If you get the expected result, then you're done; otherwise you may have to revise your guess and check again.

This method of guess-and-check is most useful in looking for what could be the result of applying the chain rule. According to the chain rule

$$\frac{d}{dx}(f(g(x))) = \underbrace{f'}_{\text{derivative of outside}}(\overbrace{g(x)}^{\text{inside}}) \cdot \underbrace{g'(x)}_{\text{derivative of inside}}$$

Thus any function which is the result of applying the chain rule will be the product of two factors, the 'derivative of the outside' and the 'derivative of the inside.' If you recognize a function of this form, its antiderivative is $f(g(x))$.

☐ **Example 1** Find $\int 3x^2 \cos(x^3)\, dx$.

Solution. The function $3x^2 \cos(x^3)$ looks like the result of applying the chain rule: there is an 'inside' function x^3 and its derivative $3x^2$ appearing as a factor. So let's guess $\sin(x^3)$ for the antiderivative.

Differentiating to check gives $\frac{d}{dx}(\sin(x^3)) = \cos(x^3) \cdot (3x^2)$. Since this is what we began with, we have

$$\int 3x^2 \cos(x^3)\,dx = \sin(x^3) + C.$$

❑

The basic idea behind this method, trying to find an 'inside' function whose derivative appears as a factor, works even when the derivative is missing a constant factor, as shown in the next example.

❑ **Example 2** Find $\int te^{(t^2+1)}\,dt$.

Solution. For our guess, we'll pick $e^{(t^2+1)}$, since taking the derivative of an exponential results in the reappearance of the exponential, together with other terms from the chain rule. Now we check:

$$\frac{d}{dt}(e^{(t^2+1)}) = (e^{(t^2+1)}) \cdot 2t.$$

Our original guess was off by a factor of 2. No problem: just re-guess $\frac{1}{2}e^{(t^2+1)}$. Then check again:

$$\frac{d}{dt}\left(\frac{1}{2}e^{(t^2+1)}\right) = \frac{1}{2}e^{(t^2+1)} \cdot 2t = e^{(t^2+1)} \cdot t.$$

Thus

$$\int te^{(t^2+1)}\,dt = \frac{1}{2}e^{(t^2+1)} + C.$$

❑

❑ **Example 3** Find $\int x^3\sqrt{x^4+5}\,dx$.

Solution. Here the inside function is x^4+5, and its derivative appears as a factor (with the exception of the missing 4). So the integrand we have is more or less of the form

$$g'(x)\sqrt{g(x)}$$

with $g(x) = x^4+5$. Thus we might guess that an antiderivative is

$$\frac{(g(x))^{\frac{3}{2}}}{\frac{3}{2}} = \frac{(x^4+5)^{\frac{3}{2}}}{\frac{3}{2}}.$$

Let's check and see:

$$\frac{d}{dx}\left(\frac{(x^4+5)^{\frac{3}{2}}}{\frac{3}{2}}\right) = \frac{3}{2}\frac{(x^4+5)^{\frac{1}{2}}}{\frac{3}{2}} \cdot 4x^3 = 4x^3(x^4+5)^{\frac{1}{2}},$$

so $\frac{(x^4+5)^{\frac{3}{2}}}{\frac{3}{2}}$ is too big by a factor of 4. The correct antiderivative is

$$\frac{1}{4}\frac{(x^4+5)^{\frac{3}{2}}}{\frac{3}{2}} = \frac{1}{6}(x^4+5)^{\frac{3}{2}}.$$

Thus

$$\int x^3\sqrt{x^4+5}\,dx = \frac{1}{6}(x^4+5)^{\frac{3}{2}} + C.$$

As a final check:

$$\frac{d}{dx}\left(\frac{1}{6}(x^4+5)^{\frac{3}{2}}\right) = \frac{1}{6}\cdot\frac{3}{2}(x^4+5)^{\frac{1}{2}}\cdot 4x^3 = x^3(x^4+5)^{\frac{1}{2}}.$$

❑

As we have seen in the preceding examples, antidifferentiating a function often involves "correcting for" constant factors: if differentiation produces an extra factor of 4, antidifferentiation will require a factor of 1/4.

The Method of Substitution

There is a formalization of guess-and-check—called *substitution*—that helps to find antiderivatives when the integrand is complicated.

$$\boxed{\begin{array}{l} \text{let } w = \text{"inside function"} \\ dw = w'(x)\,dx. \end{array}}$$

Let's redo the three examples that we just did by guess-and-check.

❑ **Example 4** Find $\displaystyle\int 3x^2\cos(x^3)\,dx$.

Solution. As before, we look for an inside function whose derivative appears; in this case x^3. We let $w = x^3$. Then $dw = w'(x)\,dx = 3x^2\,dx$. The original integrand can now be completely rewritten in terms of the new variable w:

$$\int 3x^2\cos(x^3)\,dx = \int \cos\underbrace{(x^3)}_{w}\cdot\underbrace{3x^2\,dx}_{dw} = \int \cos w\,dw = \sin w + C = \sin(x^3) + C.$$

By changing the variable to w, the function becomes simpler ($\cos w$) and can be antidifferentiated more easily. The final step, after antidifferentiating, is to convert back to the original variable x. ❑

Why do we write dw as though it were a separate quantity?

Since the indefinite integral, $\int f(x)\,dx$, represents the set of antiderivatives of f and the symbol dx tells us that we're antidifferentiating with respect to x, you might think it makes no sense to write

dx without the integral symbol "$\int$" and an integrand $f(x)$ preceding it. You might also wonder why dw is substituted for $w'(x)\,dx$. Here's why this notation is useful.

Suppose F is an antiderivative of f, so $F' = f$ and $\int f(w)\,dw = F(w) + C$. Suppose w is a function of x, so $w = w(x)$. By the chain rule:

$$\frac{d}{dx}[F(w)] = \frac{d}{dw}[F(w)] \cdot \frac{dw}{dx} = F'(w) \cdot w'(x)$$
$$= f(w) \cdot w'(x) = f(w(x)) \cdot w'(x)$$

Thus $F(w)$ is an antiderivative of $f(w(x)) \cdot w'(x)$ with respect to x, so

$$\int [f(w(x)) \cdot w'(x)]\,dx = F(w) + C.$$

But we also know $\int f(w)\,dw = F(w) + C$, so we must have

$$\int \underbrace{f(w(x))}_{\text{inside function}} \cdot w'(x)\,dx = \int f(w)\,dw.$$

On the left is the original integral in terms of x; on the right is the new integral in terms of w. Thus when we substitute w for the inside function on the left, we must set $dw = w'(x)\,dx$ also.

Let's revisit the second example that we did by guess-and-check:

☐ **Example 5** Find $\int te^{(t^2+1)}\,dt$.

Solution. Here the inside function is $t^2 + 1$, so we'll try

$$w = t^2 + 1.$$

Then

$$dw = w'(t)\,dt = 2t\,dt.$$

Notice that the original integrand has only $t\,dt$, and not $2t\,dt$. But extra multiplicative constants are not a problem. We simply write

$$\frac{1}{2}\,dw = t\,dt$$

and then substitute:

$$\int te^{(t^2+1)}\,dt = \int e^{\overbrace{(t^2+1)}^{w}} \cdot \underbrace{t\,dt}_{\frac{dw}{2}} = \int e^w \frac{1}{2}\,dw = \frac{1}{2}\int e^w\,dw = \frac{1}{2}e^w + C = \frac{1}{2}e^{(t^2+1)} + C.$$

This gives the same answer as we found using guess-and-check. ☐

Now we'll redo the third of the examples that we solved previously by guess-and-check.

◻ **Example 6** Find $\int x^3 \sqrt{x^4 + 5}\, dx$.

Solution. The inside function is $x^4 + 5$, with derivative $4x^3$. But the original problem has an x^3, not a $4x^3$ in it. However, when the only thing missing is a constant factor, we can still use substitution. As before, we let w be the inside function:

$$w = x^4 + 5,$$

so

$$dw = w'(x)\, dx = 4x^3\, dx,$$

giving

$$\frac{1}{4} dw = x^3\, dx.$$

Thus

$$\int x^3 \sqrt{x^4 + 5}\, dx = \int \sqrt{w}\, \frac{1}{4} dw = \frac{1}{4} \int w^{\frac{1}{2}}\, dw = \frac{1}{4} \cdot \frac{w^{\frac{3}{2}}}{\frac{3}{2}} + C = \frac{1}{6}(x^4 + 5)^{\frac{3}{2}} + C.$$

Once again, we get the same result as with guess-and-check. ◻

Warning: As you saw in the previous examples, you can apply the substitution method when a *constant* factor is missing from the derivative of the inside function. However, you may not be able to use substitution if anything other than a constant factor is missing. For example, to find

$$\int x^2 \sqrt{x^4 + 5}\, dx$$

setting $w = x^4 + 5$ does you no good because you can't easily get $x^2\, dx$ in terms of w. The moral of the story is that to use substitution, you want to have the derivative of the inside function, *to within a constant factor*, in the integrand.

◻ **Example 7** Find $\int e^{\cos \theta} \sin \theta\, d\theta$.

Solution. This time we let $w = \cos \theta$, since its derivative is $-\sin \theta$, and there is a factor of $\sin \theta$ in the integrand. Then

$$dw = w'(\theta)\, d\theta = -\sin \theta\, d\theta$$

and so

$$-dw = \sin \theta\, d\theta.$$

Thus

$$\int e^{\cos \theta} \sin \theta\, d\theta = \int e^w (-dw) = (-1) \int e^w\, dw = -e^w + C = -e^{\cos \theta} + C.$$

◻

You might be wondering why we didn't put $(-1)\int e^w\,dw = -e^w - C$ in the previous example. Since the constant C is arbitrary, it doesn't really matter whether we add or subtract it, but the convention is always to add it onto whatever antiderivative you have calculated. If we think of $-e^{\cos\theta}$ as one of the antiderivatives of $e^{\cos\theta}\sin\theta$, then we write

$$-e^{\cos\theta} + C$$

to remind ourselves that there are lots of other antiderivatives.

☐ **Example 8** Find $\displaystyle\int \frac{e^t}{1 + e^t}\,dt.$

Solution. If you observe that the derivative of $1 + e^t$ is e^t, then $w = 1 + e^t$ is a good choice. Then $dw = e^t\,dt$, so that

$$
\begin{aligned}
\int \frac{e^t}{1 + e^t}\,dt = \int \frac{1}{1 + e^t}\,e^t\,dt = \int \frac{1}{w}\,dw \;&=\; \ln|w| + C \\
&=\; \ln|1 + e^t| + C \\
&=\; \ln(1 + e^t) + C \qquad \text{since } (1 + e^t) \text{ is always positive.}
\end{aligned}
$$

Since the numerator is $e^t\,dt$, you might have wanted to try $w = e^t$. This substitution leads to the integral $\displaystyle\int \frac{dw}{1 + w}$, which is not much better than the original integral. The fact that $w = 1 + e^t$ does work makes it clear that you may have to make several attempts before finding a substitution that works.   ☐

Notice the pattern in the previous example: having a function in the denominator and its derivative in the numerator leads to a natural logarithm. The next example follows the same pattern.

☐ **Example 9** Find $\int \tan\theta\,d\theta.$

Solution. Recall that $\tan\theta = \dfrac{\sin\theta}{\cos\theta}$. If $w = \cos\theta$, then $dw = -\sin\theta\,d\theta$, so

$$\int \tan\theta\,d\theta = \int \frac{\sin\theta}{\cos\theta}\,d\theta = \int -\frac{dw}{w} = -\ln|w| + C = -\ln|\cos\theta| + C.$$

☐

Some people prefer the substitution method over guess-and-check since it is more systematic, but both methods achieve the same result. For simple problems, guess-and-check can be faster.

❖ Exercises for Section 6.4

1. (a) Find the derivatives of $\sin(x^2 + 1)$ and $\sin(x^3 + 1)$.

   (b) Use your answer to (a) to find antiderivatives of:

   i. $x\cos(x^2 + 1)$ ii. $x^2\cos(x^3 + 1)$

   (c) Find the antiderivatives of

   i. $x\sin(x^2 + 1)$ ii. $x^2\sin(x^3 + 1)$

For each of the functions in Problems 2–8, find $\int f(x)\,dx$:

2. $f(x) = 2x\cos(x^2)$

3. $f(x) = \sin 3x$

4. $f(x) = \sin(2 - 5x)$

5. $f(x) = e^{\sin x}\cos x$

6. $f(x) = \dfrac{x}{x^2 + 1}$

7. $f(x) = \dfrac{1}{3\cos^2 2x}$

8. $f(x) = \sin x\left(\sqrt{2 + 3\cos x}\right)$

9. For each of the integrals in Problems 12–23, state the substitution you would use but do not perform the integration.

10. For each of the integrals in Problems 24–33, state the substitution you would use but do not perform the integration.

11. For each of the integrals in Problems 34–40, state the substitution you would use but do not perform the integration.

Find the integrals in Problems 12–42. Remember, you can check your answers.

12. $\displaystyle\int xe^{-x^2}\,dx$

13. $\displaystyle\int y(y^2 + 5)^8\,dy$

14. $\displaystyle\int t^2(t^3 - 3)^{10}\,dt$

15. $\displaystyle\int x(x^2 - 4)^{\frac{7}{2}}\,dx$

16. $\displaystyle\int \frac{dy}{y + 5}$

17. $\displaystyle\int (2t - 7)^{73}\,dt$

18. $\displaystyle\int x(x^2 + 3)^2\,dx$

19. $\displaystyle\int (x^2 + 3)^2\,dx$

20. $\displaystyle\int \frac{1}{\sqrt{4 - x}}\,dx$

21. $\displaystyle\int \sin\theta(\cos\theta + 5)^7\,d\theta$

22. $\displaystyle\int x^2 e^{x^3 + 1}\,dx$

23. $\displaystyle\int \sin^3\alpha\cos\alpha\,d\alpha$

24. $\displaystyle\int \sqrt{\cos 3t}\,\sin 3t\,dt$

25. $\displaystyle\int \frac{(\ln z)^2}{z}\,dz$

26. $\displaystyle\int \sin^6\theta\cos\theta\,d\theta$

27. $\displaystyle\int \sin^6(5\theta)\cos(5\theta)\,d\theta$

28. $\displaystyle\int \frac{\cos\sqrt{x}}{\sqrt{x}}\,dx$

29. $\displaystyle\int \frac{e^t + 1}{e^t + t}\,dt$

30. $\displaystyle\int \frac{1 + e^x}{\sqrt{x + e^x}}\,dx$

31. $\displaystyle\int \frac{y}{y^2 + 4}\,dy$

32. $\displaystyle\int \frac{e^x}{2 + e^x}\,dx$

33. $\displaystyle\int \frac{e^{\sqrt{y}}}{\sqrt{y}}\,dy$

34. $\displaystyle\int \tan 2x\,dx$

35. $\displaystyle\int \frac{x + 1}{x^2 + 2x + 19}\,dx$

36. $\displaystyle\int \frac{x\cos(x^2)}{\sqrt{\sin(x^2)}}\,dx$

37. $\displaystyle\int x^2(1 + 2x^3)^2\,dx$

38. $\displaystyle\int y^2(1 + y)^2\,dy$

39. $\displaystyle\int \frac{t}{1 + 3t^2}\,dt$

40. $\displaystyle\int \frac{(t + 1)^2}{t^2}\,dt$

41. $\displaystyle\int \frac{e^x - e^{-x}}{e^x + e^{-x}}\,dx$

42. $\displaystyle\int (2x + 1)e^{x^2}e^x\,dx$

[Hint: Rewrite $e^{x^2}e^x = e^{?}$.]

43. Find $\displaystyle\int 4x(x^2 + 1)\,dx$ using two methods:

   (a) Do the multiplication first, then antidifferentiate.

(b) Now do the integral using the substitution $w = x^2 + 1$.

(c) Explain why the two expressions formed in parts (a) and (b) are different. Are they both correct?

44. Throughout much of this century, the yearly consumption of electricity in the U.S. has been increasing exponentially at a continuous rate of 7%. Assuming this trend continues, and that the electrical energy consumed in 1900 was 1.4 million megawatt-hours (a megawatt hour is a measure of electrical energy),

(a) Write an expression for electricity consumption as a function of time, t, measured in years since 1900.

(b) Find the average yearly electrical consumption throughout this century.

(c) During what year was electrical consumption closest to the average for the century?

(d) Without doing the calculation for (c), how could you have predicted which half of the century the answer would be in?

45. If we assume that wind resistance is proportional to velocity, then the velocity, v, of a body of mass m falling vertically is given by

$$v = \frac{mg}{k}\left(1 - e^{-\frac{k}{m}t}\right),$$

where g is the acceleration due to gravity and k is a constant. Find the height, h, above the surface of the earth as a function of time. Assume the body starts at height h_0.

6.5   Integration by Substitution—Part II

In this section we'll see how to use substitution to evaluate *definite* integrals, using the Fundamental Theorem of Calculus. Then we'll look at some examples where it is not easy to see the chain rule, but the method of substitution works nonetheless.

Definite Integrals by Substitution.

☐ **Example 1**   Compute $\int_0^2 xe^{x^2}\, dx$.

Solution. To evaluate a definite integral using the Fundamental Theorem of Calculus, we first need to find an antiderivative of $f(x) = xe^{x^2}$. The inside function is x^2, so let $w = x^2$. Then $dw = 2x\, dx$, so $\frac{1}{2}\, dw = x\, dx$. Thus

$$\int xe^{x^2}\, dx = \int e^w \frac{1}{2}\, dw = \frac{1}{2}e^w + C = \frac{1}{2}e^{x^2} + C.$$

Now we find the definite integral

$$\int_0^2 xe^{x^2}\, dx = \frac{1}{2}e^{x^2}\bigg|_0^2 = \frac{1}{2}(e^4 - e^0) = \frac{1}{2}(e^4 - 1).$$

There is another way to look at the same problem. After we established that

$$\int xe^{x^2}\, dx = \frac{1}{2}e^w + C,$$

our next two steps were to replace w by x^2, and then x by 2 and 0. We could have directly replaced the original limits of integration, $x = 0$ and $x = 2$, by the corresponding w limits. Since $w = x^2$, the w limits are $w = 0^2 = 0$ and $w = 2^2 = 4$, so we get

$$\int_{x=0}^{x=2} xe^{x^2}\, dx = \frac{1}{2}\int_{w=0}^{w=4} e^w\, dw = \frac{1}{2}e^w\Big|_0^4 = \frac{1}{2}\left(e^4 - e^0\right) = \frac{1}{2}(e^4 - 1).$$

As you would expect, both methods give the same answer. $\qquad\square$

To use substitution to find definite integrals, either:

- compute the indefinite integral, expressing an antiderivative in terms of the original variable, and then evaluate the result at the original limits,

or

- convert the original limits to new limits in terms of the new variable and do not convert back to the original variable.

Example 2 Evaluate $\displaystyle\int_0^{\frac{\pi}{4}} \frac{\tan^3 \theta}{\cos^2 \theta}\, d\theta.$

Solution. To use substitution, we must decide what w should be. There are two possible inside functions, $\tan\theta$ and $\cos\theta$. Now

$$\frac{d}{d\theta}(\tan\theta) = \frac{1}{\cos^2\theta} \quad \text{and} \quad \frac{d}{d\theta}(\cos\theta) = -\sin\theta,$$

and since the integral contains a factor of $\frac{1}{\cos^2\theta}$ but not of $\sin\theta$, we let $w = \tan\theta$ and $dw = \frac{1}{\cos^2\theta}\, d\theta.$ When $\theta = 0$, $w = \tan 0 = 0$, and when $\theta = \frac{\pi}{4}$, $w = \tan\frac{\pi}{4} = 1$, so

$$\int_0^{\frac{\pi}{4}} \frac{\tan^3\theta}{\cos^2\theta}\, d\theta = \int_0^{\frac{\pi}{4}} (\tan\theta)^3 \cdot \frac{1}{\cos^2\theta}\, d\theta = \int_0^1 w^3\, dw = \frac{1}{4}w^4\Big|_0^1 = \frac{1}{4}.$$

$\qquad\square$

Example 3 Evaluate $\displaystyle\int_1^3 \frac{dx}{5-x}.$

Solution. Let $w = 5 - x$, so $dw = -dx$. When $x = 1$, $w = 4$, and when $x = 3$, $w = 2$, so

$$\int_1^3 \frac{dx}{5-x} = \int_4^2 \frac{-dw}{w} = -\ln|w|\Big|_4^2 = -(\ln 2 - \ln 4) = \ln\left(\frac{4}{2}\right) = \ln 2 \approx 0.69.$$

Notice that we write the limit $w = 4$ at the bottom, even though it is larger than $w = 2$, because $w = 4$ corresponds to the lower limit $x = 1$. $\qquad\square$

More Complex Substitutions

In the examples of substitution we have seen so far, we had to guess what to let w be and then we had to be lucky enough to have dw (or some constant multiple of it) just lying around elsewhere in the problem, waiting for us. What if we weren't so lucky? It turns out that we can often get somewhere by letting w be some messy expression contained inside, say, a cosine or under a root, even if we cannot see immediately how such a substitution helps. So go ahead and try!

☐ **Example 4** Find $\int (x+7)\sqrt[3]{3-2x}\,dx$.

Solution. Here, instead of the derivative of the inside function (which is -2), we have the factor $(x+7)$. However, substituting $w = 3-2x$ turns out to help. Then $dw = -2\,dx$, so $(-1/2)\,dw = dx$. Now we must convert everything to w, including $x+7$. Well, if $w = 3 - 2x$, then $2x = 3 - w$, so $x = 3/2 - w/2$, and therefore we can write $x + 7$ in terms of w. Thus

$$
\begin{aligned}
\int (x+7)\sqrt[3]{3-2x}\,dx &= \int \left(\frac{3}{2} - \frac{w}{2} + 7\right)\sqrt[3]{w}\left(-\frac{1}{2}\right)dw \\
&= -\frac{1}{2}\int \left(\frac{17}{2} - \frac{w}{2}\right)w^{1/3}\,dw \\
&= -\frac{1}{4}\int (17 - w)\,w^{1/3}\,dw \\
&= -\frac{1}{4}\int \left(17w^{1/3} - w^{4/3}\right)dw \\
&= -\frac{1}{4}\left(17\frac{w^{4/3}}{4/3} - \frac{w^{7/3}}{7/3}\right) + C \\
&= -\frac{1}{4}\left(\frac{51}{4}(3-2x)^{4/3} - \frac{3}{7}(3-2x)^{7/3}\right) + C.
\end{aligned}
$$

Looking back over the solution, you can see the reason this substitution works is that it converts $\sqrt[3]{3-2x}$, the messiest part of the problem, to $\sqrt[3]{w}$, which can be integrated. ☐

☐ **Example 5** Find $\int \sqrt{1 + \sqrt{x}}\,dx$.

Solution. This time, the derivative of the inside function is nowhere to be seen. However, try $w = 1 + \sqrt{x}$. Then $w - 1 = \sqrt{x}$, so $(w-1)^2 = x$. Therefore $2(w-1)\,dw = dx$. We then have

$$
\begin{aligned}
\int \sqrt{1 + \sqrt{x}}\,dx &= \int \sqrt{w}\,2(w-1)\,dw \\
&= 2\int w^{1/2}(w-1)\,dw \\
&= 2\int (w^{3/2} - w^{1/2})\,dw \\
&= 2\left(\frac{2}{5}w^{5/2} - \frac{2}{3}w^{3/2}\right) + C \\
&= 2\left(\frac{2}{5}(1+\sqrt{x})^{5/2} - \frac{2}{3}(1+\sqrt{x})^{3/2}\right) + C.
\end{aligned}
$$

☐

Notice that the substitution in the previous example again converts the inside of the messiest function into something simple. In addition, since the derivative of the inside function is not waiting for us, we have to solve for x so that we can get dx entirely in terms of w.

Substitution can be used to express one definite integral in terms of another, even in situations where we don't know what the indefinite integral is.

☐ **Example 6** Derive a formula for the area of the ellipse

$$\frac{x^2}{a^2} + \frac{y^2}{b^2} = 1$$

assuming $a, b > 0$.

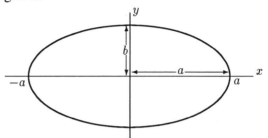

Figure 6.15: Ellipse: $\frac{x^2}{a^2} + \frac{y^2}{b^2} = 1$
(for Example 6)

Solution. Solving for y gives

$$y = \pm b\sqrt{1 - \frac{x^2}{a^2}}$$

where the positive square root gives the upper half of the ellipse and the negative square root gives the lower. (See Figure 6.15.) Thus the area of the ellipse is given by

$$\text{Area} = 2\int_{-a}^{a} b\sqrt{1 - \frac{x^2}{a^2}}\, dx$$

Substitution allows us to convert this integral to one we can compute. Let $w = \frac{x}{a}$, so $dx = a\, dw$. When $x = -a$, $w = -1$, and when $x = a$, $w = 1$. Thus

$$2\int_{-a}^{a} b\sqrt{1 - \frac{x^2}{a^2}}\, dx = 2\int_{-1}^{1} b\sqrt{1 - w^2}\, a\, dw = 2ab\int_{-1}^{1}\sqrt{1 - w^2}\, dw$$

Now the integral $\int_{-1}^{1}\sqrt{1 - w^2}\, dw$ is the area of half a circle of radius 1 (because $y = \sqrt{1 - w^2}$ is the top half of the circle $w^2 + y^2 = 1$). Thus

$$\int_{-1}^{1}\sqrt{1 - w^2}\, dw = \frac{\pi}{2}$$

so

$$\text{Area of ellipse} = 2ab\int_{-1}^{1}\sqrt{1 - w^2}\, dw = \pi ab.$$

(Alternatively, calculating $\int_{-1}^{1}\sqrt{1 - w^2}\, dw$ using left hand sums with 100 subdivisions gives the value of this integral as $1.57\ldots$, which agrees with $\frac{\pi}{2}$ to two decimal places.) ☐

❖ Exercises for Section 6.5

1. Use substitution to express the following integrals as a multiple of $\int_a^b \frac{1}{w}\, dw$ for some a and b. Then evaluate the integrals.

   (a) $\int_0^1 \frac{x}{1+x^2}\, dx$. [Hint: try $w = 1 + x^2$.]   (b) $\int_0^{\frac{\pi}{4}} \frac{\sin x}{\cos x}\, dx$.

For Problems 2–7, use substitution to calculate the definite integrals.

2. $\int_0^{\frac{1}{2}} \cos \pi x\, dx$

3. $\int_1^8 \frac{e^{\sqrt[3]{x}}}{\sqrt[3]{x^2}}\, dx$

4. $\int_0^2 \frac{x}{(1+x^2)^2}\, dx$

5. $\int_{-1}^{e-2} \frac{1}{t+2}\, dt$

6. $\int_1^4 \frac{\cos \sqrt{x}}{\sqrt{x}}\, dx$

7. $\int_{-\frac{\pi}{4}}^{\frac{\pi}{4}} \cos^2 \theta \sin^5 \theta\, d\theta$

   [Hint: $\cos^2 \theta + \sin^2 \theta = 1$.]

For Problems 8–21, evaluate the definite integrals. Some of the integrands have elementary antiderivatives and their integrals can be calculated using the Fundamental Theorem of Calculus, perhaps after a substitution; others you will have to calculate using left and right hand sums. Whenever possible, use the Fundamental Theorem.

8. $\int_1^3 \frac{1}{x}\, dx$

9. $\int_{-1}^3 (x^3 + 5x)\, dx$

10. $\int_1^2 \frac{\sin t}{t}\, dt$

11. $\int_0^\pi \sin \theta (\cos \theta + 5)^7\, d\theta$

12. $\int_{-1}^1 \frac{1}{1+y^2}\, dy$

13. $\int_0^1 x(1+x^2)^{20}\, dx$

14. $\int_0^1 \frac{x}{1+5x^2}\, dx$

15. $\int_0^{\pi/12} \sin 3\alpha\, d\alpha$

16. $\int_1^2 \frac{x^2+1}{x}\, dx$

17. $\int_4^1 x\sqrt{x^2+4}\, dx$

18. $\int_0^1 \frac{1}{x^2+2x+1}\, dx$

19. $\int_0^{\frac{1}{\sqrt{2}}} \frac{x}{\sqrt{1-x^4}}\, dx$

20. $\int_{-2}^0 \frac{2x+4}{x^2+4x+5}\, dx$

21. $\int_{\frac{1}{4}}^1 \sin \frac{1}{t}\, dt$

In Problems 22–26, explain why each pair of antiderivatives are really, despite their apparent dissimilarity, essentially different expressions of the same problem. (They are not *identical*; they are expressed using different symbols and some are multiples of their partners–but they are, for the most part, the same problem. Why?) You do not need to evaluate the integrals, just explain ...

22. $\int \frac{e^x\, dx}{1+e^{2x}}$ and $\int \frac{\cos x\, dx}{1+\sin^2 x}$

23. $\int \frac{\ln x}{x}\, dx$ and $\int x\, dx$

24. $\int (\sin x)^3 \cos x\, dx$ and $\int (x^3+1)^3 x^2\, dx$

25. $\int \sqrt{x+1}\, dx$ and $\int \frac{\sqrt{1+\sqrt{x}}}{\sqrt{x}}\, dx$

26. $\int e^{\sin x} \cos x\, dx$ and $\int \frac{e^{\arcsin x}}{\sqrt{1-x^2}}\, dx$

27. Find $\displaystyle\int \frac{dx}{x^2+4x+5}$ by completing the square and using the substitution $x+2=\tan\theta$.

28. Integrate:

    (a) $\displaystyle\int \frac{1}{\sqrt{x}}\,dx$

    (b) $\displaystyle\int \frac{1}{\sqrt{x+1}}\,dx$

    (c) $\displaystyle\int \frac{1}{\sqrt{x}+1}\,dx$

29. Find the area under $f(x)=xe^{x^2}$ between $x=0$ and $x=2$.

30. Find the average value of $f(x)=\dfrac{1}{x+1}$ on the interval $x=0$ to $x=2$. Sketch a graph showing the function and the average value.

31. (a) Find $\displaystyle\int \sin\theta\cos\theta\,d\theta$

    (b) You probably solved (a) by making the substitution $w=\sin\theta$ or $w=\cos\theta$. (If not, go back and do it that way.) Now find $\int \sin\theta\cos\theta\,d\theta$ by making the *other* substitution.

    (c) There is yet another way of finding this integral that involves the trigonometric identities

    $$\sin 2\theta = 2\sin\theta\cos\theta \text{ and } \cos 2\theta = \cos^2\theta - \sin^2\theta.$$

    Find $\int \sin\theta\cos\theta\,d\theta$ using one of these identities and then the substitution $w=2\theta$.

    (d) You should now have three different expressions for the indefinite integral $\int \sin\theta\cos\theta\,d\theta$. Are they really different? Are they all correct? Explain the reasons for your answer.

32. If we assume that wind resistance is proportional to the square of velocity, the velocity, v, of a falling body is given by

$$v = \sqrt{\frac{g}{k}}\left(\frac{e^{t\sqrt{gk}} - e^{-t\sqrt{gk}}}{e^{t\sqrt{gk}} + e^{-t\sqrt{gk}}}\right).$$

Use the substitution $w=e^{t\sqrt{gk}}+e^{-t\sqrt{gk}}$ to find the height, h, of the body above the surface of the earth as a function of time. Assume the body starts at a height h_0.

6.6   Integration by Parts

We have already pointed out that there were no rules for antidifferentiation like the product, quotient, and chain rules for differentiation. There is, however, a method called *integration by parts*, that is based on the product rule.

☐ **Example 1** Find $\displaystyle\int xe^x\,dx$.

Solution. We are looking for a function whose derivative is xe^x. If we think about how the product rule works, we might be led to guess xe^x, because we know when we take the derivative we will get two terms, one of which will be xe^x:

$$\frac{d}{dx}(xe^x) = \frac{d}{dx}(x)e^x + x\frac{d}{dx}(e^x) = e^x + xe^x$$

Of course, our guess is wrong because of the extra e^x. Maybe we can adjust our guess by subtracting from xe^x something to cancel out the extra e^x. This leads us to try $xe^x - e^x$. Let's check it:

$$\frac{d}{dx}(xe^x - e^x) = \frac{d}{dx}(xe^x) - \frac{d}{dx}(e^x) = e^x + xe^x - e^x = xe^x.$$

It works, so $\displaystyle\int xe^x \, dx = xe^x - e^x + C.$ ❑

❑ **Example 2** Find $\displaystyle\int \theta \cos \theta \, d\theta.$

Solution. We guess the antiderivative is $\theta \sin \theta$, and use the product rule to check:

$$\frac{d}{d\theta}(\theta \sin \theta) = \frac{d(\theta)}{d\theta} \sin \theta + \theta \frac{d}{d\theta}(\sin \theta) = \sin \theta + \theta \cos \theta.$$

To correct for the extra $\sin \theta$ term, we must subtract from our original guess something whose derivative is $\sin \theta$. Since $\frac{d}{d\theta}(\cos \theta) = -\sin \theta$, we'll try:

$$\frac{d}{d\theta}(\theta \sin \theta + \cos \theta) = \frac{d}{d\theta}(\theta \sin \theta) + \frac{d}{d\theta}(\cos \theta) = \sin \theta + \theta \cos \theta - \sin \theta = \theta \cos \theta.$$

Thus $\displaystyle\int \theta \cos \theta \, d\theta = \theta \sin \theta + \cos \theta + C.$ ❑

The General Case

We can formalize the process illustrated in these two examples in the following way. We begin with the product rule:

$$\frac{d}{dx}(uv) = u'v + uv'$$

where u' and v' are the derivatives of u and v. We rewrite this as:

$$uv' = \frac{d}{dx}(uv) - u'v$$

and then integrate both sides:

$$\int uv' \, dx = \int \frac{d}{dx}(uv) \, dx - \int u'v \, dx.$$

Since an antiderivative of $\frac{d}{dx}(uv)$ is just uv, we get the formula for

---

Integration by Parts

$$\int uv' \, dx = uv - \int u'v \, dx.$$

This formula is useful when the integrand can be viewed as a product and when the integral on the right hand side is easier than that on the left.

In effect, we were using integration by parts in the previous two examples. In Example 1, we let $xe^x = (x) \cdot (e^x) = uv'$, then $u = x$ and $v' = e^x$. Thus $u' = 1$ and $v = e^x$ so

$$\int xe^x \, dx = (x)(e^x) - \int (1)(e^x) \, dx = xe^x - e^x + C.$$

So uv represents our first guess and $\int u'v \, dx$ represents the correction to our guess.

Notice what would have happened in Example 1 if we took $u = x$ and $v = e^x + C_1$. Then

$$
\begin{aligned}
\int xe^x \, dx &= x(e^x + C_1) - \int (e^x + C_1) \, dx \\
&= xe^x + C_1 x - e^x - C_1 x + C \\
&= xe^x - e^x + C, \qquad \text{as before.}
\end{aligned}
$$

Therefore you see that it is not necessary to take a general antiderivative for v—any antiderivative will do.

You may wonder what would have happened in Example 1 if we had picked u and v' the other way around. If $u = e^x$ and $v' = x$, then $u' = e^x$ and $v = x^2/2$. The formula for integration by parts then gives

$$\int xe^x \, dx = \frac{x^2}{2} e^x - \int \frac{x^2}{2} \cdot e^x \, dx,$$

which is true, but hardly helpful since the integral on the right is worse than the one on the left. The moral of the story is that you want to choose u and v' to make the integral on the right easier than that on the left.

Advice on how to choose u and v'

- Whatever you let v' be, you'd better be able to find v.

- It helps if u' is 'simpler' than u (or at least no more complicated than u).

- It helps if v is 'simpler' than v' (or at least no more complicated than v').

If we pick $v' = x$ in Example 1, then $v = x^2/2$, which is certainly 'worse' than v'.

There are some examples which don't look like good candidates for integration by parts, because they don't appear to involve products, but for which the process works well. Such examples often involve $\ln x$ and the inverse trig functions. Here is one:

☐ **Example 3** Find $\displaystyle\int_2^3 \ln x \, dx$.

∗ Int. by parts works well for ln x and inverse trig. Functions

Solution. This doesn't look like a product unless you write $\ln x = (1)(\ln x)$. Then we might say $u = 1$ so $u' = 0$ (which certainly makes things 'simpler'). But if $v' = \ln x$, what is v? (If we knew, we wouldn't need integration by parts.) Let's try the other way: if $u = \ln x$, $u' = \frac{1}{x}$ and if $v' = 1$, $v = x$, so

$$\int_2^3 \ln x \, dx = \int_2^3 (1)(\ln x) \, dx = (\ln x)(x)\Big|_2^3 - \int_2^3 \frac{1}{x} x \, dx$$

$$= x \ln x \Big|_2^3 - \int_2^3 1 \, dx = (x \ln x - x)\Big|_2^3$$

$$= 3 \ln 3 - 3 - 2 \ln 2 + 2 = 3 \ln 3 - 2 \ln 2 - 1.$$

Notice that when doing a definite integral by parts, you must remember to put the limits of integration (here 2 and 3) on the uv term (here $x \ln x$) as well as on the integral $\int u'v \, dx$.

☐ **Example 4** Find $\int x^6 \ln x \, dx$.

Solution. View $x^6 \ln x$ as uv' where $u = \ln x$ and $v' = x^6$. (Notice we did not let $u = x^6$, $v' = \ln x$, as you might have expected!) Then $v = \frac{1}{7}x^7$ and $u' = \frac{1}{x}$ so integration by parts gives us:

$$\int x^6 \ln x \, dx = \int (\ln x)x^6 \, dx = (\ln x)\left(\frac{1}{7}x^7\right) - \int \frac{1}{7}x^7 \cdot \frac{1}{x} \, dx$$

$$= \frac{1}{7}x^7 \ln x - \frac{1}{7}\int x^6 \, dx$$

$$= \frac{1}{7}x^7 \ln x - \frac{1}{49}x^7 + C.$$

In Example 4, we did not choose $v' = \ln x$, because it is not immediately clear what v is. In fact, we had to use integration by parts in Example 3 to find out what the antiderivative of $\ln x$ was. Also, using $u = \ln x$, as we have done, gives $u' = \frac{1}{x}$, which most people would consider 'simpler' than $u = \ln x$. This shows that u does not have to be the factor written on the left (here, x^6). However, if you do choose $u = x^6$ and $v' = \ln x$, the integration by parts still works—it's just less pleasant.

☐ **Example 5** Find $\int x^2 \sin 4x \, dx$.

Solution. If we let $v' = \sin 4x$, then $v = -\frac{1}{4}\cos 4x$, which is no worse that v'. Also letting $u = x^2$, we get $u' = 2x$, which is 'simpler' than $u = x^2$. Using integration by parts:

$$\int x^2 \sin 4x \, dx = x^2\left(-\frac{1}{4}\cos 4x\right) - \int 2x\left(-\frac{1}{4}\cos 4x\right) dx$$

$$= -\frac{1}{4}x^2 \cos 4x + \frac{1}{2}\int x \cos 4x \, dx.$$

The trouble is we still have to grapple with $\int x \cos 4x \, dx$. This can be done by using integration by parts again with a new u and v, namely $u = x$ and $v' = \cos 4x$:

$$\begin{aligned}
\int x\cos 4x\, dx &= x\left(\frac{1}{4}\sin 4x\right) - \int 1\cdot\frac{1}{4}\sin 4x\, dx \\
&= \frac{1}{4}x\sin 4x - \frac{1}{4}\cdot\left(-\frac{1}{4}\cos 4x\right) + C \\
&= \frac{1}{4}x\sin 4x + \frac{1}{16}\cos 4x + C.
\end{aligned}$$

Thus,

$$\begin{aligned}
\int x^2\sin 4x\, dx &= -\frac{1}{4}x^2\cos 4x + \frac{1}{2}\int x\cos 4x\, dx \\
&= -\frac{1}{4}x^2\cos 4x + \frac{1}{2}\left(\frac{1}{4}x\sin 4x + \frac{1}{16}\cos 4x + C\right) \\
&= -\frac{1}{4}x^2\cos 4x + \frac{1}{8}x\sin 4x + \frac{1}{32}\cos 4x + C_1
\end{aligned}$$

where C_1 is the arbitrary constant $\frac{1}{2}C$. $\qquad\square$

In Example 6 on page 393, in order to find the area of an ellipse we evaluated the integral

$$2\int_{-a}^{a} b\sqrt{1-\frac{x^2}{a^2}}\, dx$$

by substituting $\frac{x}{a} = w$. We will now show how to compute this integral by a different substitution.

☐ **Example 6** Use the substitution $x = a\sin\theta$ to evaluate

$$2b\int_{-a}^{a}\sqrt{1-\frac{x^2}{a^2}}\, dx.$$

Solution.

If $\quad x = a\sin\theta,\qquad\qquad$ then $\quad dx = a\cos\theta\, d\theta.$
When $\quad x = -a,\quad -a = a\sin\theta\quad$ so $\quad \sin\theta = -1\quad$ and $\quad \theta = -\frac{\pi}{2}.$
When $\quad x = a,\quad a = a\sin\theta\quad$ so $\quad \sin\theta = 1\quad$ and $\quad \theta = \frac{\pi}{2}.$

In addition,

$$\sqrt{1-\frac{x^2}{a^2}} = \sqrt{1-\sin^2\theta} = \sqrt{\cos^2\theta}.$$

For $-\pi/2 \le \theta \le \pi/2$, $\cos\theta$ is non-negative, so $\sqrt{\cos^2\theta} = \cos\theta$ and

$$
\begin{aligned}
2b \int_{-a}^{a} \sqrt{1 - \frac{x^2}{a^2}}\, dx &= 2b \int_{-\frac{\pi}{2}}^{\frac{\pi}{2}} \sqrt{1 - \sin^2\theta} \cdot a\cos\theta\, d\theta \\
&= 2ab \int_{-\frac{\pi}{2}}^{\frac{\pi}{2}} \cos\theta \cdot \cos\theta\, d\theta \\
&= 2ab \int_{-\frac{\pi}{2}}^{\frac{\pi}{2}} \cos^2\theta\, d\theta.
\end{aligned}
$$

Using integration by parts with $u = \cos\theta$, $v' = \cos\theta$ gives $u' = -\sin\theta$, $v = \sin\theta$, so we get

$$
\int \cos^2\theta\, d\theta = \cos\theta\sin\theta + \int \sin^2\theta\, d\theta
$$

Substituting $\cos^2\theta = 1 - \sin^2\theta$ leads to

$$
\begin{aligned}
\int \cos^2\theta\, d\theta &= \cos\theta\sin\theta + \int (1 - \cos^2\theta)\, d\theta \\
&= \cos\theta\sin\theta + \int 1\, d\theta - \int \cos^2\theta\, d\theta
\end{aligned}
$$

Looking at the right side, we see our original integral has 'boomeranged' back to us. If we move it to the left and divide by 2, we get

$$
\begin{aligned}
2 \int \cos^2\theta\, d\theta &= \cos\theta\sin\theta + \int 1\, d\theta \\
\int \cos^2\theta\, d\theta &= \frac{1}{2}\cos\theta\sin\theta + \frac{1}{2}\int 1\, d\theta
\end{aligned}
$$

Therefore the definite integral is given by

$$
\begin{aligned}
2ab \int_{-\frac{\pi}{2}}^{\frac{\pi}{2}} \cos^2\theta\, d\theta &= 2ab \left(\frac{1}{2}\cos\theta\sin\theta \Big|_{-\frac{\pi}{2}}^{\frac{\pi}{2}} \right) + ab \int_{-\frac{\pi}{2}}^{\frac{\pi}{2}} 1\, d\theta \\
&= ab(0 - 0) + ab\theta \Big|_{-\frac{\pi}{2}}^{\frac{\pi}{2}} = ab \left[\frac{\pi}{2} - \left(-\frac{\pi}{2} \right) \right] \\
&= \pi ab, \quad \text{as before.}
\end{aligned}
$$

$\square$

❖ Exercises for Section 6.6

1. (a) Find formulas for the derivatives of the following functions:

    i.) $x\cos x$ ii.) $\cos 2x$ iii.) $x^2\cos x$ iv.) $x\ln x$

    (b) Use your answers from part (a) to find antiderivatives for

    v.) $\ln x$ vi.) $\sin 2x$ vii.) $x\sin x$ viii.) $x^2\sin x$

2. By writing $\arctan x = (1) \cdot (\arctan x)$, find $\int \arctan x \, dx$.

Find the indefinite integrals in Problems 3–28.

3. $\int t e^{5t} \, dt$

4. $\int t^2 e^{5t} \, dt$

5. $\int p e^{-0.1p} \, dp$

12. $\int \dfrac{z}{e^z} \, dz$

13. $\int (\theta + 1) \sin(\theta + 1) \, d\theta$

14. $\int \sin^2 \theta \, d\theta$

15. $\int \cos^2(3\alpha + 1) \, d\alpha$

16. $\int q^5 \ln 5q \, dq$

17. $\int \dfrac{\ln x}{x^2} \, dx$

6. $\int y \ln y \, dy$

7. $\int x^3 \ln x \, dx$

8. $\int t \sin t \, dt$

18. $\int y \sqrt{y + 3} \, dy$

19. $\int \dfrac{y}{\sqrt{5 - y}} \, dy$

20. $\int (t + 2)\sqrt{2 + 3t} \, dt$

21. $\int \dfrac{t + 7}{\sqrt{5 - t}} \, dt$

22. $\int (\ln t)^2 \, dt$

23. $\int x(\ln x)^4 \, dx$

9. $\int t^2 \sin t \, dt$

10. $\int \theta^2 \cos 3\theta \, d\theta$

11. $\int (z + 1)e^{2z} \, dz$

24. $\int \arctan 7z \, dz$

25. $\int x \arctan x^2 \, dx$

26. $\int \arcsin w \, dw$

27. $\int x^3 e^{x^2} \, dx$

28. $\int x^5 \cos x^3 \, dx$

29. In Problem 14 above, you evaluated $\int \sin^2 \theta \, d\theta$ by integration by parts. (If you didn't do it by parts, do so now!) Redo this integral by changing the form of the integrand using the identity

$$\sin^2 \theta = \frac{1 - \cos 2\theta}{2}.$$

Explain any differences in the form of the answer obtained by the two methods.

30. Compute $\int \cos^2 \theta \, d\theta$ in two different ways and explain any differences in the form of your answers. (The identity

$$\cos^2 \theta = \frac{1 + \cos 2\theta}{2}$$

may be useful.)

31. Use integration by parts twice to find $\int e^x \sin x \, dx$.

32. Use integration by parts twice to find $\int e^\theta \cos \theta \, d\theta$.

33. Use the results of Problems 31 and 32 to find $\int x e^x \sin x \, dx$.

34. Use the results from Problems 31 and 32 to find $\int \theta e^\theta \cos \theta \, d\theta$.

35. Show that $\int x^n e^x \, dx = x^n e^x - n \int x^{n-1} e^x \, dx$.

36. Show that $\int x^n \sin ax \, dx = -\frac{1}{a} x^n \cos ax + \frac{n}{a} \int x^{n-1} \cos ax \, dx$.

37. Show that $\int x^n \cos ax \, dx = \frac{1}{a} x^n \sin ax - \frac{n}{a} \int x^{n-1} \sin ax \, dx$.

38. Show that $\int \cos^n x \, dx = \frac{1}{n} \cos^{n-1} x \sin x + \frac{n-1}{n} \int \cos^{n-2} x \, dx$.

Evaluate the integrals in Problems 39–47 both symbolically (as in $\ln(3\pi)$) and numerically ($\ln(3\pi) \approx$ 2.243):

39. $\int_1^5 \ln t \, dt$

40. $\int_0^{10} ze^{-z} \, dz$

41. $\int_3^5 x \cos x \, dx$.

42. $\int_1^3 t \ln t \, dt$

43. $\int_0^5 \ln(1+t) \, dt$

44. $\int_0^1 \arctan y \, dy$

45. $\int_0^1 x \arctan x^2 \, dx$

46. $\int_0^1 \arcsin z \, dz$

47. $\int_0^1 u \arcsin u^2 \, du$

48. Using Riemann sums, find an approximate value for $\int_1^2 \ln x \, dx$. Find using antiderivatives $\int_1^2 \ln x \, dx$. Explain in words why your answers verify the Fundamental Theorem of Calculus.

49. Integrating $e^{ax} \sin b x$ by parts yields a result of the form

$$\int e^{ax} \sin bx \, dx = e^{ax}(A \sin bx + B \cos bx) + C.$$

(a) Find the constants A and B in terms of a and b. [Hint: Don't actually perform the integration by parts.]

(b) Evaluate $\int e^{ax} \cos bx \, dx$ by modifying the result in part (a). Again, it is not necessary to perform integration by parts, as the result is of the same form as that in (a).

50. During a surge in the demand for electrical power, the rate, r, at which power is used can be approximated by

$$r = te^{-at},$$

where t is the time in hours and a is a positive constant.

(a) Find the total energy, E, used in the first T hours. Give your answer as a function of a.

(b) What happens to E as $T \to \infty$?

☐ **Example 4** Find $\int (x^2 + 3) \ln x \, dx$.

Solution. Formula III-13 only applies to functions of the form $x^n \ln x$, so we'll have to multiply out and separate into two integrals.

$$\int (x^2 + 3) \ln x \, dx = \int x^2 \ln x \, dx + 3 \int \ln x \, dx.$$

Now we can use formula III-13 on each integral separately, to get

$$\int (x^2 + 3) \ln x \, dx = \frac{x^3}{3} \ln x - \frac{x^3}{9} + 3(x \ln x - x) + C.$$

☐

Part IV of the table contains reduction formulas for the antiderivatives of $\cos^n x$ and $\sin^n x$, which can be obtained by integration by parts. When n is a positive integer, formulas IV-17 and IV-18 can be used repeatedly to reduce the power n until it is 0 or 1.

☐ **Example 5** Find $\int \sin^6 \theta \, d\theta$.

Solution. Use IV-17 repeatedly:

$$\int \sin^6 \theta \, d\theta = -\frac{1}{6} \sin^5 \theta \cos \theta + \frac{5}{6} \int \sin^4 \theta \, d\theta$$

$$\int \sin^4 \theta \, d\theta = -\frac{1}{4} \sin^3 \theta \cos \theta + \frac{3}{4} \int \sin^2 \theta \, d\theta$$

$$\int \sin^2 \theta \, d\theta = -\frac{1}{2} \sin \theta \cos \theta + \frac{1}{2} \int 1 \, d\theta.$$

Calculate $\int \sin^2 \theta \, d\theta$ first and use this to find $\int \sin^4 \, d\theta$; then calculate $\int \sin^6 \, d\theta$. Putting this all together we get

$$\int \sin^6 \theta \, d\theta = -\frac{1}{6} \sin^5 \theta \cos \theta - \frac{5}{24} \sin^3 \theta \cos \theta - \frac{15}{48} \sin \theta \cos \theta + \frac{15}{48} \theta + C.$$

☐

There are reduction formulas that work for negative powers of $\sin x$ and $\cos x$ in IV-19 and IV-21. Here is an example that uses IV-21:

☐ **Example 6** Find $\int \frac{1}{\cos^5 x} \, dx$.

Solution. Use IV-21 twice to get the exponent down to 1:

$$\int \frac{1}{\cos^5 x} \, dx = \frac{1}{4} \frac{\sin x}{\cos^4 x} + \frac{3}{4} \int \frac{1}{\cos^3 x} \, dx$$

$$\int \frac{1}{\cos^3 x} \, dx = \frac{1}{2} \frac{\sin x}{\cos^2 x} + \frac{1}{2} \int \frac{1}{\cos x} \, dx.$$

Now use IV-22 to get

$$\int \frac{1}{\cos x}\, dx = \frac{1}{2} \ln \left| \frac{(\sin x) + 1}{(\sin x) - 1} \right| + C.$$

Putting this all together gives

$$\int \frac{1}{\cos^5 x}\, dx = \frac{1}{4} \frac{\sin x}{\cos^4 x} + \frac{3}{8} \frac{\sin x}{\cos^2 x} + \frac{3}{16} \ln \left| \frac{(\sin x) + 1}{(\sin x) - 1} \right| + C.$$

❑

The last item in part IV of the table is not a formula at all: it is, instead, advice on how to antidifferentiate products of integer powers of $\sin x$ and $\cos x$. There are various techniques to choose, depending on the nature (odd or even, positive or negative) of the exponents.

❑ **Example 7** Find $\int \cos^3 t \sin^4 t\, dt$.

Solution. Here the exponent of $\cos t$ is odd, so IV-23 recommends making the substitution $w = \sin t$. Then $dw = \cos t\, dt$. To make this work, we'll have to separate off one of the cosines to be dw. Also, the remaining even power of $\cos t$ can be rewritten in terms of $\sin t$ by using $\cos^2 t = 1 - \sin^2 t = 1 - w^2$ so that

$$\int \cos^3 t \sin^4 t\, dt = \int \cos^2 t \sin^4 t \cos t\, dt$$

$$= \int (1 - w^2) w^4\, dw = \int (w^4 - w^6)\, dw$$

$$= \frac{1}{5} w^5 - \frac{1}{7} w^7 + C = \frac{1}{5} \sin^5 t - \frac{1}{7} \sin^7 t + C.$$

❑

❑ **Example 8** Find $\int \cos^2 x \sin^4 x\, dx$.

Solution. In this example, both exponents are even. The advice given in IV-23 is to convert to all sines or all cosines. We'll convert to all sines by substituting $\cos^2 x = 1 - \sin^2 x$, and then we'll multiply out the integrand:

$$\int \cos^2 x \sin^4 x\, dx = \int (1 - \sin^2 x) \sin^4 x\, dx = \int \sin^4 x\, dx - \int \sin^6 x\, dx.$$

In Example 5 you found $\int \sin^4 x\, dx$ and $\int \sin^6 x\, dx$. Put them together to get:

$$\int \cos^2 x \sin^4 x\, dx = -\frac{1}{4} \sin^3 x \cos x - \frac{3}{8} \sin x \cos x + \frac{3}{8} x$$

$$-(-\frac{1}{6} \sin^5 x \cos x - \frac{5}{24} \sin^3 x \cos x - \frac{15}{48} \sin x \cos x + \frac{15}{48} x) + C$$

$$= \frac{1}{6} \sin^5 x \cos x - \frac{1}{24} \sin^3 x \cos x - \frac{3}{48} \sin x \cos x + \frac{3}{48} x + C.$$

❑

The last two parts of our table are concerned with quadratic functions: **Part V** has expressions with quadratic denominators; **Part VI** contains square roots of quadratics. The quadratics that appear in these formulas are of the form $x^2 \pm a^2$ or $a^2 - x^2$, or in factored form $(x - a)(x - b)$, where a and b are constants. Many integrands with quadratics in them are either of one of these forms or can be coerced into one of them by completing the square or factoring.

Using Factoring:

☐ **Example 9** Find $\displaystyle\int \frac{3x + 7}{x^2 + 6x + 8}\, dx$.

Solution. In this case the denominator factors

$$x^2 + 6x + 8 = (x + 2)(x + 4).$$

Now in V-27 we let $a = -2$, $b = -4$, $c = 3$, and $d = 7$, to obtain:

$$\int \frac{3x + 7}{x^2 + 6x + 8}\, dx = \frac{1}{2}(\ln|x + 2| - (-5)\ln|x + 4|) + C.$$

❑

Using **completing the square** to rewrite the quadratic in the form $w^2 + a^2$:

☐ **Example 10** Find $\displaystyle\int \frac{1}{x^2 + 6x + 14}\, dx$.

Solution. By completing the square, we get:

$$
\begin{aligned}
x^2 + 6x + 14 &= (x^2 + 6x + 9) - 9 + 14 \\
&= (x + 3)^2 + 5.
\end{aligned}
$$

Let $w = x + 3$. Then $dw = dx$ and so the substitution gives

$$
\begin{aligned}
\int \frac{1}{x^2 + 6x + 14}\, dx = \int \frac{1}{w^2 + 5}\, dw &= \frac{1}{\sqrt{5}} \arctan \frac{w}{\sqrt{5}} + C \\
&= \frac{1}{\sqrt{5}} \arctan \frac{x + 3}{\sqrt{5}} + C
\end{aligned}
$$

where the antidifferentiation uses V-24 with $a^2 = 5$.

❑

Further Examples: Transforming an Integrand to Fit the Table

Some antidifferentiation problems will arise in forms that look nearly the same as the formulas in our table, but many will not. In order to use the table, you'll often need to manipulate or reshape integrands to fit entries in the table. The kinds of manipulation that tend to be useful are expansion, factoring, long division, completion of the square, or substitution.

☐ **Example 11** Find $\displaystyle\int (x^3 + 5)^2 \, dx$.

Solution. Note that you can't use substitution here: letting $w = x^3 + 5$ doesn't work, since there is no $dw = 3x^2 \, dx$ in the integrand. What will work is expansion: simply multiply out the square: $(x^3 + 5)^2 = x^6 + 10x^3 + 25$. Then use I-1:

$$\int (x^3 + 5)^2 \, dx = \int x^6 \, dx + 10 \int x^3 \, dx + 25 \int 1 \, dx = \frac{1}{7}x^7 + 10 \cdot \frac{1}{4}x^4 + 25x + C.$$

☐

☐ **Example 12** Find $\displaystyle\int \frac{2x^2 - 3x + 7}{2x + 5} \, dx$.

Solution. You might be tempted to try to factor the numerator, or to complete the square. But a good rule of thumb when integrating a rational function whose numerator has a degree greater than or equal to that of the denominator is to do *long division*. This results in a polynomial plus a simpler rational function as a remainder.

If we perform long division, we get

$$\frac{2x^2 - 3x + 7}{2x + 5} = x - 4 + \frac{27}{2x + 5}.$$

Then use I-1 and I-2:

$$
\begin{aligned}
\int \frac{2x^2 - 3x + 7}{2x + 5} \, dx &= \int \left(x - 4 + \frac{27}{2x + 5} \right) dx \\
&= \frac{x^2}{2} - 4x + 27 \int \frac{dx}{2x + 5} \\
&= \frac{x^2}{2} - 4x + 27 \int \frac{1}{w} \left(\frac{1}{2} dw \right) \qquad \text{(where } w = 2x + 5) \\
&= \frac{x^2}{2} - 4x + \frac{27}{2} \ln|w| + C \\
&= \frac{x^2}{2} - 4x + \frac{27}{2} \ln|2x + 5| + C.
\end{aligned}
$$

☐

☐ **Example 13** Find $\displaystyle\int \frac{x^2}{x^2 + 4} \, dx$.

Solution. Since the degree of the numerator and denominator are equal, start with long division: $\dfrac{x^2}{x^2 + 4} = 1 - \dfrac{4}{x^2 + 4}$. Then by V-24 with $a = 2$ we get:

$$\int \frac{x^2}{x^2 + 4} \, dx = \int 1 \, dx - 4 \int \frac{1}{x^2 + 4} \, dx = x - 4 \cdot \frac{1}{2} \arctan \frac{x}{2} + C.$$

☐

☐ **Example 14** Find $\int e^t \sin(5t + 7)\, dt$.

Solution. This looks pretty similar to II-8. To make the correspondence more complete, let's try the substitution $w = 5t + 7$. Then $dw = 5\, dt$, so $dt = \frac{1}{5}\, dw$. Also, $t = \frac{(w-7)}{5}$. Then the integral becomes

$$\int e^t \sin(5t + 7)\, dt \;=\; \int e^{\frac{w-7}{5}} \sin w\, \frac{dw}{5}$$

$$=\; \frac{e^{-\frac{7}{5}}}{5} \int e^{\frac{w}{5}} \sin w\, dw \qquad \text{(since } e^{\frac{w-7}{5}} = e^{\frac{w}{5}} e^{-\frac{7}{5}} \text{ and } e^{-\frac{7}{5}} \text{ is a constant)}.$$

And now we can use II-8 with $a = (1/5)$ and $b = 1$ to write

$$\int e^{\frac{w}{5}} \sin w\, dw = \frac{1}{\left(\frac{1}{5}\right)^2 + 1^2} e^{\frac{w}{5}} \left(\frac{\sin w}{5} - \cos w \right) + C$$

so

$$\int e^t \sin(5t + 7)\, dt \;=\; \frac{e^{-\frac{7}{5}}}{5} \left(\frac{25}{26} e^{\frac{5t+7}{5}} \left(\frac{\sin(5t + 7)}{5} - \cos(5t + 7) \right) \right) + C$$

$$=\; \frac{5e^t}{26} \left(\frac{\sin(5t + 7)}{5} - \cos(5t + 7) \right) + C.$$

☐

❖ Exercises for Section 6.7

1. For each of the integrals in Problems 5–13, write down which formula in the table of integrals applies, but do not do the integration.

2. For each of the integrals in Problems 14–25, write down which formula in the table of integrals applies, and what substitution you would use (if one is necessary). Do not do the integration.

3. For each of the integrals in Problems 26–38, write down which formula in the table of integrals applies and what substitution you would use (if one is necessary). Do not do the integration.

4. For each of the integrals in Problems 39–50, write down which formula in the table of integrals applies, and what substitution you would use (if one is necessary). Do not do the integration.

For Problems 5–13, antidifferentiate:

5. $\displaystyle\int x^3 e^{2x}\, dx$

6. $\displaystyle\int \sin 3\theta \cos 5\theta\, d\theta$

7. $\displaystyle\int \sin 3\theta \sin 5\theta\, d\theta$

8. $\displaystyle\int e^{-3\theta} \cos \theta\, d\theta$

9. $\displaystyle\int x^5 \ln x\, dx$

10. $\displaystyle\int \sin w \cos^4 w\, dw$

11. $\displaystyle\int \frac{1}{\cos^3 x}\, dx$

12. $\displaystyle\int \sin^4 x\, dx$

13. $\displaystyle\int \frac{1}{3 + y^2}\, dy$

For Problems 14–25, antidifferentiate, using the table of integrals and a substitution if necessary.

14. $\displaystyle\int x^2 e^{3x}\,dx$

15. $\displaystyle\int x^4 e^{3x}\,dx$

16. $\displaystyle\int x^2 e^{x^3}\,dx$

17. $\displaystyle\int y^2 \sin 2y\,dy$

18. $\displaystyle\int \cos 2y \cos 7y\,dy$

19. $\displaystyle\int e^{5x} \sin 3x\,dx$

20. $\displaystyle\int x^3 \sin x^2\,dx$

21. $\displaystyle\int z e^{2z^2} \cos(2z^2)\,dz$

22. $\displaystyle\int u^5 \ln(5u)\,du$

23. $\displaystyle\int y^7 \ln(y^3)\,dy$

24. $\displaystyle\int \cos^4 3y\,dy$

25. $\displaystyle\int z \sin^3(z^2)\,dz$

For Problems 26–38, antidifferentiate, using the table of integrals and a substitution if necessary.

26. $\displaystyle\int \frac{1}{\sqrt{2-x^2}}\,dx$

27. $\displaystyle\int \frac{1}{\sqrt{1-9x^2}}\,dx$

28. $\displaystyle\int \frac{1}{1+4x^2}\,dx$

29. $\displaystyle\int \frac{1}{\sin^2 2\theta}\,d\theta$

30. $\displaystyle\int \frac{1}{\sin^3 3\theta}\,d\theta$

31. $\displaystyle\int \frac{1}{\cos^4 7x}\,dx$

32. $\displaystyle\int \frac{1}{\sqrt{9-4x^2}}\,dx$

33. $\displaystyle\int \frac{x}{\sqrt{9-4x^2}}\,dx$

34. $\displaystyle\int \cos^2 \theta \sin^2 \theta\,d\theta$

35. $\displaystyle\int \sin^3 3\theta \cos^2 3\theta\,d\theta$

36. $\displaystyle\int \tan^4 x\,dx$

37. $\displaystyle\int \frac{dz}{z(z-3)}$

38. $\displaystyle\int \frac{dy}{4-y^2}$

For Problems 39–50, antidifferentiate, using the table of integrals and a substitution if necessary.

39. $\displaystyle\int \frac{1}{1+(z+2)^2}\,dz$

40. $\displaystyle\int \frac{dy}{\sqrt{4+(1+y)^2}}$

41. $\displaystyle\int \frac{1}{\sqrt{2-(x+1)^2}}\,dx$

42. $\displaystyle\int \frac{1}{x^2+4x+3}\,dx$

43. $\displaystyle\int \frac{1}{x^2+4x+4}\,dx$

44. $\displaystyle\int \frac{1}{y^2+4y+5}\,dy$

45. $\displaystyle\int \frac{x^3+3}{x^2-3x+2}\,dx$

46. $\displaystyle\int \frac{x^2}{x^2+6x+13}\,dx$

47. $\displaystyle\int \sqrt{x^2+8x+7}\,dx$

48. $\displaystyle\int \frac{4y+9}{3y+y^2}\,dy$

49. $\displaystyle\int \frac{5z-13}{6-5z+z^2}\,dz$

50. $\displaystyle\int \frac{t^2+1}{t^2-1}\,dt$

In Problems 51–63, find the definite integrals by using antiderivatives and the Fundamental Theorem of Calculus. Check your answers numerically, using left- and right-hand sums.

51. $\displaystyle\int_0^2 \frac{1}{4+x^2}\,dx.$

52. $\displaystyle\int_\pi^{2\pi} (y^2+3)\cos 3y\,dy$

53. $\displaystyle\int_0^1 \frac{dx}{x^2+2x+5}$

54. $\displaystyle\int_{\pi/4}^{\pi/3} \frac{dx}{\sin^3 x}$

55. $\displaystyle\int_{-3}^{-1} \frac{dx}{\sqrt{x^2+6x+10}}$

56. $\displaystyle\int_0^3 \frac{1+x}{2+x^2}\,dx$

57. $\displaystyle\int_0^2 \cos\theta\sin 2\theta\,d\theta$

58. $\displaystyle\int_0^2 \sqrt{4-x^2}\,dx.$ [You can evaluate this by thinking about its geometric meaning.]

59. $\displaystyle\int_0^1 \sqrt{4-x^2}\,dx$

60. $\displaystyle\int_0^1 \sqrt{3-x^2}\,dx$

61. $\displaystyle\int_1^2 (x-2x^3)\ln x\,dx$

62. $\displaystyle\int_{-\pi}^{\pi} \sin 5x\cos 6x\,dx$

63. $\displaystyle\int_{-\pi}^{\pi} \sin 5x\cos 5x\,dx$

64. The voltage, V, in an electrical outlet in our homes is given as a function of time, t, by the function $V_0\cos(120\pi t)$ where V is in volts and t in seconds, and V_0 is a constant representing the maximum voltage.

   (a) What is the average value of the voltage over 1 second?

   (b) When we talk about 'average voltage,' we usually mean the 'root mean square' voltage, $\overline{V}$, defined by $\overline{V}=\sqrt{\text{average of }V^2}$. Find $\overline{V}$ in terms of V_0.

   (c) The standard voltage in an American house is 110 volts, meaning that $\overline{V}=110$. What is V_0?

65. An economist studying the rate of production, $R(t)$, of oil in a new oil well has proposed the following model:
$$R(t) = A + Be^{-t}\sin(2\pi t)$$
where t is the time in years, A is the average rate (a constant), and B is the 'variability' coefficient (a constant).

   (a) Find the total amount of oil produced in the first N years of operation. (Take N to be an integer.)

   (b) Find the average amount of oil produced per year over the first N years (N an integer).

   (c) From your answer to (b), find the average amount of oil produced per year as $N\to\infty$.

   (d) Looking at the function $R(t)$, explain how you might have predicted your answer to (c) without doing any calculations.

   (e) Do you think it is reasonable to expect this model to hold over a very long period? Why or why not?

66. (a) Show how you can use the fact that $\dfrac{1}{x^2 - x} = \dfrac{1}{x - 1} - \dfrac{1}{x}$ to find $\displaystyle\int \dfrac{1}{x^2 - x}\,dx$.

    (b) Show how your answer to part (a) agrees with the answer you get by using the integral tables.

67. (a) Show how combining terms in the expression $\dfrac{2}{x} + \dfrac{1}{x + 3}$ yields $\dfrac{3x + 6}{x^2 + 3x}$, and use this to evaluate $\displaystyle\int \dfrac{3x + 6}{x^2 + 3x}\,dx$.

    (b) Show how your answer to (a) agrees with the answer you get by using the integral tables.

68. Show that for all integers m and n, with $m \neq \pm n$, $\displaystyle\int_{-\pi}^{\pi} \sin m\theta \sin n\theta\,d\theta = 0$.

69. Show that for all integers m and n, with $m \neq \pm n$, $\displaystyle\int_{-\pi}^{\pi} \cos m\theta \cos n\theta\,d\theta = 0$.

6.8 Approximating Definite Integrals

The methods of the last few sections allow us to get 'exact' answers for definite integrals in a variety of special cases. In fact, most real-world applications of calculus don't require exact answers. If the problem is to predict the total amount of fuel used by the Space Shuttle, or the length of a steel cable supporting a bridge, we may only require a certain number of decimal places of accuracy. Even when we can use the Fundamental Theorem to find an 'exact' answer, the precision may be meaningless if the integrand is modeled on data that was measured in an inexact way.

We can approximate a definite integral numerically using left- and right-hand Riemann sums. The purpose of the next two sections is to introduce better methods for approximating definite integrals—better in the sense that they give more accurate results with less work than that required to find the left- and right-hand sums. The methods work so well that in most practical situations definite integrals are evaluated numerically, *instead* of by antidifferentiation and the Fundamental Theorem.

The General Riemann Sum

Remember, a left-hand Riemann sum of $f(x)$ over an interval $a \leq x \leq b$ with n subdivisions is a sum of terms like $f(z_i)\Delta x$, where z_i is chosen to be the left-hand point in each subinterval. The right-hand Riemann sum is the same except that we choose z_i to be the right-hand point. There is nothing special about the left or right endpoints of the subintervals—we can get a Riemann sum that approximates the definite integral by choosing *any* point z_i in each subinterval. This is called a *general Riemann sum* and written:

$$\sum_{i=1}^{n} f(z_i)\Delta x = f(z_1)\Delta x + \cdots + f(z_n)\Delta x,$$

where, as before, Δx is the length of an individual subinterval[3], so that $\Delta x = \dfrac{b-a}{n}$. Then z_1 is the point you chose from the first subinterval, z_2 is the point you chose from the second, and so on. The reason for considering Riemann sums other than left- and right-hand sums is that other choices of points within the subinterval can yield better approximation methods.

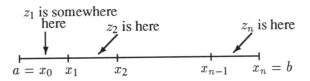

z_1 is somewhere here
z_2 is here
z_n is here

$a = x_0 \quad x_1 \quad\quad x_2 \quad\quad\quad\quad x_{n-1} \quad x_n = b$

The Midpoint Rule

An important example of this more general Riemann sum is when z_i is chosen to be the *midpoint* of each interval. For example, in approximating $\int_1^2 f(x)\,dx$ by a Riemann sum with two subdivisions, you could divide the interval $1 \le x \le 2$ into two pieces and then take $z_1 = 1.25$, the midpoint of the first piece, and $z_2 = 1.75$, the midpoint of the second piece (see Figure 6.17).

The Riemann sum is then

$$f(1.25)0.5 + f(1.75)0.5.$$

Figure 6.17 shows that evaluating f at the midpoint of each subdivision usually gives a better approximation to the area under the curve than evaluating f at either end. For this particular f, for instance, you can see that each rectangle both overshoots and undercuts the exact area, and the amounts of overshoot and undercut are nearly equal. Thus, the area of the rectangles in Figure 6.17 will be closer to the area under the curve than the left- or right-hand sums.

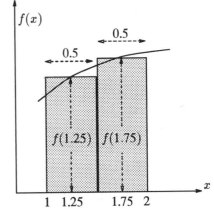

Figure 6.17: Midpoints with two subdivisions.

In fact, we shall see that this new midpoint Riemann sum *is* generally a better approximation to the definite integral than the left- or right-hand sum with the same number of subdivisions n.

---

[3]It is also possible to have unequal subdivisions, although we will not consider them.

So far we have three rules for choosing the points $z_1, \ldots, z_n$ in a Riemann sum:

1. use the left endpoint of each subinterval

2. use the right endpoint of each subinterval

3. use the midpoint of each subinterval.

These three methods are called, not surprisingly, the *left (rectangle) rule, right (rectangle) rule*, and the *midpoint (rectangle) rule*. We'll use $\text{LEFT}(n)$, $\text{RIGHT}(n)$ and $\text{MID}(n)$ to denote the results obtained by using these rules with n subdivisions.

☐ **Example 1** For $\int_1^2 \frac{1}{x}\, dx$, compute LEFT(2), RIGHT(2) and MID(2), and compare your answers with the true value of the integral.

Solution.

$$\text{LEFT}(2) = f(1)(0.5) + f(1.5)(0.5) = \frac{1}{1}(0.5) + \frac{1}{1.5}(0.5) = 0.8333\ldots$$

$$\text{RIGHT}(2) = f(1.5)(0.5) + f(2)(0.5) = \frac{1}{1.5}(0.5) + \frac{1}{2}(0.5) = 0.5833\ldots$$

$$\text{MID}(2) = f(1.25)(0.5) + f(1.75)(0.5) = \frac{1}{1.25}(0.5) + \frac{1}{1.75}(0.5) = 0.6857\ldots$$

All three Riemann sums in this example are approximating

$$\int_1^2 \frac{1}{x}\, dx = \ln x \Big|_1^2 = \ln 2 - \ln 1 = \ln 2 = 0.6931\ldots.$$

The left and right rules give quite bad approximations with only two subdivisions, but the midpoint rule is already fairly close to the true answer.

It is easy to see why the left and right rules are so bad; since $f(x) = \frac{1}{x}$ is decreasing from 1 to 2, the left rule overestimates on each subdivision, and the right rule underestimates (see Figure 6.18a). On the other hand, the midpoint rule approximates with a rectangle on each subdivision that is partly above and partly below the graph, so the errors tend to balance out (see Figure 6.18b).

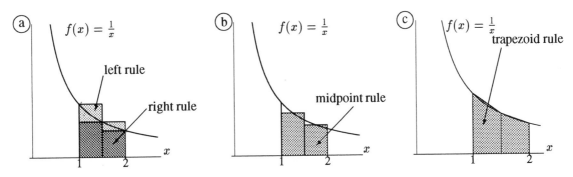

Figure 6.18: Left, right, midpoint and trapezoid rules for $\int_1^2 \frac{1}{x}\,dx$.

The Trapezoid Rule

We have just seen how the midpoint rule can have the effect of balancing out the errors of the left and right rules. There is another way of balancing these errors: why not average the left and right rules? This rule is called the *trapezoid rule*

$$\text{TRAP}(n) = \frac{\text{LEFT}(n) + \text{RIGHT}(n)}{2}.$$

The trapezoid rule averages the values of f at the left and right endpoints of each subinterval and multiplies by Δx. This is the same as approximating the area under the graph of f in each subinterval by a trapezoid (see Figure 6.19).

☐ **Example 2** For $\int_1^2 \frac{1}{x}\,dx$, compare the trapezoid rule with 2 subdivisions with the left, right, and midpoint rules.

Solution.

In the previous example we got LEFT(2)= 0.8333 ... and RIGHT(2) = 0.5833 ....
The trapezoid rule is the average of these, so TRAP(2)=0.7083 .... See Figure 6.18(c).
The exact value of the integral is 0.6931 ...,
so the trapezoid rule is better than the left or right rules. The midpoint rule is still the best, however, since MID(2)= 0.6857 .... ☐

Figure 6.19: The trapezoid rule: area of a trapezoid.

Is the Approximation an Over- or Underestimate?

It is useful to know when a rule is producing an overestimate and when it is producing an underestimate. In Chapter 3 we saw when f is increasing

$$\text{LEFT}(n) \leq \int_a^b f(x)\, dx \leq \text{RIGHT}(n),$$

when f is decreasing

$$\text{RIGHT}(n) \leq \int_a^b f(x)\, dx \leq \text{LEFT}(n).$$

The Trapezoid Rule: If the graph of the function is concave down, then each trapezoid in the trapezoid rule will lie below the graph, so the trapezoid rule will provide an underestimate, as in Figure 6.20(a). If the graph is concave up, the trapezoid rule will provide an overestimate, as in Figure 6.20(b).

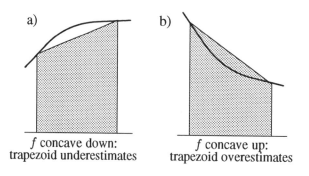

a) f concave down:
trapezoid underestimates

b) f concave up:
trapezoid overestimates

Figure 6.20: Error in the trapezoid rule.

The Midpoint Rule: To understand the relationship between the midpoint rule and concavity, take a rectangle whose top intersects the curve at the midpoint of the interval. In addition, draw a tangent to the curve at the midpoint; this gives a trapezoid. See Figure 6.21. (This is *not* the same trapezoid as in the trapezoid rule; this one does not touch the curve at the endpoints of the interval.) The midpoint rectangle and the new trapezoid have the same area, because the shaded triangles in Figure 6.21 are congruent and have the same area. Hence, if the graph of the function is concave down, the midpoint rule overestimates; if the graph is concave up, the midpoint rule underestimates (See Figure 6.22). Thus

if the graph of f is concave down

$$\text{TRAP}(n) \leq \int_a^b f(x)\, dx \leq \text{MID}(n).$$

if the graph of f is concave up

$$\text{MID}(n) \leq \int_a^b f(x)\, dx \leq \text{TRAP}(n).$$

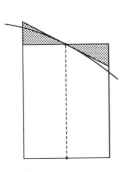

Figure 6.21: Midpoint rectangle and trapezoid with same area

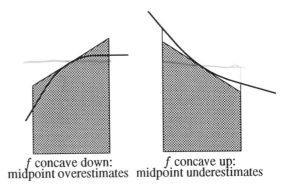

f concave down: midpoint overestimates f concave up: midpoint underestimates

Figure 6.22: Error in the midpoint rule.

How do we know if n is large enough?

Any of these rules can be made more accurate by taking n larger.[4] We usually start by taking larger and larger values of n until the decimals in the answer start to settle down; it is then likely (though there's no guarantee) that we're close to the true value. For example, since the last two values in Table 6.2 agree to 3 decimal places, it is probably safe to conclude that $\int_0^{2.5} \sin(t^2)\, dt = 0.43$ to two decimal places.

n	TRAP(n)	MID(n)
2	1.2292	0.0192
10	0.4572	0.4169
50	0.4316	0.4300
250	0.4306	0.4305

Table 6.2: Trapezoid and midpoint rules for $\int_0^{2.5} \sin(t^2)\, dt$.

If f is monotonic, or doesn't change concavity, we can be certain how close our estimates are to the true value. If f is monotonic over the range of integration, we know that the true value of the integral is trapped between LEFT(n) and RIGHT(n), so we can just increase n until LEFT(n) and RIGHT(n) are sufficiently close together. Similarly, if f doesn't change concavity, the true value of the integral will be trapped between MID(n) and TRAP(n), and we can just increase n until the difference between them is very small.

For a function that is not monotonic, or does change concavity, we can often split up the interval of intergration in such a way that the behavior does not change on each subinterval, as in the next example.

---

[4]Increasing n increases the accuracy of the approximation until n is so large that round-off error comes into play.

☐ **Example 3** Use the midpoint and trapezoid rules to find upper and lower estimates for the integral

$$\int_0^1 e^{-z^2}\, dz.$$

Solution.
The graph of $f(z) = e^{-z^2}$ appears to change concavity between 0 and 1. Differentiating f shows that

$$f'(z) = -2ze^{-z^2}$$
$$f''(z) = 2(2z^2 - 1)e^{-z^2}$$

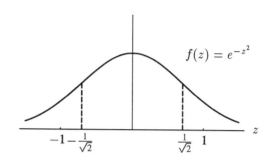

So the points of inflection are at $z = \pm\frac{1}{\sqrt{2}}$. Since the concavity changes between 0 and 1, we do not know whether the midpoint or the trapezoid rules give over- or underestimates for $\int_0^1 e^{-z^2}\, dz$. If, however, we split the interval at $z = \frac{1}{\sqrt{2}}$, we can be sure the midpoint rule overestimates the integral on $\left[0, \frac{1}{\sqrt{2}}\right]$ (where the integrand is concave down) and the trapezoid rule overestimates on $\left[\frac{1}{\sqrt{2}}, 1\right]$ (where the integrand is concave up), and vice versa for the underestimate. Using $n = 20$ for $\int_0^{\frac{1}{\sqrt{2}}} e^{-z^2}\, dz$ gives TRAP(20) $= 0.604928$ and MID(20) $= 0.605062$, so:

$$0.604928 < \int_0^{\frac{1}{\sqrt{2}}} e^{-z^2}\, dz < 0.605062.$$

Similarly for $\int_{\frac{1}{\sqrt{2}}}^1 e^{-z^2}\, dz$, we have TRAP(20) $= 0.141809$ and MID(20) $= 0.141805$:

$$0.141805 < \int_{\frac{1}{\sqrt{2}}}^1 e^{-z^2}\, dz < 0.141809.$$

Putting these two results together

$$0.604928 + 0.141805 < \int_0^{\frac{1}{\sqrt{2}}} e^{-z^2}\, dz + \int_{\frac{1}{\sqrt{2}}}^1 e^{-z^2}\, dz < 0.605062 + 0.141809$$

gives

$$0.746733 < \int_0^1 e^{-z^2}\, dz < 0.746871.$$

Thus we can say that $\int_0^1 e^{-x^2}\, dx = 0.747$ to 3 decimal places. If more accuracy is needed, we simply increase the value of n and repeat the calculation. ☐

❖ Exercises for Section 6.8

1. Table 6.3 contains approximations to $\int_0^4 \sqrt{100 + x^3} \, dx$. Fill in the rest of the table with values rounded to 4 digits to the right of the decimal point.

	$n = 1$	$n = 2$	$n = 4$
Left			
Right	51.2250		
Trap		43.5909	
Mid			

Table 6.3: Approximations of $\int_0^4 \sqrt{100 + x^3} \, dx$.
(for Problem 1)

Estimate the integrals in Problems 2–7 with $N = 10$, 100, and 1000 using left sums, right sums, trapezoids and midpoint sums. State whether each sum is an over- or under-estimate.

2. $\int_0^{\frac{1}{2}} (\sin \theta)^{\frac{3}{2}} \, d\theta$

3. $\int_1^2 \sqrt{x} e^x \, dx$

4. $\int_0^{\frac{\pi}{2}} \sqrt{3 + \cos \theta} \, d\theta$

5. $\int_0^1 \sin \frac{\theta^2}{2} \, d\theta$

6. $\int_0^{\frac{\pi}{2}} e^{-\sin x} \, dx$

7. $\int_0^3 \sqrt{1 + x^3} \, dx$

8. (a) Estimate $\int_0^1 \frac{1}{1 + x^2} \, dx$ by subdividing the interval into eight parts using

   i. the left-hand Riemann sum,

   ii. the right-hand Riemann sum,

   iii. the trapezoidal rule.

   (b) Since the exact value of the integral is $\pi/4$, you can estimate the value of π. Explain why your first estimate is too large, your second estimate is too small.

9. Consider the following integrals:

$$\text{i)} \int_1^{10} \ln x \, dx \qquad \text{ii)} \int_0^4 e^x \, dx$$

   (a) For each integral, find LEFT(32), RIGHT(32), and TRAP(32). Also find the exact value of each integral.

   (b) For each integral, arrange LEFT(32), RIGHT(32), TRAP(32) and the true value in ascending order. Explain, using diagrams, how you could predict this ordering without doing the calculations in part (a).

10. The graph of g is shown to the right. Listed below are the results from using the left, right, trapezoid and midpoint rules to approximate $\int_0^1 g(t)\,dt$, using the same number of subdivisions for each rule.

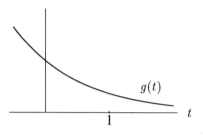

$$0.601, \quad 0.632, \quad 0.633, \quad 0.664$$

(a) Which approximation is which?

(b) Between which two approximations does the true value of the integral lie?

11. Given the velocity-time data to the right, estimate the total distance traveled from time $t = 0$ to time $t = 6$ using LEFT, RIGHT, and TRAP.

t	0	1	2	3	4	5	6
v	3	4	5	4	7	8	11

12. (a) Show that $\int_0^1 \sqrt{2 - x^2}\,dx = \dfrac{\pi}{4} + \dfrac{1}{2}$. [Hint: Break up the area under $y = \sqrt{2 - x^2}$ from $x = 0$ to $x = 1$ into two pieces: a sector of a circle and a right triangle.]

(b) Approximate $\int_0^1 \sqrt{2 - x^2}\,dx$ for $n = 5$ using the left, right, trapezoid and midpoint rules. Compute the error in each case using the answer to (a), and compare the errors.

13. (a) Find the exact value of $\int_0^{2\pi} \sin\theta\,d\theta$.

(b) Explain, using pictures, why the MID(1) and MID(2) approximations to this integral give the exact value.

(c) Does MID(3) give the exact value of this integral? How about MID(n)? Explain.

14. (a) Explain why the the area of the trapezoid in Fig. 6.23(a) is $h \cdot \dfrac{(l_1 + l_2)}{2}$.

(b) Let A be the area under the function sketched in Fig. 6.23(b). On a graph similar to Fig. 6.23(b), sketch areas representing each of the following quantities:

$$E = h \cdot f(0)$$

$$F = h \cdot f(h)$$

$$R = h \cdot f\left(\frac{h}{2}\right)$$

$$C = h \cdot \frac{f(0) + f(h)}{2} = \frac{E + F}{2}$$

$$N = \frac{h}{2} \cdot \frac{f(0) + f\left(\frac{h}{2}\right)}{2} + \frac{h}{2} \cdot \frac{f\left(\frac{h}{2}\right) + f(h)}{2} = \frac{R + C}{2}.$$

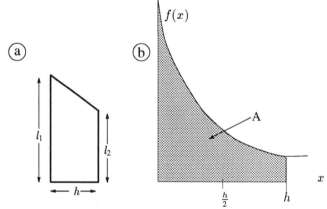

Figure 6.23: Approximate the shaded area A.

(c) Put the values A, E, F, R, C, and N in increasing order.

(d) Which is the better approximation to A: E or F?

(e) Which is the better approximation to A: R or C?

Problems 15–19 involve approximating $\int_a^b f(x)\,dx$.

15. Show that $\text{RIGHT}(n) = \text{LEFT}(n) + f(b)\Delta x - f(a)\Delta x$.

16. Show that $\text{TRAP}(n) = \text{LEFT}(n) + \frac{1}{2}\left(f(b) - f(a)\right)\Delta x$.

17. Show that $\text{LEFT}(2n) = \frac{1}{2}\left(\text{LEFT}(n) + \text{MID}(n)\right)$.

18. Using a computer or calculator, verify that the equations given in Problems 15 and 16 hold for $\int_1^2 \left(\frac{1}{x}\right)\,dx$, when $n = 10$.

19. Suppose that $a = 2$, $b = 5$, $f(2) = 13$, $f(5) = 21$ and that $\text{LEFT}(10) = 3.156$ and $\text{MID}(10) = 3.242$. Use the equations given in Problems 15–17 to compute $\text{RIGHT}(10)$, $\text{TRAP}(10)$, $\text{LEFT}(20)$, $\text{RIGHT}(20)$, and $\text{TRAP}(20)$.

20. The width, in feet, at various points along the fairway of a hole on a golf course is given in Figure 6.24. If one pound of fertilizer covers 200 square feet, estimate the number of pounds needed to fertilize the fairway.

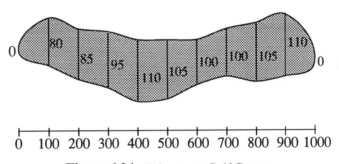

Figure 6.24: Fairway on Golf Course

(for Problem 20)

21. Estimate

$$\sum_{k=1}^{100,000} \frac{1}{k}$$

to the closest integer. [Hint: Work with left- or right-hand sums of the function $f(x) = 1/x$.]

6.9 Approximation Errors and Simpson's Rule

Whenever we compute an approximation, we are always concerned about the error, namely the difference between the exact answer and the approximation. We never know the exact error; if we did, we would also know the exact answer! What we want is some bound on the error and some idea of how much work would be involved in making the error smaller. The way to do this with definite integrals is quite practical: do the computation once for some moderate size of n, then do the computation again increasing n by a factor of 2 or 10 and see how many decimal places agree in the two answers. Continue increasing n until the desired decimal places "stabilize".

What is most interesting is that the errors are not random and follow a definite pattern as n increases. The errors for some methods are much smaller than those for others, and the errors for the midpoint and trapezoid methods are related to each other in a way that suggests an even better method called Simpson's rule.

We will work with the example $\int_1^2 \frac{1}{x}\, dx$, because we know the value of this integral to about 10 decimal places just by punching $\ln 2$ on a calculator. By studying the error in a case where we do know the exact answer, we will learn how the error behaves for other definite integrals where we may not know the answer exactly.

Error in Left and Right Rules

Let us see what happens to the error in the left and right rules as we increase n. We will increase n each time by a factor of 5 starting at $n = 2$. Since we already know what $\ln 2$ is (at least to ten digits), we can compute the error in approximating $\ln 2$ by the left- and right-hand Riemann sums for each n. The results are in Table 6.4. A negative error indicates the Riemann sum is less than $\ln 2$.

n	Error	
	Left rule	Right rule
2	0.1402	−0.1098
10	0.0256	−0.0244
50	0.0050	−0.0050
250	0.0010	−0.0010

Table 6.4: Errors for the left and right rule approximation to $\int_1^2 \frac{1}{x}\, dx = \ln 2 \approx 0.6931471806$.

One thing stands out: the errors for the left and right rules have opposite signs but are approximately equal in magnitude. This should come as no surprise. Since $f(x) = 1/x$ decreases as we move from left to right in the interval $[1, 2]$, the left rule will always overestimate and the right rule will always underestimate. Moreover, since the graph of f is almost a straight line in each subinterval, the left rule overestimates by about the same amount (geometrically a 'triangle' above the curve) as the right rule underestimates (a 'triangle' under the curve). See Figure 6.25. If we had not already decided to use the trapezoid rule, we might have been led to invent it by the observation that the errors in the left and right rule are approximately equal in magnitude and opposite in sign. The best way to try to get the errors to cancel is to average the left and right rule—this average is the trapezoidal rule.

There is another pattern to the errors in Table 6.4. If we compute the *ratio* of the errors when $n = 2$ versus $n = 10$, when $n = 10$ versus $n = 50$, and when $n = 50$ versus $n = 250$, as in Table 6.5, we see that the error, in both the left and right rules, decreases by a factor of about 5 as n increases by a factor of 5.

There is nothing special about 5; the same holds for any factor. In particular, to get 1 extra digit of accuracy in any calculation, we must make the error $\frac{1}{10}$ as big, so we must increase n by a factor of 10. Thus: *For the left or right rules, each extra digit of accuracy requires about 10 times the work.*

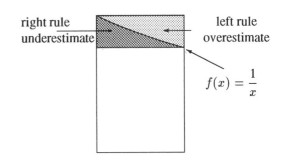

Figure 6.25: Errors in left and right sums

The calculator used to produce these tables took about 4 seconds to compute the left rule for $n = 50$, and this yields $\ln 2$ to two digits. To get three correct digits, n would need to be around 500 and the time would be 40 seconds. Four digits requires $n = 5000$ and 400 seconds.

Ten digits requires $n = 5 \times 10^9$ and 4×10^8 seconds, which is more than 12 years! The fastest computers would still take 4 seconds for ten digits, 40 seconds for eleven digits, 400 seconds for twelve digits, and 4×10^{10} seconds (or about 1200 years) for twenty digits. Clearly the error for the left and right rules does not decrease fast enough as n increases for practical computations accurate to more than a few digits.

	Ratio of Errors	
	Left rule	Right rule
2 vs. 10	5.47	4.51
10 vs. 50	5.10	4.90
50 vs. 250	5.02	4.98

Table 6.5: The ratio of the errors as n increases for $\int_1^2 \frac{1}{x}\, dx$.

Error in Trapezoid and Midpoint Rule

Now we will look at errors in the trapezoid and the midpoint rules. Table 6.6 shows the trapezoid and midpoint rules produce much better approximations to $\int_1^2 \frac{1}{x}\, dx$ than the left and right rules.

Again there is a pattern to the errors. The midpoint rule is noticeably better than the trapezoid rule; the error for the midpoint rule, in absolute value, seems to be about half the error of the trapezoid rule. To see why, remember from the last section that the trapezoid rule uses the line joining the endpoints of each subdivision to form the approximating trapezoid, whereas the midpoint rule uses a tangent line at the midpoint. Since the tangent line is the best linear approximation, we would expect the midpoint rule to do better. To visualize the difference, compare the shaded areas representing the errors of each rule in Figure 6.26. Also, notice in Table 6.6 that the errors for the two rules have opposite signs. This is due to the concavity of the curve: if the curve is concave down, the midpoint rule overestimates and the trapezoid rule underestimates, and vice versa if the curve is concave up. See Figure 6.26.

n	Error	
	Trapezoid rule	Midpoint rule
2	0.0152	−0.0074
10	0.00062	−0.00031
50	0.0000250	−0.0000125
250	0.0000010	−0.0000005

Table 6.6: The errors for the trapezoid and midpoint
rules for $\int_1^2 \frac{1}{x}\,dx$.

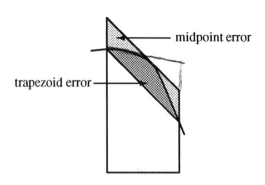

Figure 6.26: Errors in the midpoint and trapezoid
rules

In addition to comparing the trapezoid and midpoint rules with each other, we are interested in how the errors behave for each individual rule as n increases. Table 6.7, gives the ratios of the errors for each rule, as we did previously for the left and right rules.

	Ratio of Errors	
	Trapezoid rule	Midpoint rule
2 vs. 10	24.33	23.84
10 vs. 50	24.97	24.95
50 vs. 250	25.00	25.00

Table 6.7: The ratios of the errors as n increases for
$\int_1^2 \frac{1}{x}\,dx$.

For both rules we see that as n increases by a factor of 5, the error decreases by a factor of about $25 = 5^2$. In fact, it can be shown that this squaring relationship holds for any factor, so increasing n by a factor of 10 will decrease the error by a factor of about $100 = 10^2$. And reducing the error by a factor of 100 is equivalent to adding 2 more decimal places of accuracy to the result. In other words: *In the trapezoid or midpoint rules, each extra* two *digits of accuracy requires about 10 times the work.*

This result shows the advantage of the midpoint and trapezoid methods over the left and right methods: less additional work needs to be done to get another decimal place of accuracy. The calculator used to produce these tables took about 4 seconds again to compute the midpoint rule for $\int_1^2 \frac{1}{x}\,dx$ with $n = 50$ and this gets 4 digits correct. Thus to get 6 digits would take $n = 500$ and 40 seconds, to get 8 digits would take 400 seconds, and to get 10 digits would take 4000 seconds or about one hour. That is still not great, but it is certainly better than the 12 years required by the left or right rules.

Simpson's Rule

Still more improvement is possible. If you think about how the errors compare for the trapezoid and midpoint rules, you may agree that a combination of these two rules might be better yet. If the trapezoid error has the opposite sign and about twice the magnitude of the midpoint error, then we could try a weighted average of the two rules with the midpoint rule weighted twice the trapezoid rule. The result is called *Simpson's Rule*[5]:

$$\text{SIMP}(n) = \frac{2 \cdot \text{MID}(n) + \text{TRAP}(n)}{3}$$

In Table 6.8, we give the errors for Simpson's rule. Notice how much smaller the errors are than the previous errors. Of course, it is a little unfair to compare Simpson's rule at $n = 50$, say, with the previous rules because Simpson's rule must compute the value of f at both the midpoint and the endpoints of each subinterval, and hence involves evaluating the function at twice as many points.

We know by our previous analysis, however, that even if we did compute the other rules at $n = 100$ to compare with Simpson at $n = 50$, the other errors would only decrease by a factor of 2 for the left and right rules and by a factor of 4 for the trapezoid and midpoint rules.

Table 6.8 also gives the ratios of the errors. This time as n increases by a factor of 5, the

n	Error	Ratio
2	0.0001067877	550.15
10	0.0000001940	632.27
50	0.0000000003	

Table 6.8: The errors for Simpson's rule and the ratios of the errors.

errors decrease by a factor of about 600 or about 5^4. Again this behavior holds for any factor, so increasing n by a factor of 10 decreases the error by a factor of about 10^4. In other words: *In Simpson's Rule, each extra four digits of accuracy requires about 10 times the work.*

This is a great improvement over either the midpoint or trapezoid rules, which only give two extra digits of accuracy when n is increased by a factor of 10. Simpson's rule is so efficient that our benchmark problem of computing $\ln 2$ to ten digits almost falls by the wayside on the first whack. We get nine digits correct with $n = 50$ in about 8 seconds on our calculator. Doubling n will decrease the error by a factor of about $2^4 = 16$ and hence will get the tenth digit. The total time is 16 seconds, which is pretty good.

In general, Simpson's rule achieves a reasonable degree of accuracy when using relatively small values of n, and is a good choice for an all-purpose approximation method.

Alternate View of the Trapezoid and Simpson's Rule

Our approach to approximating $\int_a^b f(x)\,dx$ numerically has been empirical: try something, see how the error behaves, then try to improve it. We can also develop the various rules for numerical integration by making better and better approximations to the integrand, f. The left, right, and midpoint rectangular rules are all examples of approximating f by a constant (flat) function on each subinterval. The trapezoid rule is obtained by approximating f by a linear function on each subinterval. It turns out that Simpson's rule can, in the same spirit, be obtained by approximating f by a quadratic function on each subinterval. The details are given in Problems 20 and 21 on page 430.

How the Error Depends on the Integrand

There are other factors besides the size of n which affect the size of the error in each of the rules. Instead of looking at how the error behaves as we increase n, let's leave n fixed and imagine trying

---

[5]You should be aware that many books and computer programs use slightly different terminology for Simpson's rule; what we call $n = 50$, they call $n = 100$.

our approximation methods on different functions. We observe that the error in the left or right rule depends on how steeply the graph of f rises or falls. A steep curve makes the triangular regions missed by the left or right rectangles tall and hence large in area. Thus the error in the left or right rules depends on the size of the derivative of f (see Figure 6.27).

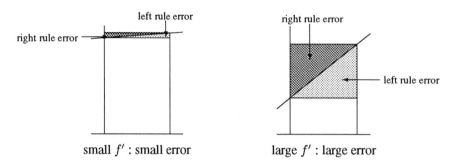

Figure 6.27: The error in the left and right rules depends on the steepness.

From Figure 6.28, we can see that the errors in the trapezoid and midpoint rules depend on how much the curve is bent up or down. In other words, the concavity, and hence the size of the second derivative of f, has an effect on the errors of these two rules. Finally, it can be shown that the error in Simpson's rule depends on the size of the *fourth* derivative of f, written $f^{(4)}$.

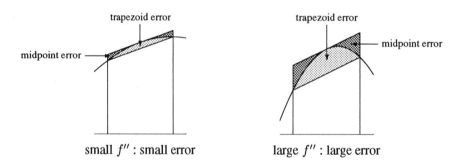

Figure 6.28: The error in the trapezoid and midpoint rules depends on how bent the curve is.

Summary of Errors

Left and Right Rules

- Errors are approximately proportional to $1/n$.

- For a given n, the errors for the left and right rules are approximately equal in absolute value and opposite in sign.

- For a given n, the size of the error depends on the size of f'.

Midpoint and Trapezoid Rules

- Errors are approximately proportional to $1/n^2$.

- For a given n, the midpoint error is about half the size of the trapezoid error and opposite in sign.

- For a given n, the size of the error depends on the size of f''.

Simpson's Rule

- Errors are approximately proportional to $1/n^4$.

- For a given n, the size of the error depends on the size of the fourth derivative, $f^{(4)}$.

Conclusion: The definite integrals $\int_a^b f(x)\,dx$ can be computed quickly and accurately in most cases using Simpson's rule. The only trouble occurs when f' or a higher derivative of f fails to exist or gets very large in the interval $a \leq x \leq b$.

❖ Exercises for Section 6.9

1. Compute SIMP(2), the value for Simpson's rule with $n = 2$, for Example 1, on page 416, and compare your value with the true value.

2. Approximate the value of $\displaystyle\int_1^2 \frac{1}{x}\,dx$ using Simpson's rule for $n = 10$, and compute the error.

For Problems 3-10, use Simpson's rule with various values of n to evaluate the definite integrals with error less than 0.001. Explain how you know you have reached a large enough value of n.

3. $\displaystyle\int_0^1 \frac{1}{4+x^2}\,dx$

7. $\displaystyle\int_0^1 \sqrt{1+x^4}\,dx$

4. $\displaystyle\int_0^2 e^{\sin t}\,dt$

8. $\displaystyle\int_0^{10} \ln(z^2+1)\,dz$

5. $\displaystyle\int_0^3 \sin^2 x\,dx$

9. $\displaystyle\int_0^1 \cos(t^2)\,dt$

6. $\displaystyle\int_0^3 \cos^2 \theta\,d\theta$

10. $\displaystyle\int_1^2 \frac{\sin x}{x}\,dx$

11. (a) What is the exact value of $\displaystyle\int_0^1 7x^6\,dx$?
    (b) Find LEFT(5), RIGHT(5), TRAP(5), MID(5), and SIMP(5), and compute the error for each.
    (c) Repeat (b) with $n = 10$ (instead of $n = 5$).
    (d) For each rule in (b), compute the ratio of the error for $n = 5$ divided by the error for $n = 10$. Are the values what you expect? Why?

12. Consider Simpson's rule approximations to $\displaystyle\int_0^2 (x^3 + 3x^2)\,dx$.

    (a) What is the exact value of this integral?
    (b) Find SIMP(n) for $n = 2, 4$, and 100. What do you notice?

13. (a) Suppose a certain computer takes 2 seconds to compute a certain definite integral accurate to 4 digits to the right of the decimal point, using the left rectangle rule. How long will it take to get 8 digits correct using the left rectangle rule? How about 12 digits? 20 digits (give answers in years)?

   (b) Same as (a) but this time assume it is the trapezoidal rule which is being used throughout.

14. Suppose that for a certain definite integral, LEFT(10) = 0.38745 and LEFT(20) = 0.36517. Estimate the actual error for LEFT(10) (and thereby the actual value of the integral) by assuming that the error is reduced by a factor of 2 in going from LEFT(10) to LEFT(20).

15. Suppose for a certain definite integral that MID(10) = 35.619 and MID(20) = 35.415. Estimate the actual error for MID(10) assuming that the error for MID(10) is reduced by a factor of 4 in going to MID(20).

Are the statements in Problems 16–19 true or false? Why?

16. The midpoint rule approximation to $\int_0^1 (y^2 - 1)\, dy$ is always smaller than the exact value of the integral.

17. The trapezoid rule approximation is never exact.

18. If LEFT(2) $< \int_a^b f(x)\, dx$, then LEFT(4) $< \int_a^b f(x)\, dx$.

19. If $0 < f' < g'$ everywhere, then the error in approximating $\int_a^b f(x)\, dx$ by LEFT(n) is less than the error in approximating $\int_a^b g(x)\, dx$ by LEFT(n).

20. Suppose that $a < b$ and that m is the midpoint $m = (a+b)/2$. Let $h = b - a$. We wish to show that if f is a quadratic polynomial, so $f(x) = Ax^2 + Bx + C$, then:

$$\int_a^b f(x)\, dx = \frac{h}{3}\left(\frac{f(a)}{2} + 2f(m) + \frac{f(b)}{2}\right).$$

   (a) Prove that this equation holds for the functions $f(x) = 1$, $f(x) = x$, and $f(x) = x^2$.

   (b) Use (a) and the "Facts about Sums and Constant Multiples of the Integrand" from Section 6.2, page 369, to prove that the equation holds for any quadratic function f.

21. Here is a scheme for approximating $\int_a^b f(x)dx$. Divide up the interval $a \leq x \leq b$ into n equal subintervals. On each subinterval approximate f by a quadratic function that agrees with f at both endpoints and at the midpoint of the subinterval.

    (a) Explain why the integral of f on the subinterval $[x_i, x_{i+1}]$ is approximately equal to the expression

    $$\frac{h}{3}\left(\frac{f(x_i)}{2} + 2f(m_i) + \frac{f(x_{i+1})}{2}\right),$$

    where m_i is the midpoint of the subinterval, $m_i = (x_i + x_{i+1})/2$. See Problem 20.

    (b) Show that if we add up these approximations for each subinterval, we get Simpson's rule:

    $$\int_a^b f(x)dx \approx \frac{2 \cdot \text{MID}(n) + \text{TRAP}(n)}{3}.$$

6.10   Improper Integrals

Our original definition of the definite integral $\int_a^b f(x)\,dx$ assumed that the interval $a \leq x \leq b$ was of finite size and that f was continuous, except perhaps at a few points, and bounded everywhere. Integrals that arise in applications don't necessarily have these nice properties. In this section, we investigate a class of integrals, called *improper* integrals, in which one limit of integration is infinite or the integrand is unbounded. We will usually consider those with positive integrands since they are the most common. As an example, to estimate the mass of the Earth's atmosphere, you might calculate an integral which sums the mass of the air up to different heights. In order to represent the fact that the atmosphere doesn't end at a specific height, you can let the upper limit of integration get larger and larger, or tend to infinity.

One Type of Improper Integral: When the Limit of Integration is Infinite

Let's examine an example of an improper integral: $\int_1^\infty \frac{1}{x^2}\,dx$. This notation is shorthand for the idea of computing a definite integral over $1 \leq x \leq b$ and then taking the limit as b approaches infinity. We simply compute the definite integral $\int_1^b \frac{1}{x^2}\,dx$ and let $b \to \infty$:

$$
\begin{aligned}
\int_1^\infty \frac{1}{x^2}\,dx &= \lim_{b\to\infty} \int_1^b \frac{1}{x^2}\,dx \\
&= \lim_{b\to\infty} \left. -x^{-1} \right|_1^b \\
&= \lim_{b\to\infty} \left(-\frac{1}{b} + \frac{1}{1}\right) = 1.
\end{aligned}
$$

If we think of this in terms of areas, it may seem strange that the region whose area is computed by $\int_1^\infty \dfrac{1}{x^2}\,dx$ extends from $x = 1$ infinitely far to the right. How could it have finite area? See Figure 6.29(a). Well, what our limit computations above are saying is that

$$\text{when } b = 10: \quad \int_1^{10} \frac{1}{x^2}\,dx \;=\; -\frac{1}{x}\Big|_1^{10} = -\frac{1}{10} + 1 = 0.9$$

$$\text{when } b = 100: \quad \int_1^{100} \frac{1}{x^2}\,dx \;=\; -\frac{1}{100} + 1 = 0.99$$

$$\text{when } b = 1000: \quad \int_1^{1000} \frac{1}{x^2}\,dx \;=\; -\frac{1}{1000} + 1 = 0.999$$

and so on. In other words, as b gets larger and larger, the area between $x = 1$ and $x = b$ tends to 1. See Figure 6.29(b). Thus, it does make sense to declare that $\int_1^\infty \dfrac{1}{x^2}\,dx = 1$. We say that $\int_1^\infty \dfrac{1}{x^2}\,dx$ *converges* to 1.

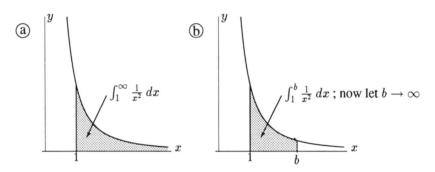

Figure 6.29: Area Representation of Improper Integral

Of course, in another example, we might not get a finite limit as b gets larger and larger. In this case we say the improper integral grows without bound or *diverges*.

If $\displaystyle\lim_{b\to\infty} \int_a^b f(x)\,dx$ is a finite number, we say that $\displaystyle\int_a^\infty f(x)\,dx$ **converges** and define

$$\int_a^\infty f(x)\,dx = \lim_{b\to\infty} \int_a^b f(x)\,dx.$$

Otherwise, we say that $\displaystyle\int_a^\infty f(x)\,dx$ **diverges**. We define $\displaystyle\int_{-\infty}^b f(x)\,dx$ similarly.

☐ **Example 1** Decide whether the improper integral $\displaystyle\int_1^\infty \frac{1}{\sqrt{x}}\,dx$ converges or not.

Solution. We have

$$\int_1^\infty \frac{1}{\sqrt{x}}\,dx = \lim_{b\to\infty}\int_1^b x^{-\frac{1}{2}}\,dx$$

$$= \lim_{b\to\infty} 2x^{\frac{1}{2}}\Big|_1^b$$

$$= \lim_{b\to\infty}(2b^{\frac{1}{2}} - 2).$$

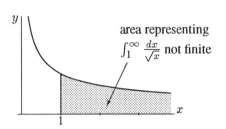

area representing $\int_1^\infty \frac{dx}{\sqrt{x}}$ not finite

Figure 6.30: $\int_1^\infty \frac{1}{\sqrt{x}}\,dx$ diverges

Clearly, there is no finite limit as $b \to \infty$. Thus we say the integral $\int_1^\infty \frac{1}{\sqrt{x}}\,dx$ grows without bound or *diverges*. We have shown that the area in Figure 6.30 is not finite. ❑

What is the difference between the functions $\frac{1}{x^2}$ and $\frac{1}{\sqrt{x}}$ that makes the area under the graph of $\frac{1}{x^2}$ approach 1 while the area under $\frac{1}{\sqrt{x}}$ grows very large? Both functions approach 0 as x grows, so as b grows larger smaller bits of area are being added to the definite integral. The difference between the functions is subtle: the values of the function $\frac{1}{\sqrt{x}}$ *don't shrink fast enough* for the integral to have a finite value. Of the two functions, $\frac{1}{x^2}$ drops to 0 much faster than does $\frac{1}{\sqrt{x}}$, and this feature keeps the area under $\frac{1}{x^2}$ from growing beyond 1.

❑ **Example 2** Find $\int_0^\infty e^{-5x}\,dx$.

Solution.

$$\int_0^\infty e^{-5x}\,dx = \lim_{b\to\infty}\int_0^b e^{-5x}\,dx = \lim_{b\to\infty} -\frac{1}{5}e^{-5x}\Big|_0^b = \lim_{b\to\infty}\left(-\frac{1}{5}e^{-5b} + \frac{1}{5}\right)$$

Since $e^{-5b} = \frac{1}{e^{5b}}$, it tends to 0 as b approaches infinity, and therefore

$$\int_0^\infty e^{-5x}\,dx = 0 + \frac{1}{5} = \frac{1}{5}.$$

Since e^{5x} grows very rapidly, we expect that e^{-5x} will approach 0 rapidly. The fact that the area approaches $\frac{1}{5}$ instead of growing without bound is a consequence of the speed with which e^{-5x} approaches 0. ❑

❑ **Example 3** Determine for which values of the exponent, p, the improper integral $\int_1^\infty \frac{1}{x^p}\,dx$ diverges.

Solution. For $p \neq 1$,

$$\int_1^\infty \frac{1}{x^p}\,dx = \lim_{b\to\infty}\int_1^b x^{-p}\,dx = \lim_{b\to\infty}\frac{1}{-p+1}x^{-p+1}\Big|_1^b = \lim_{b\to\infty}\left(\frac{1}{-p+1}b^{-p+1} - \frac{1}{-p+1}\right).$$

The important question is whether the exponent of b is positive or negative. If it is negative, as b approaches infinity, b^{-p+1} approaches 0. If the exponent is positive, then b^{-p+1} grows without bound as b approaches infinity. Thus the integral converges if $-p + 1 < 0$, or $p > 1$. It diverges if $p < 1$.

What happens if $p = 1$? In this case, we get

$$\int_1^\infty \frac{1}{x}\, dx = \lim_{b \to \infty} \ln x \Big|_1^b = \lim_{b \to \infty} \ln b - \ln 1.$$

Since $\ln b$ becomes arbitrarily large as b approaches infinity, the integral grows without bound. We conclude that $\int_1^\infty \frac{1}{x^p}\, dx$ diverges precisely when $p \leq 1$. For $p > 1$ the integral has value

$$-\left(\frac{1}{-p+1}\right) = \frac{1}{p-1}.$$ $\square$

Application to Work

If the force acting on an object at a distance r from the origin is given by $F(r)$, then we define the *work* done against the force to move the object from $r = a$ to $r = b$ by

$$W = \int_a^b F(r)\, dr$$

(Note: This assumes the force and direction of motion are parallel.) In the next example, we calculate the work done against an electric force. Chapter 7, page 489, looks at the concept of work in more detail.) The electric force, F, between two charged particles is given by *Coulomb's Law*:

$$F = k\frac{q_1 q_2}{r^2}$$

where q_1 and q_2 are the magnitudes of the charges of the particles, r is their distance apart, and k is a constant whose value is 9×10^9 if q_1 and q_2 are in coulombs, r is in meters and F is in newtons. The force is attractive if the charges are opposite, and repulsive if the charges have the same sign.

◻ **Example 4** A hydrogen atom consists of a proton and an electron, with opposite charges of magnitude 1.6×10^{-19} coulombs. Find the work which must be done to take a hydrogen atom apart (i.e. to move the electron from its orbit to an infinite distance from the proton). Assume that the initial distance between the electron and the proton is the Bohr radius, $R_B = 5.3 \times 10^{-11}$ meter.

Solution. Since $F = k\frac{q_1 q_2}{r^2}$ and $W = \int_a^b F(r)\, dr$, and we are moving from $r = R_B$ to $r = \infty$:

$$\begin{aligned} W &= \int_{R_B}^\infty k\frac{q_1 q_2}{r^2}\, dr = kq_1 q_2 \lim_{b \to \infty} \int_{R_B}^b \frac{1}{r^2}\, dr \\ &= kq_1 q_2 \lim_{b \to \infty} -\frac{1}{r}\Big|_{R_B}^b = kq_1 q_2 \lim_{b \to \infty} \left(-\frac{1}{b} + \frac{1}{R_B}\right) = \frac{kq_1 q_2}{R_B}. \end{aligned}$$

With the units we are using, W is measured in joules, and therefore

$$W = \frac{(9 \times 10^9)(1.6 \times 10^{-19})^2}{5.3 \times 10^{-11}} \approx 4.34 \times 10^{-18} \text{ joules.}$$

This is the amount of work done in lifting a speck of dust 0.000000025 inch off the ground. (In other words, not much!) For comparison, the work required to bring two 1 coulomb charges of the same sign from infinity to within 1 meter of one another is about equal the work needed to lift 1 million elephants 6 inches off the ground. ❑

What happens if the limits of integration are $-\infty$ and ∞? In this case, we break the integral at any point and write the original integral as a sum of two new improper integrals. In other words:

We can use any (finite) number c to define

$$\int_{-\infty}^{\infty} f(x)\,dx = \int_{-\infty}^{c} f(x)\,dx + \int_{c}^{\infty} f(x)\,dx.$$

If *either* of the two new improper integrals diverges, we say the original integral diverges. Only if both of the new integrals have a finite value do we add the values to get a finite value for the original integral.

It is not hard to show that the above definition does not depend on the choice for c. Can you do it?

Another Type of Improper Integral: When the Integrand becomes Infinite

There is another way for an integral to be improper. The interval may be finite but the function may be unbounded at some points in the interval. For example, consider $\int_{0}^{1} \frac{1}{\sqrt{x}}\,dx$. Since the graph of $y = \frac{1}{\sqrt{x}}$ has a vertical asymptote at $x = 0$, the region between the graph, the x-axis, and the lines $x = 0$ and $x = 1$ is unbounded. Instead of extending to infinity in the horizontal direction as in the previous improper integrals, this region extends to infinity in the vertical direction. See Figure 6.31(a). We can handle this improper integral in the same way as before: we compute $\int_{a}^{1} \frac{1}{\sqrt{x}}\,dx$ for values of a slightly larger than 0 and look at what happens as a approaches 0 from the positive side. (This is written as $a \to 0^{+}$.)

$$\int_{0}^{1} \frac{1}{\sqrt{x}}\,dx = \lim_{a \to 0^+} \int_{a}^{1} \frac{1}{\sqrt{x}}\,dx = \lim_{a \to 0^+} 2x^{\frac{1}{2}}\Big|_{a}^{1} = \lim_{a \to 0^+} (2 - 2a^{\frac{1}{2}}) = 2.$$

Geometrically, what we have done is to calculate the finite area between $x = a$ and $x = 1$ and take the limit as a tends to 0 from the right. See Figure 6.31(b). Since the limit exists, we say that the integral *converges* to 2. If the limit did not exist, we would say the improper integral *diverges*.

❑ **Example 5** Investigate the convergence of $\int_{0}^{2} \frac{1}{(x-2)^2}\,dx$.

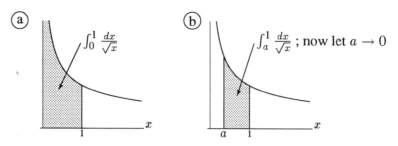

Figure 6.31: Area Representation of Improper Integral

Solution. This is an improper integral since the integrand tends to infinity as x approaches 2, and is undefined at $x = 2$. Since the trouble is at the right endpoint, we replace the upper limit by b, and let b tend to 2 from the left. This is written $b \to 2^-$, with the "$-$" signifying that 2 is approached from below. See Figure 6.32.

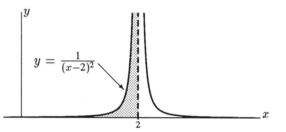

Figure 6.32: Graph of $\int_0^2 \frac{1}{(x-2)^2}\, dx$

$$\int_0^2 \frac{1}{(x-2)^2}\, dx = \lim_{b\to 2^-} \int_0^b \frac{1}{(x-2)^2}\, dx = \lim_{b\to 2^-} (-1)(x-2)^{-1}\Big|_0^b = \lim_{b\to 2^-} \left(-\frac{1}{(b-2)} - \frac{1}{2}\right).$$

Therefore, since $\lim_{b\to 2^-} \left(-\frac{1}{b-2}\right)$ does not exist, the integral diverges. $\qquad\square$

Suppose $f(x)$ tends to infinity as x approaches b.
If $\lim_{c\to b^-} \int_a^c f(x)\, dx$ is a finite number, we say that $\int_a^b f(x)\, dx$ **converges** and define

$$\int_a^b f(x)\, dx = \lim_{c\to b^-} \int_a^c f(x)\, dx$$

Otherwise, we say that $\int_a^b f(x)\, dx$ **diverges**.

It can also be the case that an integral is improper because the integrand tends to infinity *inside* the interval of integration rather than at the endpoints. In this case, we break the given improper integral into two (or more) improper integrals so that the integrand tends to infinity only at endpoints. In other words,

If $f(x)$ tends to infinity as x approaches some point c in the interval $[a, b]$, then we define

$$\int_a^b f(x)\,dx = \int_a^c f(x)\,dx + \int_c^b f(x)\,dx.$$

If *either* of the two new improper integrals diverges, we say the original integral diverges. Only if *both* of the new integrals have a finite value do we add the values to get a finite value for the original integral.

☐ **Example 6** Investigate the convergence of $\int_{-1}^2 \frac{1}{x^3}\,dx$.

Solution. See the graph in Figure 6.33. The trouble spot is $x = 0$, rather than $x = -1$ or $x = 2$. To handle this situation, we break the given improper integral into two other improper integrals which have $x = 0$ as one of the endpoints:

$$\int_{-1}^2 \frac{1}{x^3}\,dx = \int_{-1}^0 \frac{1}{x^3}\,dx + \int_0^2 \frac{1}{x^3}\,dx.$$

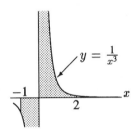

Figure 6.33: Graph of $\int_{-1}^2 \frac{1}{x^3}\,dx$

We can now use the previous technique to evaluate the new integrals, if they converge. In this example, both of the new integrals diverge:

$$\int_0^2 \frac{1}{x^3}\,dx = \lim_{a\to 0+} \left. -\frac{1}{2}x^{-2}\right|_a^2 = \lim_{a\to 0+} \frac{1}{2}\frac{1}{a^2} - \frac{1}{8}$$

so the limit does not exist. A similar computation shows that $\int_{-1}^0 \frac{1}{x^3}\,dx$ also diverges. Thus the original integral diverges.

It is easy to miss an integral which is improper because the integrand tends to infinity inside the interval. For example, it is seriously incorrect to say that $\int_{-1}^2 \frac{1}{x^3}\,dx = \left. -\frac{1}{2}x^{-2}\right|_{-1}^2 = -\frac{1}{8} + \frac{1}{2} = \frac{3}{8}$.

☐

☐ **Example 7** Find $\int_0^6 \frac{1}{(x-4)^{\frac{2}{3}}}\,dx.$

Solution. Figure 6.34 shows that the trouble spot is at $x = 4$, so we break the integral at $x = 4$ and consider the separate parts.

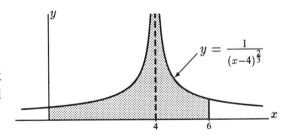

Figure 6.34: Graph of $\int_0^6 \frac{1}{(x-4)^{\frac{2}{3}}}\,dx$

$$\int_0^4 \frac{1}{(x-4)^{\frac{2}{3}}}\, dx = \lim_{b \to 4^-} 3(x-4)^{\frac{1}{3}}\Big|_0^b = \lim_{b \to 4^-}\left(3(b-4)^{\frac{1}{3}} - 3(-4)^{\frac{1}{3}}\right) = 3(4)^{\frac{1}{3}}.$$

Similarly,

$$\int_4^6 \frac{1}{(x-4)^{\frac{2}{3}}}\, dx = \lim_{a \to 4^+} 3(x-4)^{\frac{1}{3}}\Big|_a^6 = \lim_{a \to 4^+}\left(3 \cdot 2^{\frac{1}{3}} - 3(a-4)^{\frac{1}{3}}\right) = 3(2)^{\frac{1}{3}}.$$

Since both of these integrals converge, the original integral converges:

$$\int_0^6 \frac{1}{(x-4)^{\frac{2}{3}}}\, dx = 3(4)^{\frac{1}{3}} + 3(2)^{\frac{1}{3}} \approx 8.54.$$

❏

Finally, there is a question of what to do when an integral is improper at both endpoints. In this case, we just break the integral at any interior point of the interval. The original integral diverges if either or both of the new integrals diverge.

❑ **Example 8** Investigate $\int_0^\infty \frac{1}{x^2}\, dx$.

Solution. This integral is improper both because the upper limit is ∞ and because the function is undefined at $x = 0$. We break the integral into two parts at, say, $x = 1$. We know by Example 3 that $\int_1^\infty \frac{1}{x^2}\, dx$ has a finite value, but the other part $\int_0^1 \frac{1}{x^2}\, dx$ diverges since:

$$\int_0^1 \frac{1}{x^2}\, dx = \lim_{a \to 0^+} -x^{-1}\Big|_a^1 = \lim_{a \to 0^+}\left(\frac{1}{a} - 1\right).$$

Therefore $\int_0^\infty \frac{1}{x^2}\, dx$ diverges as well.

❏

❖ Exercises for Section 6.10

Calculate the values of the integrals in Problems 1–24, if they converge:

1. $\int_1^\infty e^{-2x}\, dx$

2. $\int_1^\infty \frac{x}{4+x^2}\, dx$

3. $\int_0^\infty \frac{x}{e^x}\, dx$

4. $\int_{-\infty}^0 \frac{e^x}{1+e^x}\, dx$

5. $\int_\pi^\infty \sin y\, dy$

6. $\int_{-\infty}^\infty \frac{dz}{z^2+25}$

7. $\int_{\frac{\pi}{4}}^{\frac{\pi}{2}} \frac{\sin x}{\sqrt{\cos x}}\, dx$

8. $\int_0^4 \frac{dx}{\sqrt{16-x^2}}$

9. $\int_{-1}^1 \frac{1}{v}\, dv$

10. $\int_1^\infty \frac{1}{x^2+1}\, dx$

11. $\int_1^\infty \frac{1}{\sqrt{x^2+1}}\, dx$

12. $\int_0^1 \frac{x^4+1}{x}\, dx$

13. $\int_1^\infty \dfrac{y}{y^4+1}\,dy$

14. $\int_{16}^{20} \dfrac{1}{y^2-16}\,dy$

15. $\int_0^4 \dfrac{1}{u^2-16}\,du$

16. $\int_0^1 \dfrac{\ln x}{x}\,dx$

17. $\int_2^\infty \dfrac{dx}{x\ln x}$

18. $\int_0^2 \dfrac{1}{\sqrt{4-x^2}}\,dx$

19. $\int_0^\pi \dfrac{1}{\sqrt{x}}e^{-\sqrt{x}}\,dx$

20. $\int_3^\infty \dfrac{dx}{x(\ln x)^2}$

21. $\int_1^2 \dfrac{dx}{x\ln x}$

22. $\int_7^\infty \dfrac{dy}{\sqrt{y-5}}$

23. $\int_4^\infty \dfrac{dx}{x^2-1}$

24. $\int_4^\infty \dfrac{dx}{(x-1)^2}$

25. Find the area under the curve $y = \dfrac{1}{\cos^2 t}$ between $t = 0$ and $t = \frac{\pi}{2}$.

26. Suppose a function h is defined by $h(x) = \frac{1}{x\sqrt{x}} - \frac{1}{16}, 0 < x \le 4$, and $h(x) = \frac{1}{x^2}, x > 4$;

   (a) Evaluate $\int_0^\infty h(x)\,dx$.

   (b) Is h differentiable at $x = 4$? If not, why not? If so, find $h'(4)$.

27. (a) Find the indefinite integral $\displaystyle\int \dfrac{dz}{z^2-z}$.

   (b) Sketch a graph of the integrand function $f(z) = \dfrac{1}{z^2-z}$. For what value(s) of z is $f(z)$ undefined?

   (c) Find $\displaystyle\int_3^\infty \dfrac{dz}{z^2-z}$, if it exists.

   (d) Find $\displaystyle\int_1^3 \dfrac{dz}{z^2-z}$, if it exists.

   (e) Find a formula for $\displaystyle\int_a^x \dfrac{dz}{z^2-z}$, and explain for what values of a and x the formula holds.

28. The gamma function is defined for all $x > 0$ by the rule

$$\Gamma(x) = \int_0^\infty t^{x-1}e^{-t}\,dt.$$

   (a) Find $\Gamma(1)$ and $\Gamma(2)$.

   (b) Integrate by parts with respect to t to show that, for positive n,

$$\Gamma(n+1) = n\Gamma(n).$$

   (c) Find a simple expression for $\Gamma(n)$ for positive integers n.

29. (a) For what values of p does the integral

$$\int_0^e x^p \ln x \, dx$$

converge? What is the value of the integral when it converges?

(b) For what value of p does the integral

$$\int_e^\infty x^p \ln x \, dx$$

converge? What is the value of the integral when it converges?

For Problems 30–32, use the definition of work above Example 4 on page 434.

30. Find the work done in separating opposite electric charges of magnitude 1 coulomb. Assume the charges are initially 1 meter apart and that one is moved infinitely far away from the other.

31. Suppose two electrons, initially 1 meter apart, are moved closer together.

(a) Write an integral representing the work done if the charges are moved to a small distance a apart.

(b) What happens as a tends to zero?

32. The force acting on a rocket ship due to the Earth's gravity is $F(r) = 16,000r^{-2}$ lbs. when it is r thousand miles above the center of the Earth.

(a) Write a definite integral for the amount of work that must be done for the rocket to "escape" the Earth's gravity, i.e., to move from the surface of the Earth (which is at $r = 4$) to $r = \infty$.

(b) Evaluate the integral. (This quantity is the *gravitational potential energy* of the rocket.)

33. The rate, r, at which people get sick during an epidemic of the flu can be approximated by

$$r = 1000te^{-0.5t}$$

where r is measured in people/day and t is measured in days since the start of the epidemic.

(a) Sketch a graph of r as a function of t.

(b) When are people getting sick fastest?

(c) How many people get sick altogether?

6.11   More on Improper Integrals

Making Comparisons

Sometimes it is too difficult to find an exact value of an improper integral by antidifferentiation. Still, it may be possible to get an estimate of the value of the integral, or to determine that it grows without bound. The key is to *compare* the given integral to one whose behavior you already know. Let's look at an example.

☐ **Example 1** Determine whether $\int_1^\infty \frac{1}{\sqrt{x^3+5}}\,dx$ converges or not.

Solution. First, let's see what this integrand does as $x \to \infty$. For large x, the 5 becomes insignificant besides the x^3, so

$$\frac{1}{\sqrt{x^3+5}} \approx \frac{1}{\sqrt{x^3}} = \frac{1}{x^{3/2}}.$$

Now, $\int_1^\infty \frac{1}{x^{3/2}}\,dx$ converges, by Example 3 on page 433, so we expect our integral to converge too. In order to confirm this, we observe that $x^3 \leq x^3 + 5$, so

$$\frac{1}{\sqrt{x^3+5}} \leq \frac{1}{\sqrt{x^3}}.$$

Thus

$$\int_1^b \frac{1}{\sqrt{x^3+5}}\,dx \leq \int_1^b \frac{1}{\sqrt{x^3}}\,dx.$$

See Figure 6.35.

The values of both integrals increase as b approaches infinity, since both integrands are positive. We know that the right-hand integral converges, using $p = 3/2$ in the result of Example 3, page 433:

$$\int_1^\infty \frac{1}{\sqrt{x^3}}\,dx = \int_1^\infty \frac{1}{x^{\frac{3}{2}}}\,dx = \frac{1}{\frac{3}{2}-1} = 2.$$

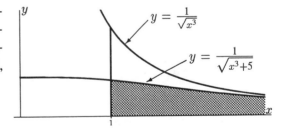

Figure 6.35: Graph showing $\int_1^\infty \frac{1}{\sqrt{x^3+5}}\,dx \leq \int_1^\infty \frac{dx}{\sqrt{x^3}}$

Since $\int_1^b \frac{1}{\sqrt{x^3+5}}\,dx$ increases as b approaches infinity but is always smaller than $\int_1^\infty \frac{1}{x^{3/2}}\,dx = 2$, we know $\int_1^\infty \frac{1}{\sqrt{x^3+5}}\,dx$ must have a finite value ≤ 2. Thus $\int_1^\infty \frac{dx}{\sqrt{x^3+5}}$ converges, and it converges to a value less than or equal to 2. ☐

Notice that we first looked at the behavior of the integrand as $x \to \infty$. This is useful because the convergence or divergence of the integral is determined by what happens as $x \to \infty$. The integrand can do any wild thing it likes for x in a finite interval (except perhaps be unbounded) without affecting the convergence of the integral. In general, we have:

The Comparison Test for $\displaystyle\int_a^\infty f(x)\,dx.$

Making a comparison usually involves two stages:

- guessing, by looking at the behavior of the integrand for large x, whether the integral converges or not. (This is the "behaves like" principle.)

- confirming the guess by comparison, usually with a power function $(g(x) = \frac{1}{x^p})$ or an exponential $(g(x) = e^{-ax},\ a > 0)$.

☐ **Example 2** Decide whether $\displaystyle\int_4^\infty \frac{dt}{(\ln t) - 1}$ converges or diverges.

Solution. Since $\ln t$ grows without bound as $t \to \infty$, the -1 is eventually going to be insignificant in comparison to $\ln t$. Thus, as far as convergence is concerned

$$\int_4^\infty \frac{1}{(\ln t) - 1}\,dt \quad \text{will behave like} \quad \int_4^\infty \frac{1}{\ln t}\,dt.$$

What does $\int_4^\infty \frac{1}{\ln t}\,dt$ do? Since $\ln t$ grows very slowly, $\frac{1}{\ln t}$ goes to zero very slowly, and so the integral probably doesn't converge. Specifically, $\ln t < t$ and $\ln t > 0$ for $t \geq 4$, so $\frac{1}{\ln t} > \frac{1}{t}$. Since $\int_4^b \frac{1}{t}\,dt$ grows without bound as $b \to \infty$ (see Example 3, page 433), $\int_4^b \frac{1}{\ln t}\,dt$ must too.

Here is the more precise argument. We know that $\ln t < t$ for all positive t, and we know that a positive fraction is increased by making its denominator smaller, so:

$$\frac{1}{(\ln t) - 1} > \frac{1}{\ln t} > \frac{1}{t}.$$

Since these inequalities hold throughout the interval between the limits of integration, we can say for all $b \geq 4$,

$$\int_4^b \frac{1}{(\ln t) - 1}\,dt \geq \int_4^b \frac{1}{t}\,dt.$$

By Example 3, page 433 (with $p = 1$), the values of $\int_1^b \frac{1}{t}\,dt$, and hence of $\int_4^b \frac{1}{t}\,dt$, increase without bound. Therefore $\int_4^\infty \frac{1}{(\ln t) - 1}\,dt$ diverges. ☐

How Do You Know What To Compare With?

In Examples 1 and 2, we investigated the convergence of an integral by comparing it with an easier integral. How did we pick the easier integral? Indeed, how do we know if the integrand we start with is to be larger or smaller than the one we use for comparison?

Which Way should the Inequalities Go?

If $f(x)$ is positive:
To show convergence, $f(x)$ needs to be *smaller* than one that converges.
To show divergence, $f(x)$ needs to be *larger* than one that diverges.

If $f(x)$ is negative:
Work with $-f(x)$, which is positive.

Finding the right integrand to compare with is a matter of trial-and-error, guided by any information you get by looking at your own integrand as $x \to \infty$. You want your comparison integrand to be easy and, in particular, to have a simple antiderivative.

Useful Integrals for Comparison

- $\int_1^\infty \dfrac{1}{x^p}\,dx$ converges for $p > 1$, diverges for $p \le 1$.

- $\int_0^1 \dfrac{1}{x^p}\,dx$ converges for $p < 1$, diverges for $p \ge 1$.

- $\int_0^\infty e^{-ax}\,dx$ converges for $a > 0$.

Of course, you can use any function for comparison, provided you can determine its behavior easily.

☐ **Example 3** Investigate the convergence of $\int_1^\infty \dfrac{(\sin x) + 3}{\sqrt{x}}\,dx$.

Solution. Since it looks difficult to find an antiderivative of this function, we try comparison. What happens to this integrand as $x \to \infty$? Since $\sin x$ oscillates between -1 and 1, this integrand oscillates between $\frac{2}{\sqrt{x}}$ and $\frac{4}{\sqrt{x}}$.

What do $\int_1^\infty \frac{2}{\sqrt{x}}\,dx$ and $\int_1^\infty \frac{4}{\sqrt{x}}\,dx$ do? As far as convergence is concerned, they certainly do the same thing, and whatever that is, our integral does it too. The important thing to notice is that $\sqrt{x}$ grows very slowly, so $\frac{1}{\sqrt{x}}$ gets small slowly – which means convergence is unlikely. Since $\sqrt{x} = x^{\frac{1}{2}}$, the result in the box above (with $p = \frac{1}{2}$) tells us that $\int_1^\infty \frac{dx}{\sqrt{x}}$ diverges. So we believe our original integral diverges and therefore we will compare it with an integral having a smaller integrand. Looking back we see the smaller integrand is $\frac{2}{\sqrt{x}}$, so we use the fact that:

$$\frac{(\sin x) + 3}{\sqrt{x}} \geq \frac{-1 + 3}{\sqrt{x}} = \frac{2}{\sqrt{x}}.$$

Since this inequality holds throughout the interval of integration, we can say for all $b \geq 1$,

$$\int_1^b \frac{(\sin x) + 3}{\sqrt{x}} \, dx \geq \int_1^b \frac{2}{\sqrt{x}} \, dx.$$

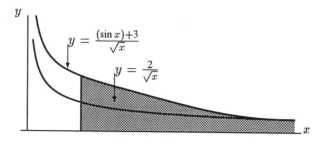

Figure 6.36: Graph showing $\int_1^\infty \frac{2}{\sqrt{x}} \, dx < \int_1^\infty \frac{(\sin x) + 3}{\sqrt{x}} \, dx$

See Figure 6.36. Now since $\int_1^\infty \frac{2}{\sqrt{x}} \, dx$ diverges, $\int_1^\infty \frac{\sin x + 3}{\sqrt{x}} \, dx$ does as well. $\qquad\square$

Notice that there are two possible comparisons we could have made in Example 3:

$$\frac{(\sin x) + 3}{\sqrt{x}} \geq \frac{-1 + 3}{\sqrt{x}} \qquad \text{or} \qquad \frac{(\sin x) + 3}{\sqrt{x}} \leq \frac{1 + 3}{\sqrt{x}}.$$

Since both $\int_1^\infty \frac{2}{\sqrt{x}} \, dx$ and $\int_1^\infty \frac{4}{\sqrt{x}} \, dx$ diverge, only the first comparison is useful. Knowing our integral is *smaller* than a divergent integral is of no help whatsoever!

The next example shows what to do if your comparison does not hold throughout the interval of integration.

$\square$ **Example 4** Show $\int_1^\infty e^{-\frac{x^2}{2}} \, dx$ converges to a finite value.

Solution. We know that $e^{-\frac{x^2}{2}}$ goes very rapidly to zero as $x \to \infty$, so we expect this integral may converge. To make a comparison for $\int_1^\infty e^{-\frac{x^2}{2}} \, dx$, the first thing that comes to mind is probably $\int_1^\infty e^{-x} \, dx$, because e^{-x} has an elementary antiderivative and $\int_1^\infty e^{-x} \, dx$ converges.

If the comparison is to be useful, e^{-x} had better be the larger. What is the relationship between $e^{-\frac{x^2}{2}}$ and e^{-x}? We know that for $x \geq 2$,

$$x \leq \frac{x^2}{2} \qquad \text{so} \qquad -\frac{x^2}{2} \leq -x,$$

and so, for $x \geq 2$

$$e^{-\frac{x^2}{2}} \leq e^{-x}.$$

But since this inequality only holds for $x \geq 2$, we must split our original interval of integration into two pieces:

$$\int_1^\infty e^{-\frac{x^2}{2}} \, dx = \int_1^2 e^{-\frac{x^2}{2}} \, dx + \int_2^\infty e^{-\frac{x^2}{2}} \, dx.$$

Now $\int_1^2 e^{-\frac{x^2}{2}} \, dx$ is finite (after all, it is a bounded function over a finite interval) and $\int_2^\infty e^{-\frac{x^2}{2}} \, dx$ is finite by comparison with $\int_2^\infty e^{-x} \, dx$. Therefore $\int_1^\infty e^{-\frac{x^2}{2}} \, dx$ is the sum of two finite pieces and therefore must be finite. $\qquad\square$

Using Numerical Methods for Improper Integrals

☐ **Example 5** Approximate the value of $\int_1^\infty e^{-\frac{x^2}{2}}\, dx$.

Solution. Unfortunately $f(x) = e^{-\frac{x^2}{2}}$ has no elementary antiderivative. We will have to use numerical methods like the trapezoidal or midpoint rule, but how do we do this for an integral over an infinite interval? What we will do is compute $\int_1^c e^{-\frac{x^2}{2}}\, dx$ for some value of c which is large enough to make the part we missed, $\int_c^\infty e^{-\frac{x^2}{2}}\, dx$, so small that we don't care about it. We estimate the part we missed by comparison with e^{-x}:

$$\int_c^\infty e^{-\frac{x^2}{2}}\, dx \leq \int_c^\infty e^{-x}\, dx = \lim_{b\to\infty} -e^{-x}\Big|_c^b = e^{-c}, \qquad \text{for } c \geq 2.$$

Suppose we pick $c = 10$. Then $e^{-10} < 0.00005$, so we know that $\int_{10}^\infty e^{-\frac{x^2}{2}}\, dx < 0.00005$. Therefore $\int_1^\infty e^{-\frac{x^2}{2}}\, dx$ and $\int_1^{10} e^{-\frac{x^2}{2}}\, dx$ differ by less than 0.00005. If we compute $\int_1^{10} e^{-\frac{x^2}{2}}\, dx$ by the midpoint rule, the approximate value will be an underestimate, while the trapezoidal approximation will be an overestimate because the graph of $e^{-\frac{x^2}{2}}$ is concave up for $x > 1$. Using 200 subdivisions, we obtain

$$\begin{aligned} \text{MID}(200) &= 0.39764; \\ \text{TRAP}(200) &= 0.39779. \end{aligned}$$

Thus

$$0.39764 \leq \int_1^{10} e^{-\frac{x^2}{2}}\, dx \leq 0.39779.$$

Since $0 < \int_{10}^\infty e^{-\frac{x^2}{2}}\, dx \leq e^{-10} < 0.00005$, we have

$$0.39764 + 0 < \int_1^{10} e^{-\frac{x^2}{2}}\, dx + \int_{10}^\infty e^{-\frac{x^2}{2}}\, dx < 0.39779 + 0.00005.$$

Therefore

$$0.39764 < \int_1^\infty e^{-\frac{x^2}{2}}\, dx < 0.39784.$$

Thus we have $\int_1^\infty e^{-\frac{x^2}{2}}\, dx \approx 0.398$.

Notice that we didn't begin to do numerical approximation in this example until after we had ascertained that the integral converged. ☐

❖ Exercises for Section 6.11

In Problems 1–14 decide if the improper integral converges or diverges. Explain your reasoning.

1. $\displaystyle\int_{50}^{\infty} \frac{dz}{z^3}$

2. $\displaystyle\int_{1}^{\infty} \frac{dx}{x^3+1}$

3. $\displaystyle\int_{2}^{\infty} \frac{d\theta}{\sqrt{\theta^3+1}}$

4. $\displaystyle\int_{1}^{\infty} \frac{dx}{1+x}$

5. $\displaystyle\int_{0}^{\infty} \frac{dy}{1+e^y}$

6. $\displaystyle\int_{-1}^{5} \frac{dt}{(t+1)^2}$

7. $\displaystyle\int_{1}^{\infty} \frac{2+\cos\phi}{\phi^2} d\phi$

8. $\displaystyle\int_{0}^{\infty} \frac{dz}{e^z+2^z}$

9. $\displaystyle\int_{0}^{\pi} \frac{2-\sin\phi}{\phi^2} d\phi$

10. $\displaystyle\int_{-\infty}^{\infty} \frac{du}{1+u^2}$

11. $\displaystyle\int_{1}^{\infty} \frac{du}{u+u^2}$

12. $\displaystyle\int_{4}^{\infty} \frac{3+\sin\alpha}{\alpha} d\alpha$

13. $\displaystyle\int_{1}^{\infty} \frac{d\theta}{\sqrt{\theta^2+1}}$

14. $\displaystyle\int_{0}^{1} \frac{d\theta}{\sqrt{\theta^3+\theta}}$

15. Does the integral $\displaystyle\int_{1}^{\infty} \frac{x}{e^{-x}+x} dx$ converge? Why or why not?

Estimate the values of the integrals in Problems 16–17 correct to two decimal places by integrating the functions on your calculator or computer for large values of the upper bound of integration.

16. $\displaystyle\int_{1}^{\infty} e^{-x^2} dx$

17. $\displaystyle\int_{0}^{\infty} e^{-x^2} \cos x \, dx$

18. The function

$$f(x) = ae^{-\frac{x^2}{2}}$$

is useful in statistics. Find the value of a (to three decimal places) that makes

$$\int_{-\infty}^{\infty} f(x)\, dx = 1.$$

19. Statisticians often use the function

$$g(x) = ae^{-\frac{(x-k)^2}{2}}.$$

(a) To three decimal places, what value of a should be chosen to ensure that

$$\int_{-\infty}^{\infty} g(x)\, dx = 1?$$

(b) Is your answer the same as or different from your answer to Problem 18? Explain why.

20. (a) Find an upper bound for

$$\int_3^\infty e^{-x^2}\, dx.$$

   [Hint: $e^{-x^2} \le e^{-3x}$ for $x \ge 3$.]

   (b) For any positive n, generalize the result of part (a) to find an upper bound for

$$\int_n^\infty e^{-x^2}\, dx$$

   by noting that $nx \le x^2$ for $x \ge n$.

21. Do the following integrals converge? If so, give an upper bound for the value of the integral.

   (a) $\displaystyle\int_1^\infty \frac{2x^2+1}{4x^4+4x^2-2}\,dx$

   (b) $\displaystyle\int_1^\infty \left(\frac{2x^2+1}{4x^4+4x^2-2}\right)^{\frac{1}{4}}\,dx$

22. For what values of p does the integral $\displaystyle\int_2^\infty \frac{dx}{x(\ln x)^p}$ converge?

23. For what values of p does the integral $\displaystyle\int_1^2 \frac{dx}{x(\ln x)^p}$ converge?

24. In Planck's Radiation Law, we encounter the integral

$$\int_1^\infty \frac{dx}{x^5(e^{1/x}-1)}.$$

   (a) Explain why the tangent line approximation to e^t at $t=0$ tells us that for all t

$$1+t \le e^t.$$

   (b) Substituting $t = \dfrac{1}{x}$, show that for all x

$$e^{1/x}-1 > \frac{1}{x}.$$

   (c) Use the comparison test to show that the original integral converges.

25. In this problem you will look at two possible ways to approximate (to four decimal places) the value of the integral $\int_1^\infty e^{-x^2/2}\,dx$. You will compute $\int_1^b e^{-x^2/2}\,dx$ using midpoint rectangles with n subdivisions. Since the original integral is improper, you need to let $b \to \infty$. In addition, you need to let $n \to \infty$. The question is: Which should go to infinity first?

   (a) First fix b and increase n. Start with $b = 3$ and let $n = 20, 50, 100, \ldots$ until your result stabilizes. Take the value to which your result has stabilized as the value of the integral with $b = 3$. Now increase $b = 4, 5, \ldots$ and for each new b value increase n as before. Continue until successive values of b give the same answer. Take the value you get as the value of $\int_1^\infty e^{-x^2/2}\,dx$.

   (b) In this part, fix n and increase b. Start with $n = 20$ and increase $b = 3, 4, 5, \ldots$ What happens? You may have to use very large values of b before the results stabilize.

   (c) Explain why the method in (a) gives a reasonable way of approximating the improper integral while the method in (b) does not.

6.12   Notes on Constructing Antiderivatives

We have seen several techniques for finding an antiderivative F of a function f. The fact remains, however, that most of the time F cannot be found easily. Suppose f is an elementary function, i.e., a combination of constants, powers of x, $\sin x$, $\cos x$, e^x, and $\ln x$. Then you have to be lucky to find an antiderivative F which is also an elementary function. But if F is not an elementary function, how can we be sure that it exists at all? After all, elementary functions are the only functions for which we have formulas. In this section we will learn to use the definite integral to construct antiderivatives, and, along the way, we will learn how to create genuinely new functions.

Construction of Antiderivatives using a Slope Field

In Section 6.3 we saw how to graph an antiderivative, F, of the function f in Figure 6.37(a) with the property that $F(0) = 0$. Since $F' = f$, the graph of F will be increasing where f is positive, and decreasing where f is negative. Also, it will slope steeply when f is large (positive or negative) and gently when f is small, and it will be horizontal when f is zero. Using this information, we can generate the lower graph in Figure 6.37(b). This graph starts at $(0,0)$, since $F(0)$ was supposed to be 0.

What about other antiderivatives? They all have the same slope for a particular value of x and so are the same shape as F, but shifted up or down. Thus, they all have the form $F(x) + C$, for some constant C. If you imagine the whole family of functions $F(x) + C$ as C varies, you get a pile of graphs stacked up on top of each other, all of the same shape but at different heights above the x-axis. Figure 6.37(b) shows $F(x) + 1$ as well as $F(x)$.

So far, all we have is a rough sketch of the antiderivative. To get a more accurate graph of F, we should be more careful about making F have the right slope at every point. The slope of F at any point (x, y) on its graph should be $f(x)$, since $F'(x) = f(x)$. We can arrange this as follows:

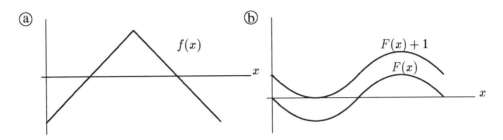

Figure 6.37: A function f and two antiderivatives. Note that $F' = f$.

at the point (x, y) in the plane, draw a small line segment with slope $f(x)$. Do this at many points. If $f(x) = x$, you get Figure 6.38.

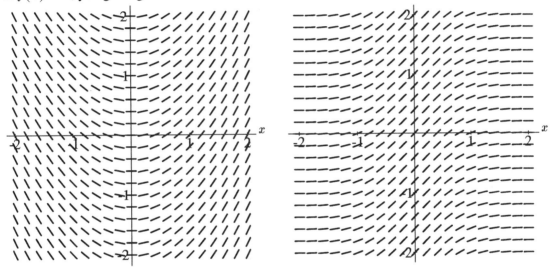

Figure 6.38: Slope Field of $f(x) = x$.   **Figure 6.39**: Slope Field of e^{-x^2}.

We call such a diagram a *slope field* (or sometimes a *direction field* or a *tangent field*). Notice how the lines in Figure 6.38 seem to be arranged in a parabolic pattern. This is because the general antiderivative of x is $x^2/2 + C$, so the lines are all the tangent lines to the family of parabolas $y = x^2/2 + C$. This suggests a way of finding antiderivatives graphically even if we can't write down a formula for them; plot the slopes and see if they suggest the graph of an antiderivative. For example, if you do this with $f(x) = e^{-x^2}$, which is one of the functions that does not have an elementary antiderivative, you get Figure 6.39.

You can see the ghost of the graph of a function lurking behind the slopes in Figure 6.39; in fact there is a whole stack of them. If you move across the plane in the direction suggested by the slope field at every point, you will trace out a curve. The slope field will be tangent to the curve everywhere, so this is the graph of an antiderivative of e^{-x^2}.

This construction is important because it shows that the antiderivative of e^{-x^2} does exist—the graph defines it. In addition, this construction is useful because the graph allows us to see the qualitative behavior of the antiderivative.

Construction of Antiderivatives Using the Definite Integral.

The previous discussion may have convinced you that an antiderivative of e^{-x^2} exists, but it does not give you an easy way of evaluating the antiderivative. If it were an elementary function, we could evaluate it using a calculator, but it isn't. However, we know from the Fundamental Theorem of Calculus that if F is an antiderivative of e^{-x^2}, then

$$F(b) - F(a) = \int_a^b e^{-t^2}\, dt$$

so

$$F(x) - F(0) = \int_0^x e^{-t^2}\, dt.$$

All we have done here is to replace the usual b in the upper limit with an x and set $a = 0$. Suppose we are looking for the antiderivative that satisfies $F(0) = 0$. Then the $F(0)$ drops out of the above equation, and we get

$$F(x) = \int_0^x e^{-t^2}\, dt.$$

This is a formula for F! F is a function because for any value of x, there is a unique value for $F(x)$. For any fixed x, we can calculate $F(x)$ to any accuracy we like by using numerical methods such as the trapezoid rule or Simpson's rule. For example, we can show that

$$F(2) = \int_0^2 e^{-t^2}\, dt = 0.88208\ldots.$$

Notice that our expression for F is not an elementary function; we have *created* a new function using the definite integral. Our next theorem says that this method of constructing antiderivatives works in general. This means that if we define F by

$$F(x) = \int_a^x f(t)\, dt$$

then F must be an antiderivative of f. This can happen in one of two ways. Either we already know what the antiderivatives of f are, in which case F turns out to be one of them. Or, if f has no elementary antiderivatives, F defines an antiderivative of f for you.

Construction Theorem for Antiderivatives *If f is a continuous function, and if a is any number, then the function F defined by*

$$F(x) = \int_a^x f(t)\, dt$$

is an antiderivative of f.

Justification of the Construction Theorem for Antiderivatives: We will look at the two cases separately:

Case 1: If we already know an antiderivative of f, which we may call G, then the Fundamental Theorem of Calculus tells us that

$$G(x) - G(a) = \int_a^x f(t)\, dt.$$

By the definition of F, we have

$$F(x) = \int_a^x f(t)\,dt.$$

Therefore,

$$G(x) - G(a) = F(x),$$

so F and G differ by a constant (namely $G(a)$), and so F and G are *both* antiderivatives of f.

Case 2: Suppose we know no antiderivative of f. Then our task is to show that F, defined by this integral, is an antiderivative of f. We want to show that $F'(x) = f(x)$. In fact the argument we will use works even if we already know an antiderivative. By the definition of the derivative,

$$F'(x) = \lim_{h \to 0} \frac{F(x+h) - F(x)}{h}.$$

Let's suppose f is positive and h is positive. Then we can visualize

$$F(x) = \int_a^x f(t)\,dt$$

and

$$F(x+h) = \int_a^{x+h} f(t)\,dt$$

as areas, which leads to representing $F(x+h) - F(x) = \int_x^{x+h} f(t)\,dt$ as a difference of two areas. From Figure 6.40, we see that $F(x+h) - F(x)$ is roughly a rectangle of height $f(x)$ and width h (shaded darker in Figure 6.40), so

$$F(x+h) - F(x) \approx f(x)h.$$

Because f is continuous, as h approaches 0, the height of this rough rectangle approaches $f(x)$. That is, as $h \to 0$

$$\frac{F(x+h) - F(x)}{h} \to f(x).$$

A similar argument holds if h is negative and $h \to 0$, so

$$F'(x) = \lim_{h \to 0} \frac{F(x+h) - F(x)}{h} = f(x).$$

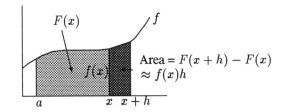

Figure 6.40: Justification of the Construction Theorem for Antiderivatives.

Using the Construction Theorem for Antiderivatives.

The construction theorem enables us to write down antiderivatives of functions that do not have elementary antiderivatives. For example, an antiderivative of $\frac{\sin x}{x}$ is

$$F(x) = \int_0^x \frac{\sin t}{t}\,dt.$$

This is an honest-to-goodness function, whose values you can calculate to any degree of accuracy using left- or right-hand sums, or the more efficient methods of Section 6.8. If you were a manufacturer of calculators, you might even put a button for it on your calculator. In fact, this particular function already has a name: it is called the *sine-integral* and it is denoted $\mathrm{Si}(x)$.

❑ **Example 1** Construct a table of values of Si(x) for $x = 0, 1, 2, 3$.

Solution.

Using Riemann sums, we calculate values of Si(x) = $\int_0^x \frac{\sin t}{t} dt$ given in Table 6.9. Since the integral is improper at $t = 0$, we took the lower limit as 0.00001 instead of 0.

x	0	1	2	3
Si(x)	0	0.95	1.61	1.85

Table 6.9: Values of Si(x) for $x = 0, 1, 2, 3$.

❑

The reason the sine-integral has a name is that scientists have found a use for it in optics. For people who use it all the time, it is just another elementary function, like sine or cosine and tables of its values have been compiled. In addition to knowing all the derivatives that we know, i.e., the derivatives of e^x, $\sin x$, $\cos x$, and $\ln x$, such people also have in their heads the derivative

$$\frac{d}{dx} \text{Si}(x) = \frac{\sin x}{x},$$

and can use this fact with the other differentiation rules.

❑ **Example 2** Find the derivative of x Si(x).

Solution. Using the product rule,

$$
\begin{aligned}
\frac{d}{dx}(x \, \text{Si}(x)) &= \left(\frac{d}{dx} x\right) \text{Si}(x) + \left(\frac{d}{dx} \text{Si}(x)\right) x \\
&= 1 \cdot \text{Si}(x) + \frac{\sin x}{x} x \\
&= \text{Si}(x) + \sin x.
\end{aligned}
$$

❑

If we were to add Si(x) to our list of elementary functions we would be able to do more indefinite integrals using the methods of this chapter.

❑ **Example 3** Find $\int \frac{\sin(t^2)}{t} dt$.

Solution. Make the substitution $t = \sqrt{w}$, giving $dt = \frac{1}{2\sqrt{w}} dw$ and $w = t^2$, so

$$\int \frac{\sin(t^2)}{t} dt = \int \frac{\sin w}{\sqrt{w}} \frac{1}{2\sqrt{w}} dw = \frac{1}{2} \int \frac{\sin w}{w} dw = \frac{1}{2} \text{Si}(w) + C = \frac{1}{2} \text{Si}(t^2) + C.$$

❑

❖ Exercises for Section 6.12

For each of the functions in Problems 1-3, let $F(x) = \int_0^x f(t)\, dt$. Draw a graph showing $F(x)$ as a function of x.

1.

2.

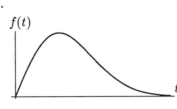

3.

4. Make a table of values for the function

$$I(x) = \int_0^x \sqrt{t^4 + 1}\, dt$$

for $x = 0, 0.5, 1.0, 1.5, 2.0$.

5. (a) Continue the table of values for $\text{Si}(x) = \int_0^x \dfrac{\sin t}{t}\, dt$, in Example 1, page 452 for $x = 4, 5$.
   (b) Why is $\text{Si}(x)$ decreasing between $x = 4$ and $x = 5$?

6. (a) Match the following functions with their slope fields as shown in Figure 6.41.

   i. $f(x) = e^{x^2}$ ii. $f(x) = e^{-2x^2}$ iii. $f(x) = e^{-\frac{x^2}{2}}$

   iv. $f(x) = e^{-0.5x} \cos x$ v. $f(x) = \dfrac{1}{(1 + 0.5 \cos x)^2}$ vi. $f(x) = -e^{-x^2}$

   (b) On each slope field, sketch the antiderivative, F, with $F(0) = 0$.

Find the derivatives in Problems 7–12 .

7. $\dfrac{d}{dx} \displaystyle\int_0^x \sqrt{3 + \cos(t^2)}\, dt$

10. $\dfrac{d}{dt} \displaystyle\int_t^\pi \cos(z^3)\, dz$

8. $\dfrac{d}{dx} \displaystyle\int_1^x (1 + t)^{200}\, dt$

11. $\dfrac{d}{dx} \displaystyle\int_x^1 \ln t\, dt$

9. $\dfrac{d}{dx} \displaystyle\int_{0.5}^x \arctan(t^2)\, dt$

12. $\dfrac{d}{dx} \left[\text{Si}(x^2) \right]$

13. Using a calculator or computer, draw the slope field of e^{-x^2}. Use it to sketch the antiderivative $F(x)$ of e^{-x^2} that satisfies $F(0) = 0$. Estimate $\lim_{x \to \infty} F(x)$.

14. (a) Sketch a graph of $f(t) = \dfrac{\sin t}{t}$.

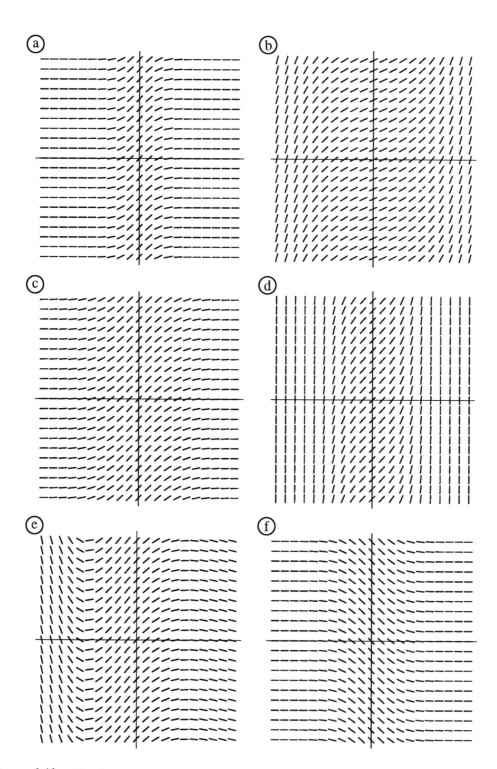

Figure 6.41: What functions are associated with these slope fields? (All show $-3 \le x \le 3$ and $-3 \le y \le 3$)

(for Problem 6)

(b) What does your graph tell you about the behavior of $\text{Si}(x)$, for $x > 0$? Is $\text{Si}(x)$ always increasing or always decreasing? Does $\text{Si}(x)$ cross the x-axis for $x > 0$?

(c) By drawing the slope field for $f(t) = \dfrac{\sin t}{t}$, decide whether $\lim\limits_{x \to \infty} \text{Si}(x)$ exists.

15. Let $F(x)$ be the antiderivative of $\sin(x^2)$ satisfying $F(0) = 0$.

(a) Describe any general features of the graph of F that you can deduce by looking at the graph of $\sin(x^2)$ in Figure 6.42.

(b) By drawing a slope field (using a calculator or computer), sketch a graph of F. Does F ever cross the x-axis in the region $x > 0$? Does $\lim\limits_{x \to \infty} F(x)$ exist?

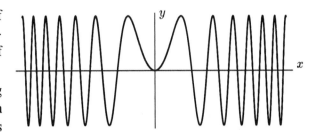

Figure 6.42: Graph of $y = \sin(x^2)$

(for Problem 15)

16. Consider the integral

$$F(x) = \int_1^x (t + 1)^2 \cos 2t \, dt.$$

(a) Evaluate the definite integral using the table of integrals and the Fundamental Theorem of Calculus.

(b) Verify the Construction Theorem by finding $\frac{dF}{dx}$ from your answer to (a) above, and by finding $\frac{dF}{dx}$ from the Construction Theorem.

Problems 17–21 concern the *error function*, $\text{erf}(x)$, defined by

$$\text{erf}(x) = \frac{2}{\sqrt{\pi}} \int_0^x e^{-t^2} \, dt.$$

17. Find the derivative of $x \, \text{erf}(x)$.

18. Find the derivative of $\text{erf}(\sqrt{x})$.

19. Find the indefinite integral

$$\int x e^{-x^4}\, dx$$

   in terms of the error function.

20. Find an expression for the cumulative probability distribution function:

$$F(x) = \frac{1}{\sqrt{2\pi}} \int_0^x e^{-\frac{t^2}{2}}\, dt$$

   in terms of $\mathrm{erf}(x)$.

21. Find an expression for

$$\frac{1}{\sqrt{2\pi}} \int_{x_1}^{x_2} e^{-\frac{t^2}{2}}\, dt$$

   in terms of $\mathrm{erf}(x)$.

22. (a) Sketch a (relatively simple) smooth curve, $y = p(x)$, satisfying the following:
    - $|p(x)| \le 0.5$ for all x.
    - $p(0) = 0.5$ and $p(3) = -0.5$.
    - $p(x) \to 0$ as $x \to \infty$.

    (b) Sketch the curve $y = q(x)$, where q is the antiderivative of p which satisfies $q(0) = 3$. [Hint: What does $|p(x)|$ being small imply for the graph of q?]

    (c) Now sketch the curve $y = r(x)$ where r is the antiderivative of q satisfying $r(0) = 0$.

    (d) Rewrite the restriction on $|p(x)|$ as a restriction on $r(x)$. In your own words, what does this restriction imply for the curve $y = r(x)$ drawn in (c)?

6.13 Miscellaneous Exercises for Chapter 6

Are the statements in Problems 1–3 true or false? Why?

1. If F and G are two different antiderivatives of $f(x) = 3x^2$, then their graphs never intersect.

2. $\int_0^2 x\, dx = 2 \int_0^1 x\, dx$.

3. $\int_0^x f(t)\, dt$ and $\int_1^x f(t)\, dt$ differ by a constant.

4. Write between one and two pages on the two different methods (Riemann sums and antiderivatives) of integration you have seen. Explain the idea behind each one, using an example to illustrate your discussion. Lastly, explain how they are related.

   One way to approach this is to imagine that you're trying to explain to a friend who has just missed a week of class what the 'big idea' of integration was.

For Problems 5–7, find $f(x)$.

5. $f'(x) = 2x^3 - \dfrac{1}{x} - \dfrac{1}{x^2}$

6. $f'(x) = e^x + x^e$

7. $f'(x) = e^\pi + \dfrac{1}{\sqrt{x^3}}$

For Problems 8–9 the graph of $f'(x)$ is given. Sketch a possible graph for $f(x)$. Mark the points $x_1 \ldots x_4$ on your graph and label local maxima, local minima and points of inflection.

8.

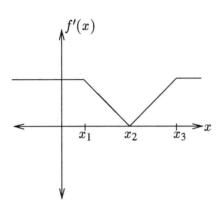

9.

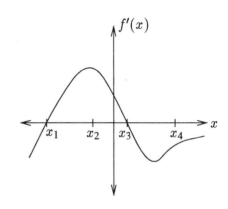

10. Sketch the derivative and antiderivative of the function in the figure to the right. Make the antiderivative go through the origin.

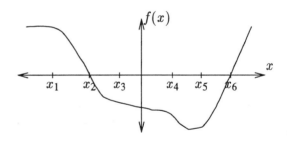

11. (a) Using a substitution, find $\displaystyle\int \frac{dx}{\sqrt{1 - 4x^2}}$ from the table of integrals.

(b) Use your answer to (a) to find $\displaystyle\int_0^{\frac{\pi}{8}} \frac{dx}{\sqrt{1 - 4x^2}}$.

(c) Check your answer to (b) by numerical integration.

12. (a) Explain why you can rewrite x^x as: $x^x = e^{x \ln x}$.

(b) Use your answer to (a) to find $\dfrac{d}{dx}(x^x)$.

(c) Find $\displaystyle\int x^x(1 + \ln x)dx$.

13. Suppose the function f is defined by $f(x) = x^2$ for $0 \le x \le 1$ and $f(x) = 2 - x$ for $1 \le x \le 2$. Compute $\displaystyle\int_0^2 f(t)\,dt$.

For Problems 14–38, evaluate the following integrals. For Problems 35–38, find the definite integrals using the Fundamental Theorem of Calculus, and check your answers numerically.

14. $\displaystyle\int \sin x \left(\sqrt{2 + 3\cos x}\right) dx$

15. $\displaystyle\int x \cos x\,dx$

16. $\displaystyle\int x^2 e^{2x}\,dx$

17. $\displaystyle\int x\sqrt{1 - x}\,dx$

18. $\displaystyle\int x \ln x\,dx$

19. $\displaystyle\int y \sin y\,dy$

20. $\displaystyle\int te^{t^2}\,dt$

21. $\displaystyle\int \frac{dz}{z^2 + z}$

22. $\displaystyle\int \frac{\cos\sqrt{y}}{\sqrt{y}}\,dy$

23. $\displaystyle\int (\ln x)^2\,dx$

24. $\displaystyle\int \ln(x^2)\,dx$

25. $\displaystyle\int e^{0.5 - 0.3t}\,dt$

26. $\displaystyle\int \cos^3 2\theta \sin 2\theta\,d\theta$

27. $\displaystyle\int \frac{5x + 6}{x^2 + 4}\,dx$

28. $\displaystyle\int x\sqrt{4 - x^2}\,dx$

29. $\displaystyle\int \sqrt{4 - x^2}\,dx$

30. $\displaystyle\int \frac{1}{\cos^2 z}\,dz$

31. $\displaystyle\int \cos^2 \theta\,d\theta$

32. $\displaystyle\int \frac{(u + 1)^3}{u^2}\,du$

33. $\displaystyle\int \tan(2x - 6)\,dx$

34. $\displaystyle\int \sqrt{2x^2 - 5}\,dx$

35. $\displaystyle\int_1^3 \ln(x^3)\,dx$

36. $\displaystyle\int_1^e (\ln x)^2\,dx$

37. $\displaystyle\int_{-\pi}^{\pi} e^{2x} \sin 2x\,dx$

38. $\displaystyle\int_{-\frac{\pi}{4}}^{\frac{\pi}{4}} x^3 \cos x^2\,dx$

In Problems 39–42, explain why the following pairs of antiderivatives are really, despite their apparent dissimilarity, essentially different expressions of the same problem. (They are not *identical*: they are expressed using different symbols and some are multiples of their partners—but they are, for the most part, the same problem. Why?) You do not need to evaluate the integrals, just explain....

39. $\displaystyle\int \frac{dx}{x^2 + 4x + 4}$ and $\displaystyle\int \frac{x}{(x^2 + 1)^2}\,dx$

40. $\displaystyle\int \frac{1}{\sqrt{1 - x^2}}\,dx$ and $\displaystyle\int \frac{x\,dx}{\sqrt{1 - x^4}}$

41. $\displaystyle\int \frac{x}{1 - x^2}\,dx$ and $\displaystyle\int \frac{1}{x \ln x}\,dx$

42. $\displaystyle\int \frac{x}{x + 1}\,dx$ and $\displaystyle\int \frac{1}{x + 1}\,dx$

43. Integrate:

   (a) $\displaystyle\int \frac{e^x}{1 + e^x}\,dx$

   (b) $\displaystyle\int \frac{e^x}{1 + e^{2x}}\,dx$

   (c) $\displaystyle\int \frac{1}{1 + e^x}\,dx$

44. (a) Find the average value of the following functions over one cycle:

   i. $f(t) = \cos t$ ii. $g(t) = |\cos t|$ iii. $k(t) = (\cos t)^2$

   (b) Put the averages you have just found in ascending order. Explain clearly, using words and graphs, why the averages should be expected to come out in the order they did.

45. In 1990, humans generated 1.4×10^{20} Joules of energy through the combustion of petroleum. All of the Earth's petroleum together would generate approximately 10^{22} Joules. Assuming the use of energy generated by petroleum combustion will increase by 2% each year, how long will it be before all of our petroleum resources are used up?

46. Suppose the graph of f is in Figure 6.43 and its formula is:
$$f(x) = \begin{cases} -x + 1, & \text{for } 0 \le x \le 1; \\ x - 1, & \text{for } 1 < x \le 2. \end{cases}$$

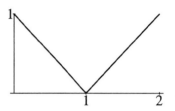

    (a) Find a function F such that $F' = f$ and $F(1) = 1$.

    (b) Use geometry to show the area under the graph of f above the x-axis between $x = 0$ and $x = 2$ is equal to $F(2) - F(0)$.

    (c) Use parts (a) and (b) to verify the Fundamental Theorem of Calculus.

Figure 6.43: Verify the Fundamental Theorem of Calculus

47. A store has an inventory of Q units of a certain product at time $t = 0$. The store sells the product at the steady rate of $\frac{Q}{A}$ units per week, and exhausts the inventory in A weeks.

    (a) Find a formula $f(t)$ for the amount of product in inventory at time t. Graph $f(t)$.

    (b) Find the average inventory level during the period $0 \le t \le A$ in two ways: graphically and by antidifferentiation. Check your answer by common sense.

Problems 48–49 concern Figure 6.44.

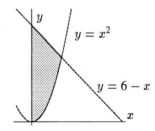

48. Find the average (vertical) height of the shaded area to the right.

49. Find the average (horizontal) width of the shaded area to the right.

Figure 6.44: Figure for Problems 48–49.

50. The curves $y = \sin x$ and $y = \cos x$ cross each other infinitely often. What is the area of the region bounded by these two curves between two consecutive crossings?

For Problems 51–59, decide if the integral converges or diverges. If an integral converges, find its value.

51. $\displaystyle\int_4^\infty \frac{dt}{t^{\frac{3}{2}}}$

52. $\displaystyle\int_0^\infty w e^{-w}\, dw$

53. $\displaystyle\int_{10}^\infty \frac{dx}{x \ln x}$

54. $\displaystyle\int_2^\infty \frac{1}{4+z^2}\, dz$

55. $\displaystyle\int_{10}^\infty \frac{1}{4-z^2}\, dz$

56. $\displaystyle\int_{-1}^1 \frac{1}{x^3}\, dx$

57. $\displaystyle\int_{-\frac{\pi}{4}}^{\frac{\pi}{4}} \tan\theta\, d\theta$

58. $\displaystyle\int_0^{\frac{\pi}{2}} \frac{1}{\sin\phi}\, d\phi$

59. $\displaystyle\int_{-5}^{10} \frac{dt}{\sqrt{t+5}}$

For Problems 60–64, decide if the integral converges or diverges. If in integral converges, find its value, or give a bound for the value.

60. $\displaystyle\int_0^{\frac{\pi}{4}} \tan 2\theta\, d\theta$

61. $\displaystyle\int_0^\infty \frac{\sin^2\theta}{\theta^2+1}\, d\theta$

62. $\displaystyle\int_0^\pi \tan^2\theta d\theta$

63. $\displaystyle\int_0^1 (\sin x)^{-\frac{3}{2}}\, dx$

64. $\displaystyle\int_1^\infty \frac{x}{x+1}\, dx$

65. Estimate the values of the following integrals correct to two decimal places by integrating the functions on your calculator for larger and larger values of the upper bound of integration.

   (a) $\displaystyle\int_0^\infty \frac{\sqrt{x}}{e^x}\, dx$ (b) $\displaystyle\int_1^\infty \ln\left(\frac{e^x+1}{e^x-1}\right)\, dx$

66. (a) Find the following indefinite integrals:

   i. $\displaystyle\int \ln x\, dx$ ii. $\displaystyle\int (\ln x)^2\, dx$ iii. $\displaystyle\int (\ln x)^3\, dx$ iv. $\displaystyle\int (\ln x)^4\, dx$

   (b) Find the following improper definite integrals:

   i. $\displaystyle\int_0^1 \ln x\, dx$ ii. $\displaystyle\int_0^1 (\ln x)^2\, dx$ iii. $\displaystyle\int_0^1 (\ln x)^3\, dx$ iv. $\displaystyle\int_0^1 (\ln x)^4\, dx$

   (c) Guess a formula for $\int_0^1 (\ln x)^n\, dx$, where $n \geq 0$

67. Find each of the integrals

$$\int_0^\infty e^{-x}\, dx, \qquad \int_0^\infty x e^{-x}\, dx, \qquad \int_0^\infty x^2 e^{-x}\, dx, \qquad \int_0^\infty x^3 e^{-x}\, dx,$$

   and hence guess the value of $\displaystyle\int_0^\infty x^n e^{-x}\, dx$.

68. Suppose you estimate $\int_0^{0.5} f(x)dx$ by the trapezoid and midpoint rules with 100 steps. Use a sketch to explain the relation between the two estimates and the true value of the integral if

   (a) $f(x) = 1 + e^{-x}$

   (b) $f(x) = e^{-x^2}$

   (c) $f(x)$ is a straight line.

69. Show that if f is linear, namely $f(x) = k + mx$, then the average value of f on the interval $a \leq x \leq b$ is just $(f(a) + f(b))/2$. Is this what you would expect? Explain.

70. Suppose for a certain definite integral that TRAP(10) = 4.6891 and TRAP(50) = 4.6966. Estimate the actual error for TRAP(10) (and thereby the actual value of the integral) by assuming that the error is reduced by a factor of 25 in going from TRAP(10) to TRAP(50).

71. Suppose for a definite integral that SIMP(5) = 7.41562 and SIMP(10) = 7.41738. Estimate the actual value of the integral by using the fact that the error is reduced by a factor of 16 in going from SIMP(5) to SIMP(10).

72. Let $F(x) = \int_0^x \sin 2t \, dt$.

    (a) Evaluate $F(\pi)$.

    (b) Draw a sketch to explain geometrically why the answer to (a) is correct.

    (c) For what values of x is $F(x)$ positive? negative?

73. Let

$$F(x) = \int_2^x \frac{1}{\ln t} \, dt \ \ \text{for } x \geq 2.$$

    (a) Find $F'(x)$.

    (b) Is F increasing or decreasing? What can you say about the concavity of its graph?

    (c) Sketch a graph of $F(x)$.

74. Use the fact that $e^x \geq 1 + x$ for all values of x and the formula

$$e^x = 1 + \int_0^x e^t \, dt$$

to show that

$$e^x \geq 1 + x + x^2/2$$

for all positive values of x.

Generalize this idea to get inequalities involving higher degree polynomials.

75. Use the fact that $\cos x \leq 1$ for all x and repeated integration to show that

$$\cos x \leq 1 - \frac{x^2}{2!} + \frac{x^4}{4!}.$$

Chapter 7

USING THE DEFINITE INTEGRAL

In Chapter 6, we focused on different ways of evaluating integrals. In Chapter 3 we saw how a definite integral can represent an area, an average value, or a total change. In this chapter we will use definite integrals to solve problems in geometry, physics, economics and probability. In the first section of this chapter, we look at the integral as an area. In each of the remaining sections, we use the same method to represent a quantity as a definite integral: we chop up the quantity and approximate it by a Riemann sum.

7.1   The Definite Integral as an Area

Often, simply knowing that you can think of some quantity as a definite integral is helpful, even if you don't evaluate the integral. In particular, you can represent the quantity you are interested in as an area, and hence visualize its magnitude. The next example shows how looking at areas allows you to compare the magnitude of two quantities.

☐ **Example 1**   Two cars start from rest at a traffic light and accelerate for two minutes. Figure 7.1 shows their velocities as a function of time. a) Which car is ahead after one minute? b) Which car is ahead after two minutes?

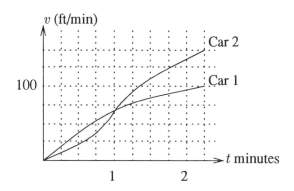

Figure 7.1: Velocities of two cars. Which is ahead when?

Solution.

(a) For the first minute, Car 1 goes faster than Car 2, and therefore Car 1 must be ahead at the end of one minute.

(b) At the end of two minutes, the situation is less clear, since Car 1 has been going faster for the first minute and Car 2 for the second. However, if $v = f(t)$ is the velocity of a car after t minutes, then we know that

$$\text{Distance traveled in two minutes } = \int_0^2 f(t)\,dt,$$

since the integral of velocity is distance traveled. This definite integral may also be interpreted as the area under the graph of f between 0 and 2. Since the area representing the distance traveled by Car 2 is clearly larger than the area for Car 1 (see Figure 7.1), we know that Car 2 has traveled farther than Car 1.

☐

☐ **Example 2**   A car starts at noon and travels with the velocity shown in Figure 7.2. A truck starts at 1 pm from the same place and travels at a constant velocity of 50 mph.

(a) How far away is the car when the truck starts?

(b) During the period when the car is ahead of the truck, when is the distance between them greatest and what is that greatest distance?

(c) When does the truck overtake the car and how far have both traveled then?

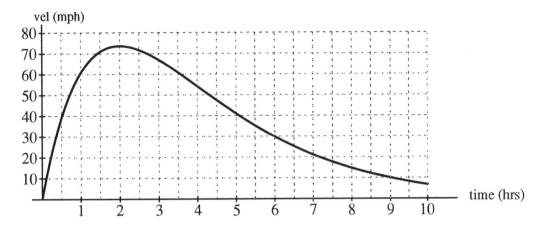

Figure 7.2: Velocity of Car for Example 2

Solution. To find distances from the velocity graph, we use the fact that if t is the time measured from noon, and v is the velocity,

$$\left(\begin{array}{c} \text{Distance traveled} \\ \text{by car at time } T \end{array} \right) = \int_0^T v \, dt = \left(\begin{array}{c} \text{Area under velocity} \\ \text{graph between 0 and } T \end{array} \right).$$

The truck's motion can be represented on the same graph by the horizontal line $v = 50$, starting at $t = 1$. The distance traveled by the truck is then the rectangular area under this line, and the distance between the two is the difference between these areas. One small square on the graph corresponds to moving at 10 mph for $\frac{1}{2}$ hour, i.e. to a distance of 5 miles.

(a) The distance traveled by the car when the truck starts is represented by the shaded area in Figure 7.3, which totals about 7 squares or about 35 miles.

(b) The car starts ahead of the truck, and the distance between them increases as long as the velocity of the car is greater than the velocity of the truck. Later, when the truck's velocity is greater than the car's, the truck will start to gain on the car, and the distance will start to decrease.

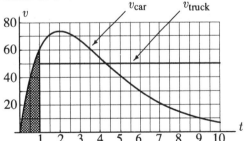

Figure 7.3: Shaded area = distance traveled by car (noon–1pm)

Thus the distance between the car and the truck will increase as long as $v_{car} > v_{truck}$, and will decrease when $v_{car} < v_{truck}$. Therefore, the maximum distance occurs when $v_{car} = v_{truck}$, i.e. when $t \approx 4.3$ hours (at about 4:18 pm). See Figure 7.4.

The distance traveled by the car is the area under the v_{car} graph between $t = 0$ and $t = 4.3$; the distance traveled by the truck is the area under the v_{truck} line between $t = 1$ (when it started) and $t = 4.3$. The distance between the car and truck is represented by the shaded area in Figure 7.4, which is approximately

$$35 \text{ miles} + 50 \text{ miles} = 85 \text{ miles}.$$

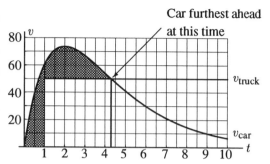

Figure 7.4: Shaded area = distance by which car is ahead at 4:18 pm

(c) The truck overtakes the car when both have traveled the same distance. This occurs when the area under the curve up to that time equals the area under the line up to that time. Since the areas under the curve and the line overlap (see Figure 7.5), they are equal when the heavily shaded area equals the lightly shaded area (which we know is about 85 miles). This happens when $t \approx 8.3$ hours, or about 8:18 pm.

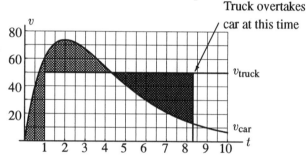

Figure 7.5: Which is ahead: the car or the truck?

❑

❖ Exercises for Section 7.1

1. For the two cars in Example 1, page 464, estimate

   (a) the distances moved by Car 1 and Car 2 during the first minute.
   (b) the time at which the two cars have gone the same distance.

2. Consider the car and the truck in Example 2 on Page 464.

   (a) How fast is the distance between the car and the truck increasing or decreasing at 3 pm?
   (b) What is the practical significance (in terms of distances) of the fact that the car's velocity is maximum at about 2 pm?

3. Consider the car and the truck in Example 2 on Page 464, but suppose the truck starts at noon. (Everything else remains the same.)

   (a) Sketch a new graph showing the velocities of both car and truck against time.
   (b) How many times do the two graphs intersect? What does each intersection mean in terms of the distance between the two?

For each region in Problems 4–6, write a definite integral which represents its area. Evaluate the integral to derive a formula for the area.

4. A rectangle with base b and height h:

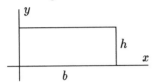

5. A right triangle of base b and height h:

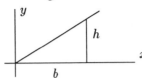

6. A circle of radius r:

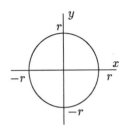

7. A car moves along a straight line with velocity, in ft/sec, given by

$$v(t) = 6 - 2t \quad \text{for } t \geq 0.$$

   (a) Describe the car's motion in words. (When is it moving forwards, backwards, and so on?)

   (b) Suppose the car's position is measured from its starting point. When is it furthest to the right? To the left?

   (c) Find s, the car's position measured from its starting point, as a function of time.

8. A car moves along a straight line with velocity given by

$$v(t) = 2 + 10t \text{ ft/sec}$$

   for $0 \leq t \leq 10$ seconds.

   (a) Graph $v(t)$ for $0 \leq t \leq 10$ and explain how you can find the total distance the car has traveled between $t = 0$ and $t = 10$ seconds geometrically (i.e., from the graph). Find this distance using the formula for the area of a trapezoid.

   (b) Find the function $s(t)$ that gives the position of the car as a function of time. Explain the meaning of any new constants.

   (c) Explain how you use your function $s(t)$ to find the total distance traveled by the car between $t = 0$ and $t = 10$ seconds. Compare with your answer in part (a).

   (d) Explain how this verifies the Fundamental Theorem of Calculus.

9. The graph of $\frac{dy}{dt}$ against t is in Figure 7.6. Suppose the three shaded regions each have area 2. Given that $y = 0$ when $t = 0$, draw the graph of y against t, indicating all special features the graph might have (known heights, maxima and minima, inflection points, etc.). Pay particular attention to the relationship between the graphs. Mark $t_1, t_2, \ldots, t_5$ on the t axis.[a]

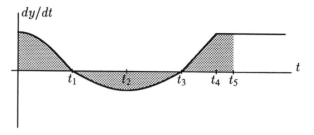

Figure 7.6: dy/dt vs. t
(for Problem 9)

[a]From *Calculus: The Analysis of Functions*, by Peter D. Taylor (Toronto: Wall & Emerson, Inc., 1992)

10. Repeat Problem 9 for the graph of $\dfrac{dy}{dt}$ given to the right. (Again the three shaded regions each have area 2.)

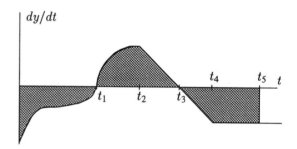

11. Water is run into a large tank through a hose at a constant rate. After 5 minutes a hole is opened in the bottom of the tank and water starts to flow out. Initially the flow rate through the hole is twice as great as the rate through the hose, but as the water level in the tank goes down, the flow rate through the hole decreases, and after another 10 minutes, the water level in the tank appears to be constant. Plot graphs of the flow rates through the hose and through the hole against time on the same pair of axes. Show how the volume of water in the tank at any time can be interpreted as an area (or the difference between two areas) on the graph. In particular, interpret the steady state volume of water in the tank.[1]

12. (a) Find the average $a(T)$ of $f(t) = \sin(t)$ between 0 and T.

   (b) Graph $a(t)$ and $f(t)$ on the same axis. Where is $a(t)$ increasing? Where is $a(t)$ smaller than $f(t)$? Where is $a(t)$ a maximum? Where is $a(t) = f(t)$? Explain.

13. Repeat Problem 12 with $f(t) = t\sin(t)$.

14. Consider the average $a(T)$ of an arbitrary function $f(t)$ between 0 and T. Explain why local minima and maxima of $a(T)$ occur at values of T where the graphs of $a(T)$ and $f(T)$ intersect.

[1]From *Calculus: The Analysis of Functions*, by Peter D. Taylor (Toronto: Wall & Emerson, Inc., 1992)

15. Figure 7.7 is a graph of the annual yield, $y(t)$ (in bushels per year), from an orchard t years after planting. The trees take about 10 years to get established, but from that point on, for the next 20 years, they give a substantial yield. After about 30 years, however, age and disease start to take their toll and the annual yield falls off.[2]

    (a) Represent on a sketch of Figure 7.7 the total yield $Y(T)$ up until time T and write an expression for $Y(T)$ in terms of $y(t)$.

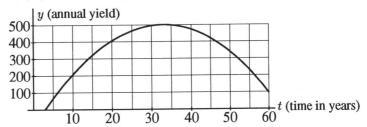

Figure 7.7: The Yield from an Orchard

(for Problem 15)

    (b) Sketch a graph of $Y(T)$ against T.

    (c) Write an expression for the average annual yield, $a(T)$, up until time T.

    (d) The important question is: when should the orchard be cut down and replanted? Assume that what you want to do is maximize your average revenue per year, and that fruit prices remain constant, so that this is achieved by maximizing your average annual yield.

        i. What condition on $Y(T)$ maximizes the average annual yield?

        ii. What condition on $y(T)$ maximizes the average annual yield? Estimate the T value which yields the maximum average annual yield.

7.2 Setting up Riemann Sums

Using the Notation

There are many situations where a quantity we want to calculate can be expressed as a definite integral. Typically this will happen when the quantity can be approximated by dividing it up into little pieces, solving the problem approximately for each little piece, and then adding up the resulting approximations. For example, the way we learned that distance traveled is the definite integral of velocity was by adding up the distances traveled over small time intervals; this yielded a Riemann sum, and the definite integral came about as the limit of that Riemann sum as the number of subdivisions went to infinity. In the examples of Section 7.1, it was not necessary to go through this process: it was enough to *know* that distance traveled can be expressed as a definite integral. However, in many of the problems we will encounter in this chapter, we will not immediately

[2]From *Calculus: The Analysis of Functions*, by Peter D. Taylor (Toronto: Wall & Emerson, Inc., 1992)

recognize the quantity we want as a Riemann sum. The key, then, is to understand how the desired quantity can be chopped up into lots of small pieces in such a way that you can calculate the desired quantity for each small piece. If adding up all these pieces gives a Riemann sum, then taking the limit as the number of subdivisions tends to infinity gives a definite integral.

Recall how we do this for a particle moving with velocity $v = f(t)$ at time t. We find the net distance traveled between $t = a$ and $t = b$ by cutting the time interval into subintervals of length Δt. Providing Δt is small, we can approximate the distance traveled during this time by assuming that the velocity is constant over the subinterval $t_i \leq t \leq t_i + \Delta t$. Let's suppose the velocity throughout the interval is $v = f(t_i)$, the velocity at the start. The distance moved during that time is approximately

$$(\text{velocity}) \times (\text{time}) \quad \text{or} \quad f(t_i)\,\Delta t.$$

Therefore, the net change in position is approximated by the sum of all of these small distances, so

$$\text{Change in Position} \approx \sum_{i=0}^{n-1} f(t_i)\,\Delta t.$$

In the limit as Δt tends to zero, this sum becomes the integral:

$$\text{Change in Position} = \lim_{\Delta t \to 0} \sum_{i=0}^{n-1} f(t_i)\,\Delta t = \int_a^b f(t_i)\,dt.$$

Informally, the $f(t)\,dt$ may be thought of as the distance traveled by the particle over the infinitesimally small time interval dt, since it is the velocity multiplied by the time. In fact, although the dt in the definite integral is no longer a number, it behaves as if it were a very small, or infinitesimal, version of Δt, and the $f(t)\,dt$ may still be thought of as a very small, or infinitesimal, distance traveled. Thus the integral may be thought of as the sum of all these small distances.

Density and How to Slice

Most of the examples in this section involve the idea of *density*. For example,

- A population density is measured in people per mile (along the edge of a road), or people per unit area (in a city), or bacteria per cm^3 (in a test-tube).

- The density of a substance (air, wood, metal, for example), is the mass of a unit volume of the substance, and is measured in, say, grams per cm^3.

We can calculate the total mass or total population using a definite integral and information about the density. To find the total quantity, we divide the region into small pieces and find the quantity from each one by multiplying the density by the size of that piece. To do this, the density must be approximately constant on each piece. Thus,

- The region should be sliced up in such a way that the density is approximately constant on each piece.

Examples Using Density

☐ **Example 1** The Massachusetts turnpike starts in the middle of Boston and heads west. The number of people living next to the pike varies as it gets further and further from town. Suppose that, x miles out of town, the population density adjacent to the pike is $P = f(x)$ people/mile. Express the total population living next to the pike within 5 miles of Boston as a definite integral.

Solution. Think about how you would calculate the population. First, you could get a rough figure for each mile of the pike. The population density at the center of Boston is $f(0)$; let's use that density for the whole first mile, giving an estimate of $f(0) \cdot 1 = f(0)$ people living in the first mile. The density one mile out is $f(1)$. Using that figure for the portion out to the second mile gives an estimate of $f(1)$ people living in the second mile. Continuing in this way, we get estimates of $f(2)$, $f(3)$, and $f(4)$ for the remaining three miles. Our estimate of the total population is

$$f(0) + f(1) + f(2) + f(3) + f(4).$$

Now, to get a more accurate estimate, you could break the distance up into tenths of miles. The population in the first 1/10 mile would be roughly the population density at the beginning of that 1/10 mile times 1/10 (since the density is in people per *mile*), or $f(0) \cdot 1/10$. The population in the second 1/10 mile would be roughly the population density at the beginning of that 1/10 mile times 1/10, or $f(0.1) \cdot 1/10$, and so on. The total estimate would be

$$\underbrace{f(0) \cdot 0.1 + f(0.1) \cdot 0.1 + f(0.2) \cdot 0.1 + \cdots + f(4.9) \cdot 0.1}_{50 \text{ terms}}.$$

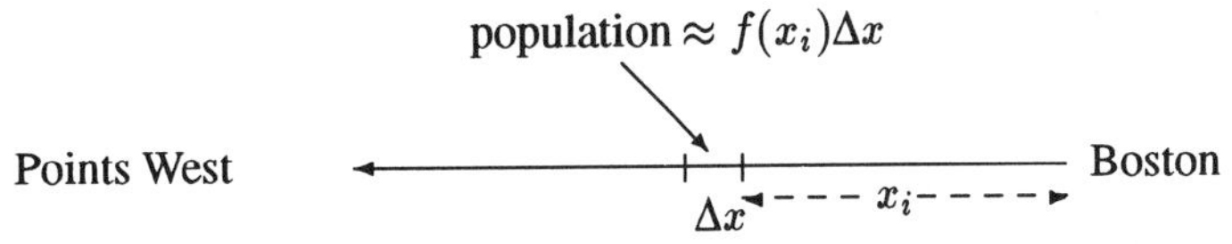

Figure 7.8: Population along the Mass Pike

This is a Riemann sum approximating the total population, with a subdivision of the interval $0 \le x \le 5$ into 50 pieces, and with $\Delta x = 0.1$. The population in each piece $x_i \le x \le x_i + \Delta x$ is roughly $f(x_i) \Delta x$, i.e., the population density times the length of the interval. See Figure 7.8. The sum of all these estimates gives the estimate

$$\text{Total Population} \approx \sum_{i=0}^{49} f(x_i) \Delta x.$$

With n subdivisions, we get

$$\text{Total Population} \approx \sum_{i=0}^{n-1} f(x_i) \Delta x.$$

Letting n tend to ∞ shows that

$$\text{Total Population} = \lim_{n\to\infty} \sum_{i=0}^{n-1} f(x_i)\,\Delta x = \int_0^5 f(x)\,dx.$$

The 5 and 0 in the limits of the integral are the upper and lower limits of the interval over which we are integrating. ☐

☐ **Example 2** The air density (in kg/m^3) h meters above the earth's surface is $P = f(h)$. Find the mass of a cylindrical column of air 2 meters wide and 25 kilometers high.

Solution. Our column of air is a circular cylinder, 2 meters wide and 25 kilometers, or 25,000 meters, high. First we should decide how we are going to slice this column into small pieces. Since the air density varies with the altitude, but remains constant horizontally, it makes sense to take horizontal slices of air. That way the density will be more or less constant over the whole slice, being close to its value at the bottom of the slice (see Figure 7.9).

A slice will be a cylinder of

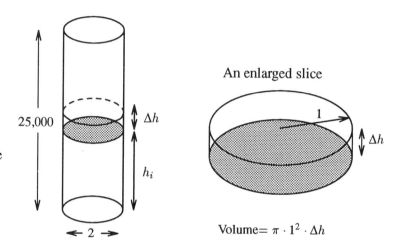

Figure 7.9: How to slice a column of air.

height Δh and width 2 meters, so its radius will be 1 m. We can find the approximate mass of the slice by multiplying its volume and the density together. If the thickness of the slice is Δh, then its volume is $\pi r^2 \cdot \Delta h = \pi 1^2 \cdot \Delta h = \pi\,\Delta h$. The density of the slice between h_i and $h_i + \Delta h$ is roughly $f(h_i)$. Thus

$$\text{Mass of Slice} \approx \text{Volume} \cdot \text{Approximate Density} = \pi\,\Delta h \cdot f(h_i).$$

If there are n slices, then adding all these up will yield a Riemann sum:

$$\text{Total Mass} \approx \sum_{i=0}^{n-1} \pi f(h_i)\,\Delta h.$$

As $n \to \infty$, this sum approximates the definite integral:

$$\text{Total Mass} = \int_0^{25000} \pi f(h)\,dh.$$

☐

☐ **Example 3** Find the mass of the column of air in Example 2 if the density of air at height h is given by

$$P = f(h) = 1.29e^{-0.000124h} \text{ kg/m}^3.$$

Solution. Using the result of the previous example, we have

$$\text{Mass} \;=\; \int_{0}^{25000} \pi 1.29 e^{-0.000124h}\, dh \;=\; \frac{1.29\pi}{0.000124}\left(\left.-e^{-0.000124h}\right|_{0}^{25000}\right)$$

$$\approx\; 32{,}683(e^{0} - e^{-0.000124(25000)}) \approx 31{,}000\,\text{kg}.$$

☐

In Example 2, we did not write down any specific Riemann sums, say with 5 or 50 subdivisions, as we did in Example 1. Rather, we thought about the general Riemann sum for the problem, in order to be able to write down the definite integral. The Δh became the dh, and the limits in the definite integral were obtained by considering the limits for the variable h. Once we decided to slice the column horizontally, we knew that a typical slice has thickness Δh and so h must be the variable in our definite integral.

☐ **Example 4**   Water leaks out of a tank through a square hole with 1 inch sides. At time t (in seconds) the velocity of water flowing through the hole is $v = g(t)$ ft/sec. Write down a definite integral that represents the total amount of water lost in the first minute.

Solution. In this problem we slice time into small intervals, because we are trying to get from a rate of change to a total amount. Since t is given in seconds, we will convert the one minute to 60 seconds. Thus we are considering water loss over the time interval $0 \le t \le 60$. We also need to convert the inches into feet. Since 1 inch = $\frac{1}{12}$ foot, the square hole has area $\frac{1}{144}$ square feet. For water flowing through a hole with constant velocity v, the amount of water which has passed

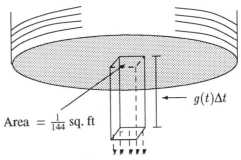

Area $= \frac{1}{144}$ sq. ft

Figure 7.10: Volume of Water passing through hole

through in some time, Δt, can be pictured as the rectangular solid in Figure 7.10, which has volume

$$\text{area}\cdot\text{height} = \text{area}\cdot\text{velocity}\cdot\text{time}.$$

In our problem, over a small time interval of length Δt, starting at time t, water flows with a nearly constant velocity $v = g(t)$ through a hole $1/144$ square feet in area. In Δt seconds, we know that

$$\text{Water Lost} \approx (1/144)g(t)\,\Delta t \text{ cubic feet.}$$

Adding the water from each subinterval gives

$$\text{Total Water Lost} \approx \sum \frac{1}{144}g(t)\,\Delta t \text{ cubic feet.}$$

from all parts

As Δt tends to 0, the sum becomes the definite integral:

$$\text{Total Water Lost} = \int_0^{60} \frac{1}{144} g(t) \, dt.$$

❑

Note that the sum represented by the $\sum$ sign is over all the small time intervals. In this example we did not bother to write down the points t_0, t_1, etc., that you need to write the Riemann sum precisely. This is unnecessary if all you want is the final expression for the definite integral.

Sometimes it is quite difficult to work out how to slice a quantity up. The key point to remember always is that you want the density, rate, or whatever, to be nearly constant within each piece, so that you will be able to write down a formula for the contribution from that piece.

❐ **Example 5** The population density in Ringsburg is a function of the distance from the city center: at r miles from the center, the density is $P = f(r)$ people per square mile. Ringsburg has a radius of 5 miles. Write down a definite integral that expresses the total population of Ringsburg.

Solution. We want to slice Rings-burg up and estimate the population on each slice. If we take straight slices, the population density will vary on each slice, since it depends on the distance from the city center. We want the population density to be pretty close to constant on each slice, so that we will be able to estimate the population by multiplying the density and the area together. We therefore take slices that are thin rings around the center, at a constant distance from the center (see Figure 7.11).

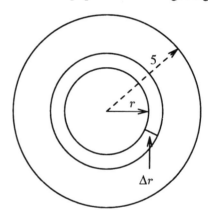

Figure 7.11: Ringsburg

The area of this ring could be computed exactly, but since it is very thin, we can approximate it quite closely by straightening it into a thin rectangle (see Figure 7.12).

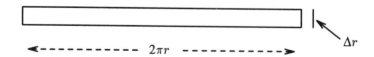

Figure 7.12: Ring from Ringsburg (straightened out)

The width of the rectangle is Δr, and its length is approximately equal to the circumference of the ring, $2\pi r$, so its area is about $2\pi r \, \Delta r$. Thus

$$\text{Population on ring} \approx \text{Density} \cdot \text{Area},$$

so

$$\text{Population on ring} \approx f(r) \cdot 2\pi r \,\Delta r.$$

Adding the contributions from each ring, we get

$$\text{Total Population} \approx \sum 2\pi r f(r) \,\Delta r.$$

So

$$\text{Total Population} = \int_0^5 2\pi r f(r) \,dr.$$

Note: You may wonder what happens if we calculate the area of the ring by subtracting the area of the inner circle (πr^2) from the area of the outer circle ($\pi(r + \Delta r)^2$), giving

$$\text{Area} = \pi(r + \Delta r))^2 - \pi r^2.$$

Multiplying out and subtracting we get

$$\begin{aligned} \text{Area} &= \pi(r^2 + 2r\,\Delta r + (\Delta r)^2) - \pi r^2 \\ &= 2\pi r\,\Delta r + \pi(\Delta r)^2. \end{aligned}$$

This expression differs from the one we used before by the $\pi(\Delta r)^2$ term. However, as Δr becomes very small, $\pi(\Delta r)^2$ becomes much, much smaller. We say it is *second order* of smallness, since the power of the small term, Δr, is 2. In the limit as $\Delta r \to 0$, we can ignore the $\pi(\Delta r)^2$. ☐

Note that, as in each of the examples above, your Riemann sum should always be a sum of terms of the form

$$(\text{function of } x) \cdot \Delta x.$$

❖ Exercises for Section 7.2

1. A rod has length 2 meters. At a distance x from its left end, the density of the rod is given by

$$\rho(x) = 2 + 6x \text{ grams}/\text{m}.$$

   (a) Write a Riemann sum approximating the total mass of the rod.

   (b) Find the exact mass by converting the sum into an integral.

2. The *center of mass* of a rod is the point at which you would place a pivot so that the rod would balance on the pivot. To find the position of the center of mass of a rod, you must first calculate its *moment*, which is given by $\int_a^b x\rho(x)\,dx$, where a and b are the left and right endpoints of the rod, and $\rho(x)$ is its density at a distance x from a. Find the moment of the rod in Problem 1 and find its center of mass by dividing its moment by its total mass.

3. The density of cars (in cars per mile) down a 20 mile stretch of the Pennsylvania Turnpike is well approximated by

$$\rho(x) = 300\left(2 + \sin\left(4\sqrt{x + 0.15}\right)\right),$$

where x is the distance in miles from the Breezewood Toll Plaza.

   (a) Sketch a graph of this function for $0 \le x \le 20$.
   (b) Write a sum that approximates the total number of cars on this 20-mile stretch.
   (c) Find the total number of cars on the 20-mile stretch.

4. Circle City, a typical metropolis, is very densely populated near its center, and its population gradually thins out toward the city limits. In fact, its population density is $10{,}000(3 - r)$ people/square mile at distance r miles from the center.

   (a) Assuming that the population at the city limits is zero, find the radius of the city.
   (b) What is the total population of the city?

5. Greater Boston can be approximated by a semicircle of radius 8 miles with its center on the coast. Moving out away from the center along a radius, the population density is constant for the first mile. Beyond that, the density starts to decrease according to the data given in the table, where $\rho(r)$ is the population density at a distance r miles from the center.

r (miles)	0	1	2	3	4	5	6	7	8
$\rho(r)$(thousand people/mi^2)	75	75	67.5	60	52.5	45	37.5	30	22.5

   (a) Using this data and Riemann sums, estimate the total population of Boston living within the 8 mile limit.
   (b) Looking at the data, do you think your answer to (a) is an over- or an underestimate? Why?
   (c) Find a possible formula for $\rho(r)$. Use this to make another estimate of the population.

6. The density of oil in a circular oil slick on the surface of the ocean at a distance r from the center of the slick is given by $\rho(r) = \dfrac{50}{1 + r}\ \dfrac{\text{kg}}{\text{m}^2}$.

   (a) If the slick extends from $r = 0$ to $r = 10{,}000$ m, find a Riemann sum approximating the total mass of oil in the slick.
   (b) Find the exact value of the mass of oil in the slick by turning your sum into an integral and evaluating it.
   (c) Within what distance r is half the oil of the slick contained?

7. Suppose you want to find the total mass of a 3 by 5 rectangular sheet, whose density per unit area at a distance x from a particular side of length 5 is $\dfrac{1}{1+x^4}$.

   (a) Find a Riemann sum which approximates the total mass.

   (b) Find the mass to one decimal place. How do you know your answer is accurate enough?

8. You are given a cardboard figure in the shape shown to the right. The region is bounded on the left by the line $x = a$, on the right by the line $x = b$, above by $f(x)$ and below by $g(x)$. If the density $\rho(x)$ gm/cm^2 varies only with x, find an expression for the total mass of the figure, in terms of $f(x)$, $g(x)$, and $\rho(x)$.

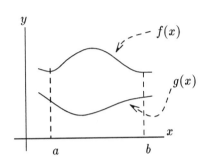

9. In the table to the right we list the density D of the Earth at a depth x below the Earth's surface. (D is measured in g/cm^3 and x is measured in km; the radius of the Earth is about 6370 km.) Find an upper and a lower bound for the Earth's mass, such that the upper bound is less than twice the lower bound. Explain your reasoning; in particular, what assumptions have you made about the density?

x	D
0	3.3
1000	4.5
2000	5.1
2900	5.6
3000	10.1
4000	11.4
5000	12.6
6000	13.0
6370	13.0

10. The exponential model for the density of the Earth's atmosphere says that if the temperature of the atmosphere were constant, then its density as a function of height, h (in meters), above the surface of the earth, would be given by

$$\rho(h) = 1.28e^{-0.000124h}\,\frac{\text{kg}}{\text{m}^3}.$$

Of course, the temperature of the atmosphere is not constant, but we will assume it is for this problem.

   (a) Write a sum that approximates the mass of the portion of the atmosphere from $h = 0$ to $h = 100$ m, i.e., the first 100 meters from sea level. Assume the radius of the Earth is 6378 km.

   (b) Find the exact answer by turning your sum in (a) into an integral. Evaluate the integral by any means.

   (c) In terms of integrals, what is the mass of the *whole* atmosphere (i.e. as $h \to \infty$)? Evaluate this integral.

11. Use your answer to Problem 10 to answer these questions.

    (a) What fraction of the mass of the Earth's atmosphere is contained within the first 100 meters?

    (b) To what height must one go such that 90% of the atmosphere's mass is contained below that height?

12. Water is flowing in a cylindrical pipe of radius 1 inch. Because water is viscous and sticks to the pipe, the rate of flow varies with the distance from the center. The speed of the water at distance r inches from the center is $10(1 - r^2)$ inches per second. What is the rate (in cubic inches per second) at which water is flowing through the pipe?

7.3   Applications to Geometry

We have already seen how the definite integral can be used to compute the areas of certain regions. In this section, we will show how the definite integral can also be used to compute volumes of solids and lengths of curves. The strategy we will employ to make these computations is always the same: divide the solid (or curve) into small pieces whose volume (or length) we can easily approximate, and then sum the contributions of all the pieces, giving a Riemann sum that approximates the total volume (or length). As the number of terms in the sum tends to infinity, the approximation gets better, and the sum approaches a definite integral. The value of the definite integral is therefore the total volume of the solid (or the total length of the curve).

Volumes of Given Cross-Section

When calculating the volume of a solid using Riemann sums, we chop the solid up into smaller pieces whose volumes we can estimate. Oftentimes, however, it's difficult to visualize the appearance of the pieces into which we slice our solid. Furthermore, we will usually have several options for how we go about our slicing.

How to Slice a Cone

Let's see how we might slice up a cone which is standing with the vertex uppermost. We'll consider two different types of slices. We could divide the cone into vertical slices, each parallel to a plane that bisects the cone and passes through its vertex. The slices will be arch-shaped (their cross-sections are actually hyperbolas); see Figure 7.13. We could also divide the cone horizontally, giving coin shaped slices; see Figure 7.14.

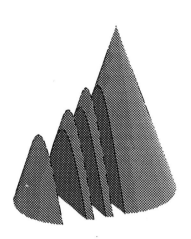

Figure 7.13: Cone Cut Into Vertical Slices   **Figure 7.14**: Cone Cut Into Horizontal Slices

To construct a Riemann sum, we probably want to cut the cone into horizontal (circular) slices, because it would be difficult to estimate the volumes of the arch-shaped slices.

☐ **Example 1** Compute the volume, in cubic feet, of the Great Pyramid of Egypt, whose base is a square 755 feet by 755 feet and whose height is 410 feet.

Solution. You may know already the equation $V = \frac{1}{3}b^2 \cdot h$ for the volume V of a pyramid with height h and base area b. We will in effect derive this formula in solving this problem.

We will view the pyramid as built up in layers starting from the base. Each layer will be square with a little bit of thickness which we denote Δh (think of Δh as meaning "a little bit of h"). The bottom layer is a square slab 755 feet by 755 feet by Δh feet. As we move up the pyramid, the layers have shorter side length. If we let s denote this side length, then the volume of each layer is approximately $s^2 \Delta h$, where s varies from 755 feet for the bottom layer to 0 feet for the top layer. (The volume is only approximately $s^2 \Delta h$ because the sides of the slab are not exactly vertical.) See Figure 7.15.

The total volume of the pyramid is the sum of the volumes, $s^2 \Delta h$, of each layer. Since each layer is at a different height h, we must express s as a function of h, so that each term in the sum depends on h alone. If we cut open the pyramid along a plane through the top point of the pyramid and perpendicular to a side of the base, we obtain the triangular cross-section drawn in the left-hand picture in Figure 7.15. By similar triangles, we get $\frac{s}{755} = \frac{410-h}{410}$. Thus $s = \left(\frac{755}{410}\right)(410 - h)$, and the total volume, V, is approximated by

$$V \approx \sum s^2 \Delta h = \sum \left[\left(\frac{755}{410}\right)(410 - h)\right]^2 \Delta h.$$

As the thickness of each slice tends to zero, the sum becomes a definite integral. Since h varies from

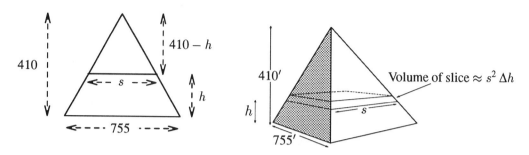

Figure 7.15: The Great Pyramid

0 to 410, the height of the pyramid, we have

$$V = \int_{h=0}^{h=410} \left[\left(\frac{755}{410} \right) (410 - h) \right]^2 dh = \left(\frac{755}{410} \right)^2 \int_0^{410} (410 - h)^2 dh$$

$$= \left(\frac{755}{410} \right)^2 \left(-\frac{(410 - h)^3}{3} \right) \Big|_0^{410} = \frac{1}{3} \left(\frac{755}{410} \right)^2 (410)^3 = \frac{1}{3}(755)^2(410)$$

$$\approx \quad 78 \text{ million cubic feet.}$$

Note that $V = \frac{1}{3}(755)^2(410) = \frac{1}{3}b^2 \cdot h$, as expected. ◻

☐ **Example 2** Find the volume of a hemisphere of radius 7 inches.

Solution. Again, you probably know the formula $\frac{4}{3}\pi R^3$ for the volume of a sphere of radius R. We will derive this using a Riemann sum and a definite integral.

We imagine the hemisphere sitting flat on a plane and divide it into horizontal layers of thickness Δh. See Figure 7.16. Each layer is a circular slab with radius r, where r varies from 7 inches for the bottom layer to 0 inches for the top layer. Thus the volume of each layer is approximately $\pi r^2 \Delta h$, and our problem is to express r in terms of h so that we can write a Riemann sum (all of whose terms depend only on h) representing the sum of the volumes of all the separate layers. As with the pyramid, we cut the hemisphere by a plane through the top point of the hemisphere and perpendicular to the base. We get a semicircle as shown in Figure 7.17, and inside it we find a right triangle which relates the radius r of the horizontal layer to the height h above the base of the hemisphere. Using this triangle, we find that $r^2 = 7^2 - h^2$, so the volume of a circular slab at height h is

$$\pi r^2 \Delta h = \pi(7^2 - h^2)\Delta h.$$

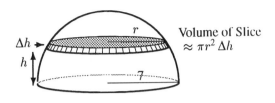

Figure 7.16: Slicing to find the volume of a hemisphere.

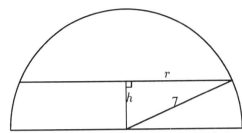

Figure 7.17: The triangle relating r and h for the hemisphere.

Summing the volumes of all slabs, we have

$$\text{Volume, } V \approx \sum \pi r^2 \, \Delta h = \sum \pi (7^2 - h^2) \, \Delta h.$$

As the thickness of each slab tends to zero, the sum becomes a definite integral. Recall that h varies from 0 to 7, the radius of the hemisphere. Thus,

$$V = \int_{h=0}^{h=7} \pi r^2 \, dh = \pi \int_0^7 (7^2 - h^2) \, dh = \pi \left(7^2 h - \frac{1}{3} h^3 \right) \Big|_0^7 = \pi \left(7^3 - \frac{1}{3} 7^3 \right) = \frac{2}{3} \pi 7^3.$$

The volume of the whole sphere is $2(\frac{2}{3}\pi 7^3) = \frac{4}{3}\pi 7^3$, as we expected. $\qquad\qquad$ $\square$

Examples 1 and 2 both had the feature that all cross-sections of the solid perpendicular to a certain direction were the same geometric figure, a square in the case of the pyramid and a circle in the case of the hemisphere. We can define solids that are much more difficult to picture, but that still have "known cross-sections" so that their volumes are easy to compute.

☐ **Example 3** Find the volume of the solid whose base is the region in the xy-plane bounded by the curves $y = x^2$ and $y = 8 - x^2$ and whose cross-sections perpendicular to the x-axis are squares with one side in the xy-plane. (See Figure 7.18.)

Solution. We view the solid as a loaf of bread, sitting on the xy-plane and made up of square slices. These slices are analogous to the square layers making up the pyramid in our first example. A typical slice is shown in Figure 7.19. The thickness of each slice is a little bit of distance along the x-axis and hence we denote it by Δx. The side length s of the square face of each slice is just the distance (in the y-direction) between

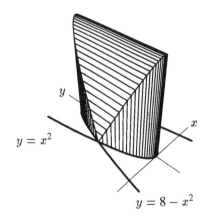

Figure 7.18: The solid for Example 3

the two curves. Therefore, in terms of x we have $s = (8 - x^2) - x^2 = 8 - 2x^2$.

Notice that this side length grows from $s = 0$ at $x = -2$ to $s = 8$ at $x = 0$, then it shrinks back to $s = 0$ at $x = 2$. The volume of the solid is obtained by summing up the volumes of all the slices. Each slice has volume $s^2 \, \Delta x = (8 - 2x^2)^2 \, \Delta x$, so the total volume is

$$V \approx \sum s^2 \, \Delta x = \sum (8 - 2x^2)^2 \, \Delta x.$$

As the thickness Δx of each slice tends to zero, the sum becomes a definite integral. Since x varies between -2 and 2, we have

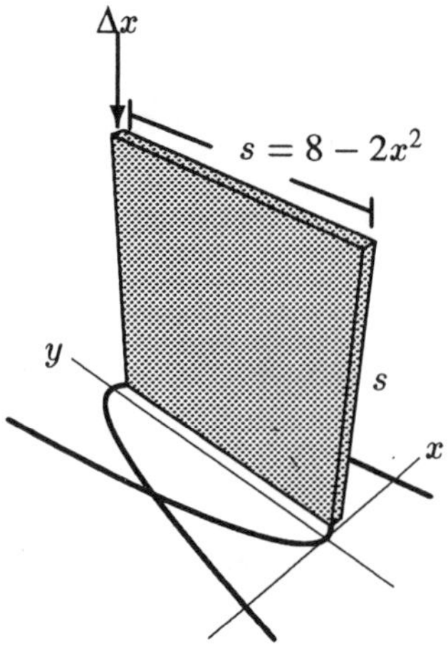

Figure 7.19: A slice of the solid for Example 3

$$
\begin{aligned}
V &= \int_{-2}^{2} (8 - 2x^2)^2 \, dx \\
&= \int_{-2}^{2} (64 - 32x^2 + 4x^4) \, dx \\
&= \left. 64x - \frac{32}{3}x^3 + 4\frac{x^5}{5} \right|_{-2}^{2} \\
&= 256 \left(1 - \frac{2}{3} + \frac{1}{5} \right) = 256 \left(\frac{8}{15} \right) \\
&\approx 136.5.
\end{aligned}
$$

We can check the plausibility of this answer by observing that the given solid is somewhat larger than the solid obtained by putting together two pyramids with 8 by 8 bases at $x = 0$ and apexes (highest points) at $x = 2$ and $x = -2$. By our first example, the total volume of these two pyramids is $2(\frac{1}{3})8^2 \cdot 2 = 256(\frac{1}{3})$, which is, as expected, less than our answer $256(\frac{8}{15})$. ☐

Volumes of Revolution

Another way to create a solid having known cross sections is to revolve a region in the plane around some line in 3-space, giving a *solid of revolution*, as in the following examples.

☐ **Example 4** The region bounded by the curve $y = e^{-x}$ and the x-axis between $x = 0$ and $x = 1$ is revolved around the x-axis. Find the volume of this solid of revolution.

Solution. Imagine the plane region divided into thin strips perpendicular to the x-axis. Each strip has width Δx. As we rotate the whole region about the x-axis, each strip sweeps out a circular slice or disk of thickness Δx (see Figure 7.20). The radius of the slice, as can be seen in Figure 7.20, is the distance from the x-axis to the curve, which is just the value of $y = e^{-x}$ on the slice. Thus the volume of the slice is $\pi y^2 \Delta x = \pi(e^{-x})^2 \Delta x$, and the total volume, V, of the solid is given by

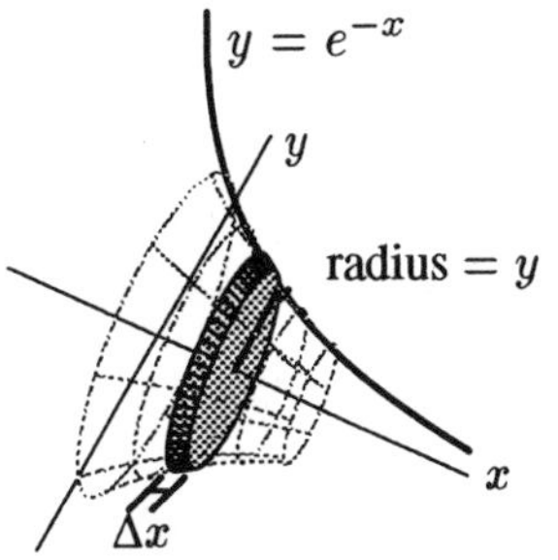

$$V \approx \sum \pi y^2 \Delta x = \sum \pi \left(e^{-x}\right)^2 \Delta x.$$

Figure 7.20: A thin strip rotated around x-axis to form a circular slice.
(Example 4)

As the thickness of each slice tends to zero, we get

$$V = \int_0^1 \pi(e^{-x})^2 \, dx = \pi \int_0^1 e^{-2x} \, dx = \pi \left(-\frac{1}{2}\right) e^{-2x} \Big|_0^1$$

$$= \pi \left(-\frac{1}{2}\right) (e^{-2} - e^0) = \frac{\pi}{2}(1 - e^{-2}) \approx 1.36.$$

□

☐ **Example 5** The region bounded by the curves $y = x$ and $y = x^2$ is rotated about the line $y = 3$. Compute the volume of the resulting solid.

Solution. Again imagine the region divided into thin vertical strips of thickness Δx, as in Figure 7.21.

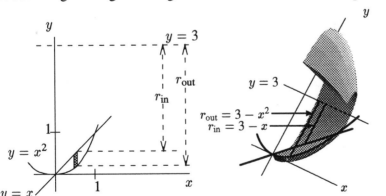

Figure 7.21: The region for Example 5

Figure 7.22: Cutaway View of Volume.
(Note Inner and Outer Radii.)

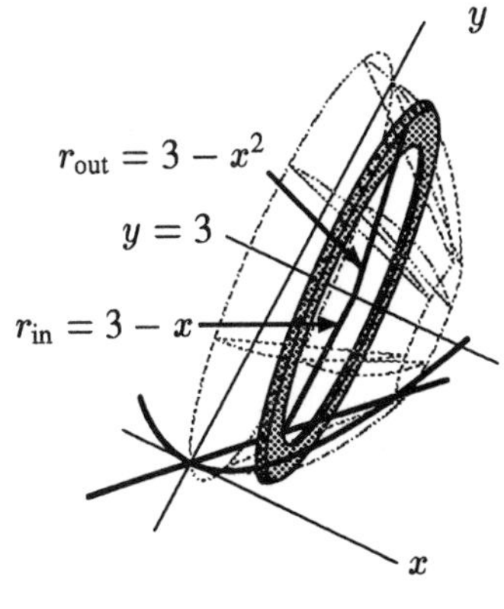

Figure 7.23: One Slice (a washer)

As each strip is rotated around the line $y = 3$, it sweeps out a washer-like slice of the total solid (see

Figure 7.22 and Figure 7.23). This washer has two radii: an inner radius r_{in}, which is the distance from the line $y = 3$ to the nearer curve $y = x$, and an outer radius r_{out}, which is the distance from the line $y = 3$ to the more distant curve, $y = x^2$. Thus the slice at x has inner radius $r_{\text{in}} = 3 - x$ and outer radius $r_{\text{out}} = 3 - x^2$. This washer may be viewed as a circular disk of radius r_{out} from which has been removed a smaller circular disk of radius r_{in}. Thus the volume of the washer is

$$\pi r_{\text{out}}^2 \Delta x - \pi r_{\text{in}}^2 \Delta x = \pi (3 - x^2)^2 \Delta x - \pi (3 - x)^2 \Delta x.$$

Adding up the volumes of all the washers, we have

$$V \approx \sum \left(\pi r_{\text{out}}^2 - \pi r_{\text{in}}^2 \right) \Delta x = \sum \left(\pi (3 - x^2)^2 - \pi (3 - x)^2 \right) \Delta x.$$

We let Δx, the thickness of each washer, tend to zero to obtain the definite integral. Since the curves $y = x$ and $y = x^2$ intersect at $x = 0$ and $x = 1$, we have

$$V = \int_0^1 \left(\pi (3 - x^2)^2 - \pi (3 - x)^2 \right) dx = \pi \int_0^1 \left((9 - 6x^2 + x^4) - (9 - 6x + x^2) \right) dx$$

$$= \pi \int_0^1 (6x - 7x^2 + x^4) dx = \pi \left(3x^2 - \frac{7x^3}{3} + \frac{x^5}{5} \right) \Big|_0^1 \approx 2.72.$$

$\square$

Arclength

Just as the definite integral can be used to compute the volumes of various solids, it can also be used to compute the lengths of various curves. Unlike the case for volumes, there is one simple formula to compute the length of any curve given as the graph of a function.

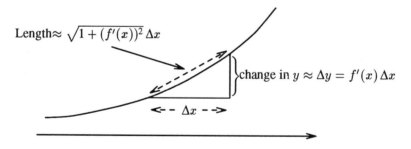

Figure 7.24: A little bit of arclength

Suppose we wish to compute the length, or the *arclength*, as it is usually called, of a curve $y = f(x)$ from $x = a$ to $x = b$, where $a < b$. As usual, we divide the curve into many little pieces; each one is approximately straight. Figure 7.24 shows that a small change Δx in the x-coordinate produces a corresponding small change in the y-coordinate of approximately $\Delta y = f'(x) \Delta x$. The length of the little piece of the curve is then approximately

$$\sqrt{(\Delta x)^2 + (\Delta y)^2} = \sqrt{(\Delta x)^2 + (f'(x) \Delta x)^2} = \sqrt{1 + (f'(x))^2} \, \Delta x.$$

Thus, for $a < b$, the arclength of the curve $y = f(x)$ from $x = a$ to $x = b$ is approximated by a Riemann sum:

$$\text{Arc Length} = L \approx \sum \sqrt{1 + (f'(x))^2} \cdot \Delta x.$$

As we let Δx tend to zero, the sum becomes the definite integral. Since x is to vary between a and b, we have

$$\boxed{\text{Arc Length} = L = \int_a^b \sqrt{1 + (f'(x))^2}\, dx.}$$

☐ **Example 6** Set up and evaluate an integral to compute the length of the curve $y = x^3$ from $x = 0$ to $x = 5$.

Solution. By the arclength formula, we have

$$L = \int_0^5 \sqrt{1 + (3x^2)^2}\, dx.$$

Although the formula for the arclength of a curve is easy to apply, the integrands it generates often have no antiderivative in terms of elementary functions. As a result, the only way to evaluate an arclength integral such as this one is by numerical methods, such as the midpoint or Simpson's Rule. In this case, Simpson's Rule for $\int_0^5 \sqrt{1 + 9x^4}\, dx$ gives 125.680669093 with 10 steps, and 125.680300379 with 20 steps. An answer of 125.68 looks reasonable. The curve starts at $(0,0)$ and goes to $(5, 125)$, so its length must be at least the length of a straight line between these points, or $\sqrt{5^2 + 125^2} = 125.10$. See Figure 7.25. Are you surprised that the length along the line and the curve are so close? ☐

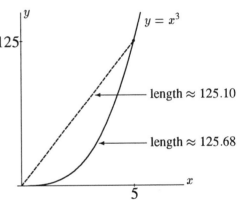

Figure 7.25: Arclength of $y = x^3$. (Note that the picture is distorted because the scales on the two axes are so different.)

❖ Exercises for Section 7.3

1. Imagine a hard-boiled egg lying on its side cut into thin slices. First think about vertical slices, and then horizontal ones. What would these slices look like? Sketch them.

2. Find the volume of a sphere of radius r by slicing up the sphere and setting up a Riemann sum that approximates the volume, and then finding the definite integral obtained when the thickness of each slice tends to zero.

3. Find, by slicing, the volume of a cone whose height is 3 cm and whose base radius is 1 cm. Slice the cone as shown in Figure 7.14, page 479.

4. Find, by slicing, a formula for the volume of a cone of height h and base radius r.

For Problems 5–7, sketch the solid obtained by rotating each region around the indicated axis. Using the sketch, show how to approximate the volume of the solid by a Riemann sum, and use this information to find the volume.

5. Region: First arch of $y = \sin x$, $y = 0$. Axis: x axis.

6. Region: $y = x^3$, $x = 1$, $y = -1$. Axis: $y = -1$.

7. Region: $y = \sqrt{x}$, $x = 1$, $y = 0$. Axis: $x = 1$.

8. When the ellipse $\frac{x^2}{a^2} + \frac{y^2}{b^2} = 1$ is rotated about the x-axis, it generates an *ellipsoid*. Compute its volume.

For Problems 9–15 consider the region bounded by $y = e^x$, the x-axis and the lines $x = 0$ and $x = 1$. Find the volume of the following solids:

9. the solid obtained by rotating the region about the x-axis.

10. the solid obtained by rotating the region about the horizontal line $y = -3$.

11. the solid obtained by rotating the region about the horizontal line $y = 7$.

12. the solid whose base is the given region and whose cross-sections perpendicular to the x-axis are squares.

13. the solid whose base is the given region and whose cross-sections perpendicular to the x-axis are semicircles.

14. the solid whose base is the given region and whose cross-sections perpendicular to the x-axis are isosceles right triangles with hypotenuses lying in the region.

15. a "pyramid" whose base is the given region and whose apex is located 10 units above the origin in the xy-plane. [Hint: each horizontal cross-section has the same shape but the x and y dimensions must decrease to 0 as the height above the xy-plane increases to 10. Thus, for example, the cross-sectional area at height 3 units should be $\left(\frac{7}{10}\right) \times \left(\frac{7}{10}\right)$ the area of the original region.]

16. Rotate the bell shaped curve $y = e^{-x^2/2}$ around the y-axis, forming a "hill" shaped solid of revolution. By slicing horizontally, find the volume of this hill.

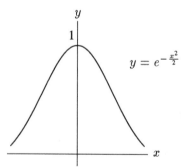

$y = e^{-\frac{x^2}{2}}$

17. The circumference of a certain tree at different heights above the ground is given in the following table:

height (feet)	0	20	40	60	80	100	120
circumference (feet)	26	22	19	14	6	3	1

Assume all horizontal cross-sections of the tree are circles. Estimate the volume of the tree using the trapezoid rule.

18. Most states expect to run out of space to put their garbage soon. In New York, solid garbage is packed into pyramid shaped dumps with square bases. (The largest such dump is on Staten Island.) A small community has a dump with a base length of 100 yards. One yard above the base, the length of the side is 99 yards; the dump can be built up to a vertical height of 20 yards. (The top point of the pyramid is never reached.) If 65 cubic yards of garbage arrive at the dump every day, how long will it be before the dump is full?

19. The hull of a certain boat has widths given by the following table. Reading across a row of the table gives widths at points 0, 10, ..., 60 feet from the front to the back at a certain level below waterline. Reading down a column of the table gives widths at levels 0, 2, 4, 6, 8 feet below waterline at a certain point from the front to the back. Use the trapezoidal rule to estimate the volume of the hull below waterline.

front of boat ⟶ back of boat

		0	10	20	30	40	50	60
	0	2	8	13	16	17	16	10
depth	2	1	4	8	10	11	10	8
below	4	0	3	4	6	7	6	4
waterline	6	0	1	2	3	4	3	2
(in feet)	8	0	0	1	1	1	1	1

For Problems 20 and 21, find the arclength of the given function from $x = 0$ to $x = 2$.

20. $f(x) = \sqrt{4 - x^2}$

21. $f(x) = \sqrt{x^3}$

22. (a) Write an integral which represents the circumference of a circle of radius r.

(b) Evaluate the integral and show that you get the answer you expect.

23. The figure to the right is a graph of the function

$$y = \frac{1}{2}(e^x + e^{-x}).$$

It is called a *catenary* and represents the shape in which a cable hangs. Find the length of this catenary between $x = -1$ and $x = 1$.

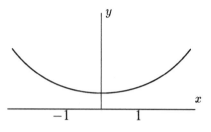

24. Compute the perimeter of the region used in Problems 9–15 with error at most 0.001.

25. Set up an integral for the circumference of an ellipse with semi-major axis $a = 2$ and semi-minor axis $b = 1$. As you try to evaluate this integral numerically (i.e., using a calculator or computer), you are faced with a problem. Describe this problem. Can you think of a way to solve it?

26. There are very few elementary functions $y = f(x)$ where arclength can be computed in elementary terms using the formula

$$\int_a^b \sqrt{1 + \left(\frac{dy}{dx}\right)^2}\, dx.$$

You have seen some such functions f in Problems 20, 21, and 23, namely, $f(x) = \frac{1}{2}(e^x + e^{-x})$, $f(x) = \sqrt{4 - x^2}$, and $f(x) = \sqrt{x^3}$. Try to find some other function that "works," i.e., a function whose arclength you can find using this formula and antidifferentiation.

27. After doing Problem 26 you may wonder what sort of functions can represent arclength. If $g(0) = 0$ and g is differentiable and increasing, then can $g(x)$, $x \geq 0$, represent arclength? That is, can we find a function $f(t)$ such that

$$\int_0^x \sqrt{1 + (f'(t))^2}\, dt = g(x)?$$

(a) Show that $f(x) = \int_0^x \sqrt{(g'(t))^2 - 1}\, dt$ works, namely that the arclength of the graph of f from 0 to x is $g(x)$, as long as $g'(x) \geq 1$.

(b) Show that if $g'(x) < 1$ for some x then $g(x)$ cannot represent the arclength of the graph of any function.

(c) Find a function f whose arclength from 0 to x is $2x$.

7.4   Applications to Physics

Although geometric problems of area, length, and volume may have been a driving force for the development of the calculus in the 17th century, it was Newton's spectacularly successful applications of the calculus to physics that most clearly demonstrated the power of this new mathematics. In this section we will apply the definite integral to problems in physics.

Work

Our first applications involve the concept of work. In physics the word 'work' has a technical meaning which is different from its everyday meaning. Physicists say that if a constant force F is applied to some object to move it a distance d, then the

> *work done*, W, is defined to be $W = F \cdot d$.

Notice that by the definition, if an object does not move, then no work is done. Thus holding a 5 pound book at the same height for an hour accomplishes no work, even though it might make you tired. You are exerting a force on the book to keep the book from falling (that is to counteract the force of gravity on the book), but you are not moving that force through a distance. This does make sense, since the book could just as easily have been sitting on a table for an hour and we wouldn't say that the table had accomplished any work. Also, if we walked across the room holding the book we would not accomplish any work on the book, since the only force we exert on the book is vertical but the motion of the book is horizontal; the distance d in the equation $W = Fd$ must be in the same direction as the force F. On the other hand, if you lift the book from the floor to a height of 3 feet off the floor, this process of lifting accomplishes $W = Fd = 5 \cdot 3 = 15$ foot-pounds of work. (See the note on units below.) Note that time has nothing to do with work. Whether the object is moved quickly or slowly is irrelevant.

In the following examples, we calculate the work done where both the force and distance moved may vary. As before, the idea is to *slice up the figure in such a way that the force is constant across each small piece*. We will calculate the work for each piece using $W = Fd$, and we will sum these pieces to approximate the total work as a Riemann sum. Letting the length of each piece tend to zero, we will obtain a definite integral that represents the total work.

Note on Units: Mass versus Weight

One further complication should be mentioned. When we talk about how much something weighs, we're talking about the force that gravity exerts on that object. A lump of iron, for example, might weigh 10 pounds on the surface of the earth. This is because the Earth's gravitational field is busily exerting 10 pounds of force on that lump of iron. If the same lump, however, is adrift in interstellar space, it is weightless, because there is no gravitational field in its vicinity to exert a force on it. Of course, the quantity of iron present hasn't changed; there's no less iron there than there was before. So we make a distinction between the *mass* of the lump (the quantity of matter it comprises) and the *weight* of that mass (the force exerted on it due to gravity).

In the British system of units, a *pound* is a unit of weight. In the metric system, however, a *kilogram* is a unit of mass, not weight. Thus knowing we have 1 kilogram of iron means that we know the mass of iron present. To figure out how much our iron weighs, we have to figure out how much force is being exerted on it by gravity. Since Newton's second law tells us that Force = Mass × Acceleration, we see that we have to multiply our iron's mass by the acceleration due to gravity, which (on the surface of the Earth) is 9.8m/sec^2. Thus our kilogram of iron weighs $1\text{kg} \cdot 9.8\text{m/sec}^2 = 9.8\text{kg m/sec}^2$, which is almost always written 9.8 newtons. (A newton $= 1\text{kg} \cdot \text{m/sec}^2$ is a unit of force or weight.)

So, if you lift a 2 kilogram book 2 meters off the floor, you have not done $(2\text{kg})(2\text{m}) = 4\text{kg} \cdot \text{m}$ worth of work! You need to instead use the book's weight (not mass) when making your calculation. Thus,

$$W = F \cdot d = \left[(2\,\text{kg})(9.8\,\text{m/sec}^2)\right](2\,\text{m}) = 39.2\ \text{newton} \cdot \text{meters}.$$

A newton · meter is usually called a joule, so you've done 39.2 joules of work in the metric system of units. In the British system the unit of work is the foot-pound, the work done when a 1 pound force moves a distance of 1 foot.

❏ **Example 1** A 28-meter chain with a mass of 20 kilograms is dangling from the roof of a building. How much work is needed to pull the chain up to the top of the building?

Solution. Since a 20 kg chain weighs $(20\,\text{kg})(9.8$ m/sec^2) = 196 newtons, it might seem the answer should be (196 n)(28 m)=5488 joules. But remember that not all the chain has to move 28 meters—links near the top move less.

Let's divide the chain into small links of length Δy, each weighing $7\,\Delta y$ newtons. (28 meters weighs 196 newtons, so one meter weighs 7 newtons.) See Figure 7.26. If Δy is small, all of this piece is hauled up approximately the same distance, namely y, against the force due to gravity of $7\,\Delta y$ newtons. Thus the work done on the small link is approximately

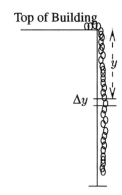

Top of Building

Figure 7.26: Chain for Example 7.26

$$7\,\Delta y \cdot y = 7y\,\Delta y.$$

The work done on the entire chain is given by the total of the work done on each piece:

$$W = \sum 7y\,\Delta y.$$

As Δy tends to zero, we obtain a definite integral. Since y varies from 0 to 28 meters, the total work is

$$W = \int_0^{28} (7y)\,dy = \frac{7}{2}y^2 \Big|_0^{28} = 2744\ \text{joules.}$$

❏

☐ **Example 2**   It is reported that the Great Pyramid of Egypt was built in 20 years. If the stone making up the pyramid weighed 200 pounds per cubic foot, find the total amount of work done in building the pyramid. Estimate how many workers were needed to build the pyramid.

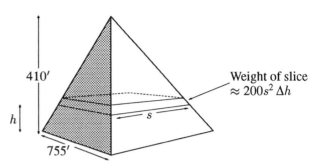

Figure 7.27: Pyramid for Example 2

Solution. We assume that the stones were originally located at the approximate height of the construction site. Imagine the pyramid built in layers as we did in Example 1, Page 479, of the previous section. Recall that the pyramid is 410 feet high and has a square base 755 feet by 755 feet.

We found by similar triangles that the layer at height h has a side length $s = \frac{755}{410}(410 - h)$. The layer at height h has a volume of $s^2 \Delta h$ and hence a weight of $200s^2 \Delta h$. Constructing this layer involved lifting a weight of $200s^2 \Delta h$ through a height of h, so the work required was $200s^2\Delta h \cdot h = 200s^2 h \Delta h$. See Figure 7.27. The total work, W, of constructing all the layers is then

$$W \approx \sum 200s^2\, h\Delta h = \sum 200 \left(\frac{755}{410}\right)^2 (410 - h)^2 h\, \Delta h.$$

As the thickness of each layer, Δh, tends to zero, we obtain a definite integral. Since h varies from 0 to 410, we have

$$W = \int_0^{410} 200s^2 h\, dh \;=\; 200\int_0^{410}\left(\frac{755}{410}\right)^2 (410 - h)^2 h\, dh$$

$$= \; 200\left(\frac{755}{410}\right)^2 \int_0^{410} (410 - h)^2 h\, dh.$$

If we make the substitution $x = 410 - h$, we avoid having to multiply out $(410 - h)^2$. Then $h = 410 - x$ and $dx = -dh$. Also $x = 410$ when $h = 0$, so the last integral becomes

$$\int_0^{410} (410 - h)^2 h\, dh \;=\; \int_{410}^{0} x^2 (410 - x)(-dx) = \int_0^{410} (410x^2 - x^3)\, dx$$

$$= \; \left(410\frac{x^3}{3} - \frac{x^4}{4}\right)\Bigg|_0^{410} = (410)^4\left(\frac{1}{3} - \frac{1}{4}\right) = 410^4\left(\frac{1}{12}\right).$$

Therefore the total work was

$$W \;=\; 200\left(\frac{755}{410}\right)^2 \int_0^{410}(410 - h)^2 h\, dh = 200\left(\frac{755}{410}\right)^2 (410)^4\left(\frac{1}{12}\right)$$

$$\approx \; 1{,}597{,}020{,}000{,}000 \approx 1.6 \times 10^{12} \text{ foot-pounds.}$$

We have calculated the total work done in building the pyramid; now we want to estimate the total number of workers needed. Let's assume every laborer worked 10 hours a day, 300 days a year, for 20 years. Assume that a typical worker lifted ten 50 pound blocks a distance of 4 feet every hour, thus performing 2000 foot-pounds of work per hour (this is a very rough estimate). Then each laborer performed $(10)(300)(20)(2000) = 12 \times 10^7$ foot-pounds of work over a twenty year period. Thus the number of workers needed was about $\frac{1.6 \times 10^{12}}{12 \times 10^7}$ or about 13,000. ❑

Escape from the Earth's Gravitational Field

If you throw an object up into the air, you expect gravity to bring it back to earth sooner or later. However, the harder you throw it, the longer you expect to have to wait before it returns. Have you ever wondered if it is possible to throw something hard enough so that it never returns to earth? We will calculate the total amount of work done to move an object infinitely far away from the earth. Amazingly enough, this total amount of work is finite, and it *is* possible to throw something hard enough that it never returns to Earth.

Previously, when we analyzed the motion of a freely falling body, we assumed that the force of gravity is constant. Now, since we are thinking of a body moving into outer space, we have to take into account the fact that the Earth's gravitational force gets weaker as you go further away from it. Newton's Law of Gravitation says that the force, F, exerted on a mass m at a distance r from the center of the Earth is given by

$$F = \frac{GMm}{r^2}$$

where $r > R$, the Earth's radius, M is the mass of the Earth and G is called the gravitational constant, whose value is about 6.67×10^{-11} if mass is measured in kg, distance in meters, and force in newtons.

❑ **Example 3**   Find the total work done to move an object of mass m infinitely far from the Earth.

Solution. The force pulling the object back to the Earth is $\dfrac{GMm}{r^2}$. The work needed to move the object a distance Δr further away is

$$\left(\frac{GMm}{r^2} \right) \Delta r.$$

Therefore the total work to move the object from the surface of the earth $r = R$ to a point "infinitely far away" is approximated by

$$W \approx \sum \frac{GMm}{r^2} \Delta r$$

where the value of r runs from $r = R$ (at the Earth's surface) to $r = \infty$ (very far away). As Δr tends to zero, we get the definite integral

$$W = \int_R^\infty \frac{GMm}{r^2} \, dr = \lim_{b \to \infty} - \frac{GMm}{r} \bigg|_R^b = \frac{GMm}{R}.$$

❑

Imagine an object shot upwards with initial velocity v. The energy of the body by virtue of its motion is called its *kinetic energy* and is equal to $\frac{1}{2}mv^2$. When the object moves upwards, it slows down, losing kinetic energy as work is done against gravity. The body moves upwards until all the kinetic energy is expended: it will stop at the height at which

$$\text{Initial Kinetic Energy} = \text{Work done against Gravity.}$$

Since the work done to move an object infinitely far from the Earth is finite, a finite starting velocity can give the object enough kinetic energy to move infinitely far away. The *escape velocity* from the Earth is defined to be the minimum vertical velocity that must be given to an object so that it is never drawn back to Earth by gravity.

☐ **Example 4** Find the escape velocity of an object from the Earth.

Solution. If v is the initial velocity, then setting the initial kinetic energy and the work done to escape infinitely far from the Earth equal we get

$$\frac{1}{2}mv^2 = \frac{GMm}{R}.$$

Solving for v gives

$$v = \sqrt{\frac{2GM}{R}}.$$

Since the mass of the earth is $M \approx 6 \times 10^{24}$ kg and its radius is $R \approx 6.4 \times 10^6$ meters, the escape velocity is about 11,000 meters/sec $\approx$ 25,000 mph, or about 30 times the speed of sound. Thus, the escape velocity is finite, though it is much higher than you can achieve by throwing or by shooting a gun (which explains why we never worry about losing tennis balls in the sky). ☐

Force and Pressure Don't do

We can use the definite integral to compute the force exerted by a liquid on a surface, such as the force of water on a dam. The idea is to get the force from the *pressure*. The pressure in a liquid is the force per unit area exerted on a small piece of area in the liquid. Two things you need to know about pressure are:

- At any point, pressure is exerted equally in all directions - up, down, sideways.

- Pressure increases with depth. (That is one of the reasons why deep sea divers have to take much greater precautions than scuba divers.)

At a depth of h feet, the pressure, exerted by the liquid and measured in pounds per square foot, is given by computing the total weight of a column of water h feet high with a base of 1 square foot. The volume of such a column of water is just h cubic feet, and since water weighs 62.4 pounds per cubic foot, the weight of the column is $62.4h$ pounds. Thus the pressure at depth h is $p = 62.4h$ pounds per square foot. See Figure 7.28. In general, no matter what the units, no matter what the liquid, the pressure, p, is proportional to the depth h:

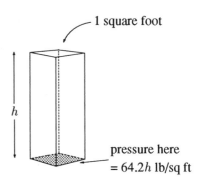

Figure 7.28: Pressure exerted by a Column of Water

$$\boxed{p = kh}$$

Next you need to know the relation between force and pressure. Provided the pressure is constant over a given area, we have

$$\boxed{\text{Force} = (\text{Pressure}) \times (\text{Area}).}$$

When the pressure isn't constant over a region, we divide the region into small pieces *in such a way that the pressure is nearly constant on each one*. Then we can use this formula on each one and a definite integral to get the force over the entire surface. Since the pressure varies with depth, this means we should divide the surface into horizontal strips, each of which is at an approximately constant depth.

◻ **Example 5** A trough 14 feet long has one rectangular vertical side 14 feet by 3 feet, one rectangular inclined side 14 feet by 5 feet, and two triangular ends with sides 3 feet, 4 feet, and 5 feet (see Figure 7.29). If the trough is filled to the top with water, compute the force on each end and each side.

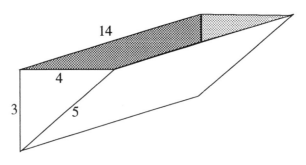

Figure 7.29: The trough for Example 5

Solution. We will do the long vertical side first. Divide the side into thin horizontal strips of width Δh. Since pressure depends only on the depth h and the strips are thin and horizontal, we can assume the pressure is the same at every point in the strip. The area of the strip is $14\,\Delta h$ and the force on the strip at depth h is the pressure at that depth times the area of the strip:

$$\text{Force on strip} = (\text{Pressure}) \times (\text{Area}) \approx (62.4h)14\,\Delta h.$$

Summing the forces on these strips, we get an expression for the total force on the vertical side:

$$F_{\substack{\text{vertical}\\\text{side}}} \approx \sum(62.4h)14\,\Delta h.$$

Letting $\Delta h \to 0$, we get a definite integral. Since h varies from 0 to 3 feet, we have

$$F_{\substack{\text{vertical}\\\text{side}}} = \int_0^3 (62.4h)14\,dh = (62.4)(14)\frac{h^2}{2}\Big|_0^3 = (62.4)(14)\left(\frac{9}{2}\right) = 3931.2 \text{ pounds.}$$

To compute the force on the inclined side, we use an inclined horizontal strip. Each strip has length 14, but its width is not Δh. By similar triangles (see Figure 7.30) the width w of the strip satisfies

$$\frac{\Delta h}{w} = \frac{3}{5}, \quad \text{so} \quad w = \frac{5}{3}\Delta h.$$

Therefore

$$\text{Area of strip} = 14w = 14\left(\frac{5}{3}\right)\Delta h,$$

so

$$\text{Force on strip} = (62.4h)(14)\left(\frac{5}{3}\right)\Delta h.$$

The total force on the inclined side is obtained by adding the forces on the strips, giving

$$F_{\substack{\text{inclined}\\\text{side}}} \approx \sum(62.4h)(14)\left(\frac{5}{3}\right)\Delta h.$$

We let $\Delta h \to 0$ to arrive at a definite integral. Again, h varies between 0 and 3 feet, so

$$F_{\substack{\text{inclined}\\\text{side}}} = \int_0^3 (62.4h)(14)\left(\frac{5}{3}\right)dh = \left(\frac{5}{3}\right)(62.4)(14)\left(\frac{9}{2}\right) = 6552 \text{ pounds.}$$

Notice that this is just $\frac{5}{3}$ the force on the vertical side.

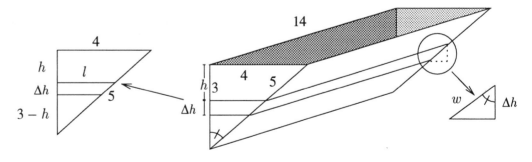

Figure 7.30: Calculating the widths of strips on the end and the inclined side

Finally, to compute the force on each triangular end, we again divide the end into horizontal strips. The width of the strip is Δh as with the vertical side, but this time the length of the strip varies from 4 feet at the top of the trough to 0 feet at the bottom. By similar triangles (see Figure 7.30) the length l of a strip is related to the depth h by

$$\frac{l}{4} = \frac{3-h}{3}, \quad \text{so} \quad l = \frac{4}{3}(3-h).$$

Thus the area of the strip is $l\,\Delta h = \frac{4}{3}(3-h)\Delta h$. As before, this means the total force is

$$F_{end} \approx \sum (62.4h)\left(\frac{4}{3}(3-h)\right)\Delta h.$$

Since h varies between 0 and 3, as $\Delta h \to 0$ we have

$$
\begin{aligned}
F_{end} &= \int_0^3 (62.4h)\left(\frac{4}{3}(3-h)\right) dh = (62.4)\left(\frac{4}{3}\right)\left(3h - \frac{h^2}{2}\right)\Big|_0^3 \\
&= (62.4)\left(\frac{4}{3}\right)\left(\frac{9}{2}\right) = 374.4 \text{ pounds.}
\end{aligned}
$$

❑

❖ Exercises for Section 7.4

1. A water tank is in the form of a right circular cylinder with height 20 ft and radius 6 ft. If the tank is half full of water, find the work required to pump all of it over the top rim. Note that water weighs 62.4 pounds per cubic foot.

2. Suppose the tank in Problem 1 is full of water. Find the work required to pump all of it to a point 10 ft above the top of the tank.

3. A rectangular water tank has length 20 feet, width 10 feet, and is 15 feet deep. If the tank is full, how much work does it take to pump all the water out? (Note that 1 cubic foot of water weighs 62.4 pounds.)

4. At a construction site a crane lifts a bucket of cement from the ground to a point 30 ft above the ground. The distance from the end of the crane arm to the ground is 75 ft. If the bucket of cement weighs 500 lb and the cable weighs 5lb/ft, find the work required to lift the cement.

5. A 2000 lb. cube of ice must be lifted 100 ft, and it is melting at a rate of 4 lb. per minute. Assume that it can be lifted at a rate of one foot every minute. Find the work needed to get the block of ice to the desired height.

6. A gas station stores its gasoline in a tank under the ground. The tank is a cylinder lying on its side with flat ends. If the radius of the cylinder is 4 feet, the length is 12 feet, and the top of the cylinder is 10 feet under the ground, find the total amount of work needed to pump the gasoline out of the tank. (Gasoline weighs 42 pounds per cubic foot.)

7. On a hot afternoon in Pétionville, Haiti, Lubonga is trying to cool off by sipping the glass of palm wine shown in Fig. 7.31. How much work does she have to do to empty the glass? [The density of the palm wine is 1.2 gm/cm³. Initially, the glass is filled to a vertical depth of 8 cm. Note that the force due to gravity acting on 1 gram is 980 dynes, and that when using these units, work is measured in ergs.]

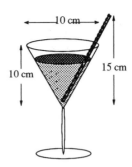

Figure 7.31: Oh, for some palm wine!

(for Problem 7)

8. How much work is required to lift a 1000 kg satellite into orbit 2×10^6 m above the surface of the earth? (The radius of the earth is 6.4×10^6 m, its mass is 6×10^{24} kg, and in these units, the gravitational constant, G, is 6.67×10^{-11}.)

9. Show that the escape velocity needed by an object to escape the gravitational influence of a spherical planet of density ρ and radius R is proportional to R and to $\sqrt{\rho}$.

10. Calculate the escape velocity of an object from the moon. The acceleration due to gravity on the moon is 1.6 m/sec², whereas on Earth it is 9.8 m/sec²; the radius of the moon is approximately 1738 km. [Hint: If g is the acceleration due to gravity, the force on an object is $F = mg = \frac{GMm}{R^2}$. See page 492.]

11. A reservoir has a dam at one end. The dam is a rectangular wall, 1000 feet long and 50 feet high. The problem is to find the total force of the water acting on the dam.

    (a) Show how to approximate this force by a Riemann sum, explaining your reasoning clearly.

    (b) Derive an integral which represents the force and evaluate it.

12. Before you pumped out any water, what was the total force on the bottom and each side of the tank in Problem 3?

13. We define the electric potential at a distance r from an electric charge q by $\frac{q}{r}$. Suppose electric charge is sprayed (with constant density σ in units of charge/unit area) onto a circular disc of radius a. Consider the axis perpendicular to the disc and through its center. Find the electric potential at the point P on this axis at a distance R from the center. (See the figure to the right.)

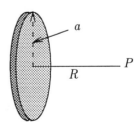

For Problems 14-15, find the kinetic energy of a rotating body, using the fact that the kinetic energy of a particle of mass m moving at a speed v is $\frac{1}{2}mv^2$. Slice the object into pieces in such a way that the velocity is approximately constant across each piece.

14. Find the kinetic energy of a rod of mass 10 kg and length 6 m rotating about an axis perpendicular to the rod at its midpoint, with an angular velocity of 2 radians per second. (Imagine a helicopter blade of uniform thickness.)

15. Find the kinetic energy of a record of mass 50 grams and radius 10 cm rotating at $33\frac{1}{3}$ revolutions per minute.

For Problems 16-17, find the gravitational force (see Page 492) between two objects. Use the fact that the gravitational attraction between particles of mass m_1 and m_2 at a distance r apart is $\frac{Gm_1m_2}{r^2}$. For each problem, slice the objects into pieces, use this formula for the pieces, and sum using a definite integral.

16. What is the force of gravitational attraction between a thin uniform rod of mass M and length l and a particle of mass m lying in the same line as the rod at a distance a from one end?

17. Two long thin rods of lengths l_1 and l_2 lie on a straight line with a gap between them of length a. Suppose their masses are M_1 and M_2, respectively, and the constant of the gravitation is G. What is the force of attraction between the rods? (Use the result of Problem 16.)

18. A car moving at speed v mph makes $25 + 0.1v$ mpg (miles per gallon) for v between 20 and 60 mph. Suppose your speed as a function of time is given by

$$v = 50\frac{t}{t+1}.$$

How many gallons do you consume between $t = 2$ and $t = 3$ (i.e., in the third hour)?

7.5 Applications to Economics

Present and Future Value

Many business deals involve payments in the future. If you buy a car or furniture, for example, you may buy it on credit and pay over a period of time.

If you are going to accept payment in the future under such a deal, you obviously need to know how much you should be paid. Being paid $100 in the future is clearly worse than being paid $100 today for many reasons. One is inflation—$100 in the future may well buy less than $100 today because prices are likely to rise. Another is the risk that you would never be repaid. Perhaps most importantly, if you are given the money today, you can do something else with it—for example, put it in the bank, invest it somewhere, or spend it. Thus, even without inflation, if you are to accept payment in the future, you would expect to be paid more to compensate for this loss of freedom. The question we will consider now is: how much more?

To simplify matters, we will only consider what you would lose by not earning interest. We will not consider the effect of inflation. Let's look at some specific numbers. Suppose you deposit $100 in an account which earns 7% interest compounded annually, so that in a year's time you will have $107. Thus, that $100 today will be worth the same as $107 a year from now.

We say that the $107 is the *future value* of the $100, and that the $100 is the *present value* of the $107. In general, we say

> - The **future value**, $\$B$, of a payment, $\$P$, is the amount to which the $\$P$ would have grown if deposited in an interest bearing bank account.
>
> - The **present value**, $\$P$, of a future payment, $\$B$, is the amount which would have to be deposited in a bank account today to have exactly $\$B$ in the account at the relevant time in the future.

As you might expect, the present value will always be smaller than the future value. In our work on compound interest in Section 1.4, we saw that with an interest rate of r, compounded annually, and a time period of t years, a deposit of $\$P$ grows to a future balance of $\$B$, where

$$B = P(1 + r)^t, \quad \text{or equivalently,} \quad P = \frac{B}{(1+r)^t}.$$

> If the interest is compounded n times a year for t years at rate r, and $\$B$ is the *future value* of $\$P$ after t years and $\$P$ is the *present value* of $\$B$, then
>
> $$B = P\left(1 + \frac{r}{n}\right)^{nt}, \quad \text{or equivalently,} \quad P = \frac{B}{\left(1 + \frac{r}{n}\right)^{nt}}.$$

Note that for a 7% interest rate, $r = 0.07$. In addition, as we saw in Section 1.4, when n tends towards infinity we say that the compounding becomes continuous, and we get

> $$B = Pe^{rt}, \quad \text{or equivalently,} \quad P = \frac{B}{e^{rt}} = Be^{-rt}.$$

☐ **Example 1**   You win the lottery and are offered the choice between $1 million in four yearly installments of $250,000 each, starting now, and a lump sum payment of $920,000 now. Assuming a 6% interest rate, compounded continuously, which should you choose?

Solution. We will do the problem in two ways. First, we assume that you pick the option with the largest present value. The first of the four $250,000 payments is made now, so it has

$$\text{Present Value} = \$250{,}000.$$

The second payment is made one year from now and so has

$$\text{Present Value} = \$250{,}000e^{-0.06(1)}.$$

Calculating the present value of the third and fourth payments similarly, we find:

Total present value of the four $250,000 payments

$$= \$250,000 + \$250,000e^{-0.06(1)} + \$250,000e^{-0.06(2)} + \$250,000e^{-0.06(3)}$$
$$\approx \$250,000 + \$235,441 + \$221,730 + \$208,818$$
$$= \$915,989.$$

Since the present value of the four payments is less than $920,000, you are better off taking the $920,000 right now.

We can also do the problem by finding which choice offers the largest future value. If you take the $920,000 now and deposit it in the bank, in three years (when you get the last $250,000 payment) it will have

$$\text{Future Value} = \$920,000e^{0.06(3)} \approx \$1,101,440.$$

Instead, if you take the four payments of $250,000, then the first payment will earn interest for three years, so after three years it is worth

$$\text{Future Value} = \$250,000e^{0.06(3)}.$$

Calculating the future value of the other payments similarly, we find:

Total future value of the four $250,000 payments

$$= \$250,000e^{0.06(3)} + \$250,000e^{0.06(2)} + \$250,000e^{0.06(1)} + \$250,000$$
$$\approx \$299,304 + \$281,874 + \$265,459 + \$250,000$$
$$= \$1,096,637.$$

The future value of the $920,000 payment is greater, so you are better off taking the $920,000 right now. Of course, since the present value of the $920,000 payment is greater than the present value of the four separate payments, you would expect the future value of the $920,000 payment to be greater than the future value of the four separate payments.

Note: If you read the fine print, you will find that many lotteries do not make their payments right away, but often spread them out, sometimes far into the future. This is to reduce the present value of the payments made – so that the value of the prizes is much less than it might first appear! ❏

Income Stream

When we consider payments made to or by an individual, we usually think of *discrete* payments, i.e., payments made at specific moments in time. However, when we consider payments made by a company, we may think of them as being *continuous*. The revenues earned by a huge corporation, for example, come in essentially all the time, and therefore they can be represented by a continuous *income stream*. Since the rate at which revenue is earned may vary from time to time, the income stream is described by

$$P(t) \text{ dollars/year.}$$

Notice that $P(t)$ is a *rate* (its units are dollars per year, for example), and that the rate depends on the time, t, usually measured in years from the present.

Present and Future Values of an Income Stream

Just as we can find the present and future values of a single payment, so we can find the present and future values of a stream of payments. As before, the future value represents the total amount of money obtained by depositing the income stream into a bank account and letting it gather interest. The present value represents the amount of money you would have to deposit today (in an interest-bearing bank account) in order to match what you would have by depositing the income stream.

When we are working with a continuous income stream, we will assume that interest is compounded continuously. The reason for this is that the approximations we are going to make (of sums by integrals) are much simpler if both payments and interest are continuous.

In order to use what we know about single payments to calculate the present and future values of an income stream, we must first divide the stream into many small payments, each of which is made at approximately one instant. Suppose payments are made at a rate of $P(t)$ dollars per year, and that we are interested in the period from now until T years in the future. If t is measured in years, we are interested in the interval $0 \leq t \leq T$. We divide it into subintervals, each of length Δt:

Assuming Δt is small, the rate, $P(t)$, at which payments are being made will not vary much within one subinterval. Thus, between t and $t + \Delta t$:

$$(\text{Total amount paid}) \approx (\text{rate of payments}) \times (\text{time}) \approx P(t)\Delta t.$$

This deposit has a period of $(T - t)$ years to earn interest, and therefore

$$\text{Future Value} \approx (P(t)\Delta t)\, e^{r(T-t)}.$$

Summing over all subintervals, we get:

$$\text{Total Future Value} \approx \sum P(t)\Delta t e^{r(T-t)}.$$

As the length of the subdivisions tends towards zero, the sum becomes an integral, and we can say:

$$\boxed{\text{Future Value} = \int_0^T P(t)e^{r(T-t)}\,dt.}$$

The same diagram shows that, measured from the present, the payment of $P(t)\Delta t$ is made t years in the future. Thus

$$\text{Present Value} \approx P(t)\Delta t e^{-rt}.$$

Summing over all subintervals gives

$$\text{Total Present Value} \approx \sum P(t)e^{-rt}\Delta t$$

and, in the limit as $\Delta t \to 0$,

$$\boxed{\text{Present Value} \;=\; \int_0^T P(t)e^{-rt}\,dt.}$$

☐ **Example 2** Find the present and future values of a constant income stream of \$100 per year over a period of 20 years, assuming an interest rate of 10% compounded continuously.

Solution.

$$\text{Present Value} \;=\; \int_0^{20} 100e^{-0.1t}\,dt = 100\left[-\frac{e^{-0.1t}}{0.1}\right]_0^{20} = 1000(1 - e^{-2}) \approx \$864.66.$$

$$\text{Future Value} \;=\; \int_0^{20} 100e^{0.1(20-t)}\,dt = \int_0^{20} 100e^2 e^{-0.1t}\,dt$$

$$= \; 100e^2\left[-\frac{e^{-0.1t}}{0.1}\right]_0^{20} = 1000e^2[1 - e^{-2}] \approx \$6389.06.$$

☐

☐ **Example 3** What is the relationship between the present and future values in the previous example? Explain this relationship.

Solution. Since

$$\text{Present Value} \;=\; 1000(1 - e^{-2}) \approx \$864.66$$
$$\text{Future Value} \;=\; 1000e^2(1 - e^{-2}) \approx \$6389.06$$

you can see that

$$(\text{Future Value}) \;=\; (\text{Present Value})e^2.$$

The reason for this is that the stream of payments is equivalent to a single payment of \$864.66 at time $t = 0$. With an interest rate of 10%, in 20 years that single payment will have grown to a future value of

$$864.66e^{0.1(20)} = 864.66e^2 \approx \$6389.02.$$

To within rounding error, this is the future value calculated previously. ☐

Supply and Demand Curves

In a free market, the quantity of a certain item produced and sold can be described by the supply and demand curves of the item. The *supply curve* shows what quantity of the item the producers will supply at different price levels. It is usually assumed that as the price increases, the quantity supplied will increase. The consumers' behavior is reflected in the *demand curve*, which shows what quantity of goods are bought at various prices. Here, an increase in price is assumed to cause a decrease in the quantity purchased.

For most producers and consumers, the price of an item has to be thought of as the independent variable, because it is set by the market as a whole, and there is nothing they can do to alter it. The quantity produced or consumed is then thought of as depending on the price. In other words, quantity is considered as a function of price. However, although in this context we think of price as the independent variable, economic tradition has placed price along the vertical axis. (The reason for this state of affairs is that economists originally took price to be the dependent variable and put it on the vertical axis. Unfortunately, when the point of view changed,

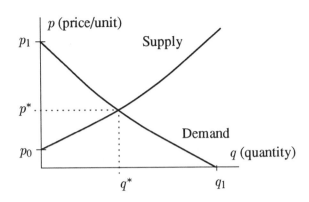

Figure 7.33: Supply and Demand Curves

the axes did not.) The higher the price, the more suppliers will produce but the less consumers will demand. Thus typical supply and demand curves are monotonic (either increasing or decreasing) and so it is in fact possible to think of either variable as a function of the other. See Figure 7.33.

☐ **Example 4** What is the economic meaning of the prices p_0, p_1 and the quantity q_1 in Figure 7.33?

Solution. The price p_0 is the vertical intercept on the supply curve, so at this price the quantity supplied is zero. In other words, unless the price is above p_0, the supplier will not produce anything.

The price p_1 is the vertical intercept of the demand curve. At this price, the quantity demanded is zero. In other words, unless the price is below p_1, consumers won't buy any of the product.

The quantity q_1 is the quantity that would be demanded if the price were zero—or the quantity which could be given away if the item were free. ☐

It is assumed that in the absence of other forces, the market will settle down to the *equilibrium price* and *quantity*, p^* and q^*, where the graphs cross. This means that at the equilibrium point, a quantity q^* of an item will be produced and sold for a price of p^* each.

Consumer and Producer Surplus

Notice that at equilibrium, a number of consumers have bought the item at a lower price than they would have been willing to pay. (For example, there are some consumers who would have been willing to pay prices up to p_1.) Similarly there are some suppliers who would have been willing to produce the good at a lower price (down to p_0, in fact). We define:

- the **consumer surplus** to measure the total amount saved by consumers by buying the good at the equilibrium price, rather than at the price they were willing to pay.

- the **producer surplus** to measure the additional revenue earned by suppliers by selling at the equilibrium price, rather than at what they would have been willing to accept.

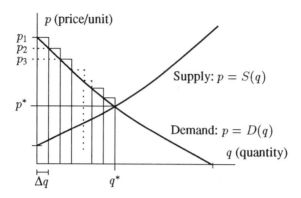

Figure 7.34: Calculation of Consumer Surplus

Suppose that all consumers buy the good at the maximum price they are willing to pay. Divide the interval from 0 to q^* into subintervals of length Δq. Figure 7.34 shows that a quantity Δq are sold at a price of about p_1, another Δq are sold for a slightly lower price of about p_2, the next Δq for a price of about p_3, and so on. Thus the consumers' total expenditure is about

$$p_1\Delta q + p_2\Delta q + p_3\Delta q \ldots = \sum p_i\Delta q.$$

If D is the demand function given by $p = D(q)$, and if all consumers who were willing to pay more than p^* paid as much as they were willing, then as $\Delta q \to 0$, we would have

$$\begin{pmatrix} \text{Consumer} \\ \text{Expenditure} \end{pmatrix} = \int_0^{q^*} D(q)dq = \begin{pmatrix} \text{Area under Demand} \\ \text{Curve from 0 to } q^* \end{pmatrix}.$$

Now if all goods are sold at the equilibrium price, the consumers' actual expenditures is p^*q^*, the area of the rectangle between the axes and the lines $q = q^*$ and $p = p^*$. Thus

$$\begin{pmatrix} \text{Consumer} \\ \text{Surplus} \end{pmatrix} = \left(\int_0^{q^*} D(q)dq \right) - p^*q^* = \begin{pmatrix} \text{Area under Demand} \\ \text{Curve above } p = p^* \end{pmatrix}.$$

See Figure 7.35. Similarly, if the supply curve is given by the function $p = S(q)$, we define:

$$\begin{pmatrix} \text{Producers'} \\ \text{Surplus} \end{pmatrix} = p^*q^* - \left(\int_0^{q^*} S(q)dq \right) = \begin{pmatrix} \text{Area between Supply} \\ \text{Curve and Line } p = p^* \end{pmatrix}.$$

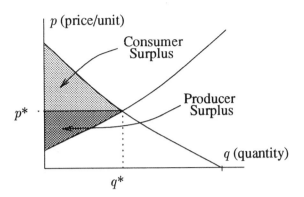

Figure 7.35: Consumer and Producer Surplus

❖ Exercises for Section 7.5

1. Draw a graph, with years on the horizontal axis, of what an income stream might look like for a company that sells sunscreen in the Northeast.

2. A company that owes your company money offers to begin repaying the debt either by making four annual payments of $5000, spread out over the next three years (one now, one in a year, one in two years, and one in three years), or by waiting and making a lump sum payment of $25,000 at the end of the three years. If you can assume an 8% annual return on the money, compounded continuously, and you use only financial considerations, which option for repayment should you choose? Justify your answer. What other considerations might you want to consider in making your decision?

3. Find the present and future values of a constant income stream of $3000 per year over a period of 15 years, assuming a 6% annual interest rate compounded continuously.

4. A certain bond is guaranteed to pay $(100 + 10t)$ dollars per year for ten years, where t is the number of years from the present. Find the present value of this income stream, given an interest rate of 5%, compounded continuously. (This is a fair price for the bond now.)

5. (a) A bank account earns 10% interest compounded continuously. At what (constant, continuous) rate must a parent deposit money into such an account in order to save $100,000 in 10 years for a child's college expenses?

   (b) If the parent decides instead to deposit a lump sum now in order to attain the goal of $100,000 in 10 years, how much must be deposited now?

6. The value of good wine increases with age. Thus, if you are a wine dealer you have the problem of deciding whether to sell your wine now, at a price of $\$P$ a bottle, or to sell it later at a higher price. Suppose that you know that the price of your wine t years from now is well approximated by $\$P(1 + 20\sqrt{t})$. Assuming continuous compounding and a prevailing interest rate of 5% per year, what is the best time to sell your wine?

7. (a) If you deposit money continuously at a constant rate of $1000 per year into a bank account that earns 5% interest, how many years will it take for the balance to reach $10,000?

   (b) How many years would it take if the account had $2000 in it initially?

8. Suppose you are manufacturing a particular item, and that the rate at which you earn a profit on the item is decreasing with time according to the formula

   $$\text{rate profit earned} = (2 - 0.1t) \text{ thousand dollars per year}$$

   after t years. (A negative profit represents a loss.) If interest is 10% compounded continuously,

   (a) write a Riemann sum approximating the present value of the total profit earned at a time T years in the future.

   (b) write an integral representing the present value of the total profit earned T years in the future. (You need not evaluate this integral.)

   (c) at what time is the present value of the stream of profits on this item maximized? What is the present value of the total profit earned up to that time?

9. In 1980, West Germany made a loan of 20 billion Deutsche Mark to the Soviet Union, to be used for the construction of a pipeline for natural gas, connecting Siberia to Western Russia, and continuing to West Germany (Urengoi – Uschgorod – Berlin). The deal was as follows: In 1985, upon completion of the pipeline, the Soviet Union would deliver natural gas to West Germany, at a constant rate, for all future times. Assuming a constant price of natural gas of 10 Pfennig (= 0.10 Deutsche Mark) per m^3, and assuming West Germany expects 10% annual interest on its investment (compounded continuously), at what rate does the Soviet Union have to deliver the gas, in billion m^3/year? Keep in mind that delivery of gas could not begin until the pipeline was completed. Thus, West Germany received no return on its investment until after five years had passed.

   (Note: A deal of this type was actually made between the two countries; obviously, the terms of the deal were much more complex than presented here. In 1981, the U.S. Government tried unsuccessfully to block the deal, saying that it compromised Western security interests.)

10. In May 1991, *Car and Driver* reported that a particularly exclusive model Jaguar, of which only 50 have been sold, goes for $980,000. If more are manufactured, 350 will be sold at a price of $560,000 each. Assuming that the demand curve is a straight line, and that $560,000 and 350 are the equilibrium price and quantity, find the consumer surplus.

11. Using Riemann sums, explain the economic significance of $\int_0^{q^*} S(q)\,dq$ to the producers.

12. Using Riemann sums, give an interpretation of the producer's surplus,

$$\int_0^{q^*} (p^* - S(q))\, dq$$

analogous to the interpretation of consumer surplus.

13. In Figure 7.33, page 503, find the regions with the following areas:

(a) $p^* q^*$

(b) $\int_0^{q^*} D(q)\, dq$

(c) $\int_0^{q^*} S(q)\, dq$

(d) $\left(\int_0^{q^*} D(q)\, dq\right) - p^* q^*$

(e) $p^* q^* - \left(\int_0^{q^*} S(q)\, dq\right)$

(f) $\int_0^{q^*} (D(q) - S(q))\, dq$

7.6 Probability and Distributions

To allocate funding for education, health care, and social security, the government needs to know how many people are in each age group. In this section we will see how to work with such information and how to represent it by a probability density function.

Suppose we have the data in Table 7.1 showing how the ages of the U.S. population were distributed in 1990.

Age Group	Percentage of Total Population
0–20	30%
20–40	31%
40–60	24%
60–80	14%
over 80	1%

Table 7.1: How Ages were Distributed in the U.S. in 1990

To represent this information graphically we use a *histogram*, putting a vertical bar above each age group in such a way that the *area* of each bar represents the percentage in that age group. The total area of all the rectangles is $100\% = 1$. We will assume that there is nobody over 100 years old, so that the last age group is 80–100. For the 0–20 age group, the base of the rectangle is 20, and we want the area to be 30%, so the height must be $30\%/20 = 1.5\%$. Notice that the vertical axis is measured in percent/year. See Figure 7.36.

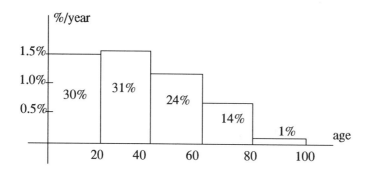

Figure 7.36: How Ages were Distributed in the U.S.in 1990

☐ **Example 1** In 1990, what percentage of the U.S. population was

(a) between 20 and 60 years old?

(b) less than 10 years old?

(c) between 75 and 80 or between 80 and 85 years old?

Solution.

(a) We add the percentages, so $31\% + 24\% = 55\%$.

(b) To find the percentage less than 10 years old, we could assume, for example, that the population was distributed evenly over the 0–20 group. (This means we are assuming that babies were born at a reasonably constant rate over the last 20 years, which is probably reasonable.) If we make this assumption, then we can say that the population less than 10 years old was about half that in that 0–20 group, that is, 15%. Notice that we get the same result by computing the area of the rectangle from 0 to 10. See Figure 7.37.

(c) To find the population between 75 and 80 years old, since 14% of Americans in 1990 were in the 60-80 group, we might apply the same reasoning and say that $\frac{1}{4}(14\%) = 3.5\%$ of the population was in this age group. This result is represented as an area in Figure 7.37. The assumption that the population was evenly distributed is not a good one here; certainly there were more people between the ages of 60 and 65 than between 75 and 80. Thus the estimate of 3.5% is certainly too high.

Again using the (faulty) assumption that ages in each group were distributed uniformly, we would find that the percentage between 80 and 85 was $\frac{1}{4}(1\%) = 0.25\%$ (See Figure 7.37). This estimate is also incorrect — there were certainly more people in the 80–85 group than, say, the 95–100 group, and so the 0.25% is too low. In addition, while the percentage of 80–85-year-olds was certainly smaller than the percentage of 75–80-year-olds, the difference between 0.25% and 3.5% (a factor of 14) is unreasonably large. We can expect the transition from one age group to the next to be smoother and more gradual.

Assuming that the distribution is smoother, we could make plausible guesses for the actual respective percentages by looking at the graph: they are probably somewhat less than 1% for 80–85-year-olds and a bit more than 1% for 75–80-year-olds. (Remember that our estimate of the percentage of 80–85-year-olds should be less than 1%, the total percentage of people over 80.)

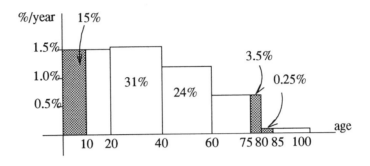

Figure 7.37: Ages in the U.S. in 1990 — Various Subgroups (Example 1)

Smoothing Out the Histogram

We could get better estimates if we had smaller age groups (each age group in Figure 7.36 is 20 years, which is quite large), or if the histogram were smoother. Suppose we have the more detailed data in Table 7.2, which leads to the new histogram in Figure 7.38. As we get more detailed information, the upper silhouette of the histogram becomes smoother — but the area of any of the bars still represents the percentage of the population in that age group. Imagine, in the limit, replacing the upper silhouette of the histogram by a smooth curve in such a way that area under the curve above one age group is the same as the area in the corresponding rectangle. The total area under the whole curve is again 100% = 1. See Figure 7.39.

If t is age in years, we define $p(t)$, the age *density function*, to be a function which 'smooths out' the age histogram. This function has the property that

$$\left(\begin{array}{c} \text{Fraction of Population} \\ \text{between ages } a \text{ and } b \end{array} \right) = \left(\begin{array}{c} \text{Area under} \\ \text{graph of } p \\ \text{between } a \text{ and } b \end{array} \right) = \int_a^b p(t)\,dt.$$

Age Group	Percentage of Total Population
0–10	15%
10–20	15%
20–30	16%
30–40	15%
40–50	13%
50–60	11%
60–70	9%
70–80	5%
80–90	1%

Table 7.2: Ages in the U.S., 1990 (more detailed)

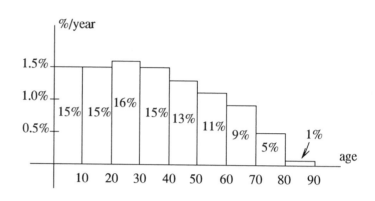

Figure 7.38: Ages in the U.S., 1990 (more detailed)

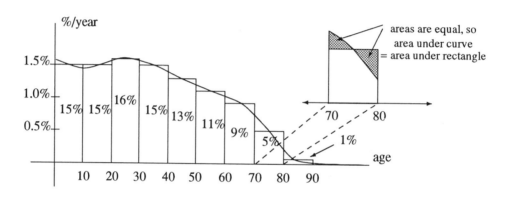

Figure 7.39: 'Smoothing Out' the Age Histogram

If a and b are the smallest and largest possible ages (say $a = 0$ and $b = 100$), so that all of the population has ages between a and b, then

$$\int_a^b p(t)dt = \int_0^{100} p(t)dt = 1.$$

What does the density function p tell us?

Notice that we have not talked about the meaning of $p(t)$ itself, but *only* of the integral $\int_a^b p(t)\, dt$. Let's look at this in a bit more detail. Suppose, for example, that $p(10) = 0.015 = 1.5\%$. This is *not* telling us that 1.5% of the population is precisely 10 years old (where 10 years old means exactly 10, not $10\frac{1}{2}$, not $10\frac{1}{4}$, not 10.1). However, $p(10) = 0.015$ does tell us that for some small interval Δt around 10, the fraction of the population with ages in this interval is approximately $p(10)\,\Delta t = 0.015\,\Delta t$. Notice also that the units of $p(t)$ are *% per year*, so $p(t)$ must be multiplied by years to give a percentage of the population.

The Density Function

In order to generalize the idea of the age distribution, let us look at the properties of a general *density function*. Suppose we are interested in how a certain characteristic, x, is distributed through a population. For example, x might be height, age, or wattage, and the population might be people, or any set of objects such as light bulbs. Then:

> If $p(x)$ is a *density function*:
>
> $$\left(\begin{array}{c} \text{Fraction of Population} \\ \text{for which } x \text{ is} \\ \text{between } a \text{ and } b \end{array} \right) = \left(\begin{array}{c} \text{Area under} \\ \text{graph of } p \\ \text{between } a \text{ and } b \end{array} \right) = \int_a^b p(x)\,dx.$$

The density function must be non-negative if its integral always gives a fraction of the population. Also, the fraction of the population with x between $-\infty$ and ∞ is 1 because the entire population has the characteristic x between $-\infty$ and ∞. In other words:

> If $p(x)$ is a *density function*:
>
> $$\int_{-\infty}^{\infty} p(x)\,dx = 1 \qquad \text{and} \qquad p(x) \geq 0 \quad \text{for all } x.$$

The function, $p(t)$, used to smooth out the age histogram, is a density function. We define $p(t) = 0$ for $t < 0$ (no one has a negative age), and $p(t) = 0$ for $t > 100$ (we assumed no one is over 100). Then

$$\int_{-\infty}^{\infty} p(t)\,dt = \int_0^{100} p(t)\,dt = 1.$$

For a general density function:

- If the characteristic x cannot be negative, the condition $\int_{-\infty}^{\infty} p(x)\,dx = 1$ is replaced by $\int_0^{\infty} p(x)\,dx = 1$.

- We do not yet assign a meaning to the value of $p(x)$ alone, but rather interpret $p(x)\,\Delta x$ as the fraction of the population with the characteristic in the interval of length Δx around x.

☐ **Example 2** The graph in Figure 7.40 shows the distribution of the number of years of education completed by adults in a population. What does the graph tell us?

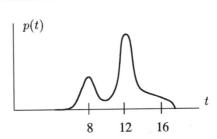

$p(t)$

8   12   16

t

Figure 7.40: Distribution of Years of Education

Solution. The fact that the peaks are at 8 and 12 years indicates that most people leave school after 8th or 12th grade, with more after 12th grade. Some complete 16 years of education.   ☐

The density function is often approximated by formulas, as in the next example.

☐ **Example 3** Find reasonable formulas representing the density function for the U.S. age distribution, assuming the function is constant at 1.5% up to age 40 and then drops linearly.

Solution. We need to construct a linear function sloping downwards from age 40 in such a way that $p(40) = 1.5\% = 0.015$ and that $\int_0^{100} p(t)dt = 1$. Since

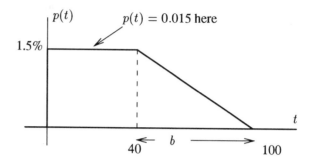

Figure 7.41: Age Density Function

$$\int_0^{100} p(t)dt = \int_0^{40} p(t)dt + \int_{40}^{100} p(t)dt = 40(0.015) + \frac{1}{2}(0.015)b = 1,$$

we have

$$\frac{0.015}{2}b = 0.4, \quad \text{giving} \quad b \approx 53.3.$$

Thus the slope of the line is $-\frac{0.015}{53.3} \approx -0.00028$, so for $40 \le t \le 93.3$,

$$p(t) - 0.015 = -0.00028(t - 40),$$
$$p(t) = 0.0263 - 0.00028t.$$

According to this way of smoothing the data, there is no one over 93.3 years old.   ☐

The Median

It is often useful to be able to give an 'average' value for a distribution. One measure that is in common use is the *median*:

> A **median** is a value T such that half the population has values of x less than (or equaling) T, and half the population has values of x greater than (or equaling) T. Thus, a median T satisfies:
>
> $$\int_{-\infty}^{T} p(x)\, dx = 0.5,$$
>
> where p is the density function. In other words, half the area under the graph of p lies to the left of T.

☐ **Example 4** Find the median age in the U.S. in 1990.

Solution. We want to find the value of T such that

$$\int_{-\infty}^{T} p(t)\, dt = \int_{0}^{T} p(t)\, dt = 0.5.$$

Since, for our model of the age distribution, $p(t) \approx 1.5\%$ up to age 40, we have

$$\text{Median} = T \approx \frac{50\%}{1.5\%} \approx 33 \text{ years}.$$

See Figure 7.42. ❑

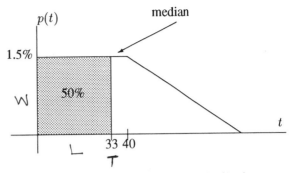

Figure 7.42: Median of Age Distribution

The Mean

Another commonly used 'average' value is the *mean*. To find the mean of N numbers, you add the numbers and divide the sum by N. For example, the mean of the numbers 1, 2, 7, and 10 is $\frac{1+2+7+10}{4} = 5$. The mean age of the entire US population is therefore defined as

$$\frac{\sum \text{ ages of all people in the U.S.}}{\text{Total number of people in the U.S.}}.$$

Calculating the sum of all the ages directly would be an enormous task; we will approximate the sum by an integral. The idea is to 'slice up' time and consider the people whose age is between t and $t + \Delta t$. How many are there?

The percentage of the population between t and $t + \Delta t$ is the area under the graph of p between these points, which is well approximated by the area of the rectangle, $p(t)\Delta t$. See Figure 7.43. If the total number of people in the population is N, then

$$\left(\begin{array}{c} \text{Number of people} \\ \text{with age between} \\ t \text{ and } t + \Delta t \end{array} \right) \approx p(t)\Delta t N.$$

The age of all of these people is approximately t:

$$\left(\begin{array}{c} \text{Sum of ages} \\ \text{of people between} \\ \text{age } t \text{ and } t + \Delta t \end{array} \right) \approx t p(t)\Delta t N.$$

Figure 7.43: Shaded Area = Percentage of Population with age between t and $t + \Delta t$

Therefore, adding and factoring out an N gives us

$$\text{Sum of ages of all people} \approx \left(\sum t p(t)\Delta t \right) N.$$

In the limit, as we allow Δt to shrink to 0, the sum becomes an integral, so as an approximation:

$$\text{Sum of ages of all people} = \left(\int_0^{100} tp(t)\,dt \right) N.$$

Therefore, with $N = $ total number of people in the U.S., and assuming no person is over 100 years old,

$$\text{Mean Age} = \frac{\text{Sum of ages of all people in U.S.}}{N} = \int_0^{100} tp(t)\,dt.$$

We can give the same argument for any density function $p(x)$. So, we define

For any[3] density function $p(x)$,

$$\textbf{Mean Value of } x = \int_{-\infty}^{\infty} xp(x)\,dx$$

It can be shown that the mean is the point on the horizontal axis where the region under the distribution graph, if it were made out of cardboard, would balance.

☐ **Example 5** Find the mean age of the U.S. population, using the density function of Example 3.

Solution. The approximate formulas for p are

$$p(t) = \begin{cases} 0.015, & \text{for } 0 \le t \le 40; \\ 0.0263 - 0.00028t & \text{for } 40 \le t \le 93.3. \end{cases}$$

Using these:

$$\begin{aligned}
\text{Mean Age} &= \int_0^{100} tp(t)\,dt = \int_0^{40} t(0.015)\,dt + \int_{40}^{93.3} t(0.0263 - 0.00028t)\,dt \\
&= 0.015\frac{t^2}{2}\Big|_0^{40} + 0.0263\frac{t^2}{2}\Big|_{40}^{93.3} - 0.00028\frac{t^3}{3}\Big|_{40}^{93.3} \\
&\approx 36 \text{ years.}
\end{aligned}$$

From the shape of the graph, we'd expect the mean to be slightly to the right of the median, as it is. See Figure 7.44.

---

[3] Provided all the relevant improper integrals converge.

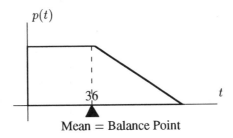

$p(t)$

36

t

Mean = Balance Point

Figure 7.44: Mean of Age Distribution

Cumulative Distribution Function

Another way of showing how ages are distributed in the U.S. is by using the

> **cumulative distribution** function, $P(t)$, defined by
>
> $$P(t) = \left(\begin{array}{c}\text{Fraction of Population}\\\text{of age less than } t\end{array}\right) = \int_0^t p(x)dx.$$
>
> Thus P is the antiderivative of p with $P(0) = 0$.

Notice that the cumulative distribution function is non-negative and *increasing* (or at least non-decreasing), since the number of people younger than age t increases as t increases. Alternatively, we can argue that $P' = p$, and p is positive (or non-negative). Thus the cumulative age distribution is a function which starts with $P(0) = 0$ and increases as t increases. $P(t) = 0$ for $t < 0$ because, when $t < 0$, there is no one whose age is less than t. The limiting value of P, as $t \to \infty$, is 1 since as t becomes very large (100 say), everyone will be younger than age t, so the fraction of people with age less than t tends towards 1. See Figure 7.45. Notice that for t less than 40, the graph of P is a straight line, because p is constant there. For $t > 40$, the graph of P levels off as p tends to 0.

To sum this up, P is non-negative and increasing, and $\lim_{t \to \infty} P(t) = 1$. In addition, $P(0) = 0$ because no one is younger than 0. As you might expect, we can generalize these ideas for any density function p.

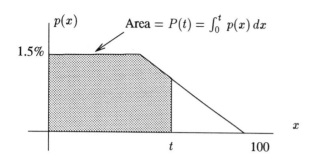
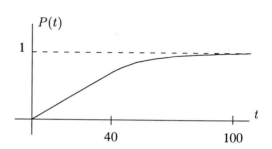

Figure 7.45: Cumulative Age Distribution Function

A **cumulative distribution function** $P(t)$ of a density function p is defined by

$$P(t) = \int_{-\infty}^{t} p(x)\, dx = \left(\begin{array}{c} \text{Fraction of population} \\ \text{having values of } x \\ \text{below } t \end{array} \right).$$

Thus P is an antiderivative of p. So $P' = p$.
Any cumulative distribution has the following properties:

- P is increasing (or non-decreasing).

- $\lim_{t\to\infty} P(t) = 1$ and $\lim_{t\to-\infty} P(t) = 0$.

- By the Fundamental Theorem of Calculus,

$$\left(\begin{array}{c} \text{Fraction of Population} \\ \text{having values of } x \\ \text{between } a \text{ and } b \end{array} \right) = \int_{a}^{b} p(x)\, dx = P(b) - P(a).$$

Probability

A density function is often called a *probability density function* (or p.d.f. for short) because we can use a density function to compute probabilities. What does the density function have to do with probability? Suppose we choose a member of the population at random. The probability, or chance, that the person is in a certain age group is equal to the fraction of the population in that age group. Thus if $p(t)$ is the density function for ages,

$$\left(\begin{array}{c} \text{Probability that} \\ \text{a person is between} \\ \text{age } a \text{ and } b \end{array} \right) = \left(\begin{array}{c} \text{Fraction of Population} \\ \text{between age } a \text{ and } b \end{array} \right) = \int_{a}^{b} p(t)\, dt.$$

Similarly, the cumulative distribution function is also called the *cumulative probability distribution*, or c.d.f. for short, because we can think of $P(t)$ as the probability that a randomly selected individual is younger than age t. In other words:

$$\left(\begin{array}{c}\text{Probability that}\\\text{a person is younger}\\\text{than age } t\end{array}\right) = \left(\begin{array}{c}\text{Fraction of Population}\\\text{younger than age } t\end{array}\right) = P(t) = \int_0^t p(x)\, dx.$$

☐ **Example 6** Suppose you want to analyze the fishing industry in a small town. Each day, the boats bring back at least two tons of fish, and never more than eight tons. Using the probability density function describing the daily catch in Figure 7.46, find the corresponding cumulative probability distribution function and explain its meaning. What is the probability that the catch will be between five and seven tons?

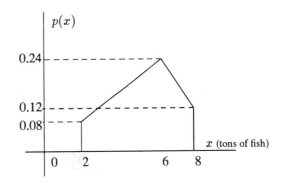

Figure 7.46: Probability Density of Daily Catch

Solution. The cumulative probability distribution function is given by $P(t) = \int_2^t p(x)\, dx$. By finding the equations of the lines that determine the probability density function, we have

$$p(x) = \begin{cases} 0.04x & \text{for } 2 \le x \le 6; \\ -0.06x + 0.6 & \text{for } 6 \le x \le 8. \end{cases}$$

Thus, for $2 \le t \le 6$,

$$P(t) = \int_2^t 0.04x\, dx = 0.04\frac{x^2}{2}\bigg|_2^t = 0.02t^2 - 0.08.$$

And for $6 \le t \le 8$,

$$\begin{aligned} P(t) &= \int_2^t p(x)\, dx = \int_2^6 p(x)\, dx + \int_6^t p(x)\, dx \\ &= 0.64 + \int_6^t (-0.06x + 0.6)\, dx = 0.64 + \left(-0.06\frac{x^2}{2} + 0.6x\right)\bigg|_6^t \\ &= -0.03t^2 + 0.6t + 1.36. \end{aligned}$$

$P(t)$ is the probability that the catch is less than t tons on any given day.

The probability that the catch is between 5 and 7 tons is

$$\int_5^7 p(x)\,dx = P(7) - P(5) = 0.43.$$

Graphically, this probability is the shaded area under $p(x)$ for $5 \leq x \leq 7$ in Figure 7.47.

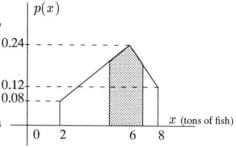

Figure 7.47: Probability Density of Daily Catch

Normal Distributions

How much rain do you expect will fall in your home town this year? If you lived in Anchorage, Alaska, the answer would be something close to 15 inches (including the snow). Of course, you don't expect exactly 15 inches. Some years there will be more than 15 inches, and some years there will be less. Most years, however, the amount of rainfall will be close to 15 inches; only rarely will it be well above or well below 15 inches.

What probability density function can we use to model the rainfall? This distribution and many other phenomena can be well-approximated by using a *normal distribution*.

A **normal distribution** has a density function of the form

$$p(x) = \frac{1}{\sigma\sqrt{2\pi}} e^{-\frac{(x-\mu)^2}{2\sigma^2}}$$

where μ and σ are constants and $\sigma > 0$. The constant μ is the *mean* of p, while σ is called the *standard deviation* of p. The standard deviation measures how tightly the observations are clustered around the mean. A small σ tells us that most of the data is close to the mean; a large σ tells us that the data is more spread out.

The factor of $\frac{1}{\sigma\sqrt{2\pi}}$ in front of the function is there to make the area under its graph equal to 1. That the factor $\sqrt{2\pi}$ is involved is one of the truly remarkable discoveries of mathematics.

The annual rainfall in Anchorage is modeled by a normal distribution with $\mu = 15$ and $\sigma = 1$. The graph in Figure 7.48 is bell-shaped, which is typical of a normal distribution.

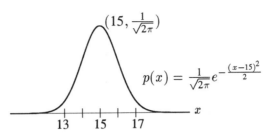

Figure 7.48: Normal Distribution with a mean of 15 and standard deviation of 1

☐ **Example 7** For Anchorage's rainfall, use the normal distribution with the density function

$$p(x) = \frac{1}{\sqrt{2\pi}} e^{-\frac{(x-15)^2}{2}},$$

to compute the fraction of the years with rainfall between (a) 14 and 16 inches, (b) 13 and 17 inches, and (c) 12 and 18 inches.

Solution.

(a) The fraction of the years with an annual rainfall between 14 and 16 inches is just $\int_{14}^{16} \frac{1}{\sqrt{2\pi}} e^{-\frac{(x-15)^2}{2}} \, dx$. Since there is no elementary antiderivative for $e^{-\frac{(x-15)^2}{2}}$, we find the integral numerically. Its value is about 0.68.

$$\left(\begin{array}{c} \text{Fraction of years with rainfall} \\ \text{between 14 and 16 inches} \end{array} \right) = \int_{14}^{16} \frac{1}{\sqrt{2\pi}} e^{-\frac{(x-15)^2}{2}} \, dx \approx 0.68.$$

(b) Finding the integral numerically again:

$$\left(\begin{array}{c} \text{Fraction of years with rainfall} \\ \text{between 13 and 17 inches} \end{array} \right) = \int_{13}^{17} \frac{1}{\sqrt{2\pi}} e^{-\frac{(x-15)^2}{2}} \, dx \approx 0.95.$$

(c)

$$\left(\begin{array}{c} \text{Fraction of years with rainfall} \\ \text{between 12 and 18 inches} \end{array} \right) = \int_{12}^{18} \frac{1}{\sqrt{2\pi}} e^{-\frac{(x-15)^2}{2}} \, dx \approx 0.997.$$

Since 0.95 is so close to 1, we expect that most of the time the rainfall will be between 13 and 17 inches a year. ☐

Notice that in the above example, the standard deviation is 1, so rainfall between 14 and 16 inches a year is within one standard deviation of the mean. Similarly, rainfall between 13 and 17 inches is within 2 standard deviations of the mean, and rainfall between 12 and 18 standard deviations is within three standard deviations of the mean. Remarkably enough, the fractions of the observations within one, two, or three standard deviations of the mean calculated in the previous example hold for any normal distribution.

Rules of Thumb for Any Normal Distribution:

- About 68% of the observations are within one standard deviation of the mean.

- About 95% of the observations are within two standard deviations of the mean.

- Over 99% of the observations are within three standard deviations of the mean.

❖ Exercises for Section 7.6

1. Consider the fishing data given in Example 6 on page 517. Show that the area under the density function in Figure 7.46 is 1. Why is this to be expected?

2. Find the mean daily catch for the fishing data in Figure 7.46, page 517.

3. IQ scores are believed to be normally distributed with mean 100 and standard deviation 15.

    (a) Write a formula for the density distribution of IQ scores.
    (b) Estimate the fraction of the population with IQ between 115 and 120.

4. Show that the area under the graph of the density function of the normal distribution

$$p(x) = \frac{1}{\sqrt{2\pi}} e^{-\frac{(x-15)^2}{2}}$$

is 1. This function has no elementary antiderivative, so you must do it numerically. Make it clear in your solution what limits of integration you used.

5. (a) Using a calculator or computer, sketch graphs of the density function of the normal distribution

$$p(x) = \frac{1}{\sigma\sqrt{2\pi}} e^{-\frac{(x-\mu)^2}{2\sigma^2}}$$

    i. for fixed μ (say $\mu = 5$) and varying σ (say $\sigma = 1, 2, 3$).
    ii. for varying μ (say $\mu = 4, 5, 6$) and fixed σ (say $\sigma = 1$).

    (b) Explain how the graphs confirm that μ is the mean of the distribution and that σ shows how closely the data is clustered around the mean.

6. The density function for the duration of telephone calls within a certain city is $p(x) = 0.4e^{-0.4x}$, where x denotes the duration in minutes of a randomly selected call.

    (a) What percentage of calls last between 1 and 2 minutes?
    (b) What percentage of calls last 1 minute or less?
    (c) What percentage of calls last 3 minutes or more?
    (d) Find the cumulative distribution function.

7. A congressional committee is investigating a defense contractor. It seems that, for the contracts this contractor is awarded, by the time the projects are completed there are almost always cost overruns. The data in Table 7.3 were presented to the committee.

Cost Overrun (C)	Fraction of Cost Overruns that are at most C
−20%	0.01
−15%	0.03
−10%	0.08
−5%	0.13
0%	0.19
10%	0.32
20%	0.50
30%	0.80
40%	0.94
50%	0.99

Table 7.3: Defense Contractor Data for Committee (for Problem 7)

(a) Plot the data with C on the horizontal axis. Is this a density function or a cumulative distribution function? Sketch the best curve you can through these points.

(b) If you think you drew a density function in (a), sketch the corresponding cumulative distribution function on another set of axes. If you think you drew a cumulative distribution function in (a), sketch the corresponding density function.

(c) Looking at the table, what is the probability that there will be a cost overrun of 50% or more? Between 20% and 50%? What appears to be the most likely cost overrun?

8. A person who travels regularly on the 9:00 am bus from Oakland to San Francisco reports that the bus is almost always at least a few minutes late, but rarely more than five minutes late. We are told further that the bus is never more than two minutes early, although it is on very rare occasions a little early.

(a) Sketch a possible density function, $p(t)$, for this situation. Shade in the region under the graph between $t = 2$ minutes and $t = 4$ minutes. Explain what this region represents.

(b) Now sketch the cumulative distribution function $P(t)$ for this situation. What measurement(s) on this graph correspond to the area shaded above? What do the inflection point(s) on your graph of P correspond to on the graph of p? How can you interpret the inflection points on the graph of P without referring to the graph of p?

9. Students at the University of California were surveyed and asked their grade point average. The distribution of GPA's is shown to the right. (GPA ranges from 0 to 4, where 2 is just passing.)

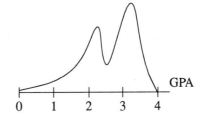

(a) Roughly what fraction of students are passing?

   (b) Roughly what fraction of the students have honor grades (a GPA above 3)?

   (c) Why do you think there is a peak around 2?

   (d) Sketch a graph of the cumulative distribution function.

10. The figure to the right shows the distribution of elevation in miles across the Earth's surface. Positive elevation denotes land above sea level; negative elevation shows land below sea level (i.e. the ocean floor).

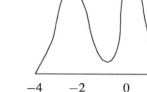

   (a) Describe in words the elevation of most of the Earth's surface.

   (b) Approximately what fraction of the Earth's surface is below sea level?

11. In southern Switzerland, most of the rain falls in the spring and fall; summers and winters are relatively dry. Sketch possible graphs for the density function and the cumulative distribution function of the rain distribution, over the course of one year. Put the date on the horizontal axis and percentage of the year's rainfall on the vertical axis.

12. Let $P(x)$ be the cumulative function for the income distribution in the U.S. in 1973 (income is measured in thousands of dollars). Some values of $P(x)$ are in the table below:

income x (in thousands)	1	4.4	7.8	12.6	20	50
$P(x)$ (in %)	1	10	25	50	75	99

   (a) What fraction of the population made between $20,000 and $50,000?

   (b) What was the median income?

   (c) Sketch a density function for this distribution. Where, approximately, does your density function have a maximum? What is the significance of this point, in terms of income distribution? How can you recognize this point on the graph of the density function and on the graph of the cumulative distribution?

13. Let v be the velocity, in meters/sec, of an oxygen molecule and let $p(v)$ be the density function of the speed distribution of oxygen molecules at room temperature. Maxwell showed that

$$p(v) = av^2 e^{\frac{-mv^2}{2kT}},$$

where $k = 1.4 \times 10^{-23}$ is the Boltzmann constant, $T = 293$ is the temperature in degrees Kelvin, and $m = 5 \times 10^{-26}$ is the mass of the oxygen molecule in kilograms.

   (a) Find the value of a.

   (b) Estimate the median and the mean speed. Find the maximum of $p(v)$.

   (c) How do your answers in (b) change as T changes?

14. If we think of an electron as a particle, the function

$$P(r) = 1 - (2r^2 + 2r + 1)e^{-2r}$$

is the cumulative distribution function of the radial distribution of the sole electron in a hydrogen atom (at ground state), where the distance r of the electron from the center of the atom is measured in Bohr radii. (1 Bohr radius $= 5.29 \times 10^{-11}$meter. Niels Bohr (1885-1962) was a Danish physicist.)

For example, $P(1) = 1 - 5e^{-2} \approx 0.32$ means that the sole electron is within 1 Bohr radius from the center of the atom 32% of the time.

(a) Find a formula for the density function of this distribution. Sketch the density function and the cumulative distribution function.

(b) Find the median distance, the mean distance, and the most likely distance of the electron from the center of the atom.

(c) The Bohr radius is sometimes called the "radius of the hydrogen atom." Why?

15. Consider a pendulum swinging through a small angle. The x-coordinate of the bob moves between $-a$ and a, as shown in the picture.

(a) Draw the probability density function for the location of the x-coordinate of the pendulum bob (i.e. neglect up and down motion). In order to do this, imagine a camera taking pictures of the pendulum at random instants. Where is the bob most likely to be found? Least likely?

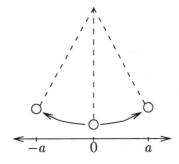

[*Hint:* Consider the speed of the pendulum at different points in its path. Is the camera more likely to take a photograph of the bob at a point on its path where the bob is moving quickly, or where it is moving slowly?]

(b) Now sketch the function

$$f(x) = \begin{cases} \frac{1}{\pi\sqrt{a^2 - x^2}} & -a < x < a; \\ 0 & |x| \geq a. \end{cases}$$

How does this graph compare with the one you drew in part (a)?

(c) Assuming that the function given in part (b) is the probability density function for the pendulum, what do you expect

$$\int_{-a}^{a} \frac{1}{\pi\sqrt{a^2 - x^2}}\, dx$$

to be? Check this by computing the integral.

(d) Does it seem reasonable, physically speaking, that $f(x)$ "blows up" at a and $-a$? Explain your answer.

7.7 Miscellaneous Exercises for Chapter 7

1. Consider a bacteria population whose birthrate, B, is given by the curve below as a function of time in hours. The birth-rate is in births per hour. The curve marked D gives the death-rate (in deaths per hour) in the same population.

   (a) Explain what the shape of each of these graphs tells you about the population.

   (b) Use the graphs to find the time at which the net rate of increase of the population is at a maximum.

   (c) Suppose at time $t = 0$ the population has size N. Sketch the graph of the total number born by time t. Also sketch the graph of the number alive at time t. Use the given graphs to find the time at which the population size is a maximum.

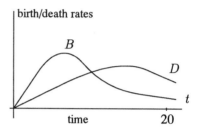

2. A student once told me the following nice rule for falling bodies: If at a certain time its velocity is u, and t seconds later its velocity is w, then the distance it has fallen during that interval is

$$s = \frac{u + w}{2}t.$$

That is, the distance can be calculated by assuming the velocity is constant over the time interval and equal to the average of the velocities at the start and finish of the interval.[4]

   (a) Using the fact that the acceleration of a falling body is constant (equal to g), draw a graph showing the velocity, v, against time, t. Assume that it started moving at time $t = 0$ with an initial velocity of zero.

   (b) How can the distance moved by time t be represented by an area on the graph?

   (c) Use area to derive the student's rule given above.

   (d) Suppose the body starts with velocity v_0 (where $v_0 > 0$). How does this change the picture? Does this change the student's rule?

   (e) So far, we have neglected air resistance. If we take air resistance into account, the equation $s = \frac{u+w}{2}t$ becomes an inequality. Explain.

---

[4]From *Calculus: The Analysis of Functions*, by Peter D. Taylor (Toronto: Wall & Emerson, Inc., 1992)

3. (a) Sketch the solid obtained by rotating the region bounded by

$$y = \sqrt{x}, \quad x = 1, \quad y = 0$$

around the line $y = 0$.

(b) Approximate its volume by Riemann sums, showing the volume represented by each term in your sum on the sketch.

(c) Now find the volume of this solid using an integral.

4. Using the region of Problem 3, find the volume when it is rotated around
   (a) the line $y = 1$ (b) y axis.

5. The probability of a transistor failing between $t = a$ months and $t = b$ months is given by

$$c \int_a^b e^{-ct} \, dt$$

for some constant c.

(a) If the probability of failure within the first six months is 10%, what is c?

(b) Given the value of c in part (a), what is the probability the transistor will fail within the second six months?

6. In this problem, you will derive the formula for the volume of a right circular cone with height l and radius of base b by rotating the line $y = ax$ from $x = 0$ to $x = l$ around the x-axis.
   (a) What value should you choose for a such that the cone will have height l and radius of base b?
   (b) Given this value of a, find the volume of the cone.

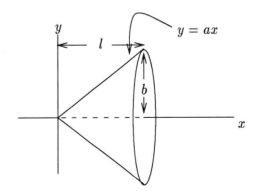

7. Mt. Shasta is a cone-like volcano whose radius at an elevation of h feet above sea level is approximately $3 \cdot \frac{10^5}{\sqrt{h}}$ feet. Its bottom is 400 feet above sea level, and its top 14,400 feet above sea level. See Figure 7.49.

   (a) Give a Riemann sum approximating the volume of Mt. Shasta.

   (b) Find the volume in cubic feet. [Note: This data about Mt. Shasta is more or less accurate. Mt. Shasta lies in northern California, and for some time was thought to be the highest point in the U.S. outside Alaska.]

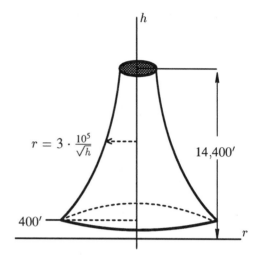

Figure 7.49: Mt. Shasta

(for Problem 7)

8. The reflector behind a car headlight is made in the shape of the parabola, $x = \frac{4}{9}y^2$, with a circular cross section, as shown in Figure 7.50.

   (a) Find a Riemann sum approximating the volume contained by this headlight.

   (b) Hence find its volume exactly.

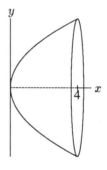

Figure 7.50: A Headlight Reflector

(for Problem 8)

9. If the circle $x^2 + y^2 = 1$ is rotated about the line $y = 3$, it forms a torus (a donut-shaped figure). Find the volume of this torus.

10. Find the volume of a generalized pyramid, whose base is a planar region of area A and whose apex is a height h above the plane containing the base. (Hint: The cross-section of the pyramid parallel to the base is of the same shape as the base, but is scaled down in area.)

For Problems 11-13, find the arc lengths.

11. $f(x) = \sqrt{1 - x^2}$ from $x = 0$ to $x = 1$

12. $f(x) = e^x$ from $x = 1$ to $x = 2$

13. $f(x) = \frac{1}{3}x^3 + \frac{1}{4x}$ from $x = 1$ to $x = 2$.

14. Find a function $f(x)$ such that its length L from (0,0) to $(x, f(x))$, $x > 0$, is given by

(a) $L = x$

(b) $L = ax, a > 0$

For the curves described in Problems 15 and 16, write the integral that gives the exact length of the curve; you need not evaluate the integral.

15. One arch of the sine curve, from $x = 0$ to $x = \pi$.

16. The ellipse with equation $(x^2/a^2) + (y^2/b^2) = 1$.

17. Find the average horizontal width of A in the figure to the right.

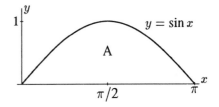

18. Marie-Jolaine walks about the xy-plane. She starts at the origin, and at time t she finds herself at the point (t, t^2).

(a) What is her speed v at time t? (Note: The speed v is defined as

$$v = \frac{ds}{dt} = \lim_{\Delta t \to 0} \frac{\Delta s}{\Delta t},$$

where s is the distance traveled; see Figure 7.51.)

(b) Find s, the distance traveled from the origin two ways: using arc length and using the relation $v = \frac{ds}{dt}$.

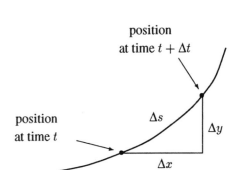

Figure 7.51: Marie-Jolaine's promenade.
(for Problem 18)

19. (a) Find the present and future values of a constant income stream of $100 per year over a period of 20 years, assuming a 10% annual interest rate compounded continuously.

(b) How many years will it take for the balance to reach $5000?

20. In 1950, an experiment was done observing the time gaps between successive cars on the Arroyo Seco Freeway. The data[5] shows that the distribution function of these time gaps was given approximately by

$$p(x) = ae^{-0.122x}$$

where x is the time in seconds and a is a constant.

   (a) Find a.

   (b) Find P, the cumulative distribution function.

   (c) Find the median and mean time gap.

   (d) Sketch rough graphs of p and P.

21. A nuclear power plant produces Strontium-90 at a rate of 1 kg/year. How much of the Strontium produced since 1971 (when the plant opened) was still around in 1992? (The half–life of Strontium-90 is 28 years.)

22. A 200 lb weight is to be hoisted from the ground to a point 20 feet high by a chain passing over a pulley. If the chain weighs 2 lb/ft, find the work done in lifting the weight.

23. Water is to be raised from a well 40 ft deep by means of a bucket attached to a rope. When the bucket is full of water it weighs 30 lb, but the bucket has a leak that causes it to lose water at the constant rate of 1/4 lb for each foot that the bucket is raised. Neglecting the weight of the rope, find the work done in raising the bucket to the top.

24. A cylindrical garbage can of depth 4 feet and radius $1\frac{1}{2}$ feet fills with rainwater up to a depth of 3 feet. How much work would be done in pumping the water up to the top edge of the can? (Recall that water weighs 62.4 pounds per cubic foot.)

25. A water tank is in the shape of a right circular cone with height 18 ft and radius 12 ft at the top. If it is filled with water to a depth of 15 ft, find the work done in pumping all of the water over the top of the tank. (Note: water weighs 62.4 pounds per cubic foot.)

---

[5]Reported by Daniel Furlough and Frank Barnes

26. What is the total force on the bottom and sides of the garbage can in Problem 24?

27. To the right you see a cross section through an apple (scale: 1 division = 1/2 inch).

    (a) Give a rough estimate for the volume of this apple (in cubic inches).

    (b) The density of "Granny Smith" apples is about 0.03 pounds per cubic inch (a little less than the density of water—as you might expect, since apples float). Estimate how much this apple would cost. (They go for 80 cents a pound.)

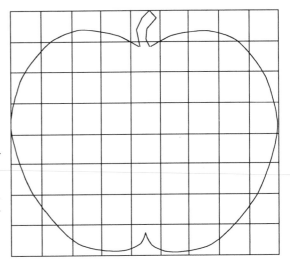

28. A cylindrical centrifuge of radius 1m and height 2m is filled with water to a depth of 1 meter (see Figure 7.52(a)). As the centrifuge accelerates, the water level rises along the wall and drops in the center; the cross-section will be a parabola. (Figure 7.52(b))

    (a) Find the equation of the cross-section in Figure 7.52(b) in terms of h, the depth of the water at its lowest point.

    (b) As the centrifuge rotates faster and faster, either water will be spilled out the top as in Figure 7.52(c), or the bottom of the centrifuge will be exposed as in Figure 7.52(d). Which happens first?

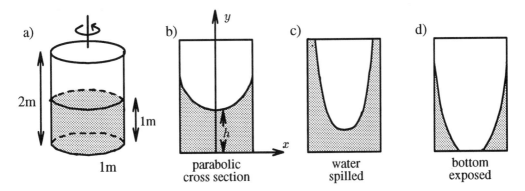

Figure 7.52: A centrifuge filled with water

29. An ancient Greek waterclock called a clepsydra is designed so that the depth of the water
decreases at a constant rate as the water runs
out a hole in the bottom. This design allows
the use of a uniform scale to mark the hours.
The tank of the clepsydra is a volume of revolution about a vertical axis. According to
Torricelli's Law, the exit speed of the water
flowing through the hole is proportional to
the square root of the depth of the water. Use
this to find the formula $y = f(x)$ for this
profile, assuming that $f(1) = 1$.

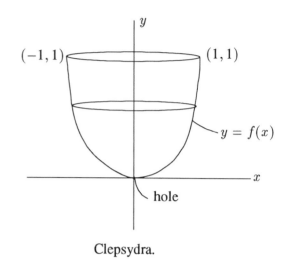

Clepsydra.

In Problems 30 and 31, you are given two objects which have the same mass M, the same radius R,
and the same angular velocity about the indicated axes (say for the sake of definiteness, 1 revolution
per minute). In each problem, say which object has the greater kinetic energy. (Recall that the kinetic
energy of a particle of mass m with speed v is $\frac{1}{2}mv^2$.) Don't actually compute the kinetic energy of
the objects to do this; just use reasoning.

30.
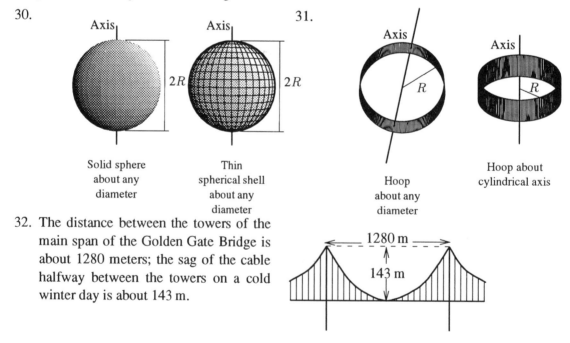

Solid sphere about any diameter

Thin spherical shell about any diameter

31.

Hoop about any diameter

Hoop about cylindrical axis

32. The distance between the towers of the
main span of the Golden Gate Bridge is
about 1280 meters; the sag of the cable
halfway between the towers on a cold
winter day is about 143 m.

1280 m

143 m

(a) How long is the cable, assuming it has a parabolic shape?

(b) On a hot summer day the cable is about 0.05% longer, due to thermal expansion. By how much does the sag increase? Assume no movement of the towers.

33. Two cylinders are inscribed in a cube of side length 2, as shown in Figure 7.54. What is the volume of the solid the two cylinders enclose? [Hint: Use horizontal slices.]

Note: The solution was known to Archimedes. The Chinese mathematician Liu Hui (third century A.D.) tried to find this volume, but he failed; he wrote a poem about his efforts calling the enclosed volume a "box-lid":

Look inside the cube
And outside the box-lid;
Though the dimension increases,
It doesn't quite fit.
The marriage preparations are complete;
But square and circle wrangle,
Thick and thin are treacherous plots,
They are incompatible.
I wish to give my humble reflections,
But fear that I will miss the correct prin-
    ciple;
I dare to let the doubtful points stand,
Waiting
For one who can expound them.

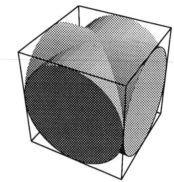

Figure 7.54: Two Cylinders for Problem 33

34. Find a formula for the volume of a bagel in terms of the diameter D of the bagel and the diameter d of the hole.

35. Consider the normal distribution

$$p(x) = \frac{1}{\sigma\sqrt{2\pi}} e^{-\frac{(x-\mu)^2}{2\sigma^2}}.$$

(a) Show that $p(x)$ is a maximum when $x = \mu$. What is that maximum value?

(b) Show that $p(x)$ has points of inflection where $x = \mu + \sigma$ and $x = \mu - \sigma$.

(c) Describe in your own words what μ and σ tell you about the distribution.

36. Triangular probability distributions are often used in business to model uncertainty. The triangular probability density function has the graph to the right. Such a distribution can be used to model a variable where only three pieces of information are available: a lower bound ($x = a$), a most likely value ($x = c$), and an upper bound ($x = b$).

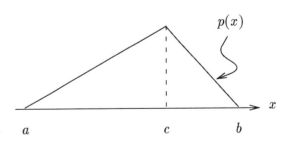

For example, say a proposed new product may cost between $6 and $10 per unit to produce, with a most likely cost of $9. Then $a = 6$, $b = 10$, $c = 9$. Thus one can write the function $p(x)$ as two lines:

$$p(x) = \begin{cases} m_1 x + b_1 & a \leq x \leq c \\ m_2 x + b_2 & c < x \leq b. \end{cases}$$

We know $p(a) = p(b) = 0$. We can find $p(c)$ and the values of m_1, m_2, b_1 and b_2 using the criterion that (Probability that $a \leq x \leq b$) = 1.

(a) Find the value of $p(c)$ geometrically.

(b) Find $p(9)$.

(c) You will now find m_1, m_2, b_1, and b_2 for the case $a = 6$, $b = 10$, $c = 9$. To do this, use the fact that $p(6) = p(10) = 0$ and the value of $p(9)$ you found in (b).

(d) What is the probability that the production cost per unit will be less than $8? (Hint: be careful in choosing the lower bound of integration.)

(e) What is the median cost?

(f) Write a formula of the cumulative probability distribution function $P(x)$ for

    (i) $6 \leq x \leq 9$,   (ii) $9 < x \leq 10$.

Sketch the graph of $P(x)$.

Chapter 8

DIFFERENTIAL EQUATIONS

In Chapter 2 we saw how the derivative of a function gives us its rate of change. In Chapter 3 we saw how to use the definite integral to compute changes in the original function from the derivative. Chapter 5 introduced antiderivatives and 'going backwards' from the derivative to the original function.

In this chapter we will start with an equation involving the derivative of a known function, and 'go backwards' to the original function. Such an equation is called a differential equation.

Although you may not have realized it, you have been solving differential equations for some time. Every time you antidifferentiate to find $\int f(x)\,dx$, you are actually solving the differential equation

$$\frac{dy}{dx} = f(x).$$

The new feature of this chapter is that the right-hand side may now be a function of y, or of both x and y, instead of just x. As before, we will see how to solve differential equations graphically first, then numerically, and, finally, in some of the special cases when it can be done, analytically.

In the course of this chapter, we will treat first order differential equations (involving first derivatives only), systems of differential equations, and some second order differential equations.

8.1   What is a Differential Equation?

Suppose you want to investigate how fast a new employee learns a complicated task. You might study how the percentage of the task the employee has mastered changes with time on the job. One theory claims that this percentage increases at a rate equal to the percentage yet to be learned. If y is the percentage mastered, the rate at which this percentage increases is $\frac{dy}{dt}$, and the percentage yet to be learned is $100 - y$. Thus the model claims that

$$\frac{dy}{dt} = 100 - y.$$

Such an equation, which gives information about the rate of change of an unknown function, is a *differential equation*. In an algebraic equation, the unknown is a number; here the unknown is a function. A differential equation tells us about the derivative of the unknown function.

Consider the function

$$y = 100 + Ce^{-t},$$

where C is any constant. Substituting this into the differential equation $\frac{dy}{dt} = 100 - y$ gives:

$$\text{left side} \quad = \quad \frac{dy}{dt} = -Ce^{-t} \quad \text{took } y' \text{ of } 100 + Ce^{-t}$$

$$\text{right side} \quad = \quad 100 - y = 100 - (100 + Ce^{-t}) = -Ce^{-t}$$

Since we get the same thing on both sides, we say $y = 100 + Ce^{-t}$ *satisfies* this differential equation.

To find the value of C, we need an additional piece of information – usually the initial value of y. If, for example, we are told that $y = 0$ when $t = 0$, substituting into

$$y = 100 + Ce^{-t}$$

shows us that

$$0 = 100 + Ce^0, \quad \text{so} \quad C = -100.$$

So the function $y = 100 - 100e^{-t}$ satisfies the differential equation *and* the condition that $y = 0$ when $t = 0$.

In this chapter we will show that any solution to this differential equation is of the form $y = 100 + Ce^{-t}$ for some constant C. Since any value of C gives a solution, we say that the *general solution* to the differential equation $\frac{dy}{dt} = 100 - y$ is the family of functions $y = 100 + Ce^{-t}$. The solution $y = 100 - 100e^{-t}$ which satisfies the differential equation together with the *initial condition* that $y = 0$ when $t = 0$ is called a *particular solution*, The differential equation and the initial condition together are called an *initial value problem*.

First, some more definitions. The differential equation

$$\frac{dy}{dt} = 100 - y$$

is called *first order*, because it involves the first derivative, but no higher derivatives. In contrast, if s is the position (in feet) of a body moving under the force of gravity and t is time (in seconds) then

$$\frac{d^2s}{dt^2} = -32.$$

This is a *second order* differential equation because it involves the second derivative of the unknown function, $s = f(t)$, but no higher derivatives.

☐ **Example 1** Show that $y = e^{2t}$ is not a solution to

$$\frac{d^2 y}{dt^2} + 4y = 0.$$

Solution. Substituting into the left side of the equation,

$$\frac{d^2 y}{dt^2} + 4y = 2(2e^{2t}) + 4e^{2t} = 8e^{2t} \neq 0.$$

Since we do not get the same quantity on each side of the equation, $y = e^{2t}$ is not a solution, i.e. $8e^{2t}$ is not identically zero. ☐

How Many Arbitrary Constants Should We Expect?

Since a differential equation involves the derivative of an unknown function, solving it usually involves antidifferentiation, which introduces arbitrary constants. The solution to a first order differential equation usually involves one antidifferentiation and one arbitrary constant (for example, the C in $y = 100 + Ce^{-t}$). Picking out one particular solution involves knowing an additional piece of information, such as an initial condition. Solving a second order differential equation generally involves two antidifferentiations and so two arbitrary constants. Consequently, finding a particular solution usually involves two initial conditions.

For example, if s is the height (in feet) of a body above the surface of the Earth at time t in seconds,

$$\frac{d^2 s}{dt^2} = -32.$$

Integrating gives

$$\frac{ds}{dt} = -32t + C_1,$$

and integrating again gives

$$s = -16t^2 + C_1 t + C_2.$$

Thus the general solution for s involves the two arbitrary constants C_1 and C_2. We can find C_1 and C_2 if we are told, for example, that the initial velocity is 100 feet per second upwards and that the initial position is 15 feet above the ground. In that case, $C_1 = 100$ and $C_2 = 15$, so

$$s = -16t^2 + 100t + 15.$$

❖ Exercises for Section 8.1

1. Show that, for any constant P_0, the function $P = P_0 e^t$ satisfies the differential equation

$$\frac{dP}{dt} = P.$$

2. Suppose you know that $Q = Ce^{kt}$ satisfies the differential equation

$$\frac{dQ}{dt} = -0.03Q.$$

   What (if anything) does this tell you about the values of C and k?

3. (a) Show that $y = \sin 2t$ satisfies

$$\frac{d^2y}{dt^2} + 4y = 0.$$

   (b) Find the value(s) of ω for which $y = \cos \omega t$ satisfies

$$\frac{d^2y}{dt^2} + 9y = 0.$$

4. (a) Show that $P = \dfrac{1}{1 + e^{-t}}$ satisfies the *logistic equation*

$$\frac{dP}{dt} = P(1 - P).$$

   (P represents the fraction of the maximum possible population in a restricted environment.)

   (b) What is the limiting value of this population, P, as time, t, tends to infinity?

5. Suppose y is the height of a hanging cable above a fixed horizontal line, as shown to the right. It can be shown that

$$\frac{d^2y}{dx^2} = k\sqrt{1 + \left(\frac{dy}{dx}\right)^2}.$$

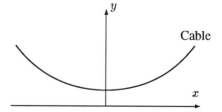

Cable

   (a) Show that $y = \dfrac{e^x + e^{-x}}{2}$ satisfies this differential equation if $k = 1$.

   (b) For general k, one solution to this differential equation is of the form

$$y = \frac{e^{Ax} + e^{-Ax}}{2A}.$$

   Substitute this expression for y into the differential equation to find A in terms of k.

6. Pick out which functions are solutions to which differential equations. (Note: Functions may be solutions to more than one equation or to none; an equation may have more than one solution.)

(a) $\dfrac{dy}{dx} = -2y$ (A) $y = 2\sin x$

(b) $\dfrac{dy}{dx} = 2y$ (B) $y = \sin 2x$

(c) $\dfrac{d^2y}{dx^2} = 4y$ (C) $y = e^{2x}$

(d) $\dfrac{d^2y}{dx^2} = -4y$ (D) $y = e^{-2x}$

7. Many of the families of curves that we have studied arise as solutions of differential equations. Match up the families of curves with the differential equations of which they are solutions.

(a) $\dfrac{dy}{dx} = \dfrac{y}{x}$ (A) $y = xe^{kx}$

(b) $\dfrac{dy}{dx} = \dfrac{y\ln y}{x}$ (B) $y = x^p$

(c) $\dfrac{dy}{dx} = \dfrac{y}{x}\left(1 + \ln\left(\dfrac{y}{x}\right)\right)$ (C) $y = e^{kx}$

(d) $\dfrac{dy}{dx} = \dfrac{y\ln y}{x\ln x}$ (D) $y = mx$

8. Match up solutions and differential equations. (Note: Each equation may have more than one solution.)

(a) $y'' - y = 0$ (A) $y = e^x$

(b) $y'' + 2y' + y = 0$ (B) $y = xe^x$

(c) $x^2y'' + 2xy' - 2y = 0$ (C) $y = x^3$

(d) $y'' - 2y' + y = 0$ (D) $y = e^{-x}$

(e) $x^2y'' - 6y = 0$ (E) $y = xe^{-x}$

                                  (F) $y = x^{-2}$

8.2   Slope Fields

In this section, you will see how to visualize a first order differential equation. Let's start with the equation

$$\frac{dy}{dx} = y.$$

You can check that the general solution to this differential equation is $y = Ce^x$, a family of exponential curves.

Any solution to this differential equation has the property that at any point in the plane, the slope of its graph is equal to its y coordinate. (That's what the equation $\frac{dy}{dx} = y$ is telling you!) This means that if the solution goes through the point $(0,1)$, its slope there is 1; when it goes through a point with $y = 4$ its slope is 4. See Figure 8.1. Another solution goes through $(0,2)$ and has slope 2 there; at the point where $y = 8$ the slope of this solution is 8.

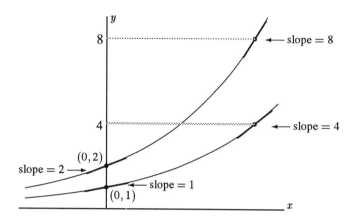

Figure 8.1: Solutions to $\frac{dy}{dx} = y$
(Note: x and y scales are different.)

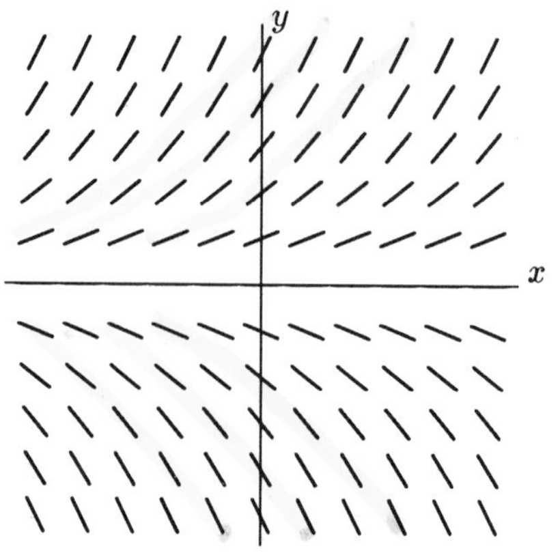

Figure 8.2: Slope field for $\frac{dy}{dx} = y$

In Figure 8.1, a small line is drawn at each of the marked points showing the slope of the curve there. Imagine drawing many of these lines, but leaving out the curves; you'll have the *slope field* for the equation $\frac{dy}{dx} = y$ shown in Figure 8.2. From the picture, you can see that above the x-axis, the slopes are all positive (because y is positive there), and they increase as you move upwards (as y increases). Below the x-axis, the slopes are all negative, and get more so as you move downwards. Notice that on a horizontal line (where y is constant) the slopes are all constant.

In the slope field you can see the ghost of the solution curve lurking. Start anywhere on the plane and move so that the slope lines are tangent to your path; you will trace out one of the solution curves. Try penciling in some solution curves on Figure 8.2, some above the x-axis, some below. The curves you draw should have the shape of an exponential function.

In most problems, we will be interested in getting the solution curves from the slope field. It may be helpful to think of the slope field as a set of signposts, pointing in the direction you should go at each point. Imagine starting anywhere in the plane: look at the slope field at that point and start to move in that direction. After a small step, look at the slope field again, and alter your direction if necessary. Continue to move across the plane in the direction the slope field points, and you'll trace out a solution curve. Notice that the solution curve is not necessarily the graph of a function, and even if it is, we may not have a formula for the function. Geometrically, solving a differential equation means finding the family of solution curves.

Let's look at another example where we do not (yet) have a formula for the solution of the differential equation.

☐ **Example 1**   Use the slope field of the differential equation

$$\frac{dy}{dx} = -\frac{x}{y}$$

to guess the form of the solution curve.

Solution. The slope field is shown in Figure 8.3. Notice that on the y-axis, where x is 0, the slope is 0. On the x-axis, where y is 0, the line segments are vertical and the slope is infinite. At the origin the slope is undefined and there is no line segment.

What do the solution curves of this differential equation look like? The slope field suggests that they look like circles centered at the origin. Later we will see how to derive the solution analytically, but even without this, we can still check that the circle is indeed a solution. Let's take the circle of radius r:

$$x^2 + y^2 = r^2$$

and differentiate implicitly, thinking of y as a function of x. Using the chain rule, we get

$$2x + 2y \cdot \frac{dy}{dx} = 0.$$

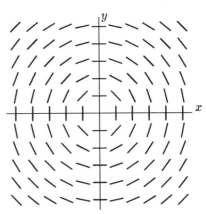

Figure 8.3: Slope field of $\frac{dy}{dx} = -\frac{x}{y}$

Solving for $\frac{dy}{dx}$ gives

$$\frac{dy}{dx} = -\frac{x}{y},$$

our differential equation.

The previous example shows that the solutions to differential equations may often be implicit functions. (Recall that implicit functions are ones which have not been "solved" for y; the dependent variable is not expressed as an explicit function of x.)

☐ **Example 2** The slope fields for $\dfrac{dy}{dt} = 2 - y$ and $\dfrac{dy}{dt} = \dfrac{t}{y}$ are shown in Figure 8.4. For each slope field:

(a) Sketch solution curves with initial conditions
   (i) $y = 1$ when $t = 0$ (ii) $y = 0$ when $t = 1$ (iii) $y = 3$ when $t = 0$

(b) For each solution curve, can you say anything about the long-run behavior of y? For example, what can you say about $\lim\limits_{t \to \infty} y$? Does $\lim\limits_{t \to \infty} y$ exist?

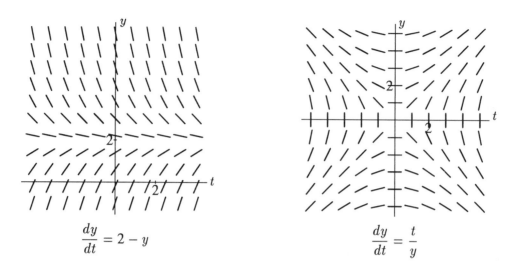

$$\frac{dy}{dt} = 2 - y \qquad\qquad \frac{dy}{dt} = \frac{t}{y}$$

Figure 8.4: Slope Fields for Example 2

Solution.

(a) See Figure 8.5.

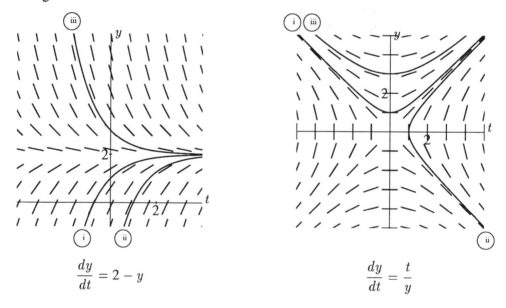

$$\frac{dy}{dt} = 2 - y \qquad\qquad \frac{dy}{dt} = \frac{t}{y}$$

Figure 8.5: Solution Curves for Example 2

(b) For $\frac{dy}{dt} = 2 - y$, all solution curves have $y = 2$ as a horizontal asymptote, so $\lim_{t \to \infty} y = 2$. For $\frac{dy}{dt} = \frac{t}{y}$, as $t \to \infty$, either $y \to t$ or $y \to -t$. $\qquad\qquad \square$

Existence and Uniqueness of Solutions

Since differential equations are used to model many real situations, the question of whether the solution is unique can have great practical importance. If we know how the velocity of a satellite is changing, can we know its velocity for all future time? If we know the initial population of a city, and we know how the population is changing, can we predict the population in the future? Common sense says yes: if we know the initial value of some quantity and we know exactly how it is changing, we should be able to figure out the future value of the quantity.

In the language of differential equations, an initial value problem (i.e., a differential equation and an initial condition) almost always has a unique solution. One way to see this is by looking at the slope field. Imagine starting at the point representing the initial condition. Through that point there will usually be a line segment pointing in the direction the solution curve must go. By following these line segments, you trace out the solution curve. Several examples with different starting points are shown in Figure 8.6. In general, at each point there is one line segment and therefore only one direction for the solution curve to go. Thus the solution curve *exists* and is *unique* provided you are given an initial point.

It can be shown that if the slope field is continuous as you move from point to point in the plane, you can be sure that the solution curve exists everywhere. Ensuring that each point has only one solution curve through it requires a slightly stronger condition.

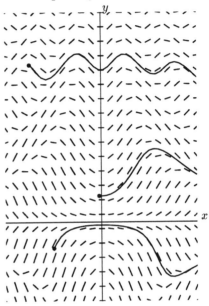

Figure 8.6: There's one and only one solution curve through each point in the plane for this slope field.

❖ Exercises for Section 8.2

1. The slope field for the equation $y' = x + y$ is shown to the right.

   (a) Carefully sketch the solutions that pass through the points
   (i) $(0,0)$ (ii) $(-3,1)$ (iii) $(-1,0)$.

   (b) From your sketch, write the equation of the solution passing through $(-1,0)$.

   (c) Verify your solution in (b) by substituting it into the differential equation.

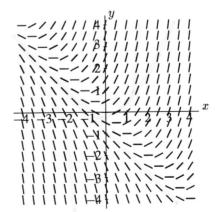

2. To the right is the slope field for the equation $y' = (\sin x)(\sin y)$.

   (a) Carefully sketch the solutions that pass through
       (i) $(-2, -2)$ (ii) $(0, \pi)$.

   (b) What is the equation of the solution, y, that
       passes through $(0, n\pi)$, where n is any integer?

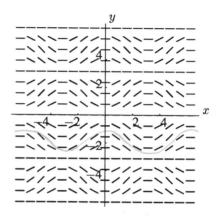

3. (a) Sketch the slope field for
        the equation $y' = x - y$
        on the attached graph at
        the points indicated.

    (b) Find the equation for
        the solution that passes
        through (1,0).

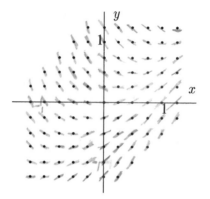

4. Figure 8.7 displays sketches of two slope fields. Sketch three solution curves for each of these
   fields.

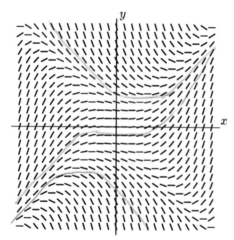 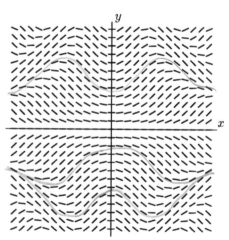

Figure 8.7: Slope Fields

(for Problem 4)

5. One of the slope fields shown in Figure 8.7 has the equation $y' = x^2 - y^2$. Which one? On this field, where is the point $(0,1)$? the point $(1,0)$? (Assume that the x and y scales are the same.) Now sketch in the line $x = 1$. Then sketch, as carefully as you can, the solution curve that passes through $(0,1)$ until it crosses $x = 1$.

6. The graph of a certain function in the xy-plane is a smooth curve. For any point (x,y) that the curve happens to pass through, the slope of the curve through that point is equal to $0.25xy$, i.e., $\frac{dy}{dx} = 0.25xy$. For example, if the curve passes through the point $(2,3)$, then the slope of the curve through $(2,3)$ is $0.25(2)(3) = 1.5$. On the graph in Figure 8.8, a short line segment through the point $(2,3)$ with a slope of 1.5 has been drawn.

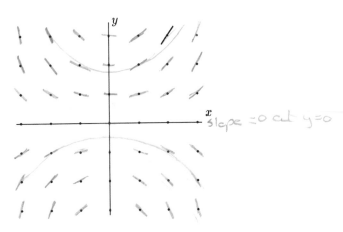

Figure 8.8: Slope Field Graph

(for Problem 6)

(a) If the curve passes through the point $(-1,2)$, what must its slope be at that point? Sketch a small line segment through the point $(-1,2)$ with such a slope. Do the same for all of the remaining points (x,y) on the grid that have integer values for x and y with $-3 \leq x \leq 3$ and $-3 \leq y \leq 3$.

(b) Sketch the curve passing through the point $(-3,-2)$. Does this curve cross the x-axis? Sketch the curve passing through the point $(-1,2)$.

(c) Guess a formula for $y = f(x)$.
   [Hint: Think of the Chain rule: What function 'differentiates to itself'? What would the inside function have to be to give a factor of $0.25x$ in the derivative?]
   Sketch your choice for $y = f(x)$ on the axes above. Does it conform to the slope field?

(d) Try multiplying the function you guessed in (c) by some number, say -2. Does this new function satisfy $\frac{dy}{dx} = 0.25xy$? Sketch its graph on the axes above. Does its graph conform to the slope field?

7. The slope field for $y' = 0.5(1+y)(2-y)$ is shown in Figure 8.9.

   (a) Plot the following points on the slope field:

      i. the origin
      ii. $(0,1)$
      iii. $(1,0)$
      iv. $(0,-1)$
      v. $(0,-5/2)$
      vi. $(0,5/2)$

   (b) Plot solution curves through the points in (a).

   (c) For which regions are all solution curves increasing? For which regions are all solution curves decreasing? When can the solution curves have horizontal tangents? Explain why, both using the slope field and using the differential equation.

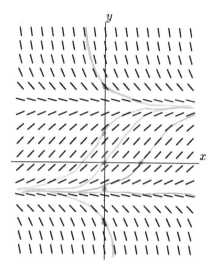

Figure 8.9: Slope Field for $y' = 0.5(1 + y)(2 - y)$ (The x and y scales are the same) (for Problem 7)

8. The Gompertz equation, which models growth of animal tumors, is $y' = -ay \ln(y/b)$, where a and b are positive constants. Write a paragraph explaining the similarities and differences between this differential equation with $a = 1$ and $b = 2$ and the equation $y' = y(2 - y)$. Use the slope fields with $0 < t < 4$ and $0 < y < 4$ in Figures 8.10 and 8.11.

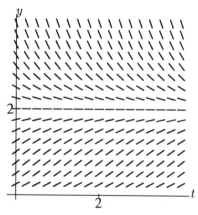

Figure 8.10: $y' = -y \ln(y/2)$ (for Problem 8)

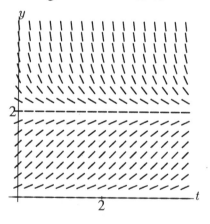

Figure 8.11: $y' = y(2 - y)$ (for Problem 8)

9. Match the following slope fields with their differential equations:

   (a) $y' = 1 + y^2$ (b) $y' = x$ (c) $y' = \sin x$
   (d) $y' = y$ (e) $y' = x - y$ (f) $y' = 4 - y$

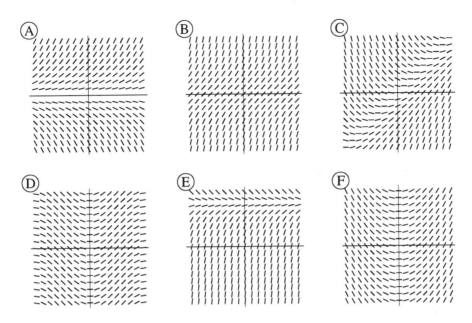

For Problems 10–15, consider a solution curve for each of the slope fields in Problem 9. What can you say about the long-run behavior of y? In other words, what can you say about $\lim_{x \to \infty} y$? You may get different limiting behavior for different starting points. In each case, your answer should discuss how the limiting behavior depends on the starting point. (Each slope field is graphed for $-5 \le x \le 5$, $-5 \le y \le 5$.)

10. Slope field (A)   12. Slope field (C)   14. Slope field (E)

11. Slope field (B)   13. Slope field (D)   15. Slope field (F)

16. In Section 4.8 on implicit differentiation, the equation $y^3 - xy = -6$ was analyzed using the derivative $\frac{dy}{dx} = \frac{y}{3y^2 - x}$. (See page 264.) View this equation for $\frac{dy}{dx}$ as a differential equation and give a rough sketch of its slope field, indicating where the slope is positive, negative, zero, or undefined (vertical). Then give a rough sketch of the solution curve through the point $(7, 2)$, the solution curve through the point $(7, 1)$, and the solution curve through the point $(7, -3)$. Notice that each of these points lies on the curve $y^3 - xy = -6$.

8.3 Euler's Method

In the previous section we saw how to sketch a solution curve to a differential equation using its slope field, whose line segments are everywhere tangent to the solution. In this section we will do the same thing numerically, by computing points on the solution curves using *Euler's method*. (Leonhard Euler was an eighteenth-century Swiss mathematician.) In some cases—but not many—we can find an equation for the solution curve; that will be the subject of the next section.

Here's the concept behind Euler's method. Think of the slope field as a set of signposts directing you across the plane. Pick a starting point (corresponding to the initial value), and calculate the slope at that point using the differential equation. This slope is a signpost telling you the direction to take. Head off a small distance in that direction. Stop and look at the new signpost. Recalculate the slope from the differential equation, using the coordinates of the new point. Change direction to correspond to the new slope, and move another small distance, and so on.

☐ **Example 1** Use Euler's method for $\dfrac{dy}{dx} = y$. Start at the point $P_0 = (0,1)$ and take $\Delta x = 0.1$.

Solution. The slope at the point $P_0 = (0,1)$ is $\frac{dy}{dx} = 1$. See Figure 8.12.

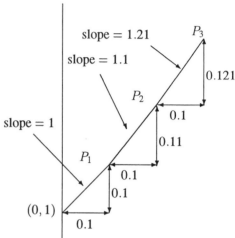

	x	y	$\Delta y = $ (slope)Δx
P_0	0	1	0.1=(1)(0.1)
P_1	0.1	1.1	0.11=(1.1)(0.1)
P_2	0.2	1.21	0.121=(1.21)(0.1)
P_3	0.3	1.331	0.1331=(1.331)(0.1)
P_4	0.4	1.4641	0.14641=(1.4641)(0.1)
P_5	0.5	1.61051	0.161051=(1.61051)(0.1)

Table 8.1: Euler's method for $\frac{dy}{dx} = y$, starting at (0,1)

Figure 8.12: Euler's method for $\frac{dy}{dx} = y$

As you move from P_0 to P_1, y will increase by Δy, where

$$\Delta y = (\text{slope at } P_0)\Delta x = 1(0.1) = 0.1,$$

so

$$y \text{ value at } P_1 = (y \text{ value at } P_0) + \Delta y = 1 + 0.1 = 1.1.$$

Thus the point P_1 is $(0.1, 1.1)$. Now, using the differential equation again, we see that

$$\text{slope at } P_1 = 1.1,$$

so if we move to P_2, then y will change by

$$\Delta y = (\text{slope at } P_1)\Delta x = (1.1)(0.1) = 0.11,$$

so

$$y \text{ value at } P_2 = (y \text{ value at } P_1) + \Delta y = 1.1 + 0.11 = 1.21.$$

Thus P_2 is $(0.2, 1.21)$. Continuing gives the results in Table 8.1. Since the solution curves of $\frac{dy}{dx} = y$ are exponentials, they are concave up, and they curve upward away from the line segments of the slope field. Therefore, in this case Euler's method produces y values which are too small.   ☐

Notice that Euler's method calculates approximate values of y for points on a solution curve; it does not give a general formula for y in terms of x. The broken line in Figure 8.12 approximates the solution curve.

☐ **Example 2** Show that Euler's method for $\dfrac{dy}{dx} = y$ starting at $(0,1)$ and using two steps with $\Delta x = 0.05$ gives $y \approx 1.1025$.

Solution. At $(0,1)$, the slope is 1 and $\Delta y = (1)(0.05) = 0.05$, so new $y = 1 + 0.05 = 1.05$. At $(0.05, 1.05)$, the slope is 1.05 and $\Delta y = (1.05)(0.05) = 0.0525$, so new $y = 1.05 + 0.0525 = 1.1025$.   ☐

In general, $\frac{dy}{dx}$ may be a function of both x and y. Euler's method still works then, as the next example shows.

☐ **Example 3** Approximate four points on the solution curve to $\frac{dy}{dx} = -\frac{x}{y}$, starting at $(0, 1)$ and using $\Delta x = 0.1$.

Solution. The results from Euler's method are in Table 8.2, along with the true y values (to two decimals) calculated from the equation of the circle $x^2 + y^2 = 1$. The fact that the curve is concave down means that the approximate y values are above the real ones. See Figure 8.13.

x	y	$\Delta y = (\text{slope})\Delta x$	true y value
0	1	$0 = (0)(0.1)$	1
0.1	1	$-0.01 = (-\frac{0.1}{1})(0.1)$	0.99
0.2	0.99	$-0.02 = (-\frac{0.2}{0.99})(0.1)$	0.98
0.3	0.97		0.95

Table 8.2: Euler's method for $\dfrac{dy}{dx} = -\dfrac{x}{y}$

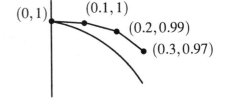

Figure 8.13: Values from Euler's method

☐

The Accuracy of Euler's Method

Let's go back to the differential equation $\frac{dy}{dx} = y$. To improve the accuracy of our Euler's method approximation, choose Δx smaller. See Figure 8.14. Let's compare the true and approximate values. The exact solution to the differential equation $\frac{dy}{dx} = y$ going through the point $(0, 1)$ is $y = e^x$, so the true values will be calculated using this function. When $x = 0.1$,

$$\text{true } y \text{ value} = e^{0.1} \approx 1.1051709.$$

From our calculations in Example 1, we know that when $x = 0.1$, the approximate y value $= 1.1$, so the error ≈ 0.005. In Example 2 we decreased Δx to 0.05. After two steps, $x = 0.1$, and we had the approximate y value $= 1.1025$, so error ≈ 0.00267. Thus halving the step size has approximately halved the error. In general:

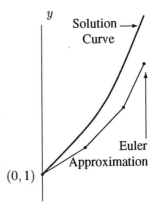

Figure 8.14: Euler's method and Solution Curve for $\frac{dy}{dx} = y$

The *error* in using Euler's method is the difference between the approximate value and the true value. If the number of steps used is n, the error is approximately proportional to $\frac{1}{n}$.

Just as there are more accurate numerical integration methods than left and right Riemann sums, so there are more accurate methods than Euler's for approximating solution curves. Euler's method will be all we need, however.

❖ Exercises for Section 8.3

1. (a) Use ten steps of Euler's method to approximate y-values for $\dfrac{dy}{dt} = \dfrac{1}{t}$, starting at $(1,0)$ and using $\Delta t = 0.1$.

    (b) Using integration, solve the differential equation to find the exact value of y at the end of these ten steps.

2. (a) Use Euler's method to approximate the value of y at $x = 1$ on the solution curve to the differential equation
$$\frac{dy}{dx} = x^3 - y^3$$
    that passes through $(0,0)$. Use $\Delta x = \frac{1}{5}$ (i.e., 5 steps).

    (b) To the right is the slope field for the differential equation, for $-2 < x < 2$ and $-2 < y < 2$. Sketch the solution that passes through $(0,0)$. Show on the graph the approximation you made in (a).

    (c) From the graph of the slope field, can you say whether your answer to (a) is an overestimate or an underestimate?

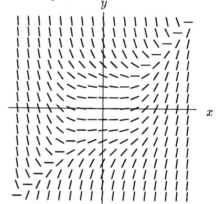

3. (a) Use Euler's method with five subintervals to approximate solution curve to the differential equation $\dfrac{dy}{dx} = x^2 - y^2$ passing through the point $(0, 1)$ and ending at $x = 1$.

    (b) Repeat this computation using ten subintervals, again ending with $x = 1$.

    In both computations, keep the approximate function values to three decimal places.

4. Why are the approximate results you obtained in Problem 3 too small? (Note: The slope field for this differential equation is one of those in Figure 8.7 on page 542.)

5. Using the fact that the error incurred using Euler's method is roughly proportional to one over the number of subintervals used, how should the errors of the five-step calculation and the ten-step calculation in Problem 3 compare? Use your answer to estimate the true value of y when $x = 1$.

6. Use Euler's method to solve
$$\frac{dB}{dt} = 0.05\,B$$
    with initial value $B = 1000$ when $t = 0$. Take:

    (a) $\Delta t = 1$ and 1 step.

    (b) $\Delta t = 0.5$ and 2 steps.

(c) $\Delta t = 0.25$ and 4 steps.

(d) Suppose B is the balance in a bank account earning interest. Explain why the result of your calculation in (a) is equivalent to compounding the interest once a year instead of continuously.

(e) Interpret the result of your calculations in (b) and (c) in terms of compound interest.

7. Consider the differential equation $\frac{dy}{dx} = f(x)$ with initial value $y(0) = 0$. Explain why using Euler's method to approximate the solution curve gives the same results as using left Riemann sums to approximate $\int_0^x f(t)\,dt$.

8. In Section 4.8 on implicit differentiation, a table of values was computed for the equation $y^3 - xy = -6$ near $x = 7$, $y = 2$. On page 264, the implicit derivative $\frac{dy}{dx} = \frac{y}{3y^2 - x}$ was evaluated at $x = 7$, $y = 2$, and then the y-values at $x = 6.8, 6.9, 7.1, 7.2$ were approximated by local linearity. Now compute the y-values for $x = 7.1, 7.2$ by Euler's method using $\Delta x = 0.05$. Compare your answer for $x = 7.2$ to the actual answer, accurate to 6 decimal places, of 2.076018.

8.4 Separation of Variables

We have seen how to sketch solution curves of a differential equation using a slope field and how to calculate approximate numerical solutions. Now we'll see how to find the equation of a solution curve exactly for certain differential equations.

First, we'll look at a familiar example, the differential equation

$$\frac{dy}{dx} = -\frac{x}{y},$$

whose solutions are the circles

$$x^2 + y^2 = C.$$

We can check that these circles are solutions by differentiation; the question now is how they were derived. The method of *separation of variables* works by putting all the x's on one side of the equation and all the y's on the other, giving

$$y\,dy = -x\,dx,$$

and then integrating each side separately:

$$\int y\,dy = -\int x\,dx,$$

$$\frac{y^2}{2} = \frac{-x^2}{2} + k,$$

thereby giving the circles we were expecting:

$$x^2 + y^2 = C \qquad \text{where } C = 2k.$$

You might worry about whether it is legitimate to integrate one side of the equation with respect to x and the other with respect to y. The reason it can be done is explained at the end of this section on page 553.

The Exponential Growth Equation

Let's use separation of variables on an equation which occurs frequently in practice:

$$\frac{dy}{dx} = ky.$$

Separating variables,

$$\frac{1}{y}\,dy = k\,dx$$

and integrating

$$\int \frac{1}{y}\,dy = \int k\,dx$$

gives

$$\ln|y| = kx + C \quad \text{for some constant } C.$$

Solving for $|y|$ leads to

$$|y| = e^{kx+C} = e^{kx}e^{C} = Ae^{kx}$$

where $A = e^{C}$, so A is positive. Thus

$$y = (\pm A)e^{kx} = Be^{kx}$$

where $B = \pm A$, so B is any nonzero constant. Even though there's no C leading to $B = 0$, B can be 0 because $y = 0$ is a solution to the differential equation. We 'lost' this solution when we divided through by y at the first step.

The general solution to $\frac{dy}{dx} = ky$ is $y = Be^{kx}$ for any B.

Thus, the differential equation $\frac{dy}{dx} = ky$ always leads to exponential growth (if $k > 0$) or exponential decay (if $k < 0$). Graphs of solution curves for some $k > 0$ are in Figure 8.15. For $k < 0$, the graphs are reflected in the y-axis.

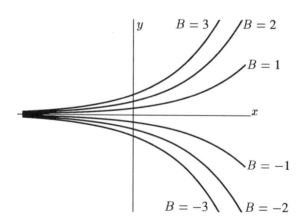

Figure 8.15: $y = Be^{kx}$, solutions to $\frac{dy}{dx} = ky$, for some fixed $k > 0$

☐ **Example 1** For $k > 0$, find and graph solutions of

$$\frac{dH}{dt} = -k(H - 20).$$

Solution. The slope field in Figure 8.16 shows the qualitative behavior of the solutions. To find the equation of the solution curves, we separate variables and integrate.

$$\int \frac{1}{H - 20} \, dH = - \int k \, dt$$

gives

$$\ln |H - 20| = -kt + C.$$

Solving for H leads to:

$$|H - 20| = e^{-kt+C} = e^{-kt} e^C = A e^{-kt}$$

or

$$H - 20 = (\pm A) e^{-kt} = B e^{-kt}$$

$$H = 20 + B e^{-kt}.$$

Graphs for $k = 1$ and $B = -10, 0, 10$, $t \geq 0$ are in Figure 8.16.

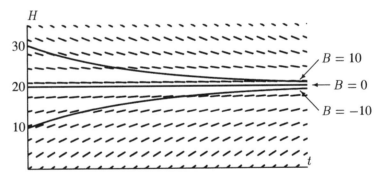

Figure 8.16: Slope Field and some Solution Curves for $\frac{dH}{dt} = -k(H - 20)$, with $k = 1$

☐

Equilibrium Solutions

The solution curves in Figure 8.16 show that as $t \to \infty$, the solutions tend towards $H = 20$. This is because $e^{-kt} \to 0$ as $t \to \infty$, so, since $H = 20 + B e^{-kt}$,

$$H \to 20 \quad \text{as } t \to \infty.$$

In the special case when $B = 0$,

$$H = 20 \quad \text{for all } t.$$

In other words, if $B = 0$, then H is constant for all t. This is called an *equilibrium solution* for H. Regardless of whether H is above or below 20, it gets closer and closer to 20 as $t \to \infty$. As a result, $H = 20$ is called a *stable* equilibrium for H. Notice that at the equilibrium, we have $\dfrac{dH}{dt} = 0$.

It is often important to know that all solutions to a particular differential equation have the same limiting behavior, as is the case here as $t \to \infty$. This means that what happens in the long run is independent of the initial condition.

In general,

- an **equilibrium solution** is constant for all values of the independent variable. The graph is a horizontal line. Equilibrium solutions can be identified by setting the derivative of the function to zero.

- an equilibrium is **stable** if a small change in the initial conditions gives a solution which tends towards the equilibrium as the independent variable tends to positive infinity.

- an equilibrium is **unstable** if a small change in the initial conditions gives a solution curve which veers away from the equilibrium as the independent variable tends to positive infinity.

☐ **Example 2** Find and sketch the solution to

$$\frac{dP}{dt} = 2P - 2Pt$$

satisfying $P = 5$ when $t = 0$.

Solution. Separating variables

$$\frac{dP}{dt} = P(2 - 2t)$$

gives

$$\int \frac{dP}{P} = \int (2 - 2t)\, dt$$

so

$$\ln |P| = 2t - t^2 + C.$$

Solving leads to

$$|P| = e^{2t - t^2 + C} = e^C e^{2t - t^2} = A e^{2t - t^2}$$

with $A = e^C$, so $A > 0$. Thus the general solution to the differential equation is

$$P = B e^{2t - t^2} \quad \text{for any } B.$$

As before, $B = 0$ gives a solution although there is no C which leads to it. To find the value of B for this example, substitute $P = 5$ and $t = 0$ giving

$$5 = Be^{2(0)-0^2} = B$$

so

$$P = 5e^{2t-t^2}.$$

The graph of this function is in Figure 8.17. Since the solution can be rewritten as

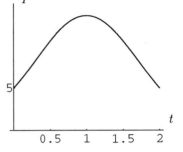

Figure 8.17: Graph of $P = 5e^{2t-t^2}$

$$P = 5e^{1-1+2t-t^2} = 5e^1 e^{-1+2t-t^2} = (5e)e^{-(1-t)^2},$$

the graph has the same shape as the graph of $y = e^{-t^2}$, the bell-shaped curve of statistics. Here, however, the maximum, normally at $t = 0$, is shifted to the right to $t = 1$. Note that the equilibrium $P = 0$ is stable, since $P(t) = Be^{2t-t^2} \to 0$ as $t \to +\infty$. ◻

Justification of Method of Separation of Variables

Suppose a differential equation can be written in the form

$$\frac{dy}{dx} = g(x)f(y).$$

Then by writing $f(y) = \frac{1}{h(y)}$, the right-hand side can be thought of as a fraction,

$$\frac{dy}{dx} = \frac{g(x)}{h(y)}.$$

If we multiply through by $h(y)$ we get

$$h(y)\frac{dy}{dx} = g(x).$$

Thinking of y as a function of x, so $y = y(x)$, and $\frac{dy}{dx} = y'(x)$, we can write the line above as

$$h(y(x)) \cdot y'(x) = g(x).$$

Now we integrate both sides with respect to x:

$$\int h(y(x)) \cdot y'(x)\, dx = \int g(x)\, dx.$$

The form of the integral on the left suggests that we use substitution. Since $y = y(x)$ and $dy = y'(x)\, dx$, we get

$$\int h(y)\, dy = \int g(x)\, dx.$$

If antiderivatives of h and g can be found, then this will give the equation of the solution curve.

Note that transforming the original differential equation,

$$\frac{dy}{dx} = \frac{g(x)}{h(y)}$$

into

$$\int h(y)\, dy = \int g(x)\, dx$$

looks as though we have treated $\frac{dy}{dx}$ as a fraction, cross multiplied and then integrated. Although that's not exactly what we've done, you may find this a helpful way of remembering the method. In fact, the dy, dx notation was introduced by Leibnitz to allow shortcuts like this. (More specifically, to make the chain rule look like cancellation.)

❖ Exercises for Section 8.4

Find the solutions to the differential equations in Problems 1–20, subject to the given initial condition.

1. $\frac{dP}{dt} = 0.02P$, $P(0) = 20$.

2. $\frac{dQ}{dt} = \frac{Q}{5}$, and $Q = 50$ when $t = 0$.

3. $\frac{dm}{dt} = 3m$, $m = 5$ when $t = 1$.

4. $\frac{dI}{dx} = 0.2I$, $I = 6$ when $x = -1$.

5. $\frac{dy}{dx} + \frac{y}{3} = 0$, $y(0) = 10$.

6. $\frac{1}{z}\frac{dz}{dt} = 5$, $z(1) = 5$.

7. $\frac{dP}{dt} = P + 4$, $P = 100$ when $t = 0$.

8. $\frac{dy}{dx} = 2y - 4$ through (2,5).

9. $\frac{dm}{dt} = 0.1m + 200$, $m(0) = 1000$.

10. $\frac{dB}{dt} + 2B = 50$, $B(1) = 100$.

11. $\frac{dz}{dt} = te^z$ through the origin.

12. $\frac{dy}{dt} = y^2(1 + t)$, $y = 2$ when $t = 1$.

13. $\frac{dy}{dx} = \frac{5y}{x}$, $y = 3$ when $x = 1$.

14. $\frac{dz}{dt} = z + zt^2$, $z = 5$ when $t = 0$.

15. $\frac{dw}{d\theta} = \theta w^2 \sin \theta^2$, $w(0) = 1$.

16. $x(x + 1)\frac{du}{dx} = u^2$, $u(1) = 1$.

17. $e^{-\cos\theta}\frac{dz}{d\theta} = \sqrt{1 - z^2}\sin\theta$, $z(0) = \frac{1}{2}$.

18. $\frac{dy}{dt} = 2^y \sin^3 t$, $y(0) = 0$.

19. $(1 + t^2)y\frac{dy}{dt} = 1 - y$, $y(1) = 0$.

20. $\frac{dy}{dx} = e^{x+y}$, and $y = 0$ when $x = 1$.

Solve the differential equations in Problems 21–26. Assume a, b and k are constants.

21. $\frac{dR}{dt} = kR$

22. $\frac{dQ}{dt} - \frac{Q}{k} = 0$

23. $\frac{dP}{dt} = P - a$

24. $\frac{dQ}{dt} = b - Q$

25. $\frac{dP}{dt} = k(P - a)$

26. $\frac{dR}{dt} = aR + b$

Solve the differential equations in Problems 27–34.

27. $\dfrac{dy}{dt} = y(2 - y), \quad y(0) = 1.$

28. $t\dfrac{dx}{dt} = (1 + 2 \ln t) \tan x$, where $t > 0$.

29. $\dfrac{dx}{dt} = \dfrac{x \ln x}{t}.$

30. $\dfrac{dy}{dt} = -y \ln(y/2), \quad y(0) = 1.$

31. $\dfrac{dy}{dt} = \dfrac{y \ln y}{t^2}.$

32. $(y\sqrt{x^3 + 1})\dfrac{dy}{dx} + x^2 y^2 + x^2 = 0.$
(Assume $x > -1$.)

33. $\dfrac{dQ}{dt} + t^2 Q^2 + Q^2 - 4t^2 - 4 = 0.$

34. $\dfrac{x}{y}\dfrac{dx}{dy} = e^{\left(\frac{x}{a}\right)^2} \ln y.$ (a is constant)

35. (a) Sketch the slope field for $y' = \frac{x}{y}$.
    (b) Sketch several solution curves.
    (c) Solve the differential equation analytically.

36. (a) Sketch the slope field for $y' = -\frac{y}{x}$.
    (b) Sketch several solution curves.
    (c) Solve the differential equation analytically.

37. Compare the slope field for $y' = \frac{x}{y}$, Problem 35, with that for $y' = -\frac{y}{x}$, Problem 36. Show that the solution curves of Problem 35 intersect the solution curves of Problem 36 at right angles.

8.5 Growth and Decay

Now we will look at a type of differential equation which occurs frequently in practice. Consider the population of a region. If there is no immigration or emigration, the rate at which the population is changing is often proportional to the population. In other words, the larger the population, the faster it is growing, because there are more people to have babies. If the population at time t is P, and its growth rate is 2%, then we know

$$(\text{rate of growth}) = 2\%(\text{current population})$$

and we can write this as

$$\frac{dP}{dt} = 0.02P.$$

The 2% growth rate is called the *relative growth rate* to distinguish it from the absolute growth rate, or rate of change of the population, $\frac{dP}{dt}$. Notice they are in different units. Since

$$\text{relative growth rate} = 2\% = \frac{1}{P}\frac{dP}{dt}$$

the relative growth rate is a number (or percent) per unit time, while

$$\text{rate of change of population} = \frac{dP}{dt}$$

is in people per unit time.

As another example, consider a bank account which advertises 7% interest compounded continuously. In Chapter 1 we introduced continuous compounding as the limiting case in which interest was added more and more often. Here we will approach continuous compounding from a different point of view, and say that continuous compounding means that, at any time, interest is being accrued at 7% of the balance at that moment:

$$\text{rate at which interest is earned} = 7\%(\text{current balance}).$$

If B is the balance at time t, and if no deposits or withdrawals are made, then the rate of change of the balance is exactly the rate the interest is earned, so

$$\text{rate of change of balance} = 7\%(\text{current balance})$$

or, in other words,

$$\frac{dB}{dt} = 0.07B.$$

Both of these equations, $\frac{dP}{dt} = 0.02P$ and $\frac{dB}{dt} = 0.07B$, are of the form

$$\frac{dP}{dt} = kP$$

for some k. As we know from the previous section:

Any solution to the equation
$$\frac{dP}{dt} = kP$$
can be written in the form
$$P = P_0 e^{kt},$$
where P_0 is the initial value of P.

□ **Example 1** A bank account earns interest continuously at a rate of 5% of the current balance per year. If the initial deposit is $1000, and if no other deposits or withdrawals are made,

(a) write the differential equation satisfied by the balance in the account;

(b) solve the differential equation and graph the solution.

Solution.

(a) In this problem we are looking for B, the balance in the account in dollars, as a function of t, time in years. At any instant, interest is being added to the account at a rate of 5% of the balance at that moment; the units of time are years. Since the addition of interest causes the balance to increase,

$$(\text{rate at which balance is increasing}) = 5\%(\text{current balance})$$

or

$$\frac{dB}{dt} = 0.05 B.$$

This is the differential equation that describes the process. Notice that it does not involve the $1000, because the initial deposit does not affect the process by which interest is earned. The $1000 is the initial condition.

(b) Solving the differential equation using separation of variables gives

$$B = B_0 e^{0.05t}$$

where B_0 is the initial value of B, so $B_0 = 1000$. Thus

$$B = 1000 e^{0.05t}$$

and this function is graphed in Figure 8.18.

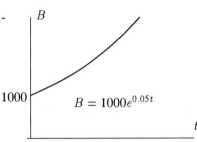

Figure 8.18: Graph of $B = 1000 e^{0.05t}$ for Example 1

❑

You may wonder how we can represent an amount of money by a differential equation, since money can only take on discrete values (you can't have fractions of a cent). In fact, the differential equation is only an approximation, but for large amounts of money, it is a pretty good approximation.

Other processes, such as radioactive decay, are described by a differential equation similar to that for population growth, but with a negative constant.

❑ **Example 2**   Radioactive carbon (Carbon-14) decays at a rate of approximately 1 part in 10,000 a year. Write and solve a differential equation for the quantity of Carbon-14 as a function of time.

Solution.

$$\left(\begin{array}{c} \text{Rate at which quantity of} \\ \text{Carbon-14 is decreasing} \end{array} \right) = \frac{1}{10,000} (\text{current quantity}).$$

If Q is the quantity of Carbon-14 at time t (in years)

$$\text{Rate at which quantity is decreasing} = -\frac{dQ}{dt}.$$

Thus

$$-\frac{dQ}{dt} = \frac{1}{10,000} Q$$

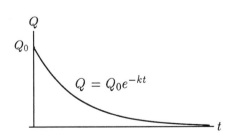

Figure 8.19: Exponential Decay

or

$$\frac{dQ}{dt} = -0.0001 Q.$$

This differential equation is of the same form as the previous one and has solution

$$Q = Q_0 e^{-0.0001t}$$

where Q_0 is the initial quantity. See Figure 8.19. ❑

☐ **Example 3** Newton's Law of Cooling says that the temperature of a hot object decreases at a rate proportional to the difference between the temperature of the object and the temperature of the surrounding air. A cup of coffee at 90°C stands in a 20°C room. After 2 minutes its temperature is 80°C.

(a) How long will it be before the coffee cools to 50°C?

(b) Sketch a graph of temperature against time.

(c) What happens if the coffee is left in the room a long time? Show this on the graph and algebraically.

Solution. We will first find the temperature of the coffee as a function of time, using the law of cooling to set up a differential equation. Newton's Law of Cooling says that for some constant α

$$(\text{rate of change of temperature}) = \alpha(\text{temperature difference}).$$

If H is the temperature of the coffee, then

$$\text{temperature difference} = H - 20$$

so

$$\frac{dH}{dt} = \alpha(H - 20).$$

What about the sign of α? If the temperature difference is positive, i.e., $H > 20$, then H is falling, making the rate of change negative. Thus α should be negative, so we'll write:

$$\frac{dH}{dt} = -k(H - 20), \qquad \text{for some } k > 0.$$

Separating variables and solving, as in Example 1 on Page 556 gives:

$$H - 20 = Be^{-kt}.$$

To find B, substitute the initial condition that $H = 90$ when $t = 0$:

$$90 - 20 = Be^{-k(0)}$$

so $B = 70$, the initial value of $(H - 20)$. Thus,

$$H - 20 = 70e^{-kt}.$$

To find k, we use the fact that after 2 minutes, the temperature is 80°C so

$$80 - 20 = 70e^{-k(2)}.$$

Dividing by 70 and taking natural logs, we get:

$$\begin{aligned} \ln\left(\frac{60}{70}\right) &= \ln(e^{-2k}) \\ -0.1542 &= -2k \\ k &\approx 0.077. \end{aligned}$$

Therefore, the temperature is given by

$$H - 20 = 70e^{-0.077t}$$

or

$$H = 20 + 70e^{-0.077t}.$$

(a) To calculate when the temperature reaches 50°C, substitute $H = 50$ and solve for t.

$$\begin{aligned} 50 &= 20 + 70e^{-0.077t} \\ \frac{30}{70} &= e^{-0.077t} \end{aligned}$$

Taking natural logs,

$$-0.8473 = -0.077t$$

gives

$$t \approx 11 \text{ minutes.}$$

(b) The graph of $H = 20 + 70e^{-0.077t}$ has a vertical intercept of $H = 90$, because the temperature of the coffee starts at 90°C. The temperature decays exponentially with $H = 20$ as the horizontal asymptote. See Figure 8.20.

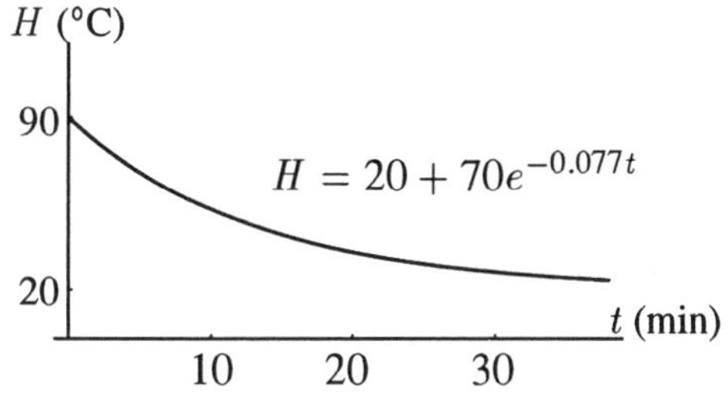

Figure 8.20: Temperature of Cooling Coffee

(c) 'In the long run' means as $t \to \infty$. The graph shows that as $t \to \infty$, $H \to 20$. Algebraically, since $e^{-0.077t} \to 0$ as $t \to \infty$,

$$H = 20 + \underbrace{70e^{-0.077t}}_{\text{goes to 0 as } t \to \infty} \longrightarrow 20$$

as $t \to \infty$. Thus, the equilibrium solution $H = 20$ is stable.

❑

Continuous versus Annual Percentage Growth Rates

If $P = P_0(1+r)^t$ with t in years, we say that r is the *annual* growth rate, while if $P = P_0e^{kt}$ we say that k is the *continuous* or *instantaneous* growth rate.

It is important to note that the constant k in the differential equation $\frac{dP}{dt} = kP$ is not the annual growth rate, but rather the *continuous* growth rate. In Example 1, with a continuous interest rate of 5%, we obtain a balance at time t of $B = B_0e^{0.05t}$, where t is measured in years. At the end of one year the balance is $B_0 e^{0.05}$. In that one year, our balance has changed from B_0 to $B_0 e^{0.05}$, that is, it has changed by the factor $e^{0.05} \approx 1.0513$. Thus the *annual* growth rate is 5.13%. This is what the bank means when it says "5% compounded continuously for an effective annual rate of 5.13%".

To be honest, most growth is measured over discrete time intervals and hence a continuous growth rate is an idealized concept. When a demographer says the population of some country is growing at the rate of 2% per year, she means just what you think: after one year the population will have increased by a factor of 1.02; and after t years the population will be given by $P = P_0(1.02)^t$. If we want to find the continuous growth rate k for the population we proceed as follows. Suppose the population is expressed as $P = P_0 e^{kt}$. At the end of one year $P = P_0 e^k$. To get 2% growth in that year we need $e^k = 1.02$. Thus $k = \ln 1.02 \approx 0.0198$. The continuous growth rate $k = 1.98\%$ is close to the annual growth rate of 2%, but it is not the same.

❖ Exercises for Section 8.5

1. The amount of arable land (land that can be used for growing crops) that we use increases as the world's population increases. Suppose $A(t)$ represents the total number of hectares of arable land in use in year t. (A hectare is about $2\frac{1}{2}$ acres.)

   (a) Explain why it is reasonable to expect $A(t)$ to satisfy the equation

   $$\frac{dA}{dt} = kA.$$

   What assumptions are you making about the world's population and its relation to the amount of arable land used?

   (b) In 1950, about 1×10^9 hectares of arable land were in use; in 1980 the figure was 2×10^9. If the total amount of arable land available is thought to be 3.2×10^9 hectares, when will it be exhausted?

2. A yam is put in a 200°C oven and heats up according to the differential equation

   $$\frac{dH}{dt} = -k(H - 200),$$

   where k is positive.

   (a) If the yam is at 20°C when it is put in the oven, solve the differential equation.

   (b) Find k using the fact that after 30 minutes the temperature of the yam is 120°C.

3. Money in a bank account grows continuously at a rate of r per year (when the interest rate is 5%, $r = 0.05$, and so on). Suppose $1000 is put into the account in 1970.

   (a) Write a differential equation satisfied by M, the amount of money in the account at time t, measured in years since 1970.

   (b) Solve the equation.

   (c) Sketch the solution until the year 2000 for interest rates of 5% and 10%.

4. (a) If $B = f(t)$ is the balance of a bank account at time t that earns interest at a rate of $r\%$, compounded continuously, what is the differential equation that describes the rate at which the balance changes? What is the constant of proportionality, in terms of r?

   (b) What is the solution to this differential equation?

   (c) Sketch the graph of $B = f(t)$ for an account that starts with $1000 and earns interest at the following rates:

   (i) 4%   (ii) 10%   (iii) 15%

5. In some chemical reactions, the rate at which the amount of a substance changes with time is proportional to the amount present. For example, this is the case as δ-glucono lactone changes into gluconic acid.

   (a) Write a differential equation satisfied by y, the quantity of δ-glucono lactone present at time t.

   (b) If 100 grams of δ-glucono lactone is reduced to 54.9 grams in one hour, how many grams will remain after 10 hours?

6. The rate at which barometric pressure decreases with altitude is proportional to the barometric pressure at that altitude. The barometric pressure is measured in inches of mercury, the altitude in feet, and the constant of proportionality is 3.7×10^{-5}. Suppose the barometric pressure at sea level is 29.92 inches of mercury.

   (a) Calculate the barometric pressure at the top of Mount Whitney, 14,500 feet (the highest mountain in the U.S. outside Alaska) and at the top of Mount Everest, 29,000 feet (the highest mountain in the world).

   (b) People cannot easily survive at a pressure below 15 inches of mercury. What is the highest altitude to which people can safely go?

7. The rate at which light is absorbed as it passes through water is proportional to the intensity, or brightness, at that point.

   (a) Find the intensity as a function of the distance the light has traveled through the water.

   (b) If 50% of the light is absorbed in 10 feet, how much is absorbed in 20 feet? 25 feet?

8. Write a differential equation describing the temperature as a function of time of a can of soda taken out of a 40°F refrigerator and left in a 65°F room. Solve the equation and sketch an approximate graph of the solution.

9. The radioactive isotope Carbon-14 is present in small quantities in all life forms, and it is constantly being replenished until the organism dies, after which it decays to stable Carbon-12 at a rate proportional to the amount present. In other words, if $C = f(t)$ is the amount of Carbon-14 present at time t, then $C' = kC$ for some constant k after the organism dies. The half-life of Carbon-14 is 5700 years, i.e., if an amount C_0 of Carbon-14 is present today, then $\frac{1}{2}C_0$ is present 5700 years from today.

   (a) Find the value of the constant k in the solution $C = f(t)$ to the differential equation $C' = kC$.

   (b) In 1988, three teams of scientists found that the Shroud of Turin, which was reputed to be the burial cloth of Jesus, contained 91% of the amount of Carbon-14 contained in freshly made cloth of the same material.[1] How old is the Shroud of Turin, according to these data?

10. Before Galileo discovered that the speed of a falling body with no air resistance is proportional to the time since it was dropped, he mistakenly conjectured that the speed was proportional to the distance it had fallen.

    (a) Assume the mistaken conjecture to be true and write an equation relating the distance fallen, $D(t)$, at time t, and its derivative.

    (b) Using your answer to part (a) and the correct initial conditions, show that D would have to be equal to 0 for all t and therefore the conjecture must be wrong.

8.6   Applications and Modeling

We often use functions to study real processes. For example, the functions which describe the gravitational attraction between two bodies, the cost of producing a good, and the oscillations of the air molecules when transmitting a musical note, have all been studied in detail. How are such functions obtained? In some cases, we fit functions to experimental data by trial and error; in other cases, we take a more theoretical approach. The more theoretical approach often leads to a differential equation whose solution is the function we want. In this section we will look at several examples which lead naturally to differential equations.

A Financial Example

In the preceding section, we saw an example in which money in a bank account was earning interest (Example 1, page 556). We will now do an example involving payments as well. Since we will use a differential equation, the payments must be made *continuously*, rather than at a single moment. In the next example, payments are made continuously at a rate of $200 per year.

---

[1]The *New York Times*, October 18, 1988

☐ **Example 1**   A bank account earns interest continuously at 5% per year and payments are made continuously out of the account at a rate of $200 per year.

(a) Write a differential equation that governs the balance in the account.

(b) Solve the differential equation, assuming an initial deposit of B_0 dollars.

(c) Sketch the solution for $B_0 = \$3000$, $\$4000$, and $\$5000$.

Solution. First, let's see how much we can learn about the solution without actually solving the differential equation. For example, we can ask if there is any initial balance B_0 which will exactly keep the bank account in equilibrium. If there's an equilibrium, the rate at which money is entering the account must balance the payments made exactly, so

$$\left(\begin{array}{c} \text{Rate interest} \\ \text{accrued} \end{array} \right) = \left(\begin{array}{c} \text{Rate payments} \\ \text{are made} \end{array} \right)$$

Remember that earning interest continuously at 5% means earning interest at a rate of 5% of whatever the balance is at the moment. If we have equilibrium, the balance starts as $\$B_0$ and remains $\$B_0$, so interest is always earned at a rate of $\$0.05 B_0$ per year. Thus, for equilibrium we must have

$$0.05 B_0 = 200$$

so

$$B_0 = \$4000.$$

Therefore, if the account starts with $4000, the interest and the payments are equal, and so the balance remains constant. Therefore $4000 is an equilibrium solution.

If, however, the initial deposit is over $4000, the interest earned will be more than the payments, and the balance in the account will increase, thereby increasing the interest still further. Thus the balance will increase faster and faster. If the initial deposit is below $4000, the interest will not be enough to meet the payments, and so the balance will decline, thereby decreasing the interest and making the balance decrease faster. Hence the balance will eventually reach 0 (and presumably the payments will stop).

(a) Now we will solve the problem analytically. As before, let B be the balance at time t, in years. To set up the differential equation, we will use the fact that

$$\left(\begin{array}{c} \text{Rate at which} \\ \text{balance is increasing} \end{array} \right) = \left(\begin{array}{c} \text{Rate interest} \\ \text{is accrued} \end{array} \right) - \left(\begin{array}{c} \text{Rate payments} \\ \text{are made} \end{array} \right).$$

Interest is earned at a rate of $0.05 B$, and payments are made at a rate of $200 per year, so we have

$$\frac{dB}{dt} = 0.05 B - 200.$$

Notice that the initial deposit does not enter into the differential equation. The equilibrium solution, $B = 4000$, is obtained by setting $\dfrac{dB}{dt} = 0$.

(b) We will solve this equation by separation of variables. It is very helpful to factor a 0.05 out of the right hand side before separating, so that the B moves over to the left hand side without a coefficient. This makes the manipulations less messy. So we write:

$$\frac{dB}{dt} = 0.05(B - 4000).$$

Separating and integrating

$$\int \frac{dB}{B - 4000} = \int 0.05 \, dt$$

gives

$$\ln|B - 4000| = 0.05t + C$$

so

$$|B - 4000| = e^{0.05t+C} = e^C e^{0.05t}$$

or

$$B - 4000 = Ae^{0.05t} \qquad \text{(where } A = \pm e^C\text{)}.$$

To find A we use the initial condition that $B = B_0$ when $t = 0$.

$$B_0 - 4000 = Ae^0 = A.$$

Substituting this value for A gives

$$B - 4000 = (B_0 - 4000)e^{0.05t}$$

or

$$B = 4000 + (B_0 - 4000)e^{0.05t}.$$

(c) If $B_0 = 4000$, then $B = 4000$, the equilibrium solution.
If $B_0 = 5000$, then $B = 4000 + 1000e^{0.05t}$.
If $B_0 = 3000$, then $B = 4000 - 1000e^{0.05t}$.
Notice that when $t \approx 27.7$, $B = 0$, so this solution holds for $0 \leq t \leq 27.7$.
The graphs of these functions are in Figure 8.21. Here $B = 4000$ is an equilibrium solution. However, if the balance starts with B_0 near, but not equal to, \$4000, then B moves further away. Thus $B = 4000$ is an unstable equilibrium.

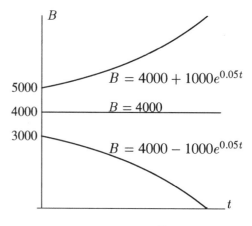

Figure 8.21: Solutions to $\frac{dB}{dt} = 0.05B - 200$

Free Fall with Air Resistance

☐ **Example 2** A freely falling body, such as a parachutist, has two forces acting on it. One is the force due to gravity, causing the body to speed up, and the other is the air resistance, causing it to slow down. Write a differential equation for velocity, v, as a function of time, t, assuming the body starts from rest and that the force due to air resistance is proportional to:

(a) v (b) $v^{\frac{3}{2}}$

Solution.

(a) If v is the velocity measured downwards and m is the mass, Newton's Second Law of Motion states

$$\text{Force} = (\text{Mass}) \times (\text{Acceleration}) = m\frac{dv}{dt}.$$

air resistance
kv (or $kv^{3/2}$)

force due to
gravity, mg

This relates the forces to the acceleration, $\frac{dv}{dt}$. Figure 8.22 shows the forces acting on the body; k is a constant of proportionality. The net downwards force is $mg - kv$, so Newton's Law becomes

$$mg - kv = m\frac{dv}{dt}$$

or

$$\frac{dv}{dt} = g - \frac{k}{m}v.$$

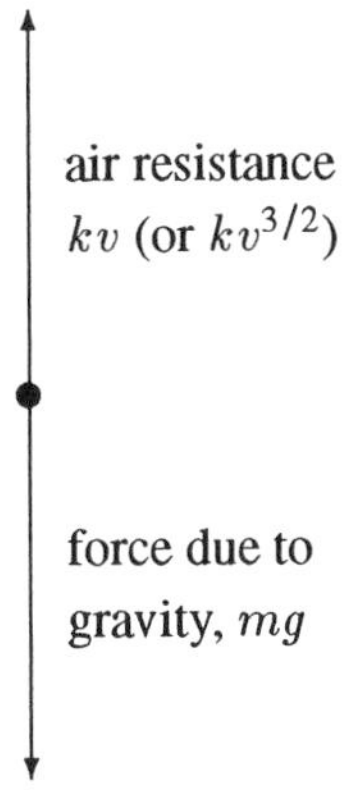

Figure 8.22: Forces acting
on a freely falling body

This can be solved by separation of variables. Again, life is easier if you factor out $-\frac{k}{m}$ on the right before separating:

$$\frac{dv}{dt} = -\frac{k}{m}\left(v - \frac{mg}{k}\right).$$

Separating and integrating gives

$$\int \frac{dv}{v - \frac{mg}{k}} = -\frac{k}{m}\int dt$$

$$\ln\left|v - \frac{mg}{k}\right| = -\frac{k}{m}t + C.$$

Solving for v:

$$\left|v - \frac{mg}{k}\right| = e^{-\frac{k}{m}t + C} = e^{C}e^{-\frac{k}{m}t}$$

$$v - \frac{mg}{k} = Ae^{-\frac{k}{m}t},$$

where A is an arbitrary constant. We find A from the condition that the body starts from rest, which means $v = 0$ when $t = 0$. Substituting:

$$0 - \frac{mg}{k} = Ae^{0}$$

gives

$$A = -\frac{mg}{k}.$$

Thus

$$v = \frac{mg}{k} - \frac{mg}{k}e^{-\frac{k}{m}t} = \frac{mg}{k}\left(1 - e^{-\frac{k}{m}t}\right).$$

The graph of v against t in Figure 8.23 shows that at the start the body is moving slowly, which means the air resistance is small, so it speeds up a lot. As the body starts to move faster, the air resistance builds until it nearly balances the gravitational force and the body speeds up very little. At this point the body has reached its *terminal velocity*, $\frac{mg}{k}$. For a parachutist, the terminal velocity is about 20 ft/sec (14 mph); without a parachute the terminal velocity of a person is about 176 ft/sec (120 mph).

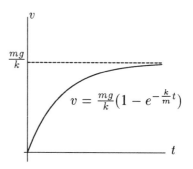

Figure 8.23: Velocity with Air Resistance $= kv$

According to the model, the body never actually reaches the terminal velocity (because $e^{-\frac{k}{m}t}$ is never actually equal to zero). However, the velocity is close to the terminal value after only a few seconds.

Notice that the terminal velocity can also be obtained from the differential equation by setting $\frac{dV}{dt} = 0$ and solving for v.

(b) If the air resistance is proportional to $v^{\frac{3}{2}}$, the net force acting on the body is $mg - kv^{\frac{3}{2}}$, so Newton's Law tells us that

$$m\frac{dv}{dt} = mg - kv^{\frac{3}{2}}.$$

This small change in the differential equation makes it much more difficult to solve analytically. If you try to proceed by separation of variables, you get

$$\int \frac{dv}{32 - 0.4v^{\frac{3}{2}}} = t + C.$$

This integral can be evaluated in closed form, but the result is such a mess that you'd be happier without it. Even if we did use this integral, we would have t in terms of v, whereas we usually want to get v as a function of t. This is a case in which getting a qualitative solution using a slope field or a numerical solution using Euler's method are the only reasonable methods we know to solve a differential equation.

Let's see how we might get a qualitative solution for the special case:

$$\frac{dv}{dt} = 32 - 0.4v^{\frac{3}{2}}$$

with v in feet per second. The slope field in Figure 8.24 shows that the solution curve starting at the origin will have a horizontal asymptote at about $v = 18$. Thus even without solving the differential equation explicitly, we can say that, under these assumptions, the speed of a falling body increases towards a terminal velocity of about 18 ft/sec.

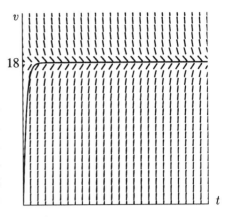

Figure 8.24: Slope field of $\frac{dv}{dt} = 32 - 0.4v^{\frac{3}{2}}$

From an analytical point of view, the equilibrium occurs where

$$\frac{dv}{dt} = 32 - 0.4v^{\frac{3}{2}} = 0,$$

giving

$$v^{\frac{3}{2}} = 80 \quad \text{or} \quad v \approx 18.57.$$

You may wonder how we know whether to assume the air resistance, or *drag force*, is proportional to the velocity, to the $\frac{3}{2}$-power of the velocity, or to some other function of the velocity. This is largely decided experimentally—by dragging different objects through various media and measuring the drag force.

Compartmental Analysis

Many biological processes are described by models similar to the model in the next example.

☐ **Example 3** A water reservoir which holds 100 million gallons supplies a city with 1 million gallons a day. The reservoir is partly refilled by a spring which provides 0.9 million gallons a day and the rest, 0.1 million gallons a day, by run-off from the surrounding land. The spring is clean, but the run-off contains salt with a concentration of 0.0001 pound per gallon. Assume that there was no salt in the reservoir initially and that the reservoir is well mixed (i.e., that the water taken by the city residents contains the concentration of salt in the tank at that instant). Find the concentration of salt in the reservoir as a function of time, assuming it remains full throughout.

Solution. In this type of problem, it is very important to distinguish between the total quantity and the concentration of salt, where

$$\text{concentration} = \frac{\text{quantity}}{\text{volume}}$$

If C is the concentration of salt (in pounds/gal) and Q is the quantity (in pounds), then since the volume of the reservoir is 100 million gals:

$$C = \frac{Q}{100 \text{ million}} \left(\frac{\text{lb}}{\text{gal}}\right).$$

The differential equation we will use is based on the relation

$$\left(\begin{array}{c} \text{rate of change of} \\ \text{quantity of salt} \end{array}\right) = (\text{rate in}) - (\text{rate out}),$$

which describes how the quantity is changing with time. We will find Q first, and then C.

To find the 'rate in,' notice that all the salt is entering through the run-off. The run-off is 0.1 million gallons per day, with each gallon containing 0.0001 pounds of salt. The rate at which salt enters the reservoir is therefore

$$\begin{aligned} \text{concentration} \times \text{volume per day} &= 0.0001 \left(\frac{\text{lb}}{\text{gal}}\right) \times 0.1 \left(\frac{\text{mil gal}}{\text{day}}\right) \\ &= 0.00001 \left(\frac{\text{mil lb}}{\text{day}}\right) = 10 \left(\frac{\text{lb}}{\text{day}}\right). \end{aligned}$$

Thus

$$\text{rate in} = 10 \text{ lb/day}$$

Salt is leaving in the water used by the city each day. The city uses a million gallons of water each day with a concentration of $\frac{Q}{100 \text{ million}} \frac{\text{lb}}{\text{gal}}$. Thus the rate at which salt is leaving is

$$\begin{aligned} \text{concentration} \times \text{volume per day} &= \frac{Q}{100 \text{ mil}} \left(\frac{\text{lb}}{\text{gal}}\right) \times 1 \left(\frac{\text{mil gal}}{\text{day}}\right) \\ &= \frac{Q}{100} \left(\frac{\text{lb}}{\text{day}}\right). \end{aligned}$$

So

$$\text{rate out} = \frac{Q}{100} \text{ lb/day}.$$

Therefore Q must satisfy

$$\frac{dQ}{dt} = 10 - \frac{Q}{100}.$$

Solving this equation by factoring out $-\frac{1}{100}$ and separating variables:

$$\frac{dQ}{dt} = -\frac{1}{100}(Q - 1000) = -0.01(Q - 1000)$$

gives

$$\int \frac{dQ}{Q - 1000} = -\int 0.01 \, dt$$

$$\ln|Q - 1000| = -0.01t + k$$

so

$$Q - 1000 = Ae^{-0.01t}.$$

There is no salt initially, so we substitute $Q = 0$ when $t = 0$:

$$0 - 1000 = Ae^0 \qquad \text{giving} \qquad A = -1000$$

Thus

$$Q - 1000 = -1000e^{-0.01t}$$

so

$$Q = 1000(1 - e^{-0.01t}).$$

Therefore

$$\text{Concentration} \;=\; C \;=\; \frac{Q}{100 \text{ million}} \;=\; \frac{1000}{10^8}(1 - e^{-0.01t}) = 10^{-5}(1 - e^{-0.01t}).$$

A sketch of concentration against time is in Figure 8.25.   ❏

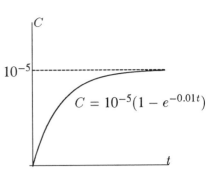

Figure 8.25: Concentration of Salt in Reservoir

❖ Exercises for Section 8.6

1. Dead leaves accumulate on the ground in a forest at 3 grams per square centimeter per year. At the same time, these leaves decompose at a continuous rate of 75% a year. Write a differential equation for the total quantity of dead leaves (per square centimeter) at time t. Sketch a solution showing that the quantity of dead leaves tends towards an equilibrium level. What is that equilibrium level?

2. According to a simple physiological model, an adult needs 20 calories per day per pound of body weight to maintain his weight. If he consumes more or less calories than those required to maintain his weight, his weight will change at a rate proportional to the difference between the number of calories consumed and the number needed to maintain his current weight; the constant of proportionality is $\frac{1}{3500}$lbs/cal. Suppose that a particular person has a constant caloric intake of I calories per day. Let $W(t)$ be his weight in pounds at time t (measured in days).

    (a) What differential equation has solution $W(t)$?

    (b) Solve this differential equation.

    (c) Draw a graph of $W(t)$ if the person starts out weighing 160 lbs and consumes 3000 calories a day. Label your axes and any intercepts and asymptotes clearly.

3. Suppose that at 1:00 p.m. one winter afternoon, there is a power failure at your house in Wisconsin, and that your heat does not work without electricity. When the power goes out, it is 68°F in your house. At 10:00 p.m., it is 57°F in the house, and you notice that it's 10°F outside.

   (a) Assuming that the temperature T, in your home obeys Newton's Law of Cooling, write the differential equation satisfied by T.

   (b) Solve the differential equation to estimate the temperature in the house when you get up at 7:00 a.m. the next morning. Should you worry about your water pipes freezing?

   (c) What assumption did you have to make in part (a) about the temperature outside? Given this (probably incorrect) assumption, would you revise your estimate up or down? Why?

4. A cylindrical barrel catches rainwater as it pours off the roof. When it stops raining, water leaks out of the barrel at a rate proportional to the square root of the depth of the water at that time. If the water level drops from 36 inches to 35 inches in 1 hour, how long will it take for all of the water to leak out of the barrel?

5. A bank account earns 5% annual interest compounded continuously. You wish to make payments out of the account at a rate of $12,000 per year (in a continuous cash flow) for 20 years.

   (a) Write a differential equation describing the balance $B = f(t)$, where t is in years.

   (b) Find the solution $B = f(t)$ to the differential equation given an initial balance of B_0 in the account.

   (c) What should the initial balance be such that the account has zero balance after precisely 20 years?

6. $1000 is put into a bank account and earns interest continuously at a rate of i per year, and in addition, continuous payments are made out of the account at a rate of $100 a year. Sketch the amount of money in the account as a function of time if the interest rate is

   (a) 5%

   (b) 10%

   (c) 15%

   In each case, you should first find an expression for the amount of money in the account at time t in years.

7. A certain commodity is currently being sold at a price of $\$p$ per unit. Over a period of time market forces will make this price tend towards the equilibrium price, which we call $\$p_0$, at which supply exactly balances demand. The rate at which the price changes is described by the Evans Price Adjustment model, which says that the rate of change in the actual market price ($\$p$) is proportional to the difference between the actual market price and the equilibrium price.

   (a) Write a differential equation for p as a function of t.

   (b) Solve for p.

   (c) Sketch solutions for various different initial prices, both above and below the equilibrium price.

   (d) What happens to p as $t \to \infty$?

8. As you know, when a course ends, students start to forget the material they have learned. One model (called the Ebbinghaus Model) assumes that the rate at which a student forgets material is proportional to the difference between the material he or she currently remembers and some constant.

   (a) Let $y = f(t)$ be the fraction of the original material remembered t weeks after the course has ended. Set up a differential equation for y. Your equation will contain two parameters.

   (b) Solve the equation and find the arbitrary constant of integration.

   (c) Describe the practical meaning (in terms of the amount remembered) of the parameter in the solution $y = f(t)$.

9. A rectangular swimming pool has dimensions in meters of 20 by 10 by 10; hence it has volume 2000 cubic meters, or 2×10^6 liters. The pool initially contains pure fresh water. At $t = 0$ minutes, salt water containing 10 grams/liter of salt is poured into the pool at a rate of 60 liters per minute. The salt water is instantly and totally mixed with the fresh water and the excess mixture is drained out of the bottom of the pool at the same rate (60 liters/minute). Let $S(t) =$ the mass of salt in the pool at time t.

   (a) Write a differential equation for the amount of salt in the pool, given that

   $$\frac{dS}{dt} = (\text{Rate at which salt enters the pool}) - (\text{Rate at which salt leaves the pool}),$$

   and, for example,

   $$\begin{pmatrix} \text{Rate at which salt} \\ \text{enters the pool} \end{pmatrix} = \begin{pmatrix} \text{Concentration of} \\ \text{salt solution} \end{pmatrix} \times \begin{pmatrix} \text{Flow rate of} \\ \text{salt solution} \end{pmatrix}$$

   $$(\text{grams/minute}) = (\text{grams/liter}) \times (\text{liters/minute})$$

   and similarly for salt leaving the pool.

   (b) Solve the differential equation to find $S(t)$.

   (c) What happens to $S(t)$ as $t \to \infty$?

10. In Brazil, interest rates increased steadily in the second half of the 1970's. As 1975 began, the interest rate was 50%, and it increased 25% a year after that (to 75% in 1976, to 100% in 1977 and so on). Assuming both that the interest rate climbed continuously and that interest accrued continuously:

  (a) Write a differential equation for the amount of money, M, in a bank account as a function of time, t, measured from 1975. Assume the only deposit was of 100,000 cruzieros on January 1, 1975.

  (b) Solve this differential equation.

  (c) How much money is in the account on January 1, 1980?

11. (Continuation of Problem 10.)

  (a) Suppose now Brazilian banks had offered a constant interest rate between 1975 and 1980. What would that interest rate have had to be to produce the same growth in the bank balance between 1975 and 1980?

  (b) Sketch a graph from 1975 into the early 1980's showing the balance in the account under both schemes, the rising and the constant interest rates.

12. A bank account that earns 10% interest compounded continuously has an initial balance of zero. Money is deposited into the account at a constant rate of $1000 per year.

  (a) Write a differential equation that describes the rate of change of the balance $B = f(t)$.

  (b) Solve the differential equation to find the balance as a function of time. Be sure to solve for any constants in your equation.

  (c) Using a Riemann sum, derive the integral that gives the balance at time t.

  (d) Evaluate the integral and show that you get the same formula as in part (b).

  (e) From your solution to the differential equation in (a), what is the formula for the balance $B = f(t)$ if the initial balance is $B_0 \neq 0$?

  (f) Show how to modify your integral in part (d) if the initial balance is B_0 and show that you get the same answer in part (e) above.

8.7   Models of Population Growth

Population projections first became important to political philosophers in the late eighteenth century. As concern for scarce resources has grown, so has interest in accurate population projections. In this section we will look at two differential equations which are used to model both human and animal population growth as well as growth processes in economics. These differential equations also have applications in epidemiology and medicine, such as modeling the spread of an infectious disease or the growth of a tumor.

When describing population growth, we often use percentages rather than absolute numbers. For example, we say the population of the world is now about 5.3 billion people and growing at about 2% a year. Thus if the population, P, is a function of time, t, we will use the relative growth rate

$$\frac{\frac{dP}{dt}}{P} \quad \text{or} \quad \frac{1}{P}\frac{dP}{dt}.$$

The Exponential Model

The first population model we will consider assumes that the relative growth rate of a population is constant, so

$$\frac{1}{P}\frac{dP}{dt} = k.$$

Rewritten as

$$\frac{dP}{dt} = kP,$$

This is a familiar differential equation whose solution is the family of exponential functions

$$P = P_0 e^{kt},$$

where P_0 is the population when t is zero.

The assumption that the relative growth rate is constant is usually a good one over a small time interval. Hence, for a suitable k, an exponential function models almost any population for a short period.

Interestingly enough, an exponential model has fitted the world's population and the population of many regions remarkably well for decades, even centuries. However, the model must break down at some point because it predicts that the population will continue to grow without bound as time goes on—and this cannot be true forever. Eventually the effects of crowding, emigration, disease, war, and lack of food will have to curb growth.

The grim predictions of this model are reflected in the ideas of Thomas Malthus, an early nineteenth century clergyman and political philosopher, who believed that, if unchecked, a population would grow geometrically (i.e., exponentially), whereas the food supply would grow arithmetically (i.e., linearly), and therefore that population growth would eventually outstrip increases in food supply.

However, for small populations the exponential model has proven itself to be quite accurate. It is only for large populations that the effects of crowding make themselves felt. In searching for an improvement, then, we should look for a model whose solution is approximately an exponential function for small values of the population, but which levels off later. That's what the next model does.

The Logistic Model

In the 1830's a Belgian mathematician, P. F. Verhulst, proposed that the relative growth rate of a population be a linearly decreasing function of P. In other words, he assumed that P satisfied the *logistic equation*:

$$\frac{1}{P}\frac{dP}{dt} = k(L - P)$$

where L is a constant, called the *carrying capacity*, which represents a limit to the population. For P small relative to L, the logistic equation becomes

$$\frac{1}{P}\frac{dP}{dt} \approx kL, \quad \text{a constant}$$

and the solution is approximately exponential. As P increases towards L, however, the relative growth rate decreases towards zero, representing the effects of crowding. Thus, as P gets closer to L, the population grows more and more slowly; if $P = L$, the growth rate is zero.

To get a picture of a solution curve for this type of differential equation, look at Figure 8.26(b), which contains the slope field for a specific case, namely

$$\frac{1}{P}\frac{dP}{dt} = 0.002(10 - P).$$

For comparison, the slope field of the equation

$$\frac{1}{P}\frac{dP}{dt} = 0.02$$

is in Figure 8.26(a). In both cases, P is in billions. You can see clearly that while the exponential function increases without bound, logistic growth levels off at $P = 10$. Looking at the logistic curve you can also see that initially it climbs like the exponential curve, but later it has an inflection point and starts to level off.

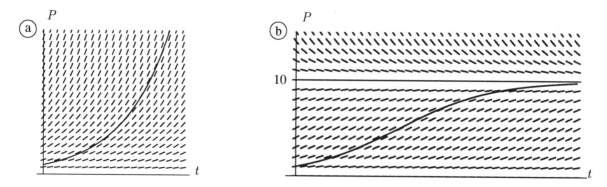

Figure 8.26: Slope Fields for (a) $\frac{1}{P}\frac{dP}{dt} = 0.02$ and (b) $\frac{1}{P}\frac{dP}{dt} = 0.002(10 - P)$

Qualitative Solution to the Logistic Equation

Even without solving the logistic equation, it is possible to locate the inflection point exactly. We'll go back to the general logistic equation and rewrite it as

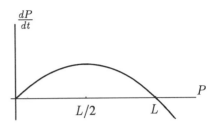

$$\frac{dP}{dt} = kP(L - P).$$

Figure 8.27: $\frac{dP}{dt} = kP(L - P)$

Now we will graph $\frac{dP}{dt}$ against P. Your graph will be a parabola, as in Figure 8.27.

You have probably never seen a graph like this before ($\frac{dP}{dt}$ graphed as a function of P, rather than of t), but we will find it helpful in analyzing the graph of P as a function of t.

In Figure 8.27 you can see that if $0 < P < L$, then $\frac{dP}{dt} > 0$ and P is increasing. If $P > L$, then $\frac{dP}{dt} < 0$ and P is decreasing. In general, a population satisfying the logistic equation will increase if P is below the carrying capacity (i.e., if $P < L$) and decrease if P is above the carrying capacity ($P > L$).

In addition, if $0 < P < L/2$, then $\frac{dP}{dt}$ is increasing, so the graph of P against t is concave up. If $L/2 < P < L$, then $\frac{dP}{dt}$ is decreasing, so the graph of P against t is concave down. Therefore, P as a function of t has an inflection point at $P = L/2$. Figure 8.28 shows the S-shaped logistic curve, also called the *sigmoid curve*, obtained if $0 < P_0 < L/2$.

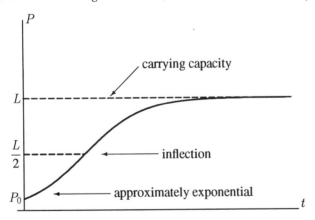

Figure 8.28: Logistic Growth with Inflection Point

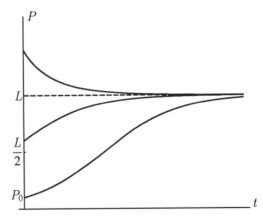

Figure 8.29: Solutions to the Logistic Equation

If $P = 0$ or $P = L$, then $\frac{dP}{dt} = 0$ and there is an equilibrium solution. (Not a very interesting one if $P = 0$.) Figure 8.29 shows that $P = 0$ is an unstable equilibrium because solutions which start near 0 move away. $P = L$, however, is a stable equilibrium.

The Analytic Solution to the Logistic Equation

We have already obtained a lot of information about logistic growth without finding a formula for the function. However, the equation can be solved by separating variables:

$$\frac{dP}{dt} = kP(L - P)$$

giving

$$\int \frac{dP}{P(L - P)} = \int k \, dt.$$

We can integrate the left side using the integral tables, or by rewriting

$$\frac{1}{P(L - P)} = \frac{1}{L}\left(\frac{1}{P} + \frac{1}{L - P}\right).$$

Thus the equation has become

$$\int \frac{1}{L}\left(\frac{1}{P} + \frac{1}{L - P}\right) dP = \int k \, dt$$

which can be integrated to give

$$\frac{1}{L}(\ln|P| - \ln|L - P|) = kt + C.$$

Multiplying through by L and using the fact that $\ln A - \ln B = \ln(\frac{A}{B})$, we have

$$\ln\left|\frac{P}{L - P}\right| = Lkt + LC.$$

Exponentiating both sides gives

$$\left|\frac{P}{L - P}\right| = e^{Lkt+LC} = e^{LC}e^{Lkt}$$

so

$$\frac{P}{L - P} = Ae^{Lkt} \quad \text{where } A = \pm e^{LC}.$$

We find A by substituting $P = P_0$ when $t = 0$, which gives

$$\frac{P_0}{L - P_0} = Ae^0 = A.$$

Thus

$$\frac{P}{L - P} = \frac{P_0}{L - P_0}e^{Lkt}.$$

Solving for P gives

$$P = \frac{LP_0e^{Lkt}}{P_0e^{Lkt} + L - P_0}.$$

Dividing top and bottom by e^{Lkt} yields

$$P = \frac{LP_0}{P_0 + (L - P_0)e^{-Lkt}}.$$

This is the formula for the logistic curve. In many cases, it is easier to argue graphically than use this formula. But if, for example, you want to calculate the exact time at which the population reaches a certain value, the formula would be quite useful.

❖ Exercises for Section 8.7

1. Assuming that Switzerland's population is growing exponentially at a continuous rate of 0.2% a year, and that its 1988 population was 6.6 million, write an expression for the population as a function of time in years. (Let $t = 0$ in 1988.)

2. (a) Consider the logistic model
$$\frac{dP}{dt} = aP - bP^2.$$
   The slope field in Fig. 8.30 is for the case $a = b = 3$. On the direction field, sketch three solution curves showing different types of behavior.
   (b) Is there a stable value of the population? If so, what is it?
   (c) Describe the meaning of the shape of the solution curves for the population: Where is P increasing? decreasing? What happens in the long run? Are there any inflection points? Where? What do they mean for the population?
   (d) Sketch a graph of $\frac{dP}{dt}$ against P. Where is $\frac{dP}{dt}$ positive? negative? zero? maximum? How do your observations about $\frac{dP}{dt}$ explain the shapes of your solution curves?

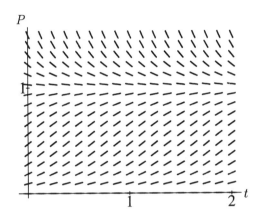

Figure 8.30: Slope field for $\frac{dP}{dt} = 3P - 3P^2$.
(for Problem 2)

3. The growth of a certain animal population is governed by the equation
$$\frac{1000}{P}\frac{dP}{dt} = 100 - P,$$
where $P(t)$ is the number of individuals in the colony at time t. The initial population is known to be 200 individuals. Sketch a graph of $P(t)$. Will there ever be more than 200 individuals in the colony? Will there ever be fewer than 100 individuals? Explain.

4. Many organ pipes in old European churches are made of tin. In cold climates such pipes can be affected with *tin pest*, where the tin becomes brittle and crumbles into a grey powder. This transformation can appear to take place very suddenly because the presence of the grey powder encourages the reaction to proceed. At the start, when there is little grey powder, the reaction proceeds slowly. Similarly, towards the end, when there is little metallic tin left, the reaction is also slow. In between, however, when there is plenty of both metallic tin and powder, the reaction can be alarmingly fast.

Suppose that the rate of the reaction is proportional to the product of the amount of tin left and the quantity of grey powder, p, present at time t. Assume also that when metallic tin is converted to grey powder, its weight does not change.

(a) Write a differential equation for p. Let the total quantity of metallic tin present originally be B.

(b) Sketch a graph of the solution $p = f(t)$ if there is a small quantity of powder initially. How much metallic tin has crumbled when it is crumbling fastest?

(c) Suppose there is no grey powder initially. (For example, suppose the tin is completely new.) What does this model predict will happen? How do you reconcile this with the fact that many organ pipes do get tin pest?

5. It is of considerable interest to policy makers to model the spread of information through a population. For example, various agricultural ministries use models to understand the spread of technical innovations or new seed types through their countries. Two models, depending on how the information is spread, are given below. Assume the population is of a constant size M.

(a) If the information is spread by mass media (TV, radio, newspapers), the rate information is spread is believed to be proportional to the number of people not having the information at that time. Write a differential equation for the number of people having the information by time t. Sketch a solution assuming that no one (except the mass media) has the information initially.

(b) If the information is spread by word of mouth, the rate of spread of information is believed to be proportional to the product of the number of people who know and the number who don't. Write a differential equation for the number of people having the information by time t. Sketch the solution for the cases in which

   i. no one,
   ii. 5% of the population,
   iii. 75% of the population,

knows initially. In each case, when is the information spreading fastest?

6. (a) The U.S. population figures for the years 1790 through 1980 appear in Table 8.3. Assuming that this population growth is well approximated by a logistic differential equation, $\frac{1}{P}\frac{dP}{dt} = a - bP$, find reasonable values for a and b.

   To do this, estimate the percentage growth rate, $\frac{1}{P}\frac{dP}{dt}$, and plot it against P. To approximate $\frac{1}{P}\frac{dP}{dt}$, calculate $\frac{1}{P}\frac{\Delta P}{\Delta t}$ from the data. Do this for about six fairly spread-out points. Draw a reasonable line (by eye) through your points, and estimate a and b.

Year	Population	Year	Population
1790	3.9	1890	62.9
1800	5.3	1900	76.0
1810	7.2	1910	92.0
1820	9.6	1920	105.7
1830	12.9	1930	122.8
1840	17.1	1940	131.7
1850	23.1	1950	150.7
1860	31.4	1960	179.0
1870	38.6	1970	205.0
1880	50.2	1980	226.5

Table 8.3: U.S. Population in Millions, 1790–1980

(for Problem 6)

(b) What does this model predict about the U.S. population in the long run?

7. (Continuation of Problem 6)

(a) You may have discovered that the logistic model does not fit the US population figures in Table 8.3 all that closely. (You can see that the points on the graph of $\frac{1}{P}\frac{dP}{dt}$ against P are not exactly on a line.) In this problem we'll try another model. This time we'll assume that $\frac{1}{P}\frac{dP}{dt}$ is a linear function of t, so you should plot $\frac{1}{P}\frac{dP}{dt}$ against t (using the same approximate values you calculated before). Put a line through them by eye, and find a and b to fit

$$\frac{1}{P}\frac{dP}{dt} = a - bt.$$

(b) When, if ever, does this model predict the U.S. population will be at its maximum?

(c) Solve the differential equation.

8.8 Two Interacting Populations

In the previous section we modeled the growth of a single population over time. We will now consider two populations which interact. For example, they may compete for food, one may prey on the other, or they may enjoy a symbiotic relationship where each helps the other. We will look at how both populations change over time and, in particular, at their long-run behavior. Does one population die out, while the other flourishes? Do both die out? Do both coexist forever? Such questions are important in interpreting the ecological balance between species.

As an example, we will model a predator-prey system using what are called the Lotka-Volterra equations. Some rather surprising behavior had been observed in predator-prey systems, and one of the early successes of the Lotka-Volterra equations was to explain that behavior.

One of the predator-prey systems about which we have long-term data is the Canadian lynx and hare. Since both animals were of interest to fur trappers, records of the Hudson Bay Company throw some light on their populations through much of the last century. These records show that the populations of both animals oscillate up and down, quite regularly, with a period of about ten years. In addition, the maximum values of the two populations did not occur at the same time; the hare was a quarter of a cycle ahead of the lynx. You will see that this behavior is predicted by differential equations.

Robins and Worms

Let's look at a simplified and idealized case in which robins are the predators and worms the prey. Suppose there are r thousand robins and w million worms. If there were no robins, the worms would increase freely according to the equation

$$\frac{dw}{dt} = aw \quad \text{where } a \text{ is a constant and } a > 0.$$

We will assume that if the robins were alone, and so had no worms for food, they would decrease according to the equation[2]

$$\frac{dr}{dt} = -br \quad \text{where } b \text{ is a constant and } b > 0.$$

Now we must model for the effect of the two populations on one another. Clearly, the presence of the robins is bad for the worms, so

$$\frac{dw}{dt} = aw - (\text{effect of robins on worms}).$$

On the other hand, the robins do better with the worms around, so

$$\frac{dr}{dt} = -br + (\text{effect of worms on robins}).$$

How exactly do the two populations interact? Let's assume the effect of one population on the other will be proportional to the number of 'encounters' (an encounter is when a robin meets and eats a worm). The number of encounters is likely to be proportional to the products of the populations because the more there are of either population, the more encounters there will be. So we will assume

$$\frac{dw}{dt} = aw - cwr \quad \text{and} \quad \frac{dr}{dt} = -br + kwr$$

where c, k are positive constants.

To analyze this system of equations, let's look at the specific example when $a = b = c = k = 1$:

$$\frac{dw}{dt} = w - wr$$

and

$$\frac{dr}{dt} = -r + wr$$

Before continuing, we need to figure out how to handle the fact that we have three variables—two populations and time—rather than the usual two. Ideally, you might hope to solve for w and r separately, getting each as a function of t. This turns out to be hard, so instead we try to find the relationship between r and w at any time. Indeed, from a biological point of view, it is often more useful to know how many of each population exist simultaneously than to know each population at particular times. Thus, for a moment, we will think of r as a function of w, and both r and w as functions of t. If we wrote $r = f(w)$ and $w = g(t)$, so that $r = f(g(t))$, then by the chain rule we could say

$$\frac{dr}{dt} = f'(g(t)) \cdot g'(t) = \frac{dr}{dw} \cdot \frac{dw}{dt},$$

---

[2]You might criticize this assumption because it predicts that the number of robins will decay exponentially, rather than die out in finite time.

so

$$\frac{dr}{dw} = \frac{dr/dt}{dw/dt}.$$

Therefore, for our example

$$\frac{dr}{dw} = \frac{-r + wr}{w - wr}.$$

The Slope Field and Equilibrium Points

You can get an idea of what solutions of this equation look like from its slope field in Figure 8.31. Probably the first thing that strikes you is that something special is going on at the point $(1, 1)$. You will also see that the solution curves appear to be closed curves centered at $(1, 1)$.

Let's look at both of these observations in more detail. At the point $(1, 1)$ there is no slope drawn because at this point we have that the rate of change of the worm popultaion with respect to time

$$\frac{dw}{dt} = 1 - (1)(1) = 0$$

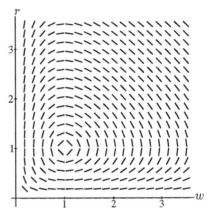

Figure 8.31: Slope field for $\dfrac{dr}{dw} = \dfrac{-r + wr}{w - wr}$

and that the rate of change of the robin population with respect to time

$$\frac{dr}{dt} = -1 + (1)(1) = 0,$$

which means that $\frac{dr}{dw}$ is undefined. In terms of worms and robins this means that if at some moment $w = 1$ and $r = 1$ (i.e., you have 1 million worms and 1 thousand robins), then w and r remain constant forever. The point $w = 1$, $r = 1$ is therefore an equilibrium solution. The slope field suggests that there are no other equilibrium points except the origin.

> At an *equilibrium point*, both w and r are constant, so
> $$\frac{dw}{dt} = 0 \qquad \text{and} \qquad \frac{dr}{dt} = 0.$$

Therefore we look for equilibrium points by solving

$$\frac{dw}{dt} = w - wr = 0$$
$$\frac{dr}{dt} = -r + rw = 0$$

which has $w = 0$, $r = 0$ and $w = 1$, $r = 1$ as the only solutions.

Trajectories in the Phase Plane

Let's look a bit more closely at the meaning of the solution curves for the populations. Remember that a point on one of the curves represents a pair of populations (w, r) which exist at the same time t (though t is not shown on the graph). At a short time later both populations will have moved to a nearby point on the curve. Thus as time passes, the point traces out the curve. The direction is marked on the curve by an arrow. See Figure 8.32. A solution curve showing the evolution of the system with time is called a *trajectory* or an *orbit*; the whole wr-plane is called the *phase plane*.

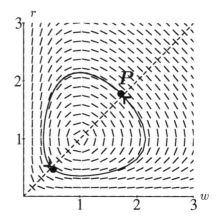

Figure 8.32: Solution Curve is Closed

Now let's think about whether the solution curves are really closed curves (i.e. they come back and meet themselves) and what this means about populations. To see why the solution curves are closed, notice that the slope field is symmetric about the line $w = r$. You can confirm this by observing that interchanging w and r does not alter the differential equation for $\frac{dr}{dw}$. This means that if you start at point P on the line $w = r$ and travel once around the point $(1, 1)$, you have to arrive back at the same point P. The reason is that the second half of your path is the reflection of the first half in the line $w = r$. See Figure 8.32. If you did not end up at P again, the second half of your path would have had a different shape from the first half.

How do we figure out which way to move on the trajectory? We need to know how w and r change as time passes, so we look at the original pair of differential equations because they involve time:

$$\frac{dw}{dt} = w - wr$$

$$\frac{dr}{dt} = -r + wr.$$

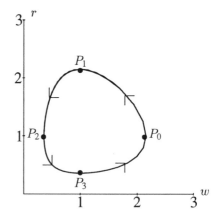

Figure 8.33: A Trajectory

At the point P_0 in Figure 8.33, $w > 1$ and $r = 1$, so

$$\frac{dr}{dt} = -r + wr = -1 + w > 0.$$

Therefore r is increasing so the point is moving counterclockwise around the closed curve.

The Populations as a Function of Time

We will now see what the trajectories tell us about the populations. Suppose we start at $t = 0$ at the point P_0 in Figure 8.33. Then suppose we move to P_1 at time t_1, to P_2 at time t_2, to P_3 at time t_3, and so on. At time t_4 we will be back at P_0 and the whole cycle will repeat. Thus *the fact that the trajectory is a closed curve means that both populations will oscillate up and down through the same values.* At P_0 the worm population is at its maximum while the robin population is at an intermediate value, which can be shown to be the average value. At P_1, the worm population has dropped to its average value, while the robins have increased to their maximum. At P_2 the worms have dropped further to their minimum while the robins have dropped to their average value, and so on. The graphs of both populations against time are shown in Figure 8.34.

Both populations oscillate with the same period, and the worms (the prey) are at their maximum a quarter of a cycle before the robins.

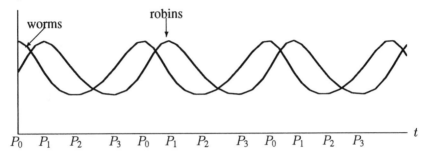

Figure 8.34: Populations of Robins (in thousands) and Worms (in millions) over Time

❖ Exercises for Section 8.8

For Problems 1–2, let x be the number of worms (in millions) and y the number of robins (in thousands) living on an island. Assume x and y satisfy the differential equations

$$\frac{1}{x}\frac{dx}{dt} = a - by$$
$$\frac{1}{y}\frac{dy}{dt} = -c + kx$$

1. Explain why this might be a reasonable way to model the interaction between the two populations. What is the meaning of the constants? Why have the signs been chosen this way?

2. Solve these equations in the two special cases when there are no robins and when there are no worms living on the island.

For Problems 3–7, assume $a = b = c = k = 1$. The slope field for

$$\frac{dy}{dx} = \frac{y(x-1)}{x(1-y)}$$

is shown in Figure 8.35, where y is the robin population in thousands and x is the worm population in millions.

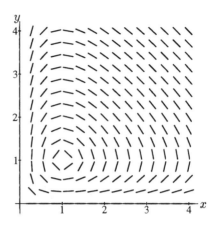

Figure 8.35: Robins and Worms

3. Describe and explain the symmetry you observe in the slope field. What consequences does this symmetry have for the solution curves?

4. Assume $x = 2$ and $y = 2$ when $t = 0$. Does the number of robins and worms increase or decrease at first? What happens in the long run?

5. For the case discussed in Problem 4, estimate the maximum and the minimum values of the robin population. How many worms are there at the time when the robin population reaches its maximum?

6. On the same axes, graph x and y (the robin and the worm populations) against time. Use initial values of 1.5 for x and 1 for y. You may do this without units for t.

7. People on our island like robins so much that they decide to import 200 robins all the way from England, to increase the initial population to $y = 2.2$ when $t = 0$. Does this make sense? Why or why not?

The differential equations in Problems 8–10 describe the rate of growth of two populations x and y (both measured in thousands) of species A and B, respectively. For each set:

(a) Describe in words what happens to the population of each species in the absence of the other.

(b) Describe in words how the species interact with one another. Can you think of reasons why the populations behave as described by the equations? Can you think of species that might interact as described by the equations?

8. $\dfrac{dx}{dt} = 0.01x - 0.05xy$

$\dfrac{dy}{dt} = -0.2y + 0.08xy$

9. $\dfrac{dx}{dt} = 0.01x - 0.05xy$

$\dfrac{dy}{dt} = 0.2y - 0.08xy$

10. $\dfrac{dx}{dt} = 0.2x$

$\dfrac{dy}{dt} = 0.4xy - 0.1y$

11. (a) Find the equilibrium points for the following system of equations

$$\frac{dx}{dt} = 15x - 3xy$$
$$\frac{dy}{dt} = -14y + 7xy$$

(b) Explain why $x = 2$, $y = 0$ is not an equilibrium point for this system.

12. (a) Find the equilibrium points for the following system of equations

$$\frac{dx}{dt} = x - 0.001x^2 - 0.005xy,$$
$$\frac{dy}{dt} = 0.02y - 3\frac{y}{x}.$$

(b) Explain why $x = 0$, $y = 0$ is not an equilibrium point for this system.

13. Two companies share the market for a new technology. They have no competition except each other. Let $A(t)$ be the net worth of one company and $B(t)$ be the net worth of the other. Assume that net worth cannot be negative, because a company goes out of business if its net worth is zero. Suppose A and B satisfy the differential equations

$$A' = 2A - AB$$
$$B' = B - AB$$

(a) What do these equations predict about the net worth of each company if the other were not present? What effect do the companies have on each other?

(b) Are there any equilibrium points? If so, what are they?

(c) Sketch a slope field for these equations (using a computer or calculator), and hence describe the different possible long-run behaviors.

14. The equations

$$\frac{dx}{dt} = y - 2xy \quad \text{and} \quad \frac{dy}{dt} = 3x - xy$$

describe the interaction between two species.

(a) Determine all the equilibrium points.

(b) Describe the behavior of the trajectories near the equilibrium points.

(c) Draw the global picture of the trajectories in the first quadrant.

15. Suppose the equations

$$\frac{dx}{dt} = x - 2xy$$
$$\frac{dy}{dt} = 4y - 2xy$$

describe the interaction between two species of animals where y is the number of thousands of animal Y, and x is the number of thousands of animal X.

(a) Which of the following best describes the relationship between X and Y?

    i. Y is the predator; X is the prey.

    ii. X is the predator; Y is the prey.

    iii. X and Y mutually benefit from each other's existence.

    iv. X and Y are competitors.

(b) Determine all the equilibrium points in the xy-phase plane.

(c) Sketch some trajectories close to the equilibrium point which is not at the origin.

16. The following systems of differential equations model the interaction of two populations x and y. In each case, answer the following two questions:

(a) What kinds of interaction do the equations describe? (Symbiosis[3], competition, predator-prey.)

(b) What happens in the long run? (For one of the systems, your answer will depend on the initial populations.) Use a calculator or computer to draw slope fields.

    i. $\dfrac{1}{x}\dfrac{dx}{dt} = y - 1,$ iii. $\dfrac{1}{x}\dfrac{dx}{dt} = y - 1 - 0.05x,$

        $\dfrac{1}{y}\dfrac{dy}{dt} = x - 1.$ $\dfrac{1}{y}\dfrac{dy}{dt} = 1 - x - 0.05y.$

    ii. $\dfrac{1}{x}\dfrac{dx}{dt} = 1 - \dfrac{x}{2} - \dfrac{y}{2},$

        $\dfrac{1}{y}\dfrac{dy}{dt} = 1 - x - y.$

---

[3]Symbiosis takes place when the interaction of two species benefits both. A favorite example is the pollination of plants by insects.

17. The concentrations of two chemicals A and B as functions of time are denoted by x and y respectively. Each alone decays at a rate proportional to its concentration. Put together, they also interact to form a third substance, at a rate proportional to the product of their concentrations. All this is expressed in the equations:

$$\frac{dx}{dt} = -2x - xy, \qquad \frac{dy}{dt} = -3y - xy.$$

   (a) Find a differential equation describing the relationship between x and y, and solve it.

   (b) Show that the only equilibrium state is $x = y = 0$. (Note that the concentrations are positive.)

   (c) Show that when x and y are positive and very small, $\dfrac{y^2}{x^3}$ is roughly constant.

   [Hint: When x is small, x is negligible compared to $\ln x$.]

   If now the initial concentrations are $x(0) = 4$, $y(0) = 8$:

   (d) Find the equation of the phase trajectory.

   (e) What would be the concentrations of each substance when they become equal?

   (f) If $x = e^{-10}$, find an approximate value for y.

8.9   Second Order Differential Equations: Oscillations

You are familiar with the equation of motion of a body falling freely under gravity:

$$\frac{d^2 s}{dt^2} = -g,$$

where s is the height of the body above ground at time t and g is the acceleration due to gravity. To solve this equation, we first integrate to get the velocity, $v = \dfrac{ds}{dt}$:

$$\frac{ds}{dt} = -gt + v_0,$$

where v_0 is the initial velocity. Then we integrate again, giving

$$s = -\frac{1}{2}gt^2 + v_0 t + s_0,$$

where s_0 is the initial height.

The differential equation $\dfrac{d^2 s}{dt^2} = -g$ is called *second order* because two is the highest order of any derivative appearing in the equation. Since getting the solution involves integrating twice, the solution involves *two* arbitrary constants, here v_0 and s_0. Thus, the general solution to a second order differential equation will be a family of functions with two parameters. Finding values for the two constants corresponds to picking a particular function out of this family.

A Mass on a Spring

Not every second order equation can be solved simply by integrating twice. Consider the following situation: a 1 kilogram mass is attached to the end of a spring hanging from the ceiling. (We will assume that the mass of the spring itself is negligible in comparison with the 1 kg mass.) See Figure 8.36.

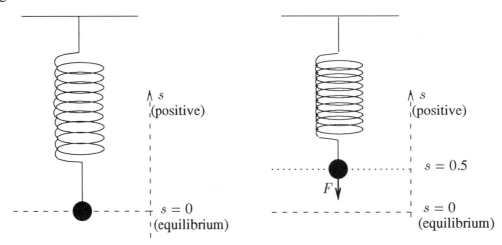

Figure 8.36: Spring and Mass in Equilibrium Position and After Upward Displacement

When the system is left undisturbed, no net force acts on the mass. The force of gravity is balanced by the force the spring exerts on the mass, and the spring is said to be in the *equilibrium position*. If you pull down on the mass, you will feel a net force pulling upwards; the further you stretch the spring, the larger the net force becomes. If instead, you push upward on the mass, the opposite happens: the net force pushes the mass down.[4]

In order to figure out how the mass moves, we need to know the exact relationship between the displacement from the equilibrium position and the net force exerted on the mass. A possible set of measurements is given in Table 8.4. In this case the relationship between the force and the displacement is given by the equation

$$F = -9s,$$

where the negative sign means that the net force is in the opposite direction to the displacement.

Displacement, s, of mass from equilibrium position (meters)	Net force, F, exerted by spring (kg·meters/sec^2)
−0.6	5.4
−0.4	3.6
−0.2	1.8
0.0	0.0
0.2	−1.8
0.4	−3.6
0.6	−5.4

Table 8.4: Data for 1 kg Mass Attached to Spring (upward is positive)

What happens if you push the mass some distance upwards, 0.5 m say, and then release it? You'd expect the mass to oscillate up and down. Let's investigate. While you were pushing on the

---

[4]Pulling down on the mass stretches the spring, increasing the tension, so the net force is upward. Pushing up the spring decreases tension in the spring, so the net force is downward.

mass, the forces acting were: gravity, the tension in the spring, and the force you were applying. When you let go, you stop applying any force, so the net force causes the mass to accelerate toward the equilibrium position. By Newton's second law of motion,

$$\text{Force} = \text{Mass} \times \text{Acceleration}.$$

Since the acceleration is $\dfrac{d^2 s}{dt^2}$, we have

$$F = 1\,\text{kg} \cdot \frac{d^2 s}{dt^2}.$$

Combining this with $F = -9s$ and dividing by 1 kg, we find

$$\frac{d^2 s}{dt^2} = -9s.$$

The motion of the mass is described by a second order differential equation! If you want to know s as a function of t, you have to solve the differential equation. Unfortunately, this equation cannot be solved just by integrating twice. The reason for this is that the differential equation involves the dependent variable s on the right-hand side as well as in the derivative, and we cannot integrate the right side without already knowing what s is.

As you might guess, if you release the mass, it oscillates up and down. In fact, later on in this section we will show that the motion of the mass is given by the following function:

$$s(t) = 0.5 \cos 3t.$$

It is easy to check (by substitution) that this function is a solution to the differential equation above. In addition, this solution satisfies the physical constraints of the problem. What are these constraints? The first constraint is that the initial displacement, $s(0)$, is 0.5 meters. Substituting, we find that

$$s(0) = 0.5 \cos 0 = 0.5\,\text{meters}.$$

The second is that at time $t = 0$, when the mass has just been released, it has not yet had time to accelerate, so its velocity is still 0:

$$v(0) = \left.\frac{ds}{dt}\right|_{t=0} = -1.5 \sin 3t \,\Big|_{t=0} = 0.$$

In the rest of this chapter, we will learn how to find exact solutions to second order differential equations like the one above. We will also learn how to approximate solutions numerically in cases where we can't find exact solutions.

Springs and Hooke's Law

The equation we got relating the force F exerted by a spring to the displacement s of the mass from the equilibrium position was an instance of *Hooke's Law*, which says that, provided s is not large enough to deform the spring,

$$F = -ks,$$

where k is the *spring constant* ($k > 0$); k depends on the physical properties of the particular spring. If there are no external forces besides gravity acting, we may combine Hooke's equation with Newton's second law, $F = m\dfrac{d^2 s}{dt^2}$, to obtain the second order differential equation which describes the motion of the mass:

$$m\frac{d^2 s}{dt^2} = -ks,$$

or equivalently,

Equation for Oscillations of a Mass on a Spring

$$\frac{d^2 s}{dt^2} = -\frac{k}{m}s.$$

Solving Differential Equations by Guess-and-Check

Guessing is the tried-and-true method for solving differential equations. It may surprise you to learn that there is no systematic method for solving most differential equations, so guesswork is often extremely important. Before looking for a general solution to the equation $\dfrac{d^2 s}{dt^2} = -\dfrac{k}{m}s$, we'll consider the special case $\dfrac{k}{m} = 1$. In Example 1, you will see how to guess specific solutions, check that they satisfy the equation, and then build the general solution.

☐ **Example 1** Find the general solution to the equation

$$\frac{d^2 s}{dt^2} = -s.$$

Solution. We want to find functions whose second derivative is the negative of the original function. We are already familiar with two functions that have this property: $s(t) = \cos t$ and $s(t) = \sin t$. We check that they are solutions by substituting:

$$\frac{d^2}{dt^2}(\cos t) = \frac{d}{dt}(-\sin t) = -\cos t,$$

and

$$\frac{d^2}{dt^2}(\sin t) = \frac{d}{dt}(\cos t) = -\sin t.$$

The remarkable thing is that starting from these two particular solutions, we can build up *all* the solutions to our equation. Here's how: if C is a constant, then $C \sin t$ and $C \cos t$ are also solutions to the differential equation. (Why?) Also, $\sin t + \cos t$ is a solution to the differential equation. (Why?) In fact, given two constants C_1, C_2, the function

$$s(t) = C_1 \cos t + C_2 \sin t$$

satisfies the differential equation, since

$$
\begin{aligned}
\frac{d^2}{dt^2}(C_1 \cos t + C_2 \sin t) &= \frac{d}{dt}(-C_1 \sin t + C_2 \cos t) \\
&= -C_1 \cos t - C_2 \sin t.
\end{aligned}
$$

It can be shown that $s(t) = C_1 \cos t + C_2 \sin t$ is the most general form of the solution. As expected, it contains two constants, C_1 and C_2. ☐

   If the differential equation represents a physical problem, then C_1 and C_2 are often determined by certain conditions of that physical problem.

☐ **Example 2**   Find the solution to

$$
\frac{d^2 s}{dt^2} = -s
$$

if the mass were displaced by a distance of s_0 and then released.

Solution. The position of the mass is given by the equation

$$
s(t) = C_1 \cos t + C_2 \sin t.
$$

We also know that the initial position is s_0; thus,

$$
s(0) = C_1 \cos 0 + C_2 \sin 0 = C_1 \cdot 1 + C_2 \cdot 0 = s_0,
$$

so $C_1 = s_0$, the initial displacement. What is C_2? To find it, we use the fact that at $t = 0$, when the spring has just been released, its velocity is 0. The velocity is the derivative of the displacement; thus,

$$
\left.\frac{ds}{dt}\right|_{t=0} = \left.(-C_1 \sin t + C_2 \cos t)\right|_{t=0} = C_1 \cdot 0 + C_2 \cdot 1 = 0,
$$

so $C_2 = 0$. Therefore the solution is $s = s_0 \cos t$. ☐

Solution to the General Spring Equation

Having found the general solution to the equation $\dfrac{d^2 s}{dt^2} = -s$, let us return to the more general equation for the motion of a spring, namely

$$
\frac{d^2 s}{dt^2} = -\frac{k}{m}s.
$$

To solve this we let $\omega = \sqrt{\frac{k}{m}}$ (why will be clear in a moment), then $\dfrac{d^2 s}{dt^2} = -\omega^2 s$ or $\dfrac{d^2 s}{dt^2} + \omega^2 s = 0$. Unfortunately, this equation no longer has $\sin t$ and $\cos t$ as solutions. For example,

$$
\frac{d^2}{dt^2}(\sin t) = -\sin t \neq -\omega^2 \sin t.
$$

All is not lost, however. We adapt the guess-and-check method as follows: We know that the functions $\sin t$ and $\cos t$ are *almost* solutions; we just have to deal with the extra ω^2 factor somehow. Ideally, we would like to be able to bring a factor of ω^2 outside in the process of differentiating the sine or cosine function twice. This would be accomplished if, each time we differentiated, we brought out a factor of ω. The chain rule suggests an answer: try the function $\sin \omega t$. Checking this:

$$\frac{d^2}{dt^2}(\sin \omega t) = \frac{d}{dt}(\omega \cos \omega t) = -\omega^2 \sin \omega t.$$

So $\sin \omega t$ is a solution, and you can probably guess that $\cos \omega t$ is a solution, too. As before, we can combine the sine and the cosine to find the general solution. Thus:

The general solution to the equation

$$\frac{d^2 s}{dt^2} + \omega^2 s = 0$$

is of the form

$$s(t) = C_1 \cos \omega t + C_2 \sin \omega t,$$

where C_1 and C_2 are arbitrary constants. (We assume that $\omega > 0$.) The period of this oscillation is

$$T = \frac{2\pi}{\omega}$$

The solution to our original equation, $\dfrac{d^2 s}{dt^2} + \dfrac{k}{m} s = 0$, is thus $C_1 \cos \sqrt{\frac{k}{m}}t + C_2 \sin \sqrt{\frac{k}{m}}t$.

Initial Value and Boundary Value Problems

A problem in which the initial position and the initial velocity are used to determine the particular solution is called an *initial value problem*. (See Example 2 on page 591.) In the following example, it is also possible to find a specific solution from a general solution given the position at two known times. In this case, the problem is known as a *boundary value problem*.

☐ **Example 3** Solve the differential equation

$$\frac{d^2 s}{dt^2} + 4s = 0,$$

given   (a) the boundary conditions $s(0) = 0$, $s(\frac{\pi}{4}) = 20$;

                    (b) the initial conditions $s(0) = 1$, $s'(0) = -6$.

Solution. Since $\omega^2 = 4$, $\omega = 2$, the general solution to the differential equation is

$$s(t) = C_1 \cos 2t + C_2 \sin 2t.$$

(a) Substituting the boundary condition $s(0) = 0$ into the general solution gives

$$s(0) = C_1 \cos(2 \cdot 0) + C_2 \sin(2 \cdot 0) = C_1 \cdot 1 + C_2 \cdot 0 = C_1 = 0.$$

We are now restricted to functions of the form $s(t) = C_2 \sin 2t$. The second condition yields the value of C_2:

$$s\left(\frac{\pi}{4}\right) = C_2 \sin\left(2 \cdot \frac{\pi}{4}\right) = C_2 = 20.$$

Therefore, the solution satisfying the boundary conditions is

$$s(t) = 20 \sin 2t.$$

(b) Now for the initial value problem. The general solution is the same: $s(t) = C_1 \cos 2t + C_2 \sin 2t$. Substituting 0 for t once again, we find

$$s(0) = C_1 \cos(2 \cdot 0) + C_2 \sin(2 \cdot 0) = C_1 = 1.$$

Differentiating $s(t) = \cos 2t + C_2 \sin 2t$ gives

$$s'(t) = -2 \sin 2t + 2C_2 \cos 2t,$$

and applying the second initial condition gives us C_2:

$$s'(0) = -2 \sin(2 \cdot 0) + 2C_2 \cos(2 \cdot 0) = 2C_2 = -6,$$

so $C_2 = -3$ and our solution is

$$s(t) = \cos 2t - 3 \sin 2t.$$

❑

More General Second Order Differential Equations

The equation we have been studying, $\dfrac{d^2 s}{dt^2} = -\omega^2 s$ or $\dfrac{d^2 s}{dt^2} + \omega^2 s = 0$, is an important example of the class of second order differential equations of the form

$$\frac{d^2 s}{dt^2} + b\frac{ds}{dt} + cs = 0.$$

Such equations are called *linear second order differential equations*. We call b and c the *coefficients* of the equation; they may be either constants or functions of t. When b and c are constants, we will say we have a *linear second order differential equation with constant coefficients*. Much of the rest of this chapter will be concerned with such equations.

What do the Graphs of Our Solutions Look Like?

Since the general solution of the equation $\frac{d^2s}{dt^2} + \omega^2 s = 0$ is of the form

$$s(t) = C_1 \cos \omega t + C_2 \sin \omega t$$

it would be useful to know what the graph of such a sum of sines and cosines looks like. We'll start with a straightforward example, $s(t) = \cos t + \sin t$, which is graphed in Figure 8.37.

Interestingly, the graph looks precisely like another sine function, and in fact it is one. If you measure the graph carefully, you will

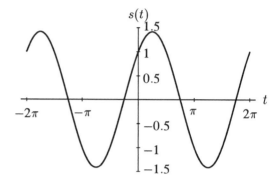

Figure 8.37: Graph of the sum of the sine and cosine: $s(t) = \cos t + \sin t$

find that its period is 2π, the same as $\sin t$ and $\cos t$, but the amplitude appears to be larger than 1, and the graph is shifted along the t axis. (This is known as a phase shift or phase angle.) If you measure carefully, you will find that the amplitude of this graph is approximately 1.414. (In fact, it's $\sqrt{2}$).

If we were to plot $C_1 \cos t + C_2 \sin t$ for a variety of C_1 and C_2, we would find that the resulting graph is always a sine with period 2π, though the amplitude and phase shift can vary. For example, the graphs of $s = 4 \cos t + 3 \sin t$ and $s = 6 \cos t - 8 \sin t$ are in Figure 8.38. Both graphs have a period of 2π, but for $C_1 = 4$, $C_2 = 3$, the amplitude is 5 and for $C_1 = 6$, $C_2 = -8$, the amplitude is 10; their phase shifts also differ.

These graphs suggest that we should be able to write the sum of a sine and a cosine as one single sine function. It turns out that a sum of sine and cosine functions of the form

$$C_1 \cos \omega t + C_2 \sin \omega t$$

can always be written in the form

$$A \sin(\omega t + \varphi)$$

for some amplitude A and phase shift φ, which tells us how the oscillation has been shifted horizontally. How do we find the constants A and φ in terms of C_1 and C_2?

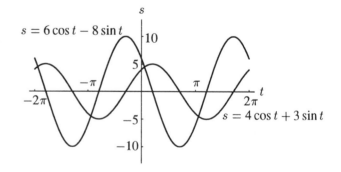

Figure 8.38: Graph of Two Sums of Sine Functions

You may already suspect the formula for A from the three functions we have graphed so far: with $C_1 = C_2 = 1$, $A = \sqrt{2}$; with $C_1 = 4$ and $C_2 = 3$, $A = 5$; and with $C_1 = 6$ and $C_2 = -8$, $A = 10$. As you will see in Problem 14 at the end of this section, it turns out that:

If $C_1 \cos \omega t + C_2 \sin \omega t = A \sin(\omega t + \varphi)$, then the *amplitude*, A, is

$$A = \sqrt{C_1^2 + C_2^2}.$$

We choose the *phase (or phase angle)* φ so that $-\pi < \varphi \le \pi$, and φ must satisfy:

$$\tan \varphi = \frac{C_1}{C_2}.$$

If $C_1 > 0$, φ is positive and if $C_1 < 0$, φ is negative.

The reason that it is often useful to write the solution to the differential equation as a single sine function $A \sin(\omega t + \varphi)$, as opposed to the sum of a sine and a cosine, is that its amplitude A and the phase shift φ are easier to recognize in this form.

☐ **Example 4** Write the function $s(t) = \cos t - \sin t$ as a single sine function. Draw its graph.

Solution. We want to find A and φ such that

$$\cos t - \sin t = A \sin(t + \varphi).$$

We know that $A = \sqrt{1^2 + (-1)^2} = \sqrt{2}$. Also, $\tan \varphi = \frac{1}{-1} = -1$, so $\varphi = -\frac{\pi}{4}$ or $\varphi = \frac{3\pi}{4}$. Since $C_1 = 1 > 0$, we take $\phi = \frac{3\pi}{4}$, giving

$$s(t) = \sqrt{2} \sin\left(t + \frac{3\pi}{4}\right)$$

as our solution. The graph of $s(t)$ is in Figure 8.39.

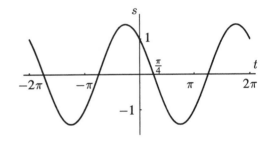

Figure 8.39: Graph of the function $s(t) = \sqrt{2} \sin(t + \frac{3\pi}{4})$

☐

Warning: Phase Shift is not the same as Horizontal Translation

Let's look at $s = \sin(3t + \frac{\pi}{2})$. If we rewrite this as $s = \sin(3(t + \frac{\pi}{6}))$, you can see from Figure 8.40 that the graph of this function is the graph of $s = \sin 3t$ shifted to the left a distance of $\frac{\pi}{6}$. (Remember that replacing x by $x - 2$ shifts a graph by 2 to the right.) But $\frac{\pi}{6}$ is *not* the phase shift; the phase shift[5] is $\frac{\pi}{2}$. From the point of view of a scientist, the important question is often not the distance the curve has shifted, but the relation between the two curves. When one is at its maximum, what is the other doing? When one is at its minimum, where is the other?

---

[5]This definition of phase shift is the one usually used in the sciences.

In this case, $s = \sin(3t + \frac{\pi}{2})$ is at its maximum when $s = \sin 3t$ is zero. In fact, $s = \sin 3t$ has its maximum a quarter of a cycle later than $s = \sin(3t + \frac{\pi}{2})$. Thus we say that $s = \sin(3t + \frac{\pi}{2})$ is a quarter of a cycle ahead of $s = \sin 3t$. (A cycle is one complete oscillation.) Since for $s = \sin t$, with period 2π, a quarter of a cycle is $\frac{2\pi}{4} = \frac{\pi}{2}$, we say that $s = \sin(3t + \frac{\pi}{2})$ is $\frac{\pi}{2}$ ahead of $s = \sin 3t$. In other words, the *phase difference* between $s = \sin(3t + \frac{\pi}{2})$ and $s = \sin 3t$ is $\frac{\pi}{2}$.

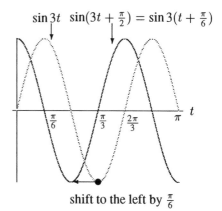

shift to the left by $\frac{\pi}{6}$

Figure 8.40: Graph showing $\sin\left(3t + \dfrac{\pi}{2}\right)$ is $\frac{\pi}{2}$ ahead of $\sin 3t$

❖ Exercises for Section 8.9

1. Check by differentiation that $y = 2\cos t + 3\sin t$ is a solution to $y'' + y = 0$.

2. Check that $y = A\cos t + B\sin t$ is a solution to $y'' + y = 0$ for any constants A and B.

3. What values of α and A make $y = A\cos \alpha t$ a solution to $y'' + 5y = 0$ such that $y'(1) = 3$?

The functions in Problems 4–6 describe the motion of a mass on a spring satisfying the differential equation $y'' = -9y$. In each case, describe in words how the motion starts when $t = 0$. For example, is the mass at the highest point, the lowest point, or in the middle? Is it moving up or down or at rest?

4. $y = 2\cos 3t$ 5. $y = -0.5\sin 3t$ 6. $y = -\cos 3t$

7. Consider the equation $y'' + 4y = 0$.

    (a) Find a general solution to this equation.

    (b) For each set of initial conditions below, find a solution which satisfies them:

        i) $y(0) = 5$, $y'(0) = 0$.   ii) $y(0) = 0$, $y'(0) = 10$.   iii) $y(0) = 5$, $y'(0) = 5$.

    (c) Sketch a graph of each of the three solutions you found in (b).

8. Each graph in Figure 8.41 below represents a solution to one of these equations:

$$\text{i) } x'' + x = 0, \qquad \text{ii) } x'' + 4x = 0, \qquad \text{iii) } x'' + 16x = 0.$$

Assuming that the t scales on the four graphs are the same, decide:

(a) For which equation each graph represents a solution.

(b) The equation of each graph.

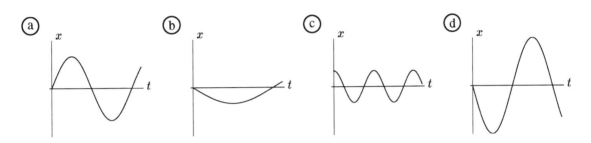

Figure 8.41: Graphs of solutions to $x'' + kx = 0$, for various values of k
(for Problem 8)

9. The following differential equations represent oscillating springs.

$$
\begin{array}{llll}
\text{(i)} & s'' + 4s = 0 & s(0) = 5, & s'(0) = 0 \\
\text{(ii)} & 4s'' + s = 0 & s(0) = 10, & s'(0) = 0 \\
\text{(iii)} & s'' + 6s = 0 & s(0) = 4, & s'(0) = 0 \\
\text{(iv)} & 6s'' + s = 0 & s(0) = 20, & s'(0) = 0
\end{array}
$$

Which differential equation represents:

(a) the spring oscillating most quickly (with the shortest period)?

(b) the spring oscillating with the largest amplitude?

(c) the spring oscillating most slowly (with the longest period)?

(d) the spring with largest maximum velocity?

10. A pendulum of length l makes an angle of x (radians) with the vertical (see Figure 8.42). When x is small, it can be shown that, approximately:

$$\frac{d^2x}{dt^2} = -\frac{g}{l}x,$$

where g is the acceleration due to gravity.

Figure 8.42: The Motion of a Pendulum

(a) Solve this equation assuming that $x(0) = 0$ and $x'(0) = v_0$.

(b) Solve this equation assuming that the pendulum starts with $x = x_0$, and is let go from this position. ("Let go" means that the velocity of the pendulum is zero when $x = x_0$. Measure t from the moment when the pendulum is let go.)

11. Look again at the pendulum motion in Problem 10. What effect does it have on x as a function of time if:

(a) x_0 is increased?

(b) l is increased?

12. Find the amplitude of $3\sin 2t + 7\cos 2t$.

13. Write $7\sin \omega t + 24\cos \omega t$ in the form $A\sin(\omega t + \varphi)$.

14. (a) Expand the $A\sin(\omega t + \varphi)$ using the trigonometric identity $\sin(x + y) = \sin x \cos y + \cos x \sin y$.

(b) Assume $A > 0$. If $A\sin(\omega t + \varphi) = C_1 \cos \omega t + C_2 \sin \omega t$, show that we must have

$$A = \sqrt{C_1^2 + C_2^2} \quad \text{and} \quad \tan \varphi = C_1/C_2.$$

Problems 15–17 show how linear second order differential equations can be used to describe electrical circuits. A charged capacitor connected to an inductor (as shown in Figure 8.43) will cause a current to flow through the inductor until the capacitor is fully discharged. The current in the inductor will, in turn, charge up the capacitor until the capacitor is fully charged once again. Suppose the capacitor has capacitance C and the inductor has inductance L.

If we assume the circuit resistance is zero, then it can be shown that the current I in the circuit satisfies the differential equation

$$L\frac{d^2 I}{dt^2} + \frac{I}{C} = 0.$$

The capacitance, C, is measured in farads. The unit for inductance is the henry, and the unit of current is the ampere. This equation is similar to the differential equations we have seen for springs based on Hooke's law.

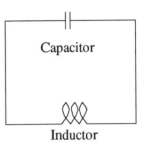

Figure 8.43: Circuit Diagram with Capacitor and Inductor (for Problems 15–17)

15. If $L = 36$ henry and $C = 9$ farad, find a formula for I if

    (a) $I(0) = 0$, $I'(0) = 2$.
    (b) $I(0) = 6$, $I'(0) = 0$.

16. Suppose we wanted to set up our circuit so that $I(0) = 0$, $I'(0) = 4$, and the maximum possible current is $2\sqrt{2}$ amperes. What should the capacitance of our capacitor be if we have an inductor with inductance 10 henry?

17. What happens to the current as t goes to infinity? What does it mean that the current is sometimes positive and sometimes negative?

8.10   Damped Oscillations and Numerical Methods

A Spring with Friction

The differential equation $\frac{d^2 s}{dt^2} = -(k/m)s$, which we used to describe the motion of a spring, disregards friction. But there is friction in every real system. For a mass on a spring, the frictional force, or drag, from air resistance increases with the velocity of the mass. We usually assume that the drag force is proportional to velocity, and so introduce a *damping term* of the form $c\frac{ds}{dt}$, where c is a constant and $\frac{ds}{dt}$ is the velocity of the mass. Remember that without damping, the differential equation was obtained from

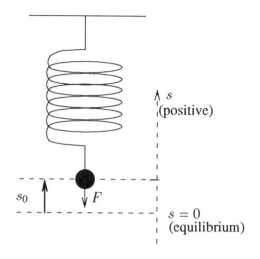

$$\text{Force} = \text{Mass} \times \text{Acceleration},$$

or

$$F = m\frac{d^2s}{dt^2}.$$

With damping, the force $-ks$ is replaced by $-ks - c\frac{ds}{dt}$, where c is positive and the $c\frac{ds}{dt}$ term is subtracted to have the effect of slowing down the motion. The new differential equation is therefore:

$$-ks - c\frac{ds}{dt} = m\frac{d^2s}{dt^2}$$

or, equivalently,

Equation for Damped Oscillations of Spring

$$\frac{d^2s}{dt^2} + \frac{c}{m}\frac{ds}{dt} + \frac{k}{m}s = 0.$$

Figure 8.44: Mass Released from Above Equilibrium

What are the Solutions to the Equation for Damped Motion?

Let's first try to get an intuitive sense of what the solution to the equation

$$\frac{d^2s}{dt^2} + \frac{c}{m}\frac{ds}{dt} + \frac{k}{m}s = 0$$

will look like.

Without damping, a mass on a spring oscillates, and the graph of its displacement against time is a sine curve. With some (but not too much) damping, the mass will still oscillate, but the size of the oscillations will decrease with time. The graph of the displacement is still a sine curve, but the amplitude decays over time. With a great deal of damping, the oscillation may be damped out entirely.

An Example

Let us look at the case where $\frac{k}{m} = 4.04$ and $\frac{c}{m} = 0.4$:

$$\frac{d^2s}{dt^2} + 0.4\frac{ds}{dt} + 4.04s = 0.$$

Suppose the mass passes through the point $s = 0.8415$ meters at time $t = 0$ sec with velocity $\frac{ds}{dt} = 0.9123$ m/sec. We can solve the above equation numerically using an adaptation of Euler's method described on Page 603, to approximate the solution of a second order differential equation and get the oscillation in Figure 8.45.

If you measure the distance between the points where the curve
crosses the t-axis, you will see
that they occur approximately every $3.14 \approx \pi$ units, and so the
position s looks like a sine function with period approximately π
and a decaying amplitude. To get
a period of π, the solution must
involve $\sin 2t$ or $\cos 2t$.

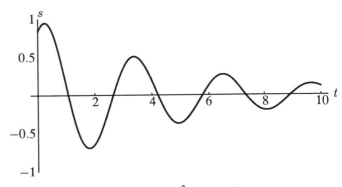

Figure 8.45: Solution to $\dfrac{d^2s}{dt^2} + 0.4\dfrac{ds}{dt} + 4.04s = 0$

Let's draw the function that
touches the peaks of the curve as
in Figure 8.46. The dotted curve
represents the decay of the oscillations; if we take a sine function with the appropriate period,
amplitude, and phase and multiply it by this decay function, we would get the function for the
position s.

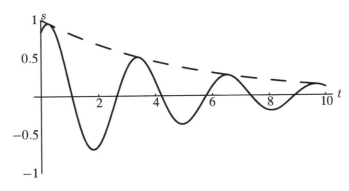

Figure 8.46: Solution with Decay Curve Superimposed

What kind of function is the decay function? The approximate values of s at the peaks (calculated using Euler's
method) are given in Table 8.5. Notice that the peaks are
equally spaced and that the ratio of the values is approximately constant: $0.5039/0.9445 \approx 0.53, 0.2688/0.5039 \approx$
0.53, etc. Hence, the decay curve seems to be exponential, and the solution to the original differential equation is
probably the product of a sine or cosine and a decaying
exponential. Thus, this solution appears to have the form

t	s
0.29	0.9445
3.43	0.5039
6.57	0.2688
9.71	0.1434

Table 8.5: Values of s at
Successive Peaks

$$s(t) = (0.53)^{t/\pi}(A\sin(2t + \varphi)).$$

Such a solution can also be written in the form

$$s(t) = e^{-0.63t}(C_1 \cos 2t + C_2 \sin 2t).$$

As you will see in the next section, this is the form if the damping is not too large; the form of the solution will be different if the damping is large.

Numerical Solutions to Second Order Differential Equations

Many second order differential equations cannot be solved analytically. Therefore, it is useful to be able to solve such equations numerically, using an extension of Euler's method. This is done by first writing the second order equation as a pair of first order equations. To illustrate the process, let's apply the technique to a differential equation that we have already solved:

$$\frac{d^2 s}{dt^2} = -s,$$

with initial conditions that at $t = 0$, $s = 0$ and $\frac{ds}{dt} = 1$. Using the methods of Section 8.9 we can show that the exact solution to this equation is

$$s = \sin t, \quad \text{so} \quad v = \frac{ds}{dt} = \cos t.$$

Now we'll solve the same equation numerically, using the phase plane.

Rewrite the second order equation as a pair of first order equations: We let $v = \frac{ds}{dt}$ be a new variable. Then our original equation $\frac{d^2 s}{dt^2} = -s$ can be rewritten as $\frac{dv}{dt} = -s$, so the system is:

$$\frac{ds}{dt} = v$$
$$\frac{dv}{dt} = -s.$$

As in Section 8.8, we have

$$\boxed{\frac{dv}{ds} = \frac{dv/dt}{ds/dt} = -\frac{s}{v}}$$

The slope field for the system of differential equations is the slope field for $\frac{dv}{ds} = -\frac{s}{v}$, shown in Figure 8.47. If we plot the exact solution, $s = \sin t$, $v = \cos t$, in the phase plane, the trajectory will be a unit circle, because

$$s^2 + v^2 = \sin^2 t + \cos^2 t = 1.$$

Euler's Method for a System of Differential Equations

Suppose that we want to approximate the solution using Euler's method with 8 steps and $\Delta t = 0.5$. For each step, we calculate $\dfrac{ds}{dt}$ and $\dfrac{dv}{dt}$ at a particular point in the phase plane and use these slopes to move to a new point in the phase plane, as illustrated in Figure 8.48.

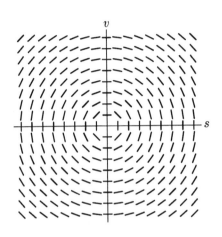

Figure 8.47: Slope Field for
$\frac{ds}{dt} = v, \frac{dv}{dt} = -s$

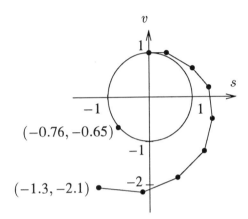

Figure 8.48: Euler's Method Trajectory
Compared with True Trajectory

Each step we take is an approximation, in the sense that it assumes that the slope field doesn't change over the step in the phase plane. As you can see in Figure 8.48 above, the Euler's method trajectory ends up deviating further and further from the true circular trajectory. The true solution to the differential equation is $s = \sin t$, and so at $t = 4$, $s = \sin 4 \approx -0.76$ and $v = \cos 4 \approx -0.65$. Euler's method with 8 steps gives $s = -1.3$ and $v = -2.1$. One way of improving the accuracy, of course, is to take smaller steps.

☐ **Example 1** Given the differential equation

$$\frac{d^2 y}{dt^2} + 2\frac{dy}{dt} + 2y = 0$$

with initial conditions that at $t = 0$, $y = 2$ and $y' = 0$, estimate the value of y and y' at $t = 1$ by using Euler's method with $n = 10$ steps ($\Delta t = 0.1$).

Solution. To rewrite the differential equation as a system of first order differential equations, let

$$\frac{dy}{dt} = v.$$

Then $\dfrac{d^2y}{dt^2} = -2\dfrac{dy}{dt} - 2y$, so

$$\frac{dv}{dt} = -2v - 2y.$$

We write $\dfrac{dv}{dy} = \dfrac{dv/dt}{dy/dt} = \dfrac{-2v - 2y}{v}$ and is the slope field for the system.

The slope field for this equation is in Figure 8.49 and is the slope field for the system. Starting at $y_0 = 2$ and $v_0 = 0$, we calculate

$$\begin{aligned}\frac{dy}{dt} &= v_0 = 0,\\\frac{dv}{dt} &= -2v_0 - 2y_0 = -4.\end{aligned}$$

Therefore, the values of y_1 and v_1 are given by

$$\begin{aligned}y_1 &= y_0 + \frac{dy}{dt}\Delta t = 2 + 0(0.1) = 2,\\v_1 &= v_0 + \frac{dv}{dt}\Delta t = 0 - 4(0.1) = -0.4.\end{aligned}$$

We now calculate the derivatives at $y_1 = 2$, $v_1 = -0.4$ for the next step:

$$\begin{aligned}\frac{dy}{dt} &= v_1 = -0.4,\\\frac{dv}{dt} &= -2v_1 - 2y_1 = 0.8 - 4 = -3.2.\end{aligned}$$

Therefore,

$$\begin{aligned}y_2 &= y_1 + \frac{dy}{dt}\Delta t = 2 - 0.4(0.1) = 1.96,\\v_2 &= v_1 + \frac{dv}{dt}\Delta t = -0.4 - 3.2(0.1) = -0.72.\end{aligned}$$

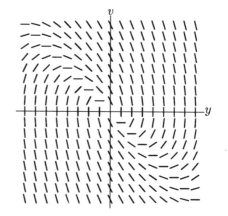

Figure 8.49: Slope Field for $dy/dt = v$, $dv/dt = -2v - 2y$

i	t_i	y_i	v_i	dy/dt	dv/dt
0	0.0	2.00	0.00	0.00	-4.00
1	0.1	2.00	-0.40	-0.40	-3.20
2	0.2	1.96	-0.72	-0.72	-2.48
3	0.3	1.89	-0.97	-0.97	-1.84
4	0.4	1.79	-1.15	-1.15	-1.28
5	0.5	1.68	-1.28	-1.28	-0.79
6	0.6	1.55	-1.36	-1.36	-0.38
7	0.7	1.41	-1.40	-1.40	-0.03
8	0.8	1.27	-1.40	-1.40	0.25
9	0.9	1.13	-1.37	-1.37	0.48
10	1.0	0.99	-1.33	-1.33	0.66

Table 8.6: Values from Euler's Method

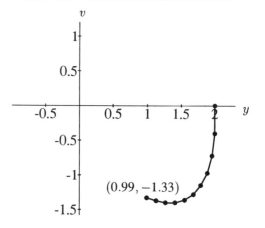

Figure 8.50: Euler Trajectory

Table 8.6 lists the values of y, v, dy/dt, and dv/dt at each step of the process, and Figure 8.50

shows the Euler trajectory. Hence, at $t = 1$, $y \approx 0.99$, and $v \approx -1.33$.

We will find the exact solution to this differential equation later, but for now let us borrow the results to check the accuracy of the numerical solution. At $t = 1$, the true value is

$$y = \frac{2}{e}(\sin 1 + \cos 1) = 1.016\ldots \qquad \text{and} \qquad v = \frac{dy}{dt}\bigg|_{t=1} = -\frac{4}{e}\sin 1 = -1.238\ldots$$

With only 10 steps, Euler's method gives solutions correct to within 3% in y and 7% in v.   ❏

❖ Exercises for Section 8.10

1. Show the first two steps of the Euler's method calculation for $\dfrac{ds}{dt} = v$, $\dfrac{dv}{dt} = -s$, starting at $s = 0$, $v = 1$, and using $\Delta t = 0.5$.

2. Continue the calculation of Problem 1 to show that after 8 steps you reach $s = -1.3$, $v = -2.1$.

3. Write a spreadsheet, calculator or computer program to generate the data in Table 8.6, Page 604.

4. (a) Write the differential equation $y'' = y$ as a system of two first order differential equations.

   (b) Starting at the point $t = 0$, $y = 1$, $y' = 1$, use Euler's method with $\Delta t = 0.25$ to approximate $y(1)$.

   (c) Check that $y = e^t$ is the exact solution to this differential equation, and satisfies the initial conditions. Compare the true value $y(1)$ to your approximation.

8.11   Solving Second Order Differential Equations

The Principle of Superposition

Suppose, as in the case of a damped spring, we are given a differential equation of the form

$$\frac{d^2 y}{dt^2} + b\frac{dy}{dt} + cy = 0,$$

where b and c are constants. As we have seen for the spring equation, if $f_1(t)$ and $f_2(t)$ satisfy this differential equation, then for any constants C_1 and C_2, the *principle of superposition* says that the function

$$y(t) = C_1 f_1(t) + C_2 f_2(t)$$

is a solution as well. It can be shown that provided $f_1(t)$ is not a multiple of $f_2(t)$, the general solution is of this form.

The Characteristic Equation

We will now look at a powerful method of solving the differential equation

$$\frac{d^2y}{dt^2} + b\frac{dy}{dt} + cy = 0.$$

We are going to try to find a solution of the form:

$$y = Ce^{rt},$$

where r may be a complex number.[6]

In order to find r, we substitute into the equation:

$$\frac{d^2y}{dt^2} + b\frac{dy}{dt} + cy = r^2Ce^{rt} + b \cdot rCe^{rt} + c \cdot Ce^{rt} = Ce^{rt}(r^2 + br + c) = 0.$$

We can divide this equation through by $y = Ce^{rt}$ so long as $C \neq 0$, because the exponential function is never zero. If $C = 0$, then $y = 0$, which is not a very interesting solution (though it is a solution). So we assume $C \neq 0$. Then $y = Ce^{rt}$ will be a solution to the differential equation if

$$r^2 + br + c = 0.$$

This quadratic equation is called the *characteristic equation* of the differential equation, and its solutions are

$$r = -\frac{1}{2}b \pm \frac{1}{2}\sqrt{b^2 - 4c}.$$

There are three different types of solution to the differential equation, depending on whether the solutions to the characteristic equation are real or complex, which is determined by the sign of $b^2 - 4c$.

The Overdamped Case: $b^2 - 4c > 0$

There are two real solutions r_1 and r_2 to the characteristic equation, and so the following two functions satisfy the differential equation:

$$C_1e^{r_1t} \quad \text{and} \quad C_2e^{r_2t}.$$

The sum of these two solutions will be the general solution to the differential equation:

The Overdamped Case: If $b^2 - 4c > 0$, the general solution to:

$$\frac{d^2y}{dt^2} + b\frac{dy}{dt} + cy = 0$$

is

$$y(t) = C_1e^{r_1t} + C_2e^{r_2t}$$

where r_1 and r_2 are the solutions to the characteristic equation.

This type of system is called *overdamped* because it occurs when there is a lot of friction in the system; for example, a spring moving in a thick fluid such as oil or molasses.

---

[6]See Section 8.12, the appendix on complex numbers.

☐ **Example 1** Suppose a spring is placed in oil, where it satisfies the differential equation

$$\frac{d^2s}{dt^2} + 3\frac{ds}{dt} + 2s = 0.$$

Solve this equation with the initial conditions $s = -0.5$ and $\dfrac{ds}{dt} = 3$ when $t = 0$.

Solution. The characteristic equation is

$$r^2 + 3r + 2 = 0,$$

with solutions $r = -1$ and $r = -2$, and so the general form of the solution to the differential equation is

$$s(t) = C_1 e^{-t} + C_2 e^{-2t}.$$

We use the initial conditions to find C_1 and C_2. At $t = 0$, we have

$$s = C_1 e^{-0} + C_2 e^{-2(0)} = C_1 + C_2 = -0.5.$$

Furthermore, since $\dfrac{ds}{dt} = -C_1 e^{-t} - 2C_2 e^{-2t}$, we have

$$\left.\frac{ds}{dt}\right|_{t=0} = -C_1 e^{-0} - 2C_2 e^{-2(0)} = -C_1 - 2C_2 = 3.$$

Solving these equations simultaneously, we find $C_1 = 2$ and $C_2 = -2.5$, so that the solution is

$$s(t) = 2e^{-t} - 2.5e^{-2t}.$$

The graph of this function is in Figure 8.51. The behavior is what you might expect: the mass passes through the equilibrium point only once when $t \approx \frac{1}{4}$, and it is so slowed by the oil that for all practical purposes, it comes to rest after a short time. The motion has been "damped out" by the oil.

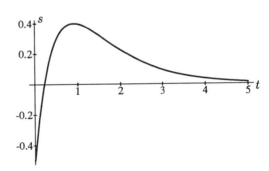

Figure 8.51: Graph of the function $s(t) = 2e^{-t} - 2.5e^{-2t}$

☐

The Critically Damped Case: $b^2 - 4c = 0$

This is the case in which the characteristic equation has only one solution, $r = -b/2$. By substitution, you can check that

---

The Critically Damped Case:

$$\frac{d^2y}{dt^2} + b\frac{dy}{dt} + \frac{b^2}{4}y = 0$$

has solution

$$y(t) = (C_1 t + C_2)e^{-\frac{b}{2}t}$$

The Underdamped Case: $b^2 - 4c < 0$

☐ **Example 2** An object of mass $m = 10$ kg is attached to a spring with spring constant $k = 20$ kg/sec^2, and the object experiences a drag force with a coefficient of $c = 20$ kg/sec (in other words, the drag force is $c \cdot \frac{ds}{dt}$). At time $t = 0$, the object is released from rest 2 meters above the equilibrium position. Write the differential equation that describes the motion.

Solution. The differential equation that describes the motion can be obtained from the following general expression for the damped motion of a spring:

$$\underbrace{m\frac{d^2s}{dt^2}}_{\text{Net Force}} = -\underbrace{ks}_{\text{Spring}} - \underbrace{c\frac{ds}{dt}}_{\text{Drag}}.$$

Substituting $m = 10$ kg, $k = 20$ kg/sec^2, and $c = 20$ kg/sec, we obtain the following differential equation for the motion:

$$\frac{d^2s}{dt^2} + 2\frac{ds}{dt} + 2s = 0.$$

At $t = 0$, the object is at rest 2 meters above equilibrium, so $s(0) = 2$ and $s'(0) = 0$. ☐

Notice that this is the same differential equation as in Example 1, except that the coefficient of drag has decreased from 3 to 2. This time, the roots of the characteristic equation have imaginary parts which lead to oscillations.

☐ **Example 3** Solve the differential equation

$$\frac{d^2s}{dt^2} + 2\frac{ds}{dt} + 2s = 0,$$

subject to $s(0) = 2$, $s'(0) = 0$.

Solution. The characteristic equation is

$$r^2 + 2r + 2 = 0 \quad \text{giving} \quad r = -1 \pm i.$$

The solution to the differential equation is

$$s(t) = A_1 e^{(-1+i)t} + A_2 e^{(-1-i)t},$$

where A_1 and A_2 are arbitrary complex numbers. The initial condition $s(0) = 2$ gives

$$\begin{aligned} 2 &= A_1 e^{(-1+i)\cdot 0} + A_2 e^{(-1-i)\cdot 0} \\ &= A_1 + A_2. \end{aligned}$$

Also,

$$s'(t) = A_1(-1+i)e^{(-1+i)t} + A_2(-1-i)e^{(-1-i)t},$$

so $s'(0) = 0$ gives

$$0 = A_1(-1+i) + A_2(-1-i).$$

Solving the simultaneous equations for A_1 and A_2 gives (after some algebra):

$$A_1 = 1 - i \quad \text{and} \quad A_2 = 1 + i.$$

The solution is therefore

$$s(t) = (1-i)e^{(-1+i)t} + (1+i)e^{(-1-i)t} = (1-i)e^{-t}e^{it} + (1+i)e^{-t}e^{-it}.$$

Using Euler's formula, $e^{it} = \cos t + i\sin t$, and $e^{-it} = \cos t - i\sin t$, we get:

$$s(t) = (1-i)e^{-t}(\cos t + i\sin t) + (1+i)e^{-t}(\cos t - i\sin t).$$

Multiplying out and simplifying, all the complex terms drop out, giving

$$\begin{aligned} s(t) &= e^{-t}\cos t + ie^{-t}\sin t - ie^{-t}\cos t + e^{-t}\sin t \\ &\quad + e^{-t}\cos t - ie^{-t}\sin t + ie^{-t}\cos t + e^{-t}\sin t \\ &= 2e^{-t}\cos t + 2e^{-t}\sin t. \end{aligned}$$

The $\cos t$ and $\sin t$ terms tell us that the solution oscillates. However the oscillations are damped extremely quickly by the factor e^{-t}. See Figure 8.52. Alternatively, you can see the behavior of the solution by rewriting it in the form

$$s(t) = 2e^{-t}(\cos t + \sin t) = 2\sqrt{2}e^{-t}\sin\left(t + \frac{\pi}{4}\right).$$

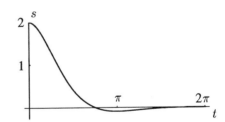

Figure 8.52: Solution $2e^{-t}\cos t + 2e^{-t}\sin t$

□

In Example 3, notice that although the coefficients A_1 and A_2 are complex, the solution, $s(t)$, is real. (You might expect this, since $s(t)$ represents a real displacement.) In general, provided the coefficients b and c in the original differential equation and the initial values are real, the solution will always be real too.

The underdamped case: If $b^2 - 4c < 0$, to solve

$$\frac{d^2y}{dt^2} + b\frac{dy}{dt} + cy = 0,$$

- find the solutions $r = \alpha \pm i\beta$ to the characteristic equation $r^2 + br + c = 0$.
- The general solution to the differential equation is, for some real C_1 and C_2,

$$y = C_1 e^{\alpha t} \cos \beta t + C_2 e^{\alpha t} \sin \beta t.$$

☐ **Example 4** Find the general solution of the equations
(a) $y'' = 9y$ (b) $y'' = -9y$.

Solution.

(a) The characteristic equation is $r^2 - 9 = 0$, so $r = \pm 3$. Thus the general solution is

$$y = C_1 e^{3t} + C_2 e^{-3t}.$$

(b) The characteristic equation is $r^2 + 9 = 0$, so $r = \pm 3i$. Thus $y = e^{3it}$ and $y = e^{-3it}$ are solutions, and the general solution is

$$y = C_1 \cos 3t + C_2 \sin 3t.$$

Notice that we have seen the solution to this equation in Section 8.9; it's the equation of undamped simple harmonic motion.

☐

❖ Exercises for Section 8.11

For Problems 1–8, find the general solution to the given differential equation.

1. $y'' + 4y' + 3y = 0$
2. $y'' + 4y' + 4y = 0$
3. $y'' + 4y' + 5y = 0$

4. $s'' - 7s = 0$
5. $s'' + 7s = 0$
6. $4z'' + 8z' + 3z = 0$

7. $\dfrac{d^2x}{dt^2} + 4\dfrac{dx}{dt} + 8x = 0$
8. $\dfrac{d^2p}{dt^2} + \dfrac{dp}{dt} + p = 0$

For Problems 9–12, solve the initial value problem.

9. $y'' + 6y' + 5y = 0$, $y(0) = 1$, $y'(0) = 0$

10. $y'' + 6y' + 5y = 0$, $y(0) = 5$, $y'(0) = 5$

11. $y'' + 6y' + 10y = 0$, $y(0) = 0$, $y'(0) = 2$

12. $y'' + 6y' + 10y = 0$, $y(0) = 0$, $y'(0) = 0$

For Problems 13–14, solve the boundary value problem.

13. $p'' + 2p' + 2p = 0$, $p(0) = 0$, $p(\pi/2) = 20$

14. $p'' + 4p' + 5p = 0$, $p(0) = 1$, $p(\pi/2) = 5$

15. Match the graphs of solutions in Figure 8.53 below with the differential equations.

   i) $x'' + 4x = 0$ ii) $x'' - 4x = 0$ iii) $x'' - 0.2x' + 1.01x = 0$ iv) $x'' + 0.2x' + 1.01x = 0$.

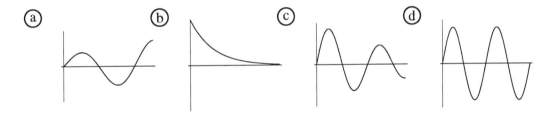

Figure 8.53: Graphs for Problem 15

Each of the differential equations below represents the oscillations of a 1 gram mass on the end of a damped spring. For Problems 16–20 below, pick the differential equation representing the system which answers the question.

   (i) $s'' + s' + 4s = 0$ (iii) $s'' + 3s' + 3s = 0$
   (ii) $s'' + 2s' + 5s = 0$ (iv) $s'' + 0.5s' + 2s = 0$

16. Which spring has the largest coefficient of damping?

17. Which spring exerts the smallest restoring force for a given displacement?

18. In which system does the mass experience the smallest frictional force for a given velocity?

19. Which oscillation has the longest period?

20. Which spring is the stiffest?

For each of the differential equations in Problems 21–23, find the values of c that make the general solution (i) overdamped, (ii) underdamped, and (iii) critically damped.

21. $s'' + 4s' + cs = 0$.

22. $s'' + 2\sqrt{2}s' + cs = 0$.

23. $s'' + 6s' + cs = 0$.

24. If $y = e^{2t}$ is a solution to the differential equation

$$\frac{d^2y}{dt^2} - 5\frac{dy}{dt} + ky = 0,$$

find the value of the constant k and the general solution to this equation.

25. Find a solution to the equation

$$\frac{d^2z}{dt^2} + \frac{dz}{dt} - 2z = 0,$$

which satisfies $z(0) = 3$ and does not tend to infinity as $t \to \infty$.

Recall the discussion of circuits in Section 8.9 on page 599. Just as springs often have a damping force which affects their motion, we can also introduce a damping force into a circuit. Problems 26–29 illustrate this situation. The damping force is caused by a resistor, which has resistance R, measured in ohms. The equation for the current I through a circuit with inductance L, capacitance C, and resistance R is given by the differential equation

$$L\frac{d^2I}{dt^2} + R\frac{dI}{dt} + \frac{1}{C}I = 0.$$

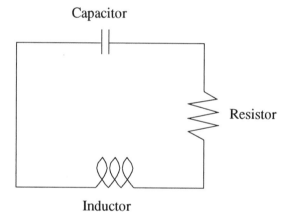

Figure 8.54: Circuit Diagram with Inductor, Capacitor, and Resistor (for Problems 26–29)

26. If $L = 1$ henry, $R = 2$ ohms, and $C = 4$ farads, find a formula for the current when

   (a) $I(0) = 0$, $I'(0) = 2$.
   (b) $I(0) = 2$, $I'(0) = 0$.

27. If $L = 1$ henry, $R = 1$ ohm, and $C = 4$ farads, find a formula for the current when

   (a) $I(0) = 0$, $I'(0) = 2$.
   (b) $I(0) = 2$, $I'(0) = 0$.
   (c) How did reducing the resistance affect the current? Compare with your solution to Problem 26.

28. If $L = 8$ henry, $R = 2$ ohm, and $C = 4$ farads, find a formula for the current when

    (a) $I(0) = 0$, $I'(0) = 2$.

    (b) $I(0) = 2$, $I'(0) = 0$.

    (c) How did increasing the inductance affect the current? Compare with your solution to Problem 26.

29. Given any positive values for R, L and C, what happens to the current as t goes to infinity?

8.12   Appendix: Complex Numbers

The quadratic equation

$$x^2 - 2x + 2 = 0$$

is not satisfied by any real number x. If you try applying the quadratic formula, you get

$$x = \frac{2 \pm \sqrt{4 - 8}}{2} = 1 \pm \frac{\sqrt{-4}}{2}.$$

Apparently, you need to take a square root of -4. But -4 doesn't have a square root, at least, not one which is a real number. Let's give it a square root.

We define the imaginary number i to be a number such that

$$i^2 = -1.$$

Using this i, we see that $(2i)^2 = -4$, so

$$x = 1 \pm \frac{\sqrt{-4}}{2} = 1 \pm \frac{2i}{2} = 1 \pm i.$$

This solves our quadratic equation. The numbers $1 + i$ and $1 - i$ are examples of complex numbers.

> A *complex number* is defined as any number that can be written in the form
>
> $$z = a + bi,$$
>
> where a and b are real numbers.

Calling the number i "imaginary" makes it sound like i doesn't really exist in the same way that real numbers exist. In some cases, it is useful to make such a distinction between real and imaginary numbers. For example, if we measure mass or position, we want our answers to be real numbers. But the imaginary numbers are just as legitimate mathematically as the real numbers are.

As an analogy, consider the distinction between positive and negative numbers. Originally, people thought of numbers only as tools to count with; their concept of "five" or "ten" was not far removed from "five arrows" or "ten stones". They were unaware that negative numbers existed at all. When negative numbers were introduced, they were viewed only as a device for solving equations like $x + 2 = 1$. They were considered "non-numbers", or, in Latin, "negative numbers". Thus, even though people started to use negative numbers, they did not view them as existing in the same way that positive numbers did. An early mathematician might have reasoned: "The number 5 exists because I can have 5 coins in my hand. But how can I have -5 coins in my hand?" Today we have an answer: "I have -5 coins" means I owe somebody 5 coins. We have realized that negative numbers are just as real as positive ones, and that in some cases, negative numbers can have physical meaning, even though they are useless for measuring length or keeping baseball scores. As we will see, complex numbers can have physical meaning as well. For example, complex numbers are used in studying wave motion in electric circuits.

Algebra of Complex Numbers

Numbers such as 0, 1, $\frac{1}{2}$, π, and $\sqrt{2}$ are called *purely real*, because they contain no imaginary components. Numbers such as i, $2i$, and $\sqrt{2}i$ are called *purely imaginary*, because they contain only the number i multiplied by a non-zero real coefficient.

Two complex numbers are called *conjugates* if their real terms are equal and if their imaginary terms are opposites. The complex conjugate of the complex number $z = a + bi$ is denoted $\overline{z}$, so

$$\overline{z} = a - bi.$$

(Note that z is real if and only if $z = \overline{z}$.) Complex conjugates have the following remarkable property: if $f(x)$ is any polynomial with real coefficients ($x^3 + 1$, say) and $f(z) = 0$, then $f(\overline{z}) = 0$. This means that if z is the solution to a polynomial equation with real coefficients, then so is $\overline{z}$.

- Adding two complex numbers is straightforward: $(a + bi) + (c + di) = (a + c) + (b + d)i$.

- Subtracting is similar: $(a + bi) - (c + di) = (a - c) + (b - d)i$.

- Multiplication works just like for polynomials, using $i^2 = -1$:

$$\begin{aligned} (a + bi)(c + di) &= a(c + di) + bi(c + di) = ac + adi + bci + bdi^2 \\ &= ac + adi + bci - bd = (ac - bd) + (ad + bc)i. \end{aligned}$$

- Powers of i: We know that $i^2 = -1$; then, $i^3 = i \cdot i^2 = -i$, and $i^4 = (i^2)^2 = (-1)^2 = 1$. Then $i^5 = i \cdot i^4 = i$, and so on. Thus we have

$$(bi)^n = b^n i^n = \begin{cases} b^n i & n = 1, 5, 9, 13, \ldots \\ -b^n & n = 2, 6, 10, 14, \ldots \\ -b^n i & n = 3, 7, 11, 15, \ldots \\ b^n & n = 4, 8, 12, 16, \ldots \end{cases}$$

- The product of a number and its conjugate is always real and non-negative:

$$z\overline{z} = (a + bi)(a - bi) = a^2 - abi + abi - b^2 i^2 = a^2 + b^2.$$

- Dividing is done by multiplying the denominator by its conjugate, thereby making it real:

$$\frac{a+bi}{c+di} = \frac{a+bi}{c+di} \cdot \frac{c-di}{c-di} = \frac{ac-adi+bci-bdi^2}{c^2+d^2} = \frac{ac+bd}{c^2+d^2} + \frac{bc-ad}{c^2+d^2} i.$$

☐ **Example 1** Compute $(2+7i)(4-6i) - i$.

Solution.

$$(2+7i)(4-6i) - i = 8 + 28i - 12i - 42i^2 - i = 8 + 15i + 42 = 50 + 15i.$$

❏

☐ **Example 2** Compute $\dfrac{2+7i}{4-6i}$.

Solution.

$$\frac{2+7i}{4-6i} = \frac{2+7i}{4-6i} \cdot \frac{4+6i}{4+6i} = \frac{8+12i+28i+42i^2}{4^2+6^2} = \frac{-34+40i}{52} = \frac{-17}{26} + \frac{10}{13}i.$$

❏

The Complex Plane and Polar Coordinates

It is often useful to picture a complex number $z = x + iy$ in the plane, with x along the horizontal axis and y along the vertical. Figure 8.55 shows the complex numbers $-2i$ and $1 + i$, $-2 + 3i$ plotted in the complex plane.

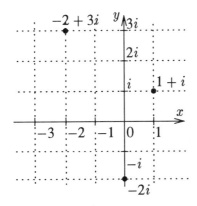

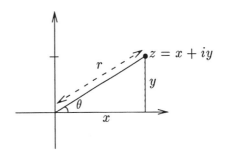

Figure 8.55: Points $-2i$, $1 + i$, $-2 + 3i$ in the Complex Plane

Figure 8.56: The point $z = x + iy$ in the complex plane, showing polar coordinates

A point z can be referred to by its *polar coordinates, r and θ.* The number r is the distance between z and the origin, and θ is the angle between the positive x-axis and the line joining z to the origin (with the convention that counterclockwise is positive). From the triangle in Figure 8.56, we have the following

---

Relation between Cartesian and Polar Coordinates

$$x = r\cos\theta \qquad r = \sqrt{x^2 + y^2}$$
$$y = r\sin\theta \qquad \tan\theta = \frac{y}{x}$$

Using polar coordinates, a complex number can be written

$$\boxed{z = x + iy = r\cos\theta + ir\sin\theta.}$$

☐ **Example 3** Express $z = -2i$ and $z = 3 + 4i$ in polar coordinates.

Solution. For $z = -2i$, the distance of z from the origin, r, is 2. (See Figure 8.55.) Also, one value for θ is $\theta = 3\pi/2$. Thus $-2i = 2\cos(3\pi/2) + i\,2(\sin 3\pi/2)$.
   For $z = 3 + 4i$, we have $x = 3$, $y = 4$, so $r = \sqrt{3^2 + 4^2} = 5$ and $\tan\theta = \frac{4}{3}$, giving, for example, $\theta \approx 0.92$. Thus $3 + 4i \approx 5\cos(0.92) + i\,5\sin(0.92)$.   ☐

☐ **Example 4** Consider a point with polar coordinates $r = 5$ and $\theta = \frac{3\pi}{4}$. What complex number does this point represent?

Solution. We see that $x = 5\cos\frac{3\pi}{4} = -\frac{5}{\sqrt{2}}$, and $y = 5\sin\frac{3\pi}{4} = \frac{5}{\sqrt{2}}$, so $z = -\frac{5}{\sqrt{2}} + i\frac{5}{\sqrt{2}}$.   ☐

Euler's Formula

Consider the point z lying on the unit circle in Figure 8.57. Writing z in polar coordinates, and using the fact that $r = 1$, we have

$$z = f(\theta) = \cos\theta + i\sin\theta.$$

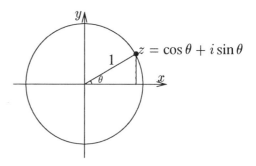

It turns out that there is a particularly beautiful and compact way of rewriting this function using complex exponentials. Let's take the derivative of f, treating i like any other constant but using the fact that $i^2 = -1$:

Figure 8.57: Point on Unit Circle

$$f'(\theta) = -\sin\theta + i\cos\theta = i\cos\theta + i^2\sin\theta.$$

Factoring out an i gives

$$f'(\theta) = i(\cos\theta + i\sin\theta) = i \cdot f(\theta).$$

As you know, the only function whose derivative is proportional to the function itself is the exponential function. In other words, if

$$g'(x) = k \cdot g(x), \quad \text{then} \quad g(x) = Ce^{kx}$$

for some constant C. In our case, we have

$$f'(\theta) = if(\theta) \quad \text{so} \quad f(\theta) = Ce^{i\theta}.$$

Now $f(0) = Ce^{i \cdot 0} = C$, and since $f(0) = \cos 0 + i \sin 0 = 1$, we must have $C = 1$ and $f(\theta) = e^{i\theta}$. Thus we have the remarkable relationship

$$\boxed{e^{i\theta} = \cos \theta + i \sin \theta.}$$

This elegant and surprising relationship was discovered by the Swiss mathematician Leonhard Euler in the eighteenth century, and it is particularly useful in solving second order differential equations. Another derivation of Euler's formula (using Taylor series) is given in Chapter 9 on page 641. For now, it allows us to write the complex number represented by the point with polar coordinates (r, θ) in the following form:

$$\boxed{z = r(\cos \theta + i \sin \theta) = re^{i\theta}.}$$

❑ **Example 5** Write the point $r = 8$, $\theta = 3\pi/4$ in Cartesian and polar form.

Solution. In Cartesian coordinates, the complex number is

$$z = 8(\cos(3\pi/4) + i \sin(3\pi/4)) = \frac{-8}{\sqrt{2}} + i\frac{8}{\sqrt{2}}.$$

Since $e^{i\theta} = \cos \theta + i \sin \theta$, using polar coordinates we have

$$z = 8e^{i\frac{3\pi}{4}}.$$

❑

Among its many benefits, the polar form of complex numbers makes finding powers and roots of complex numbers much easier. Using the polar form, $z = re^{i\theta}$, for a complex number, we may find any power of z as follows:

$$z^p = (re^{i\theta})^p = r^p e^{ip\theta}.$$

❑ **Example 6** Find a cube root of the complex number whose polar coordinates are $(8, \frac{3\pi}{4})$.

Solution. In Example 5, we saw that this complex number could be written as $z = 8e^{i\frac{3\pi}{4}}$. Thus,

$$z^{\frac{1}{3}} = \left(8e^{i\frac{3\pi}{4}}\right)^{\frac{1}{3}} = 8^{\frac{1}{3}}e^{i\frac{3\pi}{4} \cdot \frac{1}{3}} = 2e^{i\frac{\pi}{4}} = 2\left(\cos\frac{\pi}{4} + i\sin\frac{\pi}{4}\right) = 2\left(\frac{1}{\sqrt{2}} + \frac{1}{\sqrt{2}}i\right) = \sqrt{2}(1 + i).$$

❑

❖ Exercises for Section 8.12

For Problems 1–8, give the Cartesian coordinates for the points with the following polar coordinates (r, θ). The angles are measured in radians.

1. $(1, 0)$ 3. $(2, \pi)$ 5. $(5, -\frac{\pi}{6})$ 7. $(-3, \frac{\pi}{2})$

2. $(0, 1)$ 4. $(\sqrt{2}, \frac{5\pi}{4})$ 6. $(-5, -\frac{\pi}{6})$ 8. $(1, 1)$

For Problems 9–16, express the given complex number in polar form, $z = re^{i\theta}$.

9. $2i$ 11. $1 + i$ 13. 0 15. $-1 + 3i$

10. -5 12. $-3 - 4i$ 14. $-i$ 16. $5 - 12i$

For Problems 17–26, perform the indicated calculations. Give your answer in Cartesian form, $z = x + iy$.

17. $(2 + 3i) + (-5 - 7i)$ 22. $(2i)^3 - (2i)^2 + 2i - 1$

18. $(2 + 3i)(5 + 7i)$ 23. $(e^{i\pi/3})^2$

19. $(2 + 3i)^2$ 24. $\sqrt{e^{i\pi/3}}$

20. $(1 + i)^2 + (1 + i)$ 25. $(5e^{i7\pi/6})^3$

21. $(0.5 - i)(1 - \frac{i}{4})$ 26. $\sqrt[4]{10e^{i\frac{\pi}{2}}}$

By writing the complex numbers in polar form, $z = re^{i\theta}$, find a value for the quantities in Problems 27–36. Give your answer in Cartesian form, $z = x + iy$.

27. $\sqrt{i}$ 30. $\sqrt{7i}$ 33. $(-4 + 4i)^{2/3}$ 35. $(\sqrt{3} + i)^{-1/2}$

28. $\sqrt{-i}$ 31. $(1 + i)^{100}$ 34. $(\sqrt{3} + i)^{1/2}$ 36. $(\sqrt{5} + 2i)^{\sqrt{2}}$

29. $\sqrt[3]{i}$ 32. $(1 + i)^{2/3}$

37. Let $z_1 = -3 - i\sqrt{3}$ and $z_2 = -1 + i\sqrt{3}$.

   (a) Find $z_1 z_2$ and $\dfrac{z_1}{z_2}$.

   (b) Put z_1 and z_2 into polar form. Find $z_1 z_2$ and $\dfrac{z_1}{z_2}$ using the polar form, and verify that you get the same answer as in (a).

38. If the roots of the equation $x^2 + 2bx + c = 0$ are the complex numbers $p \pm iq$, find expressions for p and q in terms of a and b.

39. Every point in the complex plane can be represented by some pair of polar coordinates, but are the polar coordinates (r, θ) uniquely determined by the Cartesian coordinates (x, y)? In other words, for each pair of Cartesian coordinates, is there one and only one pair of polar coordinates for that point? Why or why not?

Are the statements in Problems 40–45 true or false? Explain your answer.

40. Any non-negative real number has a real square root.

41. For any complex number z, the product $z \cdot \bar{z}$ is a real number.

42. The square of any complex number is a real number.

43. If f is a polynomial, and $f(z) = i$, then $f(\bar{z}) = i$.

44. Any nonzero complex number z can be written in the form $z = e^w$, where w is another complex number.

45. If $z = x + yi$, where x and y are positive, then $z^2 = a + bi$ has a and b positive.

For Problems 46–50, use Euler's Formula to derive the following relationships. (Note that if a and b are real numbers, $a + bi = c + di$ means that $a = c$ and $b = d$.)

46. $\sin^2 \theta + \cos^2 \theta = 1$

47. $\sin 2\theta = 2 \sin \theta \cos \theta$

48. $\cos 2\theta = \cos^2 \theta - \sin^2 \theta$

49. $\dfrac{d}{d\theta} \sin \theta = \cos \theta$

50. $\dfrac{d^2}{d\theta^2} \cos \theta = -\cos \theta$

8.13   Miscellaneous Exercises for Chapter 8

For the differential equations in Problems 1–11, find the equation of a solution which passes through the given point.

1. $\dfrac{dP}{dt} = 0.03P + 400$, $P(0) = 0$

2. $\dfrac{dy}{dt} = y(10 - y)$, $y(0) = 1$

3. $\dfrac{df}{dx} = \sqrt{x f(x)}$, $f(1) = 1$

4. $\dfrac{dy}{dx} = \dfrac{y(3 - x)}{x(0.5y - 4)}$, $y(1) = 5$

5. $\dfrac{dy}{dx} = e^{x-y}$, $y(0) = 1$

6. $\dfrac{dk}{dt} = (1 + \ln t)k$, $k(1) = 1$

7. $2 \sin x - y^2 \dfrac{dy}{dx} = 0$, $y(0) = 3$

8. $1 + y^2 - \dfrac{dy}{dx} = 0$, $y(0) = 0$

9. $\dfrac{dy}{dx} + xy^2 = 0$, $y(1) = 1$

10. $\dfrac{dy}{dx} = \dfrac{y(100 - x)}{x(20 - y)}$, $(1, 20)$

11. $\dfrac{dy}{dx} = \dfrac{0.2y(18 + 0.1x)}{x(100 + 0.5y)}$, $(10, 10)$

For Problems 12–13, sketch solution curves with a variety of initial values for the differential equations. You do not need to find an equation for the solution.

12. $\dfrac{dy}{dt} = \alpha - y,$

where α is a positive constant.

13. $\dfrac{dw}{dt} = (w - 3)(w - 7)$

14. Consider the initial value problem

$$y' = 5 - y, \qquad y(0) = 1.$$

(a) Use Euler's method with five steps to estimate $y(1)$.

(b) Sketch the slope field for this differential equation in the first quadrant, and use it to decide if your estimate is an over- or under-estimate.

(c) Find the exact solution to the differential equation and hence find $y(1)$ exactly.

(d) Without doing the calculation, roughly what would you expect the approximation for $y(1)$ to be with ten steps?

15. Solve the differential equation

$$\frac{dy}{dx} = \frac{1}{(\cos x)(\cos y)}$$

by Euler's method, for the solution passing through $(0,0)$. Find y when $x = \frac{1}{2}$.

(a) Find the solution from Euler's method when $\Delta x = \frac{1}{2}$ (1 step),

(b) $\Delta x = \frac{1}{4}$ (2 steps),

(c) $\Delta x = \frac{1}{8}$ (4 steps).

(d) Solve the differential equation by separation of variables for y passing through $(0,0)$ and find $y(\frac{1}{2})$. Compare with your results in (a)–(c).

Find a general solution to the differential equations in Problems 16–19.

16. $y'' + 6y' + 8y = 0$

17. $9z'' + z = 0$

18. $9z'' - z = 0$

19. $x'' + 2x' + 10x = 0$

For each of the differential equations in Problems 20–21, find the values of b that make the general solution (i) overdamped, (ii) underdamped, and (iii) critically damped.

20. $s'' + bs' + 5s = 0$.   21. $s'' + bs' - 16s = 0$.

22. Explain why $s'' + bs' - s = 0$ cannot have a critically damped solution for any real number b.

23. Suppose that the rate at which a drug leaves the bloodstream and passes into the urine is proportional to the quantity of the drug in the blood at that time. If an initial dose of Q_0 is injected directly into the blood, only 20% is left in the blood after 3 hours.

    (a) Write and solve a differential equation for the quantity, Q, of the drug in the blood at time, t, measured in hours.

    (b) How much of this drug is in a patient's body after 6 hours if he is given 100 mg initially?

24. If a body at temperature T is placed in surroundings of a different temperature, then it will either gain or lose heat so as to acquire the temperature of the surroundings according to the differential equation $\frac{dT}{dt} = -k(T - A)$, where A is the the temperature of the surroundings. Here k is a positive constant that depends on the physical nature of the body.

    (a) Why is k positive?

    (b) Suppose the units in which time is measured are changed, say from days to hours, or from hours to minutes. Then k will change; how?

    (c) Suppose the body in question is a cup of hot coffee. Will k be larger or smaller if the cup itself is styrofoam or thin china ?

    (d) Suppose the coffee comes out of the coffee machine at 170 degrees, the room temperature is 70, and the constant k is 0.14 when time is measured in minutes. How long must you wait to drink the coffee if you can't stand it above 120 degrees? How soon must you drink it, if you hate coffee below 90 degrees?

25. A detective finds a murder victim at 9 a.m. The temperature of the body is measured at 90.3°F. One hour later, the temperature of the body is 89.0°F. The temperature of the room has been maintained at a constant 68°F.

    (a) Assuming the temperature, T, of the body obeys Newton's Law of Cooling, write a differential equation for T.

    (b) Solve the differential equation to estimate the time the murder occurred.

26. A bank account earns 10% annual interest, compounded continuously. Money is deposited in a continuous cash flow at a rate of $1200 per year into the account.

    (a) Write a differential equation that describes the rate at which the balance $B = f(t)$ is changing.

    (b) Solve the differential equation given an initial balance $B_0 = 0$.

    (c) Find the balance after 5 years.

27. (Continuation of Problem 26.) Now suppose the money is deposited once a month (instead of continuously) but still at a rate of $1200 per year.

    (a) Write down the sum that gives the balance after 5 years, assuming the first deposit is made one month from today, and today is $t = 0$.

    (b) The sum you wrote in part (a) is a Riemann sum approximation to the integral

    $$\int_0^5 1200e^{0.1(5-t)}dt.$$

    Determine whether it is a left sum or right sum, and determine what Δt and N are. Then use your calculator to evaluate the sum.

    (c) Compare your answer in part (b) to your answer to Problem 26(c).

28. A species of bird that only eats the rare poha berry is introduced to the Hawaiian Islands. As a result, the maximum number of birds that can be supported is 10,000. Hence, we choose to model the rate of growth of the bird's population, $\frac{dP}{dt}$, as proportional to the number of birds currently on the Islands, P, times the remaining capacity for birds, or $10,000-P$. At time $t = 0$ years, there are 10 birds on the Islands, and their population is increasing at the rate $\frac{dP}{dt} = 100$ birds per year. Assume nothing disastrous happens to the birds.

    (a) What differential equation describes the rate of growth of the bird population? Solve for the constant of proportionality from the given information.

    (b) What is the maximum rate of growth of the bird population, and at what population does it occur?

    (c) Sketch P, the solution to the differential equation, as a function of t.

    (d) Find a formula for P as a function of t.

29. The following equations describe the rates of growth of an insect and bird population in a particular region, where x is the insect population in millions at time t and y is the bird population in thousands:

    $$\frac{dx}{dt} = 3x - 0.02xy$$
    $$\frac{dy}{dt} = -10y + 0.001xy$$

    (a) Describe in words the growth of each population in the absence of the other, and describe in words their interaction.

    (b) Find the two points (x, y) at which the populations are in equilibrium.

    (c) Suppose that when the populations are at the non-trivial equilibrium, 10 thousand additional birds are suddenly introduced. Let A be the point in the phase plane representing these populations. Get a differential equation in terms of just x and y (i.e., eliminate t), and find an equation for the particular solution passing through the point A.

(d) Show that the following points lie on the trajectory in the phase plane that passes through point A:

(i) B (9646.91, 150)   (ii) C (10,000, 140.43)   (iii) D (10,361.60, 150)

(e) Sketch this trajectory in the phase plane, with x on the horizontal axis, y on the vertical. Show the equilibrium point relative to this trajectory.

(f) In what order will the points A, B, C, D be traversed? [Hint: Find $\frac{dy}{dt}$, $\frac{dx}{dt}$ at each point.]

(g) On another graph, sketch x and y versus time, t. Use the same initial value as in part (c). You do not need to indicate actual numerical values on the t-axis.

(h) Find $\frac{dy}{dx}$ at points A and C. Find $\frac{dx}{dy}$ at points B and D. [Hint: $\frac{dx}{dy} = \frac{1}{dy/dx}$. Notice that the points A and C are the maximum and minimum values of y, respectively, and B and D are the maximum and minimum values of x, respectively.]

30. Romeo is in love with Juliet, who happens (in our version of this story) to be a fickle lover. The more Romeo loves her, the more she begins to dislike him. When he hates her, her feelings for him warm up. On the other hand, his love for her grows when she loves him, and withers when she hates him. A model for their ill-fated romance is

$$\frac{dr}{dt} = Aj, \qquad \frac{dj}{dt} = -Br,$$

where A, B are positive constants, and $r(t)$ represents Romeo's love for Juliet at time t, and $j(t)$ represents Juliet's love for Romeo at time t. (Negative love is hate.)

(a) The constant on the right hand side of Juliet's equation (the one including dj/dt) has a negative sign, whereas the constant in Romeo's equation is positive. Explain why these signs follow from the story.

(b) Derive a second order differential equation for $r(t)$ and solve it. (Your equation should involve r and its derivatives, not j and its derivatives.)

(c) Express $r(t)$ and $j(t)$ as functions of t, given $r(0) = 1$ and $j(0) = 0$. Your answer will contain A and B.

(d) As you may have discovered, the outcome of the relationship is a never-ending cycle of love and hate. Find what fraction of the time they both love one another.

Chapter 9

APPROXIMATIONS

In this chapter we will see how to approximate functions by simpler functions. One way to think about this is through an analogy with numbers. Numbers can be described by exact expressions, like $\frac{1}{2}$, $\sqrt{2}$, or π, or they can be given in decimal form, which may not be exact. For example,

$$\frac{1}{2} = 0.5$$
$$\sqrt{2} = 1.4142\ldots$$
$$\pi = 3.14159\ldots.$$

The first equation is exact, but the other two are not. In the last two cases the dots indicate that there are more digits in the decimal expansion, and that the decimal on the right side of the equation is only an approximation.

In calculus, we manipulate functions instead of numbers. We learn how to add, subtract, multiply, divide, differentiate, and integrate functions. Just as rational numbers are the simplest sorts of numbers, so rational functions, and in particular polynomials, are the simplest sorts of functions. In a Taylor approximation, we construct something like a decimal expansion that enables us to approximate functions by polynomials. Fourier approximations use trigonometric functions instead of polynomials.

9.1 Taylor Polynomials

In this section, we approximate a function by polynomials. To get more accuracy, we take a higher degree polynomial. The approximation is usually good near one particular point, but often not so good further away.

Linear Approximations

We have already seen how to approximate a function using a tangent line. The tangent line at a point is the best linear approximation to the function at that point in the sense that it is the line which lies closest to the curve there. Figure 9.1 shows the graph of a function and its tangent line. Notice that the tangent line and the curve have the same slope at the point of tangency. This means they are both climbing up or down at the same rate there. As in Chapters 2 and 4, we have the equation of the

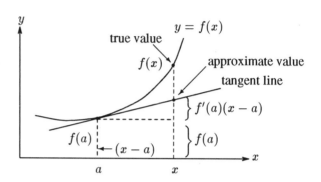

Figure 9.1: Tangent Line Approximation

Tangent Line Approximation near $x = a$:

$$y = f(a) + f'(a)(x - a)$$

☐ **Example 1** Approximate $f(x) = \cos x$, with x in radians, by its tangent line at $x = 0$.

Solution. The tangent line is just the horizontal line $y = 1$, as shown in Figure 9.2. If we take $x = 0.05$, then

$$f(x) = \cos(0.05) = 0.999,$$

which is quite close to the approximation $y = 1$. Similarly, if $x = -0.1$, then

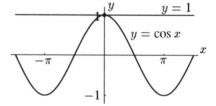

Figure 9.2: Graph of $\cos x$ and its Tangent Line at $x = 0$

$$f(x) = \cos(-0.1) = 0.995 \approx 1.$$
However, if $x = 0.4$, then
$$f(x) = \cos(0.4) = 0.921,$$

so the approximation $y = 1$ is less accurate. The further a point x is away from the original point, 0, the worse the corresponding approximation is likely to be. ☐

Notice that when $x = 0$, the approximation becomes the

Tangent Line Approximation near $x = o$:

$$y = f(0) + f'(0)x.$$

Why do this?

At this point, you are undoubtedly wondering why we should be interested in approximating the values for $\cos x$ at all. After all, you can obtain the correct values simply by punching the right keys on a calculator. On the other hand, have you ever wondered how a calculator actually generates the values for the cosine function? What would happen if you were to face some new function whose values cannot be found directly from a calculator? Some of the most effective methods for calculating such values involve approximating one function by another, simpler function, such as by a tangent line approximation.

Quadratic Approximations

Suppose we want a more accurate way of approximating $f(x) = \cos x$ near $x = 0$. The problem with using a linear approximation, as we did above, is that the curve usually bends away from the tangent line. Consequently, it makes sense to approximate the function not by a line, but by a curve which bends in the same way as the original curve. We will require that at $x = 0$ the original curve and the approximating curve have the same slope, $f'(0)$, and that they bend at the same rate, in other words that they have the same second derivative, $f''(0)$. The simplest type of function we might use for this approximation is a quadratic,

$$P(x) = C_0 + C_1 x + C_2 x^2,$$

where we must determine the values for C_0, C_1, and C_2. At $x = 0$, we want the two curves to agree, so $P(0) = f(0)$; we want the two curves to have the same slope at this point, so $P'(0) = f'(0)$; and we also want the two curves to have the same second derivative, which gives $P''(0) = f''(0)$. These three conditions will determine C_0, C_1, and C_2.

☐ **Example 2** Find the quadratic approximation to $f(x) = \cos x$ near $x = 0$.

Solution. We find C_0, C_1, C_2 using $P(0) = f(0)$, $P'(0) = f'(0)$, $P''(0) = f''(0)$.
Since

$$\begin{aligned} P(x) &= C_0 + C_1 x + C_2 x^2 & & \text{and} & f(x) &= \cos x \\ P'(x) &= C_1 + 2C_2 x & & & f'(x) &= -\sin x \\ P''(x) &= 2C_2 & & & f''(x) &= -\cos x, \end{aligned}$$

we have

$$\begin{aligned} C_0 &= P(0) = f(0) = \cos 0 = 1 & \text{so} & & C_0 &= 1 \\ C_1 &= P'(0) = f'(0) = -\sin 0 = 0 & & & C_1 &= 0 \\ 2C_2 &= P''(0) = f''(0) = -\cos 0 = -1 & & & C_2 &= -\tfrac{1}{2} \end{aligned}$$

Consequently the approximating quadratic is

$$P(x) = 1 + 0 \cdot x - \frac{1}{2}x^2 = 1 - \frac{x^2}{2}.$$

From Figure 9.3, you can see that the quadratic approximation is better than the linear approximation near $x = 0$. Let's compare the accuracy of the two approximations numerically. The linear approximation is $y = 1$. At $x = 0.4$, $\cos(0.4) = 0.921$ and $P(0.4) = 0.920$, so the quadratic approximation is a significant improvement over the linear approximation. (The error is about 0.001 instead of 0.08.) In addition, the quadratic approximation is much better than the linear approximation for small x values: $\cos(0.1) = 0.995004$ and $P(0.1) = 0.995$.

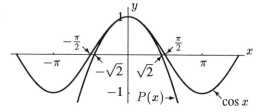

Figure 9.3: Graph of $\cos x$ and Quadratic Approximation near $x = 0$

In general, suppose we have a function $f(x)$ which is defined, along with its first two derivatives, in an interval about $x = 0$. We can construct a quadratic polynomial approximating $f(x)$ about $x = 0$ by the same steps we used above. Assume that the quadratic polynomial is

$$P(x) = C_0 + C_1 x + C_2 x^2.$$

We choose C_0, C_1, and C_2 so that the original function and the quadratic have the same value and the same first and second derivatives at $x = 0$ and obtain what is called the

> **Taylor polynomial of degree 2 about $x =$ o:**
>
> $$P(x) = f(0) + f'(0)x + \frac{f''(0)}{2}x^2$$

Higher Degree Polynomials

As a rule, over a given interval the quadratic function will be a better approximation than the linear function. However, Figure 9.3 shows that even though we have matched up the function and the quadratic in terms of their values, their slopes, and their concavity at the point $x = 0$, they usually still bend away from one another for large x. We can fix this somewhat by using an approximating polynomial of higher degree. Suppose that we want to approximate a function $f(x)$ near $x = 0$ by a polynomial of degree n:

$$f(x) \approx P_n(x) = C_0 + C_1 x + C_2 x^2 + \cdots + C_{n-1} x^{n-1} + C_n x^n.$$

We now need to find the values of the constants: C_0, C_1, $C_2, \ldots$, C_n. To do this, we require that the function and all of its first n derivatives agree with those of the polynomial at the point $x = 0$. Notice that the higher order derivatives of a function contribute more subtle information about its shape than the first two derivatives do. (For instance, the third derivative measures how fast the concavity changes.) The more derivatives that agree at $x = 0$ there are, the longer the function and the polynomial will remain close to each other.

To see how to find the constants, let's take as an example

$$f(x) \approx P_3(x) = C_0 + C_1 x + C_2 x^2 + C_3 x^3.$$

Substituting $x = 0$ gives:

$$f(0) = P_3(0) = C_0.$$

Differentiating $P_3(x)$ yields

$$P_3'(x) = C_1 + 2C_2 x + 3C_3 x^2,$$

so substituting $x = 0$ shows that

$$f'(0) = P_3'(0) = C_1.$$

Differentiating and substituting again, we get:

$$P_3''(x) = 2 \cdot 1 C_2 + 3 \cdot 2 C_3 x,$$

which gives

$$f''(0) = P_3''(0) = 2C_2,$$

so that

$$C_2 = \frac{f''(0)}{2}.$$

The third derivative, denoted by P_3''', is

$$P_3'''(x) = 3 \cdot 2 \cdot 1 C_3,$$

so that

$$f'''(0) = P_3'''(0) = 3 \cdot 2 \cdot 1 C_3.$$

and so

$$C_3 = \frac{f'''(0)}{3 \cdot 2 \cdot 1}.$$

You can imagine a similar calculation starting with $P_4(x)$, using the fourth derivative $f^{(4)}$, which would give

$$C_4 = \frac{f^{(4)}(0)}{4 \cdot 3 \cdot 2 \cdot 1},$$

and so on. Using factorial notation[1] , we write these expressions as

$$C_3 = \frac{f'''(0)}{3!}, \quad C_4 = \frac{f^{(4)}(0)}{4!}.$$

In general, for any integer k between 1 and n

$$C_k = \frac{f^{(k)}(0)}{k!},$$

---

[1]We define 3! (read *three factorial*) by $3! = 3 \cdot 2 \cdot 1 = 6$, and $4! = 4 \cdot 3 \cdot 2 \cdot 1 = 24$, etc.

where $f^{(k)}$ means the kth derivative of f and $k! = k(k-1)\cdots 2\cdot 1$. We have therefore obtained the

---

Taylor polynomial of degree n about $x = $ o for the function f:

$$f(x) \approx P_n(x)$$
$$= f(0) + f'(0)x + \frac{f''(0)}{2!}x^2 + \frac{f'''(0)}{3!}x^3 + \frac{f^{(4)}(0)}{4!}x^4 + \cdots + \frac{f^{(n)}(0)}{n!}x^n$$

☐ **Example 3** Construct the Taylor polynomial of degree 7 about $x = 0$ for the function $f(x) = \sin x$. Compare the value of the Taylor approximation with the true value of f at $x = \frac{\pi}{3}$.

Solution. We have:

$$
\begin{array}{rclcrcl}
f(x) & = & \sin x & \text{giving} & f(0) & = & 0 \\
f'(x) & = & \cos x & & f'(0) & = & 1 \\
f''(x) & = & -\sin x & & f''(0) & = & 0 \\
f'''(x) & = & -\cos x & & f'''(0) & = & -1 \\
f^{(4)}(x) & = & \sin x & & f^{(4)}(0) & = & 0 \\
f^{(5)}(x) & = & \cos x & & f^{(5)}(0) & = & 1 \\
f^{(6)}(x) & = & -\sin x & & f^{(6)}(0) & = & 0 \\
f^{(7)}(x) & = & -\cos x & & f^{(7)}(0) & = & -1.
\end{array}
$$

Using these values, we see that the Taylor polynomial approximation of degree 7 is

$$\sin x \approx P_7(x) = 0 + x + 0\cdot\frac{x^2}{2!} - \frac{x^3}{3!} + 0\cdot\frac{x^4}{4!} + \frac{x^5}{5!} + 0\cdot\frac{x^6}{6!} - \frac{x^7}{7!}$$
$$= x - \frac{x^3}{3!} + \frac{x^5}{5!} - \frac{x^7}{7!}.$$

In Figure 9.4, we show the graph of the sine curve and the approximating polynomial of degree 7. They are indistinguishable when x is close to 0. However, as we look at values of x further away in either direction, the two curves begin to diverge from one another.

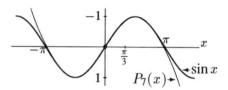

Figure 9.4: Graph of $\sin x$ and its 7th degree Taylor polynomial, $P_7(x)$

To see the accuracy of this approximation numerically, we see how well it approximates $\sin(\frac{\pi}{3}) = \sqrt{3}/2 \approx 0.8660254$. When we substitute $\frac{\pi}{3} \approx 1.0471976$ into the polynomial approximation, we obtain $P_7(\frac{\pi}{3}) \approx 0.8660213$, which is extremely accurate—to about 5 parts in a million. ☐

☐ **Example 4** Show that the Taylor polynomial of degree 8 about $x = 0$ for $f(x) = \cos x$ is:

$$f(x) \approx P_8(x) = 1 - \frac{x^2}{2!} + \frac{x^4}{4!} - \frac{x^6}{6!} + \frac{x^8}{8!}.$$

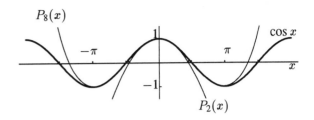

Figure 9.5: Graph of cos x and two Taylor polynomials, $P_2(x)$ and $P_8(x)$.

Solution. We can find the coefficients by the method of the previous example. Figure 9.5 shows that $P_8(x)$ is close to the cosine function for more x values than the quadratic approximation $P_2(x) = 1 - \dfrac{x^2}{2}$ that we found on page 628.

☐

☐ Example 5 Construct the Taylor polynomial of degree 10 about $x = 0$ for the function $f(x) = e^x$.

Solution. We have $f(0) = 1$. Moreover, since the derivative of e^x is equal to e^x, all the higher order derivatives will be equal to e^x. Consequently, for any $k = 1, 2, \ldots, 10$, $f^{(k)}(x) = e^x$ and $f^{(k)}(0) = e^0 = 1$. Therefore, the Taylor polynomial approximation of degree 10 is given by

$$e^x \approx P_{10}(x) = 1 + x + \frac{x^2}{2!} + \frac{x^3}{3!} + \frac{x^4}{4!} + \cdots + \frac{x^{10}}{10!}.$$

☐

To verify the accuracy of this approximation, suppose we use it to approximate $e = e^1 = 2.718281828$, correct to nine decimal places. If we substitute $x = 1$ into the approximating polynomial and evaluate it, we obtain $P_{10}(1) = 2.718281801$. Thus, this tenth degree polynomial actually yields the first seven decimal places for e. For large values of x, however, the accuracy diminishes. In Figure 9.6, we show the graph of $f(x) = e^x$ as well as the Taylor polynomials of degree $n = 0, 1, 2, 3, 4$. Notice that each successive approximation remains close to the exponential curve for a larger interval of x-values.

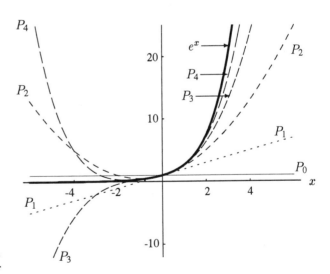

Figure 9.6: Graph of e^x and Taylor Polynomials

☐ Example 6 Construct the Taylor polynomial of degree 5 about $x = 0$ for the function $f(x) = \ln x$.

Solution. As soon as we notice that $f(0)$ is not defined, we realize there is no such polynomial. ☐

Moving Away from $x = 0$

The graph of $f(x) = \ln x$ in Figure 9.7 shows that it is not surprising that the polynomial for $\ln x$ about $x = 0$ does not exist. After all, the function does not exist for $x = 0$ or for $x < 0$. Can we find another Taylor polynomial for $f(x) = \ln x$? Although there is no tangent line at $x = 0$, the graph certainly does have tangent lines for $x > 0$.

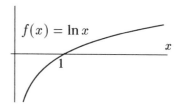

Figure 9.7: Graph of $f(x) = \ln x$

Thus, rather than constructing a polynomial approximation about $x = 0$, let's construct a polynomial centered about some other point, $x = a$. Before constructing an approximating polynomial at $x = a$, let's look at the equation of the tangent line there. Recall that since the tangent line must go through the point $(a, f(a))$ and has slope $f'(a)$, its equation is

$$y = f(a) + f'(a)(x - a).$$

Notice that if $x = a$, the second term drops out so $y = f(a)$. The $f'(a)(x - a)$ term is a correction term which gives the change in y as you move away from $x = a$.

We want the approximating polynomial to give $y = f(a)$ when $x = a$. The approximating polynomial will be set up as $f(a)$ plus correction terms which depend on the derivative, and are zero when $x = a$. This is achieved by writing the polynomial in powers of $(x - a)$ instead of powers of x:

$$f(x) \approx P_n(x) = C_0 + C_1(x - a) + C_2(x - a)^2 + \cdots + C_n(x - a)^n.$$

If we require the derivatives of the approximating polynomial and the original function to agree at $x = a$, we get the

---

Taylor polynomial of degree n about $x = a$ for the function f:

$$f(x) \approx P_n(x)$$
$$= f(a) + f'(a)(x - a) + \frac{f''(a)}{2!}(x - a)^2 + \cdots\cdots + \frac{f^{(n)}(a)}{n!}(x - a)^n$$

You can derive these coefficients in the same way as for $a = 0$. (See Problem 35, page 635.)

☐ **Example 7**   Construct the Taylor polynomial of degree 4 about $x = 1$ for the function $f(x) = \ln x$.

Solution. We have

$$
\begin{array}{lllllll}
f(x) & = & \ln x & \text{so} & f(1) & = & \ln(1) & = & 0 \\
f'(x) & = & 1/x & & f'(1) & = & 1 \\
f''(x) & = & -1/x^2 & & f''(1) & = & -1 \\
f'''(x) & = & 2/x^3 & & f'''(1) & = & 2 \\
f^{(4)}(x) & = & -6/x^4 & & f^{(4)}(1) & = & -6
\end{array}
$$

so that the Taylor polynomial is

$$\ln x \approx P_4(x) \;\; = \;\; 0 + (x-1) - \frac{(x-1)^2}{2!} + 2\frac{(x-1)^3}{3!} - 6\frac{(x-1)^4}{4!}$$
$$= \;\; (x-1) - \frac{(x-1)^2}{2} + \frac{(x-1)^3}{3} - \frac{(x-1)^4}{4}.$$

We show $\ln x$ and several Taylor polynomials in Figure 9.8. You will notice that $P_4(x)$ stays reasonably close to the function near $x = 1$, and then bends away.

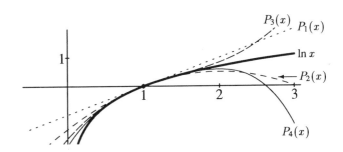

Figure 9.8: Graph of $\ln x$ and Taylor Polynomials

The moral of the story is that you can be sure that the Taylor polynomials are a good approximation only when you are "close to home" — near $x = a$. Further away, they may be good and they may not.

❖ Exercises for Section 9.1

For Problems 1–10, find the Taylor polynomials of degree n about $x = 0$.

1. $\cos x$, $\quad n = 2, 4, 6$

2. $\sqrt{1+x}$, $\quad n = 2, 3, 4$

3. $\dfrac{1}{1-x}$, $\quad n = 3, 5, 7$.

4. $\dfrac{1}{1+x}$, $\quad n = 4, 6, 8$

5. $\tan x$, $\quad n = 3, 4, 5$

6. $\arctan x$, $\quad n = 3, 4, 5$

7. $\ln(1+x)$, $\quad n = 5, 7, 9$

8. $\sqrt[3]{1-x}$, $\quad n = 2, 3, 4$

9. $\dfrac{1}{\sqrt{1+x}}$, $\quad n = 2, 3, 4$

10. $(1+x)^\alpha$, $\quad n = 2, 3, 4$ $\qquad$ (α is a constant.)

11. Show how the results of Problems 2, 3, 4, 8, and 9 are special cases of the result of Problem 10.

For Problems 12–22, find the Taylor polynomial of degree n about the given point $x = a$.

12. e^x, $a = 1$, $n = 4$

13. $\cos x$, $a = \frac{\pi}{2}$, $n = 4$

14. $\sin x$, $a = \frac{\pi}{2}$, $n = 4$

15. $\cos x$, $a = \frac{\pi}{4}$, $n = 3$

16. $\sin x$, $a = -\frac{\pi}{4}$, $n = 3$

17. $\sqrt{1 - x}$, $a = 0$, $n = 3$

18. $\sqrt{1 + x}$, $a = 1$, $n = 3$

19. $\dfrac{1}{1 + x}$, $a = 2$, $n = 4$

20. $\dfrac{1}{1 - x}$, $a = 2$, $n = 4$

21. $\ln x$, $a = 1$, $n = 6$.
[Hint: Compare with Example 7, page 632.]

22. $\ln x$, $a = 2$, $n = 4$

For Problems 23–26, suppose $P_2(x) = a + bx + cx^2$ is the second degree Taylor polynomial for the function f about $x = 0$. What can you say about the signs of a, b, c if f has the graph given below?

23. 24. 25. 26.

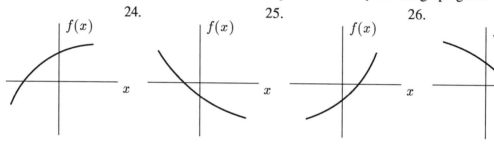

27. Show how you can use the Taylor approximation $\sin x \approx x - \dfrac{x^3}{3!}$ to explain why $\displaystyle\lim_{x \to 0} \dfrac{\sin x}{x} = 1$.

28. Use the fourth degree Taylor approximation $\cos x \approx 1 - \dfrac{x^2}{2!} + \dfrac{x^4}{4!}$ to evaluate $\displaystyle\lim_{x \to 0} \dfrac{1 - \cos x}{x^2}$.

29. Use a fourth degree Taylor approximation for e^h to evaluate the following limits. Would your answer be different if you used a Taylor polynomial of higher degree?

    (a) $\displaystyle\lim_{h \to 0} \dfrac{e^h - 1 - h}{h^2}$

    (b) $\displaystyle\lim_{h \to 0} \dfrac{e^h - 1 - h - \frac{h^2}{2}}{h^3}$

30. When we model the motion of a pendulum, we replace the differential equation

$$\frac{d^2\theta}{dt^2} = -\frac{g}{l}\sin\theta \quad \text{by} \quad \frac{d^2\theta}{dt^2} = -\frac{g}{l}\theta$$

where θ is the angle between the pendulum and the vertical. Explain why, and under what circumstances, it is reasonable to make this replacement.

31. Find the second degree Taylor polynomial for $f(x) = 4x^2 - 7x + 2$ at $x = 0$. What do you notice?

32. Find the third degree Taylor polynomial for $f(x) = x^3 + 7x^2 - 5x + 1$ at $x = 0$. What do you notice?

33. (a) Based on your observations in Problems 31–32, make a conjecture about Taylor approximations in the case when f is itself a polynomial.

    (b) Show that your conjecture is true.

34. (a) Find the Taylor polynomial approximation of degree 4 about $x = 0$ for the function $f(x) = e^{x^2}$.

    (b) Compare this result to the Taylor polynomial approximation for the function $f(x) = e^x$ about $x = 0$. What do you notice?

    (c) Use your observation in (b) to write out the Taylor polynomial approximation of degree 20 to the function in (a).

    (d) What is the Taylor polynomial approximation of degree 5 for the function $f(x) = e^{-2x}$?

35. Derive the formulas given in the box on page 632 for the coefficients of the Taylor polynomial of a function f about the point $x = a$.

9.2   Taylor Series

We have just seen how to approximate a function by Taylor polynomials. Now we will define a Taylor series, which can be thought of as a polynomial which goes on forever, and look at where such a series serves as a good approximation to the function.

In the previous section, we calculated Taylor polynomials such as those given below for $\cos x$:

$$\cos x \approx P_0(x) = 1$$

$$\cos x \approx P_2(x) = 1 - \frac{x^2}{2!}$$

$$\cos x \approx P_4(x) = 1 - \frac{x^2}{2!} + \frac{x^4}{4!}$$

$$\cos x \approx P_6(x) = 1 - \frac{x^2}{2!} + \frac{x^4}{4!} - \frac{x^6}{6!}$$

$$\cos x \approx P_8(x) = 1 - \frac{x^2}{2!} + \frac{x^4}{4!} - \frac{x^6}{6!} + \frac{x^8}{8!}$$

Here, we have a sequence of polynomials, $P_0(x)$, $P_2(x)$, $P_4(x)$, $P_6(x)$, $P_8(x)$, ..., each of which is a better approximation to $f(x) = \cos x$ than the last, at least near $x = 0$. Notice that when you go to a higher degree polynomial (say from P_6 to P_8), you add more terms ($\frac{x^8}{8!}$, for example), but the terms of lower degree don't change. Thus each polynomial includes all the previous ones. We will write the *Taylor series* for $\cos x$:

$$T(x) = 1 - \frac{x^2}{2!} + \frac{x^4}{4!} - \frac{x^6}{6!} + \frac{x^8}{8!} - \cdots$$

to represent the whole sequence of polynomials.

What Does ⋯ Mean?

The three dots indicate that the terms in the series go on forever. This is the same idea as in $\pi = 3.14\ldots$ where the three dots mean that the digits go on forever after the 4.

$$1 - \frac{x^2}{2!} + \frac{x^4}{4!}\qquad \text{is a finite polynomial of degree 4,}$$

whereas

$$1 - \frac{x^2}{2!} + \frac{x^4}{4!} - \cdots \qquad \text{is a series, and continues forever.}$$

Convergence of Taylor Series for $\cos x$

We have already seen that the Taylor polynomials are good approximations to $\cos x$ near $x = 0$. (See Figure 9.9.) Amazingly enough, for any value of x, if you take a Taylor polynomial of high enough degree, its graph is indistinguishable from the graph of the cosine near that point.

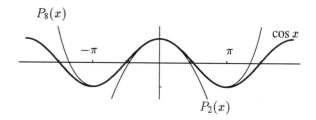

Figure 9.9: Graph of $\cos x$ and two Taylor polynomials, $P_2(x)$ and $P_8(x)$.

Let's see what happens numerically. Suppose we take $x = \pi/2$. The successive Taylor approximations to $\cos(\pi/2) = 0$ are:

$$\begin{aligned}
P_2(\pi/2) &= & 1 - (\pi/2)^2/2! & = & -0.23370\\
P_4(\pi/2) &= & 1 - (\pi/2)^2/2! + (\pi/2)^4/4! & = & 0.01997\\
P_6(\pi/2) &= & \cdots & = & -0.00089\\
P_8(\pi/2) &= & \cdots & = & 0.00002
\end{aligned}$$

and we see that the approximations converge to the correct value very rapidly. If we take a value of x somewhat further away from $x = 0$, say $x = \pi$, then $\cos\pi = -1$ and

$$\begin{aligned}
P_2(\pi) &= & 1 - (\pi)^2/2! & = & -3.93480\\
P_4(\pi) &= & \cdots & = & 0.12391\\
P_6(\pi) &= & \cdots & = & -1.21135\\
P_8(\pi) &= & \cdots & = & -0.97602
\end{aligned}$$

and we see that the rate of convergence is somewhat slower. If x were taken still further away from $x = 0$, then we need still more terms to obtain an accurate approximation. However, no matter how large x is, either positive or negative, the approximations will eventually converge to $\cos x$.

Thus, we feel justified in replacing the '$\approx$' by an '$=$', and writing

$$\cos x = T(x) = 1 - \frac{x^2}{2!} + \frac{x^4}{4!} - \frac{x^6}{6!} + \frac{x^8}{8!} - \cdots \qquad \text{for all } x.$$

We say that the Taylor series *converges* to cos x for all x, because for all x, the Taylor polynomials

$$P_1(x),\ P_2(x),\ P_3(x),\ldots,\ P_n(x),\ldots$$

converge to cos x, as $n \to \infty$.

Taylor Series for $\sin x$, $\cos x$, e^x

Similarly, we can define the Taylor series for $\sin x$ and e^x. They, too, converge for all x. Thus, for all x, we can write

$$\sin x = x - \frac{x^3}{3!} + \frac{x^5}{5!} - \frac{x^7}{7!} + \frac{x^9}{9!} - \cdots$$

$$\cos x = 1 - \frac{x^2}{2!} + \frac{x^4}{4!} - \frac{x^6}{6!} + \frac{x^8}{8!} - \cdots$$

$$e^x = 1 + x + \frac{x^2}{2!} + \frac{x^3}{3!} + \frac{x^4}{4!} + \cdots$$

These series are also called *Taylor expansions*.

Taylor Series in General

The series for $\sin x$, $\cos x$, and e^x are all examples of Taylor series. Any function f, all of whose derivatives exist at 0, has a Taylor series:

Taylor Series about $x = 0$:

$$f(x) = f(0) + f'(0)x + \frac{f''(0)}{2!}x^2 + \frac{f'''(0)}{3!}x^3 + \cdots$$

The Taylor series about $x = 0$ will in general not converge for all values of x but only for values at or near 0. In the next examples we will look at functions whose Taylor series do not converge for all x. In addition, just as we have Taylor polynomials centered at points other than 0, so (provided all the derivatives exist at $x = a$) we can have a Taylor series centered at $x = a$:

Taylor Series about $x = a$:

$$f(x) = f(a) + f'(a)(x - a) + \frac{f''(a)}{2!}(x - a)^2 + \frac{f'''(a)}{3!}(x - a)^3 + \cdots$$

Again, this Taylor series may converge only at or near $x = a$.

Intervals of Convergence

Look again at the Taylor polynomials for $\ln x$ about $x = 1$ that we derived on page 632.

Graphical Viewpoint. What happens if we attempt to improve the approximation to $\ln x$ by using higher degree Taylor polynomials? Figure 9.10 shows that for $0 < x < 2$, the polynomials fit the curve well, though outside that interval they are not close at all. In fact for $0 < x < 2$, the higher the degree of the polynomial, the better it fits the curve. Thus we say that, for $0 < x < 2$, the polynomials

$$P_5(x), P_6(x), P_7(x), \ldots, P_n(x), \ldots$$

converge to $\ln x$ as $n \to \infty$. For such x values, a higher degree polynomial will, in general, give a better approximation.

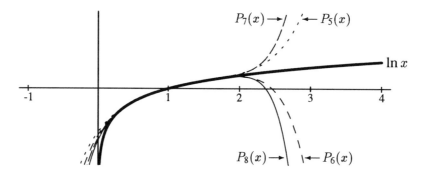

Figure 9.10: Graph of $\ln x$ and Taylor Polynomials

However, when $x > 2$, the polynomials move away from the curve and the approximations get worse as the degree of the polynomial increases. For such x values, we say the polynomials *diverge* from the original function.

Thus, the Taylor polynomial approximations about $x = 1$ are only effective as approximations to $\ln x$ for values of x between 0 and 2; beyond that, they definitely should not be used. We say the *interval of convergence* of the Taylor polynomials is $0 < x < 2$. As x gets near 0 or 2, the polynomials converge very slowly—meaning you might have to take a very high degree to get close to the correct value for $\ln x$.

Numerical Viewpoint. To illustrate these ideas, consider the successive approximations obtained from

$$\ln x = (x - 1) - \frac{(x-1)^2}{2} + \frac{(x-1)^3}{3} - \frac{(x-1)^4}{4} + \cdots + (-1)^{n-1}\frac{(x-1)^n}{n}.$$

If $x = 1.4$, so that $\ln 1.4 = 0.3364722$, then we find that

$$
\begin{array}{llll}
P_2(1.4) &=& 0.32 & \qquad P_6(1.4) &=& 0.33630 \\
P_4(1.4) &=& 0.33493 & \qquad P_8(1.4) &=& 0.33645
\end{array}
$$

and we see that the convergence is quite rapid. Similarly, we show the results of using $x = 1.9$ ($\ln 1.9 = 0.641854$) and $x = 2.3$ ($\ln 2.3 = 0.832909$) in the table to the right. Notice that for $x = 1.9$, which is inside the interval of convergence, but close the the endpoint, the approximations converge, though rather slowly.

n	$P_n(1.9)$	n	$P_n(2.3)$
2	0.495	2	0.455
5	0.69020	5	1.21589
8	0.61802	8	0.28817
11	0.65473	11	1.7171
		14	−0.70701

For $x = 2.3$, which is outside the interval of convergence, the approximations diverge and the larger the value for n, the less accurate the "approximations" become. In fact, if we continued on to a polynomial of degree 25, the contribution of the highest power term would be 28.2256; the contribution of the 100th term would be -2479335111. Thus, outside the interval of convergence, the "approximations" will oscillate ever more rapidly away from the desired value.

Thus, when we define the Taylor series for $\ln x$ about $x = 1$, we say that it converges for $0 < x < 2$, or that its interval of convergence is $0 < x < 2$, and write

$$\ln x = (x - 1) - \frac{(x-1)^2}{2} + \frac{(x-1)^3}{3} - \frac{(x-1)^4}{4} + \cdots \qquad \text{for } 0 < x < 2.$$

Outside of the interval $0 < x < 2$, the series does not represent $\ln x$. Notice that the interval of convergence is centered at $x = 1$. In general, a Taylor series about $x = a$ either converges for all x, or has an interval of convergence centered at $x = a$.

☐ **Example 1** Find the Taylor series for $\ln(1 + x)$ about $x = 0$, and its interval of convergence.

Solution. Taking derivatives of $\ln(1 + x)$ and substituting $x = 0$ leads to the Taylor series:

$$\ln(1 + x) = x - \frac{x^2}{2} + \frac{x^3}{3} - \frac{x^4}{4} + \cdots$$

Notice that this is the same series that you get by substituting $(1 + x)$ for x in the series for $\ln x$:

$$\ln x = (x - 1) - \frac{(x-1)^2}{2} + \frac{(x-1)^3}{3} - \frac{(x-1)^4}{4} + \cdots \qquad \text{for } 0 < x < 2.$$

Since the series for $\ln x$ about $x = 1$ converges for $0 < x < 2$, you shouldn't be surprised to see in Figure 9.11 that the interval of convergence for the Taylor series for $\ln(1 + x)$ about $x = 0$ is $-1 < x < 1$. Thus we write

$$\ln(1 + x) = x - \frac{x^2}{2} + \frac{x^3}{3} - \frac{x^4}{4} + \cdots \qquad \text{for } -1 < x < 1.$$

By the way, it should be clear that the series cannot possibly converge to $\ln(1 + x)$ for $x \leq -1$ since $\ln(1 + x)$ is not defined there.

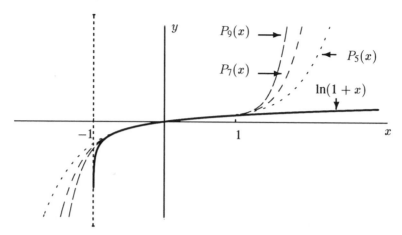

Figure 9.11: Graph of $\ln(1+x)$ and Taylor Polynomials

The Binomial Series

We will find the Taylor series about $x = 0$ for the function $f(x) = (1+x)^\alpha$, with α a constant. Taking derivatives:

$$
\begin{aligned}
f(x) &= (1+x)^\alpha & \text{so} && f(0) &= 1 \\
f'(x) &= \alpha(1+x)^{(\alpha-1)} & && f'(0) &= \alpha \\
f''(x) &= \alpha(\alpha-1)(1+x)^{(\alpha-2)} & && f''(0) &= \alpha(\alpha-1) \\
f'''(x) &= \alpha(\alpha-1)(\alpha-2)(1+x)^{(\alpha-3)} & && f'''(0) &= \alpha(\alpha-1)(\alpha-2)
\end{aligned}
$$

Thus, the third degree Taylor polynomial is

$$(1+x)^\alpha \approx P_3(x) = 1 + \alpha x + \frac{\alpha(\alpha-1)}{2!}x^2 + \frac{\alpha(\alpha-1)(\alpha-2)}{3!}x^3.$$

For a specific value of α, graphing $P_3(x)$, $P_4(x)$, and so on, shows that the Taylor polynomials converge to $f(x)$ for $-1 < x < 1$. (See Problems 22- 24, page 642.) Thus, we define the Taylor series for $f(x) = (1+x)^\alpha$ as follows:

$$(1+x)^\alpha = 1 + \alpha x + \frac{\alpha(\alpha-1)}{2!}x^2 + \frac{\alpha(\alpha-1)(\alpha-2)}{3!}x^3 + \cdots \qquad \text{for } -1 < x < 1.$$

This series is called the **Binomial series**.

As you will see in the next example, this series gives the Binomial Theorem of algebra when α is a positive integer.

☐ **Example 2**   Use the Binomial series with $\alpha = 3$ to expand $(1+x)^3$.

Solution. The series is

$$(1+x)^3 = 1 + 3x + \frac{3 \cdot 2}{2!}x^2 + \frac{3 \cdot 2 \cdot 1}{3!}x^3 + \frac{3 \cdot 2 \cdot 1 \cdot 0}{4!}x^4 + \cdots$$

The term in x^4 and all terms beyond it turn out to be zero, because the coefficient contains a factor of 0. Simplifying gives

$$(1+x)^3 = 1 + 3x + 3x^2 + x^3,$$

which is the usual binomial expansion of $(1+x)^3$. ❑

☐ **Example 3** Write down the Taylor series for $\dfrac{1}{1+x}$.

Solution. Since $\dfrac{1}{1+x} = (1+x)^{-1}$, let $\alpha = -1$. Then

$$\begin{aligned}
\frac{1}{1+x} = (1+x)^{-1} &= 1 + (-1)x + \frac{(-1)(-2)}{2!}x^2 + \frac{(-1)(-2)(-3)}{3!}x^3 + \cdots \\
&= 1 - x + x^2 - x^3 + \cdots
\end{aligned}$$

$$-1 < X < 1$$

❑

Another View of Euler's Formula

If you studied Section 8.12 on complex numbers, you may have found the expression $e^{i\theta}$ quite mysterious. What does it mean to take a number to a complex power? We can now answer this question by another method using the Taylor series for e^x:

$$e^x = T(x) = 1 + x + \frac{x^2}{2!} + \frac{x^3}{3!} + \frac{x^4}{4!} + \cdots.$$

$T(x)$ converges to e^x, for all real values of x. For complex values z, we *define* e^z to be $T(z)$.

We can now calculate $e^{i\theta}$.

$$e^{i\theta} = T(i\theta) = 1 + i\theta + \frac{i^2\theta^2}{2!} + \frac{i^3\theta^3}{3!} + \frac{i^4\theta^4}{4!} + \frac{i^5\theta^5}{5!} + \frac{i^6\theta^6}{6!} + \cdots.$$

Simplify each term, using $i^2 = -1$, $i^3 = -i$, $i^4 = 1$, $i^5 = i,\ldots$, to get

$$e^{i\theta} = 1 + i\theta - \frac{\theta^2}{2!} - i\frac{\theta^3}{3!} + \frac{\theta^4}{4!} + i\frac{\theta^5}{5!} - \frac{\theta^6}{6!} - i\frac{\theta^7}{7!} + \cdots.$$

Now group all the real and complex terms separately:

$$e^{i\theta} = \left(1 - \frac{\theta^2}{2!} + \frac{\theta^4}{4!} - \frac{\theta^6}{6!} + \cdots\right) + i\left(\theta - \frac{\theta^3}{3!} + \frac{\theta^5}{5!} - \frac{\theta^7}{7!} + \cdots\right).$$

Amazingly enough, the series for $\cos\theta$ and $\sin\theta$ appear on the right, so we have shown that

$$e^{i\theta} = \cos\theta + i\sin\theta.$$

❖ Exercises for Section 9.2

For Problems 1–6, find the Taylor series for the given function about 0.

1. $\dfrac{1}{1-x}$

2. $\sqrt{1+x}$

3. $\arctan z$

4. $\ln(1-t)$

5. $\dfrac{1}{\sqrt{1+x}}$

6. $\sqrt[3]{1-y}$

For Problems 7–13, find the Taylor series for the function about the given point $x = a$.

7. $\dfrac{1}{x}$, $\quad a = 1$

8. $\dfrac{1}{x}$, $\quad a = -1$

9. $\dfrac{1}{x}$, $\quad a = 2$

10. $\sin x$, $\quad a = \frac{\pi}{4}$

11. $\cos\theta$, $\quad a = \frac{\pi}{4}$

12. $\sin\theta$, $\quad a = -\frac{\pi}{4}$

13. $\tan x$, $\quad a = \frac{\pi}{4}$

Use Taylor series to calculate the limits in Problems 14–19.

14. $\displaystyle\lim_{\theta \to 0} \frac{\theta - \sin\theta}{\theta^3}$

15. $\displaystyle\lim_{\alpha \to 0} \frac{\tan\alpha}{\alpha}$

16. $\displaystyle\lim_{x \to 0} \frac{\sqrt{1+x} - 1}{x}$

17. $\displaystyle\lim_{h \to 0} \left(\frac{h}{\sqrt{1+h} - 1} \right)$

18. $\displaystyle\lim_{x \to \pi} \left(\frac{\sin x}{x - \pi} \right)$

19. $\displaystyle\lim_{\theta \to \frac{\pi}{2}} \frac{\cos\theta}{\left(\theta - \frac{\pi}{2}\right)}$

20. (a) Find the Taylor series for $f(x) = \ln(1 + 2x)$ about $x = 0$ by taking derivatives.

    (b) Compare your result in (a) to the series for $\ln(1 + x)$. How could you have obtained your answer to (a) from the series for $\ln(1 + x)$?

21. (a) Find the Taylor series for $f(x) = \sin(x^2)$ about $x = 0$ by taking derivatives.

    (b) Compare your result in (a) to the series for $\sin x$. How could you have obtained your answer to (a) from the series for $\sin x$?

22. By plotting the function $f(x) = \dfrac{1}{1-x}$ and several of its Taylor polynomials, estimate the interval of convergence of the series you found in Problem 1.

23. By plotting the function $f(x) = \sqrt{1+x}$ and several of its Taylor polynomials, estimate the interval of convergence of the series you found in Problem 2.

24. By plotting the function $f(x) = \dfrac{1}{\sqrt{1+x}}$ and several of its Taylor polynomials, estimate the interval of convergence of the series you found in Problem 5.

25. Suppose you know that all the derivatives of some function f exist at 0, and that the function has Taylor series:

$$x + \frac{x^2}{2} + \frac{x^3}{3} + \frac{x^4}{4} + \cdots + \frac{x^n}{n} + \cdots$$

   Find $f'(0)$, $f''(0)$, $f'''(0)$, and $f^{(10)}(0)$.

26. Suppose that you are told that the Taylor series of $f(x) = x^2 e^{x^2}$ is

$$x^2 + x^4 + \frac{x^6}{2!} + \frac{x^8}{3!} + \frac{x^{10}}{4!} \cdots$$

   Find $\dfrac{d}{dx}(x^2 e^{x^2})\Big|_{x=0}$ and $\dfrac{d^6}{dx^6}(x^2 e^{x^2})\Big|_{x=0}$.

9.3 Finding and Using Taylor Series

Finding a Taylor series for a function means finding the coefficients. Assuming the function has all its derivatives, finding the coefficients can always be done in theory, at least, by differentiation. However, in practice, the differentiation can be a very laborious business. The purpose of this section is to give you an easier way.

New Series by Substitution

Suppose you want to find the series for e^{-x^2}. You could find the coefficients by differentiation. Differentiating e^{-x^2} by the Chain rule gives $-2xe^{-x^2}$, and differentiating again gives $-2e^{-x^2} + 4x^2 e^{-x^2}$. Now, every time you differentiate you will need the Product rule, and the number of terms will grow. Thus finding, say, the 10th or 20th derivative, and finding the series for e^{-x^2} up to the terms in x^{10} or x^{20} by this method, will be tiresome (at least without a differentiating computer or calculator).

Fortunately, there's a quicker way. Recall that for all y:

$$e^y = 1 + y + \frac{y^2}{2!} + \frac{y^3}{3!} + \frac{y^4}{4!} \cdots$$

Substituting $y = -x^2$ tells us that for all x:

$$e^{-x^2} = 1 + (-x^2) + \frac{(-x^2)^2}{2!} + \frac{(-x^2)^3}{3!} + \frac{(-x^2)^4}{4!} + \cdots$$

Simplifying shows that for all x:

$$e^{-x^2} = 1 - x^2 + \frac{x^4}{2!} - \frac{x^6}{3!} + \frac{x^8}{4!} + \cdots$$

Using this method, it is easy to write down the series up to the terms in x^{10} or x^{20}.

☐ **Example 1** Find the Taylor series about $x = 0$ for $f(x) = \dfrac{1}{1 + x^2}$.

Solution. The Binomial Theorem tells us that for $-1 < y < 1$,

$$\frac{1}{1 + y} = (1 + y)^{-1} = 1 - y + y^2 - y^3 + y^4 + \cdots$$

Substituting $y = x^2$ gives, for $-1 < x < 1$,

$$\frac{1}{1 + x^2} = 1 - x^2 + x^4 - x^6 + x^8 + \cdots$$

☐

These examples show that you can get new series from old by substitution. Proving that the new series you get is the same one you would have obtained by direct calculation of the derivatives is more difficult, and is done in more advanced texts.

In Example 1, we made the substitution $y = x^2$. We can also substitute an entire series into another one, as in the next example.

☐ **Example 2** Find the Taylor series about $\theta = 0$ for $g(\theta) = e^{\sin \theta}$.

Solution. For all y, and θ, we know that

$$e^y = 1 + y + \frac{y^2}{2!} + \frac{y^3}{3!} + \frac{y^4}{4!} + \cdots$$

and

$$\sin \theta = \theta - \frac{\theta^3}{3!} + \frac{\theta^5}{5!} - \cdots$$

Let's substitute the series for $\sin \theta$ for y:

$$e^{\sin \theta} = 1 + \left(\theta - \frac{\theta^3}{3!} + \frac{\theta^5}{5!} - \cdots\right) + \frac{1}{2!}\left(\theta - \frac{\theta^3}{3!} + \frac{\theta^5}{5!} - \cdots\right)^2 + \frac{1}{3!}\left(\theta - \frac{\theta^3}{3!} + \frac{\theta^5}{5!} - \cdots\right)^3 + \cdots$$

To simplify, we multiply out and collect terms. The only constant term is the 1, and there's only one θ term. The only θ^2 term is the first term you get by multiplying out the square, and is $\dfrac{\theta^2}{2!}$. There are two contributors to the θ^3 term: the $-\dfrac{\theta^3}{3!}$ from right after the θ, and the first term you get from multiplying out the cube, which is $\dfrac{1}{3!}\theta^3$. Thus the series starts:

$$e^{\sin \theta} = 1 + \theta + \frac{\theta^2}{2!} + \left(-\frac{\theta^3}{3!} + \frac{\theta^3}{3!}\right) + \cdots$$

$$= 1 + \theta + \frac{\theta^2}{2!} + 0 \cdot \theta^3 + \cdots$$

New Series by Integration

Just as we can get new series by substitution, we can also get new series by integration. Here again, proving that the new series obtained this way is actually the Taylor series we want, and that it has the same interval of convergence, are important questions that we will not deal with.

☐ **Example 3** Find the Taylor series about $x = 0$ for arctan x from the series for $\dfrac{1}{1 + x^2}$.

Solution. We know that $\dfrac{d(\arctan x)}{dx} = \dfrac{1}{1 + x^2}$, so, using the series derived in Example 1, page 644, we have that

$$\frac{d(\arctan x)}{dx} = \frac{1}{1 + x^2} = 1 - x^2 + x^4 - x^6 + x^8 - \cdots$$

Antidifferentiating term–by–term (which turns out to be legal) gives

$$\arctan x = C + x - \frac{x^3}{3} + \frac{x^5}{5} - \frac{x^7}{7} + \frac{x^9}{9} - \cdots$$

where C is an arbitrary constant. The fact that $\arctan 0 = 0$ tells us that we must have $C = 0$, so

$$\arctan x = x - \frac{x^3}{3} + \frac{x^5}{5} - \frac{x^7}{7} + \frac{x^9}{9} - \cdots$$

☐

Applications of Taylor Series

☐ **Example 4** Use a series to estimate the numerical value of π.

Solution. Since $\arctan 1 = \frac{\pi}{4}$, we can substitute $x = 1$ into the series in Example 3, giving

$$\pi = 4 \arctan 1 = 4 \left(1 - \frac{1}{3} + \frac{1}{5} - \frac{1}{7} + \frac{1}{9} \cdots \right)$$

Unfortunately this series converges very slowly, meaning that you have to take a large number of terms to get an accurate estimate for π. The table below shows the value of the sum, S_n, obtained from the nth degree Taylor polynomial. As you see, the values of S_n do seem to converge to $\pi = 3.1415\ldots$

n	7	9	25	100	500	1000	10,000
S_n	2.895	3.340	3.218	3.122	3.138	3.140	3.141

☐

One of the most basic questions we can ask about two functions is which one is larger. The first few terms of the Taylor series for the two functions can often be used to answer this question over a small interval. If the constant terms of the two series are the same, compare the linear terms; if the linear terms are the same, compare the quadratic terms, and so on.

☐ **Example 5** By looking at the Taylor series, decide which of the following functions is largest, and which is smallest for a small positive θ.

(a) $1 + \sin \theta$ (b) $\cos \theta$ (c) $\dfrac{1}{1 - \theta^2}$

Solution. The Taylor expansion about $\theta = 0$ for $\sin \theta$ is

$$\theta - \frac{\theta^3}{3!} + \frac{\theta^5}{5!} - \frac{\theta^7}{7!} + \cdots$$

So

$$1 + \sin \theta = 1 + \theta - \frac{\theta^3}{3!} + \frac{\theta^5}{5!} - \frac{\theta^7}{7!} + \cdots$$

The Taylor expansion about $\theta = 0$ for $\cos \theta$ is

$$\cos \theta = 1 - \frac{\theta^2}{2!} + \frac{\theta^4}{4!} - \frac{\theta^6}{6!} + \cdots$$

The Taylor expansion for $\dfrac{1}{1 + \theta}$ about $\theta = 0$ is

$$\frac{1}{1 + \theta} = 1 - \theta + \theta^2 - \theta^3 + \theta^4 - \cdots$$

So, substituting $-\theta^2$ for θ:

$$\begin{aligned}
\frac{1}{1 - \theta^2} &= 1 - (-\theta^2) + (-\theta^2)^2 - (-\theta^2)^3 + (-\theta^2)^4 + \cdots \\
&= 1 + \theta^2 + \theta^4 + \theta^6 + \theta^8 + \cdots
\end{aligned}$$

For small θ, we can neglect the higher order terms in these expansions:

$$\begin{aligned}
1 + \sin \theta &\approx 1 + \theta \\
\cos \theta &\approx 1 - \frac{\theta^2}{2} \\
\frac{1}{1 - \theta^2} &\approx 1 + \theta^2
\end{aligned}$$

For all $\theta \neq 0$, we have

$$1 - \frac{\theta^2}{2} < 1 + \theta^2.$$

Also, since $\theta^2 < \theta$ for $0 < \theta < 1$, we have

$$1 - \frac{\theta^2}{2} < 1 + \theta^2 < 1 + \theta.$$

So, for small positive θ, we have

$$\cos \theta < \frac{1}{1 - \theta^2} < 1 + \sin \theta.$$

❑

❏ **Example 6** Two electrical charges of equal magnitude and opposite signs located near one another are called an electrical dipole. Suppose the charges Q and $-Q$ are a distance r apart. See Figure 9.12. The electric field, E, at the point P is given by

$$E = \frac{Q}{R^2} - \frac{Q}{(R + r)^2}.$$

Use series to investigate the behavior of the electric field far away from the dipole. Show that when R is large in comparison to r, the electric field is approximately proportional to $\frac{1}{R^3}$.

Figure 9.12: A Dipole

Solution. We will use a series approximation, but in order to do this we need to choose a variable whose value we know to be small, so that we can safely use only the first few terms. Although we know that r is much smaller than R, we do not know that r is itself small. The quantity $\frac{r}{R}$ is however very small (much smaller than 1, for example), and so we expand $\frac{1}{(R + r)^2}$ in terms of $\frac{r}{R}$ as follows:

$$\begin{aligned}
\frac{1}{(R + r)^2} &= \frac{1}{R^2(1 + \frac{r}{R})^2} = \frac{1}{R^2}\left(1 + \frac{r}{R}\right)^{-2} \\
&= \frac{1}{R^2}\left(1 + (-2)\left(\frac{r}{R}\right) + \frac{(-2)(-3)}{2!}\left(\frac{r}{R}\right)^2 + \frac{(-2)(-3)(-4)}{3!}\left(\frac{r}{R}\right)^3 + \cdots\right) \\
&= \frac{1}{R^2}\left(1 - 2\frac{r}{R} + 3\frac{r^2}{R^2} - 4\frac{r^3}{R^3} - \cdots\right)
\end{aligned}$$

So

$$\begin{aligned}
E = \frac{Q}{R^2} - \frac{Q}{(R + r)^2} &= Q\left[\frac{1}{R^2} - \frac{1}{R^2}\left(1 - 2\frac{r}{R} + 3\frac{r^2}{R^2} - 4\frac{r^3}{R^3} + \cdots\right)\right] \\
&= \frac{Q}{R^2}\left(2\frac{r}{R} - 3\frac{r^2}{R^2} + 4\frac{r^3}{R^3} - \cdots\right)
\end{aligned}$$

Notice that $\dfrac{r}{R}$ is smaller than 1, so the Binomial expansion will converge. We are interested in the electric field far away from the dipole. The quantity $\dfrac{r}{R}$ is small there, and $\left(\dfrac{r}{R}\right)^2$ and higher powers are much smaller still. Thus, we approximate by disregarding all except the first term, giving

$$E \approx \frac{Q}{R^2}\left(\frac{2r}{R}\right) \quad \text{so} \quad E \approx \frac{2Qr}{R^3}.$$

Since Q and r are constants, this means that E is nearly proportional to $\dfrac{1}{R^3}$. ❑

In the previous example, we say that E is *expanded in terms of* $\dfrac{r}{R}$, meaning that the independent variable in the expansion is $\dfrac{r}{R}$.

❖ Exercises for Section 9.3

Find the first four non-zero terms of the Taylor series about the origin for the functions in Problems 1–12.

1. e^{-x}

2. $\sqrt{1 - 2x}$

3. $\cos(\theta^2)$

4. $\dfrac{t}{1 + t}$

5. $\ln(1 - 2y)$

6. $\dfrac{1}{\sqrt{1 - z^2}}$

7. $\arcsin x$

8. $\dfrac{z}{e^{z^2}}$

9. $\phi^3 \cos(\phi^2)$

10. $\sqrt{1 + \sin\theta}$

11. $e^t \cos t$

12. $\sqrt{(1 + t)} \sin t$

For Problems 13–14, expand the quantity about the origin in terms of the variable given. Give the first four non-zero terms.

13. $\dfrac{1}{2 + x}$ in terms of $\dfrac{x}{2}$

14. $\dfrac{a}{\sqrt{a^2 + x^2}}$ in terms of $\dfrac{x}{a}$, where $a > 0$.

15. For values of y near 0, put the following functions in increasing order, using their Taylor expansions.

    (a) $\ln(1 + y^2)$

    (b) $\sin(y^2)$

    (c) $1 - \cos y$

16. The figure on the right shows the graph of the four functions below for values of x near 0. Match up graphs and formulas.

    (a) $\dfrac{1}{1-x^2}$

    (b) $(1+x)^{\frac{1}{4}}$

    (c) $\sqrt{1+\dfrac{x}{2}}$

    (d) $\dfrac{1}{\sqrt{1-x}}$

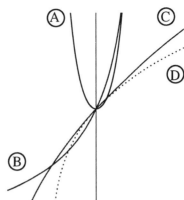

17. Consider the two functions $y = e^{-x^2}$ and $y = \dfrac{1}{1+x^2}$.

    (a) Write the Taylor expansions for the two functions about $x = 0$. What is similar about the two series? What is different?

    (b) Looking at the series, which function do you predict will be greater over the interval $(-1, 1)$? Graph both and see.

    (c) Are these functions even or odd? How might you see this by looking at the series expansions?

    (d) By looking at the coefficients, explain why it is reasonable that the series for $y = e^{-x^2}$ converges for all values of x, but the series for $y = \dfrac{1}{1+x^2}$ converges only on $(-1, 1)$.

18. One of Einstein's most amazing predictions was that light traveling from distant stars would bend around the sun on the way to earth. His calculations involved solving the equation

$$\sin \phi + b(1 + \cos^2 \phi + \cos \phi) = 0$$

where b is a very small positive constant.

    (a) Explain why the equation has a solution for ϕ which is near 0.

    (b) Expand the left hand side of the equation in Taylor series about $\phi = 0$, disregarding terms of order ϕ^2 and higher. Solve for ϕ. (Your answer will involve b.)

19. An electric dipole consists of a charge Q on the x axis at $x = 1$, and a charge $-Q$ on the x axis at $x = -1$. The electric field strength, E, at the point $x = R$ on the x axis is given (for $R > 1$) by

$$E = \frac{kQ}{(R-1)^2} - \frac{kQ}{(R+1)^2}$$

where k is a positive constant whose value depends on the units. Expand E as a series in $\frac{1}{R}$, giving the first two non-zero terms.

20. A hydrogen atom consists of an electron, of mass m, orbiting a proton, of mass M, where m is much smaller than M. The *reduced mass*, μ, of the hydrogen atom is defined by

$$\mu = \frac{mM}{m+M}$$

   (a) Show that $\mu \approx m$.

   (b) To get a more accurate approximation for μ, express μ as m times a series in $\frac{m}{M}$.

   (c) The approximation $\mu \approx m$ is obtained by disregarding all but the constant term in the series. The first order correction is obtained by including the linear term but no higher terms. If $m \approx \frac{1}{1836} M$, by what percentage does including the first order correction change the estimate $\mu \approx m$?

21. The electric potential, V, at a distance R along the axis perpendicular to the center of a charged disc with radius a and constant charge density σ, is given by

$$V = 2\pi\sigma(\sqrt{R^2 + a^2} - R).$$

Show that, for large R,

$$V \approx \frac{\pi a^2 \sigma}{R}.$$

22. Studying resonance in electric circuits leads us to consider the expression

$$\left(\omega L - \frac{1}{\omega C}\right)^2,$$

where ω is the variable and L and C are constant.

   (a) Find ω_0, the value of ω making the expression zero.

   (b) Because, in practice, ω fluctuates about ω_0, we are interested in the behavior of this expression for values of ω near ω_0. Let $\omega = \omega_0 + \Delta\omega$ and expand the expression in terms of $\Delta\omega$ up to the linear term. Give your answer in terms of ω_0 and L but not C.

23. The electric force, F, between two atoms depends on the distance, r, separating them. See Figure 9.13. A positive F represents a repulsive force, a negative F represents an attractive force.

    (a) Why does $r = a$ represent an equilibrium position?
    (b) Qualitatively, what happens to the force if the atoms start with $r = a$, and are pulled slightly further apart?
    (c) Qualitatively, what happens to the force if the atoms start with $r = a$, and are pushed slightly closer together?
    (d) Write the Taylor series for F around $r = a$.
    (e) By discarding all except the first non-zero term in this series, describe how the force between the atoms depends on the displacement from the equilibrium, when that displacement is small.

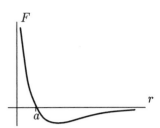

Figure 9.13: Force between atoms

9.4   The Error in a Taylor Polynomial Approximation

Whenever we make an approximation, there is nearly always an error, the difference between the exact answer (which we usually do not know) and the approximate value. In order to use the approximation intelligently, it is necessary to assess the size of the error. In general, we cannot find the exact value of the error; if we could, then we would not need an approximation. However, we can often estimate the maximum possible size of the error. This gives us a feel for how good the approximation is.

We now consider how we can assess the size of the error involved in using $P_n(x)$, the nth degree Taylor polynomial, to approximate $f(x)$ near $x = a$. The error is the difference

$$E = f(x) - P_n(x)$$

If E is positive, the approximation is smaller than the true value. If E is negative, the approximation is too large. Sometimes we are only interested in the magnitude of the error, $|E|$.

Error in $P_0(x)$

We begin with the simplest case, that of approximating $f(x)$ by the constant function $P_0(x)$ for values of x close to $x = a$. Since $P_0(x)$ means the Taylor polynomial with only the constant term,

$$P_0(x) = f(a)$$

For simplicity, suppose we consider one particular value of x, say $x = b$, which is close to $x = a$. We want to estimate how far the function is from the horizontal line $y = f(a)$ when $x = b$, as shown in Figure 9.14.

That is, we want to estimate the difference

$$E = f(b) - P_0(b) = f(b) - f(a)$$

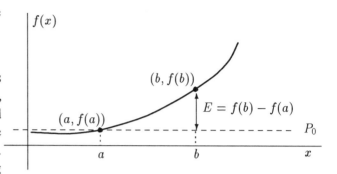

for any given value of b. The graph makes it clear that if the derivative is positive, the error is positive to the right of a and negative to the left. If the derivative were negative, the situation would be reversed. We can estimate the error by the following geometric argument.

Figure 9.14: The error in $P_0(x)$ at $x = b$

Consider the graph shown in Figure 9.15 for some function $f(x)$ on the interval $[a, b]$. Join the points on the curve where $x = a$ and $x = b$ with a line and observe that the slope of this secant line AB is given by

$$m = \frac{f(b) - f(a)}{b - a}.$$

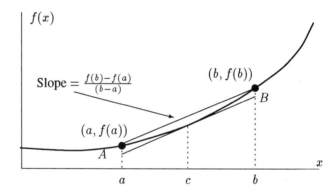

Consider now the tangent line drawn to the curve at each point between $x = a$ and $x = b$. In general, these lines will have different slopes. For the curve shown in Figure 9.15, the tangent line at $x = a$ is flatter than the secant line and so its slope is less than the number m above. Similarly, the tangent line at $x = b$ is steeper than the secant line and so its slope is greater than m. As a result, there will be at least one point between a and b where the slope of the tangent line to the curve

Figure 9.15: Estimating $f(b) - f(a)$

will be precisely the same slope of the secant line. Suppose this occurs at $x = c$. Then we must have

$$f'(c) = m = \frac{f(b) - f(a)}{b - a}.$$

Thus, it follows that

$$E = f(b) - f(a) = f'(c)(b - a)$$

and that

$$f(b) = f(a) + f'(c)(b - a).$$

Since we do not know where the point c is (except in the simplest cases), we cannot get an exact value for $f'(c)$. However, if we know the derivative $f'(x)$, then we can find its maximum value on the interval $[a, b]$ and so find the maximum possible value for the error in the approximation. In the next example, we'll estimate the error using this method in a situation in which we can also find the error exactly, allowing us to compare the two.

☐ **Example 1**   Using the formula above, estimate the error in approximating $f(x) = \cos x$ by the constant function $y = 1$ on the interval $[0, \frac{\pi}{10}]$.

Solution. The line $y = 1$ is the zero degree polynomial approximation of $\cos x$ centered at $x = 0$, since $P_0(x) = f(0) = 1$. So the error in the approximation is given by

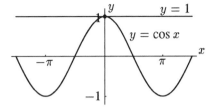

Figure 9.16: Graph of $\cos x$ and its Tangent

$$E = f(x) - P_0(x) = \cos x - 1.$$

This error is negative because $y = 1$ is an overestimate. See Figure 9.16. The error is also a function of x. However, we can estimate the magnitude of the error by noticing that $|\cos x - 1|$ is increasing on the interval $[0, \frac{\pi}{10}]$, so

$$|E| = |\cos x - 1| \leq |\cos \tfrac{\pi}{10} - 1| = |f'(c)| \cdot |\tfrac{\pi}{10} - 0|$$

with $0 < c < \frac{\pi}{10}$. Since $f'(x) = -\sin x$, this becomes

$$|\cos x - 1| \leq |-\sin c| \cdot \tfrac{\pi}{10}.$$

On the interval $[0, \frac{\pi}{10}]$, the sine function is increasing and so its greatest value occurs when $x = \frac{\pi}{10}$, so the maximum possible value of $|-\sin c|$ on this interval is $\sin(\frac{\pi}{10})$. From Problem 23 of page 288 we know that $\sin x < x$ for $x > 0$, so $\sin(\frac{\pi}{10}) < \frac{\pi}{10}$ and therefore the maximum possible error is

$$|E| = |\cos x - 1| < \tfrac{\pi}{10} \cdot \tfrac{\pi}{10} = 0.099.$$

Thus, the largest possible error is less than one tenth.

You can check this answer graphically. Figure 9.16 shows that, for $0 \leq x \leq \frac{\pi}{10}$, the largest distance between the cosine curve and the line $y = 1$ occurs at $x = \frac{\pi}{10}$, and you can read this distance off a graph on a calculator or computer. In fact, the maximum error is only $|\cos \frac{\pi}{10} - 1| = 0.049$.
☐

Error in $P_1(x)$

We will now see how to estimate the error in using the first order Taylor approximation. Suppose we consider the linear approximation to the function f near $x = a$:

$$f(x) \approx P_1(x) = f(a) + f'(a)(x - a)$$

and we again consider the particular value $x = b$. We need to estimate the size of the error

$$E = f(b) - P_1(b) = f(b) - f(a) - f'(a)(b - a).$$

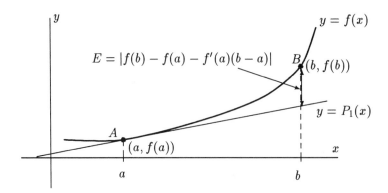

Figure 9.17: The error in $P_1(x)$ at $x = b$.

See Figure 9.17. The graph shows that the error will be positive when the function is concave up. In the case of the constant approximation the size of the error was controlled by the size of the derivative. Here it is the curviness of the graph that determines the size of the error: if the curviness is large, then the error is large. So it shouldn't be too surprising to find a second derivative in the error formula. It turns out that in this case the maximum possible error is given by

$$E = \frac{f''(c)}{2!}(b - a)^2,$$

where c is between a and b. Equivalently, since $E = f(b) - f(a) - f'(a)(b - a)$,

$$f(b) = f(a) + f'(a)(b - a) + \frac{f''(c)}{2!}(b - a)^2$$

where c is some value between a and b.

❑ **Example 2** Find the maximum possible error if the function $f(x) = e^x$ is approximated by the linear Taylor polynomial at $x = 0$ over the interval $[0, 1]$.

Solution. The error is given by

$$E = f(x) - P_1(x) = \frac{f''(c)(x - 0)^2}{2} = \frac{e^c x^2}{2}$$

where c is between 0 and x, and x is in the interval $[0, 1]$. Notice that the error is clearly positive. Since x cannot be more than 1, and since the exponential is an increasing function, the maximum value it can attain on this interval is $e^1 = e$. Since $e < 3$, we have

$$E = |E| \leq \frac{e(1)^2}{2} < \frac{3}{2} = 1.5.$$

Clearly, this is a rather large value for an error. However, it is not surprising to find that a linear approximation is not very accurate approximation to a fast-growing exponential function. ❑

Error in $P_n(x)$

If you examine the formulas above for the error term, you should notice an interesting pattern. The additional term at the end, representing the error, looks just like the next term in the Taylor series except that the derivative is evaluated at some intermediate point $x = c$ rather than at the point $x = a$. This pattern persists for Taylor polynomial approximations of any degree n. If we denote the error by E_n:

$$
\begin{aligned}
E_n &= f(b) - P_n(b) \\
&= f(b) - \left(f(a) + f'(a)(b-a) + \frac{f''(a)}{2!}(b-a)^2 + \frac{f'''(a)}{3!}(b-a)^3 + \cdots + \frac{f^{(n)}(a)}{n!}(b-a)^n \right) \\
&= \frac{f^{(n+1)}(c)}{(n+1)!}(b-a)^{n+1}.
\end{aligned}
$$

Or, equivalently

$$
f(b) = f(a) + f'(a)(b-a) + \frac{f''(a)}{2!}(b-a)^2 + \cdots + \frac{f^{(n)}(a)}{n!}(b-a)^n + \frac{f^{(n+1)}(c)}{(n+1)!}(b-a)^{n+1}.
$$

The magnitude of the error in $P_n(x)$ is given by

$$
|E| = \left| \frac{f^{(n+1)}(c)}{(n+1)!}(b-a)^{n+1} \right| \le \frac{K|b-a|^{n+1}}{(n+1)!} \quad \text{where} \quad K = \left| \begin{array}{l} \text{max. value of} \\ f^{n+1} \text{ on } [a,b] \end{array} \right|
$$

☐ **Example 3** What degree Taylor polynomial for e^x about $x = 0$ do you have to take to calculate $e = e^1$ to four decimal places?

Solution. In Section 1.12, on page 104, we defined accuracy to four decimal places to mean $|E| \le 0.00005$. Since $a = 0$ and $b = 1$, we know that

$$
|E| = \left| \frac{f^{(n+1)}(c)}{(n+1)!}(1)^{n+1} \right| = \frac{e^c}{(n+1)!}
$$

for some c between 0 and 1. The biggest this could be is if $c = 1$, in which case, since $e < 3$, we have

$$
|E| = \frac{e}{(n+1)!} < \frac{3}{(n+1)!}.
$$

This number is an upper bound on the difference between e and

$$
P(n) = 1 + 1 + \frac{1}{2} + \frac{1}{6} + \cdots + \frac{1}{n!}.
$$

n	$3/(n+1)!$
1	1.5000000000
2	0.5000000000
3	0.1250000000
4	0.0250000000
5	0.0041666667
6	0.0005952381
7	0.0000744040
8	0.0000082672
9	0.0000008267
10	0.0000000752

Table 9.1: Magnitude of Error

We tabulate the values of this upper bound in Table 9.1. Looking at the table, we can see that the difference is less than 0.00005 when $n = 8$. So taking an 8th degree Taylor polynomial should give

e to 4 decimal places. Since we already know $e \approx 2.718282$, we can check this by computing

$$1 + 1 + \frac{1}{2} + \frac{1}{6} + \frac{1}{24} + \frac{1}{120} + \frac{1}{720} + \frac{1}{5040} + \frac{1}{40320} \approx 2.718278.$$

Sure enough, this agrees with e to 4 decimal places. Of course, the error formula is most useful in situations where you don't already know the exact answer, and want some guarantee of the accuracy of your approximation. ❑

❖ Exercises for Section 9.4

1. Suppose you approximate $f(t) = e^t$ by a Taylor polynomial of degree 0 about $t = 0$ on the interval $[0, 0.5]$.

   (a) Is the approximation an overestimate or an underestimate?

   (b) Estimate the magnitude of the largest possible error. Check your answer graphically on a computer or calculator.

2. Repeat Problem 1 using the second degree Taylor approximation, $P_2(t)$, to e^t.

3. Consider the error in using the approximation $\sin \theta \approx \theta$ on the interval $[-1, 1]$.

   (a) Where is the approximation an overestimate, and where is it an underestimate?

   (b) Estimate the magnitude of the largest possible error. Check your answer graphically on a computer or calculator.

4. Repeat Problem 3 for the approximation $\sin \theta \approx \theta - \dfrac{\theta^3}{3!}$.

5. Use the methods of this section to give a bound on the magnitude of the error if we approximate $\sqrt{2}$ using the Taylor approximation of degree three for $\sqrt{1+x}$ about $x = 0$.

Use the methods of this section to show how you can estimate the magnitude of the error in estimating the quantities in Problems 6–9 using a third degree Taylor polynomial about $x = 0$.

6. $\tan 1$ 7. $0.5^{\frac{1}{3}}$ 8. $\ln(1.5)$ 9. $\frac{1}{\sqrt{3}}$

10. Give a bound for the maximum possible error for the nth degree Taylor polynomial about $x = 0$ approximating $\cos x$ on the interval $[0, 1]$. What is the bound for $\sin x$?

11. What degree Taylor polynomial about $x = 0$ do you need to calculate $\cos 1$ to four decimal places? To six decimal places? Justify your answer using the results of Problem 10.

$$X = 100^{\frac{7}{4}}$$
$$100 = X^4$$

In Section 9.2, we discussed the convergence of Taylor polynomials. In Problems 12–14, we will use the error formula to justify the convergence analytically.

12. (a) Suppose we wish to approximate e^x using Taylor polynomials. What is the error if we use the nth degree Taylor polynomial about $x = 0$ to approximate e^x? Find an expression in terms of x and n which bounds this error.

(b) Show that for a fixed x this error bound goes to 0 as n goes to ∞. In other words, if we take a high enough degree Taylor polynomial, the error approaches 0, so we can say the Taylor polynomials converge to e^x.

13. (a) Suppose we wish to approximate $\ln x$ (for $x \geq 1$) using Taylor polynomials. What is the error if we use the nth degree Taylor polynomial about $x = 1$ to approximate $\ln x$? Give a bound for the error.

(b) Show that this error bound goes to 0 as n goes to ∞ if $1 < x < 2$. Show that the size of the error cannot be bounded as n goes to ∞ if $x > 2$. In other words, the Taylor polynomials converge to $\ln x$ if $1 < x < 2$, but they do not appear to converge if $x > 2$.

14. Using the ideas of Problem 13, argue that the Taylor polynomials for $f(x) = \frac{1}{1-x}$ about $x = 0$ converge for $-1 < x < \frac{1}{2}$. You may have to consider $x < 0$ and $x > 0$ separately.

9.5   Fourier Series

We have seen how to approximate a function by a Taylor polynomial. This polynomial is usually very close to the true value of the function near one point (the point at which the Taylor polynomial is centered), but not necessarily at all close anywhere else. In other words, Taylor polynomials are a good approximation *locally*, but not necessarily *globally*. In this section, we take another approach: we approximate the function by trigonometric functions, called *Fourier approximations*. The resulting approximation may not be as close to the original function at some points as the Taylor polynomial, but in general, it is closer over a larger interval. In other words, a Fourier approximation can be a better approximation globally. In addition, unlike Taylor approximations, Fourier approximations are periodic, so they are useful for approximating periodic functions.

The sciences are full of periodic (repeating) processes, so it is natural to approximate them by periodic functions. For example, sound waves are made up of periodic oscillations of air molecules. Heartbeats, the movement of the lungs, and the electrical current that powers our homes are all periodic phenomena. Two of the simplest periodic functions are the square wave and the saw-tooth wave, shown in Figure 9.18. Electrical engineers use the square wave as the model for the flow of electricity as a switch is repeatedly flicked on and off.

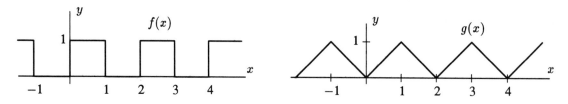

Figure 9.18: Square Wave (f) and Saw-Tooth Wave (g)

Fourier Polynomials

While we can express the square wave and the saw–tooth wave by formulas

$$f(x) = \begin{cases} 1 & 0 \le x < 1 \\ 0 & 1 \le x < 2 \\ 1 & 2 \le x < 3 \\ 0 & 3 \le x < 4 \\ \vdots & \vdots \end{cases} \qquad g(x) = \begin{cases} x & 0 \le x < 1 \\ 2 - x & 1 \le x < 2 \\ x - 2 & 2 \le x < 3 \\ 4 - x & 3 \le x < 4 \\ \vdots & \vdots \end{cases}$$

these formulas are not particularly easy to work with. Worse, the functions are not differentiable at various points. Here we will approximate such a function by a differentiable, periodic function.

Since the sine and cosine are the simplest periodic functions, these are the building blocks we will use. Because they repeat every 2π, we will assume that the function f we want to approximate repeats every 2π. (Later, we will deal with the case where f has some other period.) We will start by considering the square wave in Figure 9.19. Because of the periodicity of all the functions concerned, we only have to consider what happens in the course of a single period; the same behavior repeats in any other period.

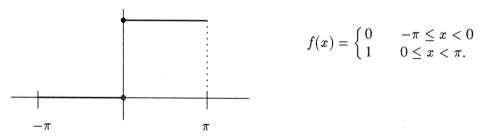

$$f(x) = \begin{cases} 0 & -\pi \le x < 0 \\ 1 & 0 \le x < \pi. \end{cases}$$

Figure 9.19: Square Wave on $[-\pi, \pi]$

We will attempt to approximate f with a sum of trigonometric functions of the form:

$$\begin{aligned} f(x) &\approx F_n(x) \\ &= a_0 + a_1 \cos x + a_2 \cos 2x + a_3 \cos 3x + \cdots + a_n \cos nx \\ &\quad + b_1 \sin x + b_2 \sin 2x + b_3 \sin 3x + \cdots + b_n \sin nx \end{aligned}$$

$$= a_0 + \sum_{k=1}^{n} a_k \cos kx + \sum_{k=1}^{n} b_k \sin kx.$$

This is known as a *Fourier polynomial of degree n* after the French mathematician Joseph Fourier, who first investigated it.[2] The coefficients a_k and b_k are the *Fourier coefficients*. Since each of the component functions $\cos kx$ and $\sin kx$, $k = 1, 2, \ldots, n$ repeats every 2π, $F_n(x)$ must repeat every 2π and so is a potentially good match for $f(x)$, which also repeats every 2π. The problem is to determine the values for the Fourier coefficients to achieve a close match between $f(x)$ and $F_n(x)$. The values we choose are

---

The Fourier Coefficients:

$$a_0 = \frac{1}{2\pi} \int_{-\pi}^{\pi} f(x)\,dx,$$

$$a_k = \frac{1}{\pi} \int_{-\pi}^{\pi} f(x) \cos kx\,dx \quad \text{for } k > 0,$$

$$b_k = \frac{1}{\pi} \int_{-\pi}^{\pi} f(x) \sin kx\,dx \quad \text{for } k > 0.$$

For a derivation of these formulas, please see the Appendix to this section. Notice that a_0 is just the average value of f over the interval $[-\pi, \pi]$.

☐ **Example 1** Construct successive Fourier polynomials for the square wave function f, with period 2π, given by

$$f(x) = \begin{cases} 0 & -\pi \le x < 0 \\ 1 & 0 \le x < \pi. \end{cases}$$

Solution. We find that

$$a_0 = \frac{1}{2\pi} \int_{-\pi}^{\pi} f(x)\,dx = \frac{1}{2\pi} \int_{-\pi}^{0} 0\,dx + \frac{1}{2\pi} \int_{0}^{\pi} 1\,dx = 0 + \frac{1}{2\pi}(\pi) = \frac{1}{2}.$$

Furthermore,

$$a_1 = \frac{1}{\pi} \int_{-\pi}^{\pi} f(x) \cos x\,dx = \frac{1}{\pi} \int_{0}^{\pi} 1 \cos x\,dx = 0$$

and

$$b_1 = \frac{1}{\pi} \int_{-\pi}^{\pi} f(x) \sin x\,dx = \frac{1}{\pi} \int_{0}^{\pi} 1 \sin x\,dx = \frac{2}{\pi}.$$

Therefore, the Fourier polynomial of degree 1 is given by

$$f(x) \approx F_1(x) = \frac{1}{2} + \frac{2}{\pi} \sin x$$

and the graph of the function and the first Fourier approximation is shown in Figure 9.20.

---

[2]The Fourier polynomials are not polynomials in the usual sense of the word.

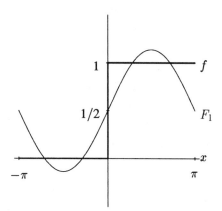

Figure 9.20: First Fourier Approximation to the Square Wave

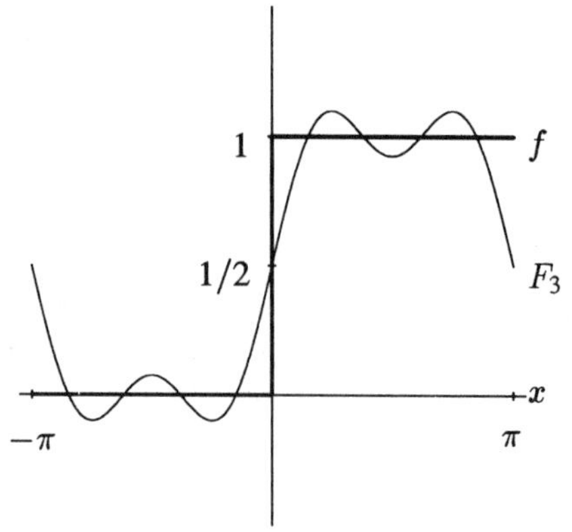

Figure 9.21: Third Fourier Approximation

We next construct the Fourier polynomial of degree 2. The coefficients a_0, a_1, b_1 are the same as before. In addition,

$$a_2 = \frac{1}{\pi} \int_{-\pi}^{\pi} f(x) \cos 2x \, dx = \frac{1}{\pi} \int_{0}^{\pi} 1 \cos 2x \, dx = 0$$

and

$$b_2 = \frac{1}{\pi} \int_{-\pi}^{\pi} f(x) \sin 2x \, dx = \frac{1}{\pi} \int_{0}^{\pi} 1 \sin 2x \, dx = 0.$$

Since $a_2 = b_2 = 0$, the Fourier polynomial of degree 2 is identical to the Fourier polynomial of degree 1. Let's look at the Fourier polynomial of degree 3:

$$a_3 = \frac{1}{\pi} \int_{-\pi}^{\pi} f(x) \cos 3x \, dx = \frac{1}{\pi} \int_{0}^{\pi} 1 \cos 3x \, dx = 0$$

and

$$b_3 = \frac{1}{\pi} \int_{-\pi}^{\pi} f(x) \sin 3x \, dx = \frac{1}{\pi} \int_{0}^{\pi} 1 \sin 3x \, dx = \frac{2}{3\pi}$$

and so the approximation is given by

$$f(x) \approx \frac{1}{2} + \frac{2}{\pi} \sin x + \frac{2}{3\pi} \sin 3x$$

The corresponding graphs are shown in Figure 9.21. This approximation is much more accurate than $F_1(x) = \frac{1}{2} + \frac{2}{\pi} \sin x$, as comparing Figure 9.21 to Figure 9.20 shows.

Without going through the details, we show the results of successively higher degree Fourier approximations in Figure 9.22, where

$$F_5(x) = \frac{1}{2} + \frac{2}{\pi} \sin x + \frac{2}{3\pi} \sin 3x + \frac{2}{5\pi} \sin 5x$$

$$F_7(x) = \frac{1}{2} + \frac{2}{\pi} \sin x + \frac{2}{3\pi} \sin 3x + \frac{2}{5\pi} \sin 5x + \frac{2}{7\pi} \sin 7x.$$

In each case, notice that the approximation matches the step-like nature of the square wave function more and more closely. ❑

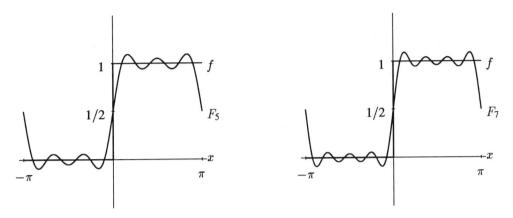

Figure 9.22: Fifth Approximation and Seventh Approximation to the Square Wave

We could have used Taylor series to approximate the square wave, provided we did not center it at a point of discontinuity. Since the square wave is a constant function on each interval, its Taylor series approximations are the constant functions: 0 or 1, depending on where your Taylor series is centered. They approximate the square wave perfectly on each piece, but they do not do a good job over the whole interval. That is what Fourier polynomials succeed in doing: they approximate a curve fairly well everywhere, rather than just near a particular point. The Fourier approximations above look a lot like square waves, so they are *globally* correct; but they do not actually give good values for individual points (at $x = 0$ they all give $1/2$, which is incorrect). Thus Fourier polynomials may not be good *local* approximations.

> Taylor polynomials give good *local* approximations to a function;
> Fourier polynomials give good *global* approximations to a function.

Fourier Series

As with Taylor polynomials, the higher the degree of the Fourier approximation, the more accurate it is. Therefore, we carry this procedure on indefinitely, and call the infinite sequence of approximations

> the **Fourier Series** for f on $[-\pi, \pi]$:
>
> $$f(x) = a_0 + a_1 \cos x + a_2 \cos 2x + a_3 \cos 3x + \cdots$$
> $$+ b_1 \sin x + b_2 \sin 2x + b_3 \sin 3x + \cdots$$
>
> where a_k and b_k are the Fourier coefficients

Thus, the Fourier series for the square wave is

$$f(x) = \frac{1}{2} + \frac{2}{\pi} \sin x + \frac{2}{3\pi} \sin 3x + \frac{2}{5\pi} \sin 5x + \frac{2}{7\pi} \sin 7x + \cdots.$$

What do we do if our function does not have period 2π?

We can easily adapt what we have done above by changing variables. Suppose we have a function f defined for $0 \le x \le b$. We will suppose that it is periodic with period b. Then let

$$t = \frac{2\pi x}{b} - \pi.$$

For instance, if f is periodic on $[0, 12]$ with period 12, then $t = \frac{2\pi x}{12} - \pi$. As x varies in the interval $0 \le x \le 12$, t varies in the interval $-\pi \le t \le \pi$. Thus we find the Fourier series in terms of t and then convert back to x's.

☐ **Example 2** Suppose $I(t)$ is the number of cases of measles per 10,000 people per month. The data points in Figure 9.23 show the variation in $f(t) = \log I(t)$ throughout the year. The curve in Figure 9.23 shows the first three terms of the Fourier series fitted to this data. Figure 9.24 contains the graphs of the constant term and the first two oscillating terms of the Fourier series plotted separately.[3] Describe what these two terms of the Fourier series tell you about measles.

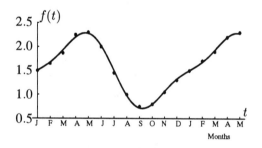

Figure 9.23: Logarithm of incidence of measles per month and Fourier series

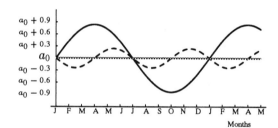

Figure 9.24: Three terms of Fourier series

Solution. Taking the log of $I(t)$ has the effect of compressing the oscillations. However, since the log of a function increases when the function increases, and decreases when it decreases, oscillations in $f(t)$ correspond to oscillations in $I(t)$.

Figure 9.24 shows that the first term in the Fourier series has period one year, (the same period as the original function); the second term has a period of six months. Reading off Figure 9.24 shows that the first term is approximately a sine with amplitude about 0.8; the second term is approximately the negative of a sine with amplitude about 0.25. Thus, for t in months ($t = 0$ in January)

$$f(t) \approx a_0 + 0.8 \sin\left(\frac{\pi}{6}t\right) - 0.25 \sin\left(\frac{\pi}{3}t\right)$$

---

[3]From Bliss and Blevins (1959), and Batschelet (1979).

where the $\frac{\pi}{6}$ and $\frac{\pi}{3}$ are introduced to make the periods 12 and 6 months respectively. We can read a_0 off the original graph of f: it is the average value of approximately 1.5. Thus

$$f(t) \approx 1.5 + 0.8 \sin\left(\frac{\pi}{6}t\right) - 0.25 \sin\left(\frac{\pi}{3}t\right)$$

The fact that this Fourier polynomial fits the original data so well (see Figure 9.23), suggests that the variation in incidence in measles comes from two sources, one with a yearly cycle and one with a half-yearly cycle. At this point the mathematics can tell us no more; we must turn to the epidemiologists for further explanation.   ☐

Appendix: Justification of the Formulas for the Fourier Coefficients

Recall that the coefficients in Taylor Series (which is a good approximation locally) are found by differentiation. In contrast, the coefficients in a Fourier Series (which is a good approximation globally) are found by integration.

We will write the entire series using a $\sum$ sign with ∞ at the top to denote that the series does not have a last term:

$$f(x) = a_0 + \sum_{k=1}^{\infty} a_k \cos kx + \sum_{k=1}^{\infty} b_k \sin kx.$$

Consider the integral:

$$\int_{-\pi}^{\pi} f(x)\,dx = \int_{-\pi}^{\pi} \left[a_0 + \sum_{k=1}^{\infty} a_k \cos kx + \sum_{k=1}^{\infty} b_k \sin kx \right] dx.$$

Splitting the integral into separate terms, we get:

$$= \int_{-\pi}^{\pi} a_0\,dx + \int_{-\pi}^{\pi} \sum_{k=1}^{\infty} a_k \cos kx\,dx + \int_{-\pi}^{\pi} \sum_{k=1}^{\infty} b_k \sin kx\,dx$$

$$= \int_{-\pi}^{\pi} a_0\,dx + \sum_{k=1}^{\infty} \int_{-\pi}^{\pi} a_k \cos kx\,dx + \sum_{k=1}^{\infty} \int_{-\pi}^{\pi} b_k \sin kx\,dx.$$

But, for $k \geq 1$, thinking of the integral as an area shows that

$$\int_{-\pi}^{\pi} \sin kx\,dx = 0 \quad \text{and} \quad \int_{-\pi}^{\pi} \cos kx\,dx = 0,$$

so all terms drop out except the first, giving

$$\int_{-\pi}^{\pi} f(x)\,dx = \int_{-\pi}^{\pi} a_0\,dx = a_0 x \Big|_{-\pi}^{\pi} = 2\pi a_0$$

and so

$$\boxed{a_0 = \frac{1}{2\pi} \int_{-\pi}^{\pi} f(x)\,dx \,.}$$

Thus a_0 is the average value of f on the interval $[-\pi, \pi]$.

To determine the values of any of the other a_k, $k = 1, 2, ..., n$, we use a rather clever method which depends on the fact that for all integers k and m

$$\int_{-\pi}^{\pi} \sin kx \cos mx \, dx = 0,$$

and that, provided $k \neq m$,

$$\int_{-\pi}^{\pi} \cos kx \cos mx \, dx = 0.$$

(See Problems 11–15 on page 666.) In addition,

$$\int_{-\pi}^{\pi} \cos^2 mx \, dx = \pi.$$

To use this, we multiply through the expression for the Fourier series by $\cos mx$, where m is any positive integer, and integrate term–by–term. (The justification for this step is omitted.) We then have

$$f(x) \cos mx = a_0 \cos mx + \sum_{k=1}^{\infty} a_k \cos kx \cos mx + \sum_{k=1}^{\infty} b_k \sin kx \cos mx.$$

When we integrate this between $-\pi$ and π, we obtain

$$\int_{-\pi}^{\pi} f(x) \cos mx \, dx = \int_{-\pi}^{\pi} \left(a_0 \cos mx + \sum_{k=1}^{\infty} a_k \cos kx \cos mx + \sum_{k=1}^{\infty} b_k \sin kx \cos mx \right) dx$$

$$= a_0 \int_{-\pi}^{\pi} \cos mx \, dx + \sum_{k=1}^{\infty} \left(a_k \int_{-\pi}^{\pi} \cos kx \cos mx \, dx \right)$$

$$+ \sum_{k=1}^{\infty} \left(b_k \int_{-\pi}^{\pi} \sin kx \cos mx \, dx \right).$$

Provided $m \neq 0$, we have $\int_{-\pi}^{\pi} \cos mx \, dx = 0$. (Think of it in terms of area.) Since $\int_{-\pi}^{\pi} \sin kx \cos mx = 0$, all the terms in the third sum are zero. Since $\int_{-\pi}^{\pi} \cos kx \cos mx \, dx = 0$ provided $k \neq m$, all the terms in the second sum are zero except when $k = m$. Thus the right-hand side reduces to one term:

$$\int_{-\pi}^{\pi} f(x) \cos mx \, dx = a_m \int_{-\pi}^{\pi} \cos mx \cos mx \, dx = \pi a_m,$$

and so

$$\boxed{a_m = \frac{1}{\pi} \int_{-\pi}^{\pi} f(x) \cos mx \, dx}$$

for each value of $m = 1, 2, ..., n$.

Using a similar line of argument, we multiply through by $\sin mx$ instead of $\cos mx$ and eventually obtain

$$\boxed{b_m = \frac{1}{\pi} \int_{-\pi}^{\pi} f(x) \sin mx \, dx}$$

for each value of $m = 1, 2, ..., n$.

❖ Exercises for Section 9.5

1. (a) For $-2\pi \le x \le 2\pi$, use a graphing calculator to sketch:

    i) $y = \sin x + \frac{1}{3} \sin 3x$

    ii) $y = \sin x + \frac{1}{3} \sin 3x + \frac{1}{5} \sin 5x$

    (b) Each of the functions in (a) is a Fourier approximation to a function whose graph is a square wave. What term would you add to the right hand side of the second function in (a) to get a better approximation to the square wave?

    (c) What is the equation of the square wave function? Is this function continuous?

2. Construct the first three Fourier approximations to the square-wave function

    $$f(x) = \begin{cases} -1 & -\pi \le x < 0 \\ 1 & 0 \le x < \pi. \end{cases}$$

    Use a calculator or computer to draw the graph of each approximation.

3. Repeat Problem 2 with the saw-tooth function

    $$f(x) = \begin{cases} -x & -\pi \le x < 0 \\ x & 0 \le x \le \pi. \end{cases}$$

For Problems 4–6, find the nth Fourier polynomial for the given functions, assuming them to be periodic with period 2π. Graph the first three approximations with the original function.

4. $g(x) = x, \quad -\pi < x \le \pi.$

5. $f(x) = x^2, \quad -\pi < x \le \pi.$

6. $h(x) = \begin{cases} 0 & -\pi < x \le 0 \\ x & 0 < x \le \pi. \end{cases}$

7. If the Fourier coefficients of f are a_k and b_k, and if the Fourier coefficients of g are c_k and d_k, show that the Fourier coefficients of $Af + Bg$ are $Aa_k + Bc_k$ and $Ab_k + Bd_k$.

8. Use the results of Problem 7 to find the Fourier coefficients of the function h in Problem 6 from those of f and g in Problems 3 and 4.

9. Suppose we have a periodic function f with period 1 defined by $f(x) = x$ for $0 \le x < 1$. Find the fourth degree Fourier polynomial for f and graph it on the interval $0 \le x < 1$. [Hint: Remember that since the period is not 2π, you will have to start by doing a substitution. Notice that the terms in the sum are not $\sin(nx)$ and $\cos(nx)$, but instead turn out to be $\sin(2\pi nx)$ and $\cos(2\pi nx)$.]

10. Suppose f has period 2 and $f(x) = x$ for $0 \le x < 2$. Find the fourth degree Fourier polynomial and graph it on $0 \le x < 2$. [Hint: See Problem 9]

For Problems 11–15 use the Table of Integrals in Chapter 6, page 404, to show the following statements are true for any positive integers k and m.

11. $\int_{-\pi}^{\pi} \sin kx \cos mx \, dx = 0.$

14. $\int_{-\pi}^{\pi} \sin kx \sin mx \, dx = 0, \quad \text{if } k \neq m.$

12. $\int_{-\pi}^{\pi} \cos kx \cos mx \, dx = 0, \quad \text{if } k \neq m.$

15. $\int_{-\pi}^{\pi} \sin^2 mx \, dx = \pi.$

13. $\int_{-\pi}^{\pi} \cos^2 mx \, dx = \pi.$

9.6 Miscellaneous Exercises for Chapter 9

Find the first four non-zero terms of the Taylor series about the origin of the functions in Problems 1–4.

1. $\sin t^2$

2. $\theta^2 \cos \theta^2$

3. $\dfrac{1}{1 - 4z^2}$

4. $\dfrac{1}{\sqrt{4 - x}}$

For Problems 5–6, expand the quantity around the origin in terms of the variable given. Give the first four non-zero terms.

5. $\dfrac{a}{a+b}$ in terms of $\dfrac{b}{a}$

6. $\sqrt{R - r}$ in terms of $\dfrac{r}{R}$, for $R > 0$

For Problems 7–9, find the second degree Taylor polynomial about the given point.

7. $\sin x, \quad x = -\frac{\pi}{4}$

8. $e^x, \quad x = 1$

9. $\ln x, \quad x = 2$

10. Find the third degree Taylor polynomial for $f(x) = x^3 + 7x^2 - 5x + 1$ at $x = 1$.

11. (a) Find $\lim\limits_{\theta \to 0} \dfrac{\sin 2\theta}{\theta}$. Explain your reasoning.

(b) Use series to explain why $f(\theta) = \dfrac{\sin 2\theta}{\theta}$ looks like a parabola near $\theta = 0$. What is the equation of the parabola?

12. Suppose x is positive but very small. Arrange the following expressions in increasing order:

$$x, \quad \sin x, \quad \ln(1+x), \quad 1 - \cos x, \quad e^x - 1, \quad \arctan x, \quad x\sqrt{1-x}.$$

13. (a) Find the Taylor series for $f(t) = te^t$ about $t = 0$.

    (b) Using your answer to (a), find a Taylor series expansion about $x = 0$ for

$$\int_0^x te^t \, dt$$

    (c) Using your answer to (b), show that

$$\frac{1}{2} + \frac{1}{3} + \frac{1}{4(2!)} + \frac{1}{5(3!)} + \frac{1}{6(4!)} + \cdots\cdots = 1$$

14. Padé approximants are rational functions used to approximate more complicated functions. In this problem, you will derive the Padé approximant to the exponential function.

    (a) Let $f(x) = \dfrac{1 + ax}{1 + bx}$, where a and b are constants. Write down the first three terms of the Taylor series for $f(x)$ about $x = 0$.

    (b) By equating the first three terms of the Taylor series about $x = 0$ for $f(x)$ and for e^x, find a and b so that $f(x)$ approximates e^x as closely as possible near $x = 0$.

15. The theory of relativity predicts that when an object moves at speeds close to the speed of light, the object appears heavier. The apparent, or relativistic, mass, m, of the object when it is moving at speed v is given by the formula

$$m = \frac{m_0}{\sqrt{1 - \frac{v^2}{c^2}}}$$

where c is the speed of light and m_0 is the mass of the object when it is at rest.

    (a) Use the formula for m to decide what values of v are possible.

    (b) Sketch a rough graph of m against v, labeling intercepts and asymptotes.

    (c) Write the first three non-zero terms of the Taylor series for m in terms of v.

    (d) For what values of v do you expect the series to converge?

16. The *gravitational field* at a point in space is the gravitational force that would be exerted on a unit mass placed there. We will assume that the gravitational field strength at a distance d away from a mass M is

$$\frac{GM}{d^2}$$

where G is constant. In this problem you will investigate the gravitational field strength, F, exerted by a system consisting of a large mass M and a small mass m, with a distance r between them. (See Figure 9.25)

Figure 9.25: Find the gravitational field strength at P

(a) Write an expression for the gravitational field strength, F, at the point P.

(b) Assuming r is small in comparison to R, expand F in a series in $\dfrac{r}{R}$.

(c) By discarding terms in $\left(\dfrac{r}{R}\right)^2$ and higher powers, explain why the series allows you to look at the field as resulting from a single particle of mass $M + m$, plus a correction term. What is the position of the particle of mass $M + m$? Why is the sign of the correction term what it is?

17. The potential energy, V, of two gas molecules separated by a distance r is given by

$$V = -V_0 \left(2 \left(\frac{r_0}{r} \right)^6 - \left(\frac{r_0}{r} \right)^{12} \right),$$

where V_0 and r_0 are positive constants.

(a) Show that if $r = r_0$, then V takes on its minimum value, $-V_0$.

(b) Write V as a series in $(r - r_0)$ up through the quadratic term.

(c) For r near r_0, show that the difference between V and its minimum value is proportional to $(r - r_0)^2$. In other words, show that $V - (-V_0) = V + V_0$ is proportional to $(r - r_0)^2$.

(d) The force, F, between the molecules is given by $F = -\dfrac{dV}{dr}$. What is F when $r = r_0$? For r near r_0, show that F is proportional to $(r - r_0)$.

18. Suppose you know that the Taylor series of $g(x) = \sin(x^2)$ is

$$x^2 - \frac{x^6}{3!} + \frac{x^{10}}{5!} - \frac{x^{14}}{7!} \cdots$$

Find $g''(0)$, $g'''(0)$, and $g^{(10)}(0)$. (There is an easy way and a hard way to do this!)

19. Consider the Taylor expansion

$$f(x) = \frac{1}{1+x} = 1 - x + x^2 - x^3 + x^4 \cdots$$

By plotting several Taylor polynomials and the function $f(x) = \frac{1}{1+x}$, confirm that the interval of convergence of this series is $-1 < x < 1$.

20. Using the fact that, for example, $\ln 4 = -\ln(\frac{1}{4})$, develop an algorithm for approximating $\ln x$ for $x > 2$ by Taylor polynomials. (Remember that the Taylor series for $\ln x$ about $x = 1$ diverges for $x > 2$.)

21. Suppose all the derivatives of g exist at $x = 0$, and that g has a critical point at $x = 0$.

    (a) Write the nth Taylor polynomial for g at $x = 0$.

    (b) What does the Second Derivative test for local maxima and minima say?

    (c) Use the Taylor polynomial to explain why the Second Derivative test works.

22. (Continuation of Problem 21). You may remember that the Second Derivative test tells us nothing when the second derivative is zero at the critical point. In this problem you will investigate that special case.
    Assume g has the same properties as in Problem 21, and that, in addition, $g''(0) = 0$. What does the Taylor polynomial tell you about whether g has a local maximum or minimum at $x = 0$?

23. Suppose you wanted to approximate π using a Taylor polynomial. You could use either the arctangent or the arcsine to give you an approximation. In this problem, you will compare the two methods.

    (a) Using the fact that $\arctan 1 = \frac{\pi}{4}$ and $\frac{d}{dx}(\arctan x) = \frac{1}{(1+x^2)}$, approximate the value of π using the third degree Taylor polynomial of $4 \arctan x$ about $x = 0$.

    (b) Using the fact that $\arcsin 1 = \frac{\pi}{2}$ and $\frac{d}{dx}(\arcsin x) = \frac{1}{\sqrt{1-x^2}}$, approximate the value of π using the third degree Taylor polynomial of $2 \arcsin x$ about $x = 0$.

    (c) Estimate the maximum error of the approximation you found in part (a).

    (d) Looking at the error formula for the arcsine approximation, explain why using the Taylor polynomials for the arctangent seems like a better idea than using the Taylor polynomial for the arcsine.

24. Find a Fourier polynomial of degree three for $f(x) = e^{2\pi x}$, for $0 \le x < 1$.

25. Use the Fourier polynomials for the square wave

$$f(x) = \begin{cases} -1 & -\pi < x \le 0 \\ 1 & 0 < x \le \pi \end{cases}$$

to explain why

$$1 - \frac{1}{3} + \frac{1}{5} - \frac{1}{7} + \cdots + (-1)^{2n+1}\frac{1}{2n+1}$$

must approach $\frac{\pi}{4}$ as $n \to \infty$.

Index

Δt, 185
arctan x
    Taylor series, 645
$\frac{dy}{dx}$ notation, 150
$\int$, no limits, 375
$\int_a^b$, 191
$\ln(1+x)$
    Taylor series, 639
$\cos x$
    Taylor series, 637
$\sin x$
    Taylor series, 637
$\sum$, 191
e, 31, 169, 235
e^x
    Taylor series, 637
$f'(x)$ notation, 140
i, 613

absolute value function, 172
acceleration, 148
    constant, 343
    relation to velocity, 158
accuracy, 104
adding sines and cosines, 595
amplitude, 595
annual percentage growth, 559
antiderivative, 346, 347, 363
    computing, 376, 390
    constructing, 448
    definite integrals, 450
    formulas for, 376
    properties of, 378
    slope fields, 448
    visualizing, 375
approximation
    error in, 194
    linear, 162, 626
    quadratic, 627
    tangent line, 163, 626
arclength, 484

circle, 77
arctangent function
    Taylor series, 645
area
    as definite integral, 191
    from integral, 186, 464
    limit of left- and right-hand sums, 186
asymptote
    horizontal, 24
    of rational functions, 92
autocatalysis, 577
average, 199
    as integral, 199

binomial series, 640
bisection method, 103
Boltzmann Distribution, 522
bound
    upper and lower, 321
boundary value, 592
boundary value problem, 343
bounded functions, 101

calculators
    round-off error, 235
Carbon-14 dating, 32
carrying capacity, 574
catenary, 488
center of mass, 476
chain rule, 245
    applications, 258
change
    total, 196
characteristic equation, 606
    critically damped case, 607
    overdamped case, 606
    underdamped case, 608
Comparison Test, 442
compartmental analysis, 567
complex number, 613
    algebra of, 614

1

complex plane, 615
conjugates, 614
definition, 613
polar representation of, 616
powers of, 617
roots of, 617
complex plane
polar coordinates, 615
composite functions, 70
compound interest, 33, 34, 36, 499, 562
rule of 70, 268
concavity, 21, 289
and error in definite integrals, 418
and second derivative, 156
concentration, 567
conjugates, 614
consumer surplus, 503
continuity, 170
continuous
at a point, 170
on an interval, 99
variable, 10
convergence
of improper integrals, 432, 435
of left- and right-hand sums, 193
of left-(right-)sums, 362
cosine function, 77
Taylor series, 637
cost
average, 318
fixed, 16, 382
marginal, 151, 313, 382
total, 311
variable, 16
cost function, 311
Coulomb's Law, 434
critical damping, 607
critical point, 279
critical value, 279
cumulative distribution, 515
cumulative distribution function, 515

damped oscillations, 307
damped spring equation, 599, 600
damping, 83, 307, 599
decay
differential equation for, 555
exponential, 23, 557
in damped spring system, 601
radioactive, 557

decreasing function, 11, 12
definite integral, 181, 190, 191, 362
approximating by LEFT(n), 185
approximating by RIGHT(n), 186
arclength, 484, 485
as area, 362, 464
as average, 199, 363
as distance, 185
by substitution, 390
constructing antiderivative, 450
consumer surplus, 504
density, 470
error in approximations, 424
force, 494
Fundamental Theorem of Calculus, 203, 364
improper, 431
mass, 472
notation, 364, 375
present and future value, 501
producer surplus, 504
properties, 366, 369, 371
relation to derivative, 204
rotation, 200
volume, 478
volumes of revolution, 482
vs. indefinite, 375
work, 492
delta t (Δt), 185
demand curve, 8, 502
density, 470
density function, 511
dependent variable, 2, 5
derivative
acceleration, 149
as limit, 129, 168
as tangent line, 132
chain rule, 245
concavity, 289
critical points, 279
definition of, 129, 140
difference quotient, 127, 168
differentiability, 171
finding algebraically, 135, 143
finding graphically, 134, 140
finding numerically, 134, 141
formulas for, 143
$\ln x$, 259
a^x, 260
constant function, 220

constant times a function, 221
exponential functions, 233
inverse trigonometric functions, 261
linear function, 220
polynomials, 228
powers, 225
sine and cosine functions, 252
sums/differences, 221
tangent function, 253
implicit functions, 263
inflection point, 289
interpretation, 144, 148, 151
linear approximation, 162
local maxima/minima, 280
test for, 281
notation, 150, 220
product rule, 239, 242
quotient rule, 242
second and concavity, 156
second derivative, 155
second derivative test, 293
units of measurement, 152
visualizing, 131
derivative function, 140
difference quotient, 11, 127, 168
differentiability, 171
differentiable, 140, 171
differentiable everywhere, 140
differential equations, 342, 346, 534
arbitrary constants, 535
characteristic equation, 606
critically damped case, 608
damped spring equation, 599
equilibrium solution, 552
Euler's method, 545
existence of solutions, 541
exponential growth, 550, 573
first order, 534
general solution to, 347, 534
growth (decay), 555
Hooke's Law, 590
implicit functions, 539
initial value problem, 348, 534, 591
logistic model, 574, 576
modeling, 562
numerical solution of second order, 602
numerical solutions of first order, 545
overdamped case, 606
particular solution, 534

second order, 535, 587, 599
separation of variables, 549
slope field, 538, 604
solution curve, 538
solution to, 346
spring equation, 590
two interacting populations, 580
underdamped case, 610
uniqueness of solutions, 541
direction field, 449
distance
from velocity, 182
visualizing from velocity graph, 183
distribution, 507
divergence
of improper integrals, 432
domain, 2
examples of, 4
dominate, 41
doubling time, 21, 23
drag, 567, 599
dummy variable, 364
dy/dx notation, 150

e, 31, 169, 235
definition of, 235
e^x
Taylor series, 637
economics
consumer surplus, 503
cost
average, 318
marginal, 313
cost function, 311
demand curve, 503
equilibrium price, 503
fixed costs, 311
future value, 498, 499, 501
marginality, 311
present value, 498, 499, 502
producer surplus, 503
revenue
marginal, 313
revenue function, 311
supply curve, 503
electric circuits, 599, 612
electric dipole, 647
ellipse
area of, 393
equilibrium

solution, 551
stable(unstable), 552
equilibrium point, 581
equilibrium price, 503
error, 104, 194
left and right rules, 424
numerical integration, 424
round-off, 235
Taylor polynomial, 651
trapezoid and midpoint rule, 425
error function, 455
escape velocity, 493
Euler's Formula, 616, 641
Euler's method, 545
accuracy, 547
extended, 602
Euler, Leonhard, 545
explicit function, 13
exponential decay, 23, 31
exponential function, 19, 20, 23, 31
applications of, 573
base e, 31
compared to power functions, 45
derivative of, 236, 260
doubling time, 21, 23
formula for, 23
half-life, 23
percentage formulas, 25
solution to differential equation, 537, 550
Taylor series, 637
exponential growth, 19, 31
exponential growth equation, 550
exponents
rules for manipulating, 26
extrapolate, 10, 163

factorial, 629
factoring, 96
family of functions(curves)
cubics, 304
damped harmonic motion, 307
exponential with asymptote, 300
normal distribution, 305
solution of differential equations, 347, 534, 537
Fibonacci sequence, 212
first derivative
inflection point, 291
maxima/minima, 281
first derivative test, 281

fixed costs, 311
force, 493
Fourier series, 657, 661
changing period, 662
square wave, 661
free falling bodies, 565
function, 2
absolute value, 172
approximation by linear, 162
average of, 199
bounded, 101
composite, 70
concave (up or down), 21
constant, 144
cost, 311
cumulative distribution, 515
decreasing, 11, 12, 144
definition, 2
density, 511
derivative, 139
domain of, 2
elementary, 383
error, 455
even, 72
explicit, 13
exponential, 19, 20, 23
derivative, 233
factoring, 96
gamma, 439
global maximum/minimum, 282
graph of, 2
implicit, 13
increasing, 10, 12, 144
inverse, 50
inverse trigonometric, 82
derivative, 260
invertible, 51
linear, 9, 10
derivative of, 220
linear approximation, 267, 626
local maxima/minima, 281
logarithm, 63
monotonic, 193
odd, 72
piecewise linear function, 174
polynomial, 88, 228
derivative, 228
power, 40, 223
derivative, 225

proportional, 5
range of, 2
rational, 92
representation, 2
revenue, 311
shift, 68
stretch, 68
sums of, 69
trigonometric, 76
derivative, 250
Fundamental Theorem of Calculus, 203, 363, 374
future value, 499, 501
formula for, 499

Galileo
law of motion, 355
gamma function, 439
gas consumption, 322
gasoline
density of, 496
global maximum/minimum, 282
finding, 283
Gompertz equation, 544
graph
area under, 197
helpful vs. non-helpful, 278
gravitation, 202, 272, 492
gravitational field, 668
growth
annual percentage, 559
differential equation for, 555
exponential, 19
guess-and-check, 383

half-life, 23
hanging cable, 488, 536
histogram, 507
Hooke's Law, 589
horizontal asymptote, 24
hyperbola, 43

i, 613
imaginary numbers, 614
implicit function, 13
differential equation, 539
improper integral, 431
comparison, 441, 442
Comparison Test, 442
numerical methods, 445
income stream, 500

increasing function, 10, 12
increment, 185
indefinite integral, 375
by substitution, 385
computing, 376
formulas for, 376
properties, 378
visualizing, 375
vs. definite, 375
independent variable, 2, 5
inertia, 353
inflection point, 289
initial value, 592
initial value problem, 343, 534
damped spring system, 608
initial conditions, 348
mass/spring system, 591
integral
approximating
RIGHT(n), 186
as average, 199
as total change, 196
definite, 181, 190, 191, 362
as area, 362
as average, 199, 363
as distance, 185
as limit, 191
as total change, 196, 362
by substitution, 390
constructing antiderivative, 450
consumer surplus, 504
definition of, 191
density, 470
error in approximations, 424
force, 494
Fundamental Theorem of Calculus, 203, 364, 374
improper, 431
mass, 472
notation, 200, 364, 375
present and future value, 501
producer surplus, 504
properties, 366, 369, 371
Riemann sum, 362
sigma ($\sum$) notation, 191
volume, 478
volumes of revolution, 482
work, 492
definite vs. indefinite, 375

improper, 431
    comparison, 441, 442
    infinite integrand, 435
    numerical methods, 445
  indefinite, 375
    by substitution, 385
integrand, 191, 362
  infinite, 435
integration
  by parts, 395
  by substitution, 385
  limits of, 191, 362
  methods
    reduction formulas, 406
  numerical methods, 414
    midpoint rule, 415
    trapezoid rule, 417
  reduction formulas, 406
  tables of, 404
  techniques
    by parts, 395
    guess-and-check method, 383
    reduction formulas, 407
    substitution method, 385, 391
    tables, 404, 406
intercept, 10
interest, 33, 34, 36, 499, 562
  rule of 70, 268
inverse functions, 50
  formulas for, 52
  graphs of, 52
  trigonometric, 82
inverse trigonometric functions, 82
  derivative of, 260
inversely proportional, 5
invertible functions, 51
iteration, 104
iterative methods, 337

joules, 490

kinetic energy, 492

LEFT(n)
  error in approximation, 424, 428
left-hand sum, 185, 190, 362
  sigma ($\sum$) notation, 191
limit, 191, 211
  improper integrals, 431
  instantaneous acceleration, 149

instantaneous velocity, 125
  left and right, 169
  left-hand, 169
  meaning of, 168
  notation, 125, 211
  of $\frac{\sin \theta}{\theta}$, 169
  of left-(right-)sums, 362
  of $(1 + h)^{1/h}$, 169
  right-hand, 169
limits of integration, 191
linear approximation, 267
linear function, 10
  approximation by, 162
  derivative, 220
  piecewise, 174
linearity
  local, 162
local linearization, 162, 163, 267
local maxima/minima, 280
  first derivative test, 281
  second derivative test, 293
logarithm function
  derivative of, 259
  graph of, 60
  Taylor polynomial approximation, 632
  Taylor series, 639
logarithms, 55
  base e (natural), 64
  base 10, 56
  compared to power functions, 60
  definition, 56
  manipulation rules, 57, 64
  natural, 31
logistic growth model, 157, 574
  analytic solution to, 576
  carrying capacity, 574
Lotka-Volterra equations, 579

marginal analysis, 151, 312
  profit, 315
marginal costs(revenues), 313
mathematical modeling, 328
maxima/minima, 293
  global, 282
  local, 280
maximizing averages, 322
mean, 513
  arithmetic vs. geometric, 325
mean value, 514
median, 512

midpoint rule, MID(n), 415, 428
    error in approximation, 419, 426
minimizing averages, 322
model
    mathematical, 2
modeling
    how to, 331
    maxima(minima), 329
moment, 476
monotonic, 193
motion
    uniform, 342
    uniformly accelerated, 343

Newton's Method, 337
    chaos, 340
    definition, 338
    failings, 340
Newton, Sir Isaac
    Law of Cooling(Heating), 558
    Law of Gravity, 492
    Second Law of Motion, 566, 600
newtons, 490
normal distribution, 305, 373, 518
    standard deviation, 520
numerical integration
    error, 424
numerical methods
    accuracy and error, 104, 417, 603
    bisection, 103, 337
    Euler's method, 545, 602
    finding derivative, 141
    improper integrals, 445
    integration
        MID(n), 414
        SIMP(n), 427
        TRAP(n), 417
        LEFT(n), 416
        MID(n), 416
        RIGHT(n), 416
        error, 424
        error and integrand, 427
    iterative, 104
    Newton's Method, 338
    second order differential equations, 602
    writing answers, 106
    zooming, 98

optimization, 319, 328
orbit, 582

overdamping, 606

parabola, 89
parameter, 13
pendulum, 598
period, 78
periodic functions
    amplitude, 78
    period of, 78
phase plane, 582
phase shift, 595
physics
    escape velocity, 493
    force, 493
    free falling body, 565
    kinetic energy, 493
    Law of Gravity, 493
    pressure, 493
    work, 489
piecewise linear function, 174
polar coordinates, 615
    relation to Cartesian, 616
polynomial functions, 88
    derivatives of, 228
polynomials
    factoring, 96
population growth, 19, 573
    carrying capacity, 574
    equilibrium population, 581
    logistic model, 574
    two interacting populations, 579
power function, 40
    compared to exponential functions, 45
    compared to logarithms, 60
    derivative of, 225, 258
    even (odd) power, 41
    formula for, 40
    fractional power, 43
    zero, negative power, 42
predation, 579
present value, 499, 501, 502
    formula for, 499
pressure, 493
probability, 373
    cumulative distribution function, 515
    density function, 511
    distribution, 507
    histogram, 507
    mean, 513, 518
    mean value, 514

median, 512
normal distribution, 518
probability density function, 516
standard deviation, 518
probability density function, 516
producer surplus, 503
product rule, 239
three dimensional, 275
profit, 312, 315
proportional, 5
inverse, 5

quadratic formula, 97
quotient rule, 242

radians, 76, 77
vs. degrees, 133
radioactive decay, 32
range, 2
examples of, 4
rate of change: average, 128
rate of change: instantaneous, 128
rational functions, 92
reduction formulas, 406
relative growth, 555
revenue
marginal, 313
total, 311
revenue function, 311
Riemann sum, 191, 362, 469
consumer surplus, 504
density, 470
force, 494
general, 414
mass, 472
present and future value, 501
producer surplus, 504
setting up, 469
volume, 478
volumes of revolution, 482
work, 492
RIGHT(n)
error in approximation, 424, 428
right-hand sum, 185, 186, 190, 362
sigma ($\sum$) notation, 191
roots
finding, 96

second derivative, 155
concavity, 156

interpretation of, 156
maxima/minima, 293
second derivative test, 293
second order
differential equations, 587
separation of variables, 549
exponential growth equation, 550
justification, 549, 553
sigma ($\sum$)
for sum, 191
sigma ($\sum$) notation, 191
sigmoid curve, 575
Simpson's rule, SIMP(n), 426, 429
sine function, 77
Taylor series, 637
sine-integral function, Si(x), 451
sine-integral, Si(x), 453, 455
slope, 10
and velocity, 123, 124
slope field, 449, 537, 574
solution curve, 538
speed, 120
spring, 588
square wave
Fourier series, 662
standard deviation, 518
standard distribution, 446
substitution, 385
into definite integrals, 390
sum
left- and right-hand, 185, 190
Riemann, 191, 362, 414
summation notation, 191
superposition, 605
principle of, 605
supply curve, 8, 502
symmetry, 72

table of integrals, 404
using table, 406
tangent field, 449
tangent function, 81
derivative of, 253
tangent line
approximation, 163, 267
slope of, 132
Taylor expansions, 637
Taylor polynomial, 626, 631
about $x = a$, 632
degree n, 630

   error, 651
   natural logarithm, 632
   second degree, 628
Taylor series, 635
   $(1 + x)^{\alpha}$, 640
   arctan x, 645
   $\ln(1 + x)$, 639
   e^x, 637
   e^{-x^2}, 644
   convergence of, 636, 638
   integration of, 645
   manipulations of, 643
   sine and cosine functions, 637
terminal velocity, 566
trajectory, 582
transcendental equation, 59
trapezoid rule, TRAP(n), 417, 428
   error in approximation, 418, 426
trigonometric functions, 76
   amplitude, 595
   cosine, 77
   derivatives of, 250
   inverse, 82
   phase shift, 78, 595
   sine, 77
   superposition of, 594
   tangent, 81
trigonometric identity, 77, 395, 401

underdamping, 608
uniformly accelerated motion, 343
unit circle, 77

variable
   dummy, 364
velocity
   and slope, 124
   average, 121, 122, 125
   definition of, 124
   from distance, 121
   instantaneous, 122
   relation to acceleration, 148, 158
   visualizing from distance graph, 122
volume, 478
   of revolution, 482

water
   weight of, 528
work, 434, 489

zooming, 98